World Scientific Series in 20th Century Physics

World Scientific Series in 20th Century Physics – Vol. 10

HOW WE LEARN; HOW WE REMEMBER: TOWARD AN UNDERSTANDING OF BRAIN AND NEURAL SYSTEMS

Selected Papers of Leon N. Cooper

Editor

Leon N. Cooper

Brown University

World Scientific

Singapore • New Jersey • London • Hong Kong

Published by

World Scientific Publishing Co. Pte. Ltd.

P O Box 128, Farrer Road, Singapore 9128

USA office: Suite 1B, 1060 Main Street, River Edge, NJ 07661

UK office: 57 Shelton Street, Covent Garden, London WC2H 9HE

Library of Congress Cataloging-in-Publication Data

Cooper, Leon N.
     How we learn, how we remember : toward an understanding of brain
and neural systems : selected papers of Leon N. Cooper / editor,
Leon N. Cooper.
         p.   cm. -- (World Scientific Series in 20th Century Physics ; vol. 10)
     Includes bibliographical references.
     ISBN 9810218141 -- ISBN 981021815X (pbk.)
     1. Memory. 2. Learning.   3. Neural networks (Neurobiology)
I. Title.   II. Series.
QP406.C665   1995
612.8'2--dc20
                                       95-37677
                                          CIP

British Library Cataloguing-in-Publication Data
A catalogue record for this book is available from the British Library.

Printed in Singapore.

DEDICATION

To Kay

for the happy times we spend together

ACKNOWLEDGEMENTS

I would like to express my appreciation to the private foundations and government agencies that have provided resources to make possible the work on which the articles in this collection are based.

I am grateful to the Alfred P. Sloan Foundation and the Ittleson Family Foundation for early support and to the Charles A. Dana Foundation whose support is vital for our current work.

I am also grateful for the support of the National Science Foundation, the Army Research Office, the Air Force Office of Scientific Research and, in particular, the office of Naval Research whose visionary Learning and Memory Program initiated in the early 1980's has been and is critical for the continuation of our program.

Above all, I would like to acknowledge the contributions of my students and colleagues whose efforts made these publications possible and whose companionship made their creation pleasurable.

CONTENTS

II.   Neural Networks

(Fortune Teller)

A Final Few Words

(Monet's Sunrise)

GENERAL INTRODUCTION

When World Scientific Publishing Inc. asked me to put together a collection of my papers, I thought, at first, these could be naturally divided into sections on various subjects on which I have worked. I was intrigued, in particular, by the section I had already titled mentally, "Orphans". It would include those papers of which I am particularly fond but which have been almost totally neglected by everyone else. I reveled in the opportunity to hector a reluctant world on the gems it was overlooking.

The more I considered the project, however, other than the chapter "Orphans", the less I liked it. It seemed too much an obituary, which I am stubbornly reluctant to write. Rather, I decided I would put together a compilation of those papers that lead directly to work in which I am currently involved. However, when one includes book chapters and written versions of various lectures, there is certain amount repetition (I sometimes feel like an eighteenth century musician, traveling from one concert appearance to another, hastily putting together new symphonies, concerti or overtures, partially constructed from previous work so that in the promotion something can be listed as new). Therefore, I have deleted sections of some articles and added occasional comments so that there is a reasonable sequence of ideas. In addition to papers I have written by myself and with colleagues, I have included some especially relevant papers on which I do not appear as an author. In all I have tried to present a more or less coherent picture of the line of research that has led us to our present investigation of the biological basis for learning and memory storage and the information processing and classification properties of neural systems.

Part I. Physiological Basis of Learning and Memory Storage

Some Properties of a Neural Model for Memory

(with J. A. Anderson, M. M. Nass, W. Freiberger, and U. Grenander)

AAAS Symposium, December 30, 1972

I had always been interested in biological problems. My project for the Westinghouse Science Talent Search, of which I was a finalist in 1947, concerned the development of strains of the bacterium bacillus subtilis more resistant to penicillin than the wild type. This choice of project was certainly influenced by the availability of a wonderful biology laboratory at the Bronx High School of Science, where I spent every possible afternoon hour when classes had finished.

My decision to choose physics was influenced by several factors. At the end of World War II, physics was very much in the public eye. The explosion of the first nuclear bomb had created a sensation. And for years I had had a burning desire to understand the theory of relativity and quantum theory, to understand the deep mysteries of the universe. And, as for many aspiring young scientists, my great idol was Albert Einstein. Einstein with his blown hair and mysterious smile; how much one wanted to understand the deep thoughts of that saint-like figure.

In my early years at Columbia College, when the decision had to be made, I chose physics. Only this way, I thought to myself, would I ever be able to participate in the deepest of all mysteries: the fundamental laws of nature, the meaning of space, time, relativity, and quantum particles. If I didn't learn early, I would never really understand. The rest, I hoped, I could somehow do later. I recall John Ward's shock at the Institute for Advanced Study in Princeton when I confided to him that after I had finished all of the problems in physics I would return to biology (having written a few operas on the way).

Thus, in the late sixties when papers on superconductivity were becoming longer, more technical, and somehow less exciting, in my peripatetic search for new objects of interest, I returned to my old love. In particular, I was fascinated by statements in various articles and books, paraphrased roughly: "Although much is known about the structure and function of individual neurons, almost nothing is known about how memory is stored and retrieved". This seemed to me particularly anomalous since, for example, so much is known about how memory is stored in computers. When I began to think about the problem, it occurred to me that memory might somehow be a collective property of very large ensembles of neurons. I had already had some success teasing subtle collective properties from large ensembles of interacting electrons and had been working on aspects of many-body theory for almost a decade, and it seemed possible that perhaps there might be some technical connection. This proved illusory, but it got me going.

So, when a young graduate student, Menasche Nass, appeared in my office and said that he was interested in working on a problem with me for his Ph. D., somehow (I don't remember exactly how) the possibility of working on a biological problem arose. I warned Menasche that among the normal risks one takes, this was really a high-risk thing to do. However, after some thought, Menasche decided he wanted to do it. It turned out he was an excellent student who accomplished a great deal in his Ph. D. thesis, unfortunately, some half a generation ahead of his time. The career options when he finished were not brilliant in theory, and Menasche decided hedidn't want to be an experimentalist. He has since pursued a successful career as a tax lawyer in Los Angeles.

As I recall, I had come upon a paper by Longuett–Higgins proposing a model for memory based on a hologram analogy. The essential notion was that rather than storing an entire item of memory in a single spot, the memory would be distributed over a region. And, as we and everyone else was saying even at that time, this would be contained somehow in the synaptic junctions between neurons.[1]

So, my first assignment to Menasche (a summer project) was to try to devise a more physiological rendering of the type of holographic memory that Longuett–Higgins had suggested. This was toward the beginning of the summer of '70. In grand graduate student style, Menasche told me, when I encountered him again in September, that he hadn't solved the problem but he had found in the literature, an extremely interesting proposal. He then showed me a paper of James Anderson, at that time a Research Associate at the Rockefeller University in New York. In fact Anderson's proposal seemed highly attractive. It was a physiologically possible distributed memory that did seem as though it could be acquired in an actual nervous system.

I contacted Jim and there began a collaboration that lasted for many years. Jim, by the way, is now Chairman of the Department of Cognitive and Linguistic Sciences at Brown University.

The following is an excerpt from the first article we wrote together. It provides a summary of some of our very early thinking.

---

[1] Synaptic junctions arose in the course of evolution as part of the solution to the problem of establishing communication between one portion of an animal and another. Once excitable membranes became available, their use in various cells such as muscles and neurons (especially in neurons) provided a means of electrical communication. When the animal becomes large one can, of course, string a single neuron from one end to the other; but this is impractical in most situations and probably risky since if the neuron is severed it is likely more difficult to replace a single long cell than a smaller, shorter one. Thus the problem of communicating between neurons arises. A straightforward means would presumably be to have a direct electrical contact and this does, in fact, occur. A seemingly less straightforward means is to transmit the information from one neuron to another by the very complex mechanism of chemical transmission: a chemical transmitter released unto the synaptic cleft, diffuses and attaches itself to special receptors on post-synaptic membrane, opens channels and produces currents in the post-synaptic dendrite. These propagate passively to the cell body of the post-synaptic neuron and, if these are sufficient, produce action potentials which travel along the axon of the post-synaptic neuron. Thus the information flow continues.

One might speculate that a great advantage of chemical transmission between neurons is the relative ease for modification of the transmission efficacy. Thus the same action potential in the pre-synaptic axon can produce a different current or response in the post-synaptic dendrite. And it is this possibility that produces the dramatic new capability of information storage.

SOME PROPERTIES OF A
NEURAL MODEL FOR MEMORY

James A. Anderson

The Rockefeller University, New York, NY 10021, USA

Leon Cooper and Menasche M. Nass

Department of Physics, Brown University,
Providence, RI 02912, USA

Walter Freiberger and Ulf Grenander

Division of Applied Mathematics, Brown University,
Providence, RI 02912, USA

2:00 pm December 30, 1972

Washington Hilton, Hemisphere Room

AAAS Symposium, Theoretical Biology and Biomathematics

ABSTRACT

A model of long-term memory, motivated by the anatomy and physiology of the mamalian central nervous system is proposed. We suggest that what is of importance to the nervous system is the collective individual activities of large numbers of individual neurons and that just such a collection of activities constitutes the memory trace. We assume that the long-term memory is the sum of such individual traces. With a minimum of some of the functions of a biological memory and that many of its properties are reminiscent of the brain. Throughout the emphasis is on consistency with the known physiology.

Enough is known about the mammalian central nervous system to allow us to suggest some general principles that a neural model for memory must satisfy.

First, the biological memory system is inherently noisy in that what is recalled is rarely (if ever) identical to what is stored and the input to the system is almost never identical to what has been stored. Also, we know that nervous activity often displays the character of noise-like processes: electrical activity, particularly in higher centers such as cortex, shows Gaussian or Poisson distributions of amplitudes or spike activity.

Second, the physical representation of memory appears to be distributed over the brain, parts of a single 'memory' presumably occuring in many different spatial locations. This property has recently been called the 'holographic' property because, by analogy to a hologram, information is not stored locally (point-to-point) as in an ordinary photograph, but globally. This distributed nature of the memory has been generally recognized since Lashley conducted his famous ablation experiments on rat cortex.

Third, there is no evidence that at any time is part of the brain left 'vacant' for future storage. Rather, most parts of the brain show continuous activity. Nor does the amount of information a human brain can stored show saturation.

Fourth, there appears to be no strong evidence in the higher centers of the mammalian central nervous system for the existence of 'pontifical' or 'decision making' neurons, although such neurons may be present in some invertebrates. Highly parallel, numerous sensory afferents excite simultaneously large populations of neurons in cortex and thalamus. Action in the mammalian central nervous system appears to involve processing by large numbers of basically rather similar neurons and the critical steps in neural processing do not appear to depend on one or only a few neurons out of this population.

Fifth, areas capable of memory storage are written over, again and again. This conclusion is strongly suggested by our previous points and is supported by the neurophysiology of cortex. To quote Sir John Eccles (1971), "Any cortical neurons does not exclusively belong to one engram ('memory') but, on the contrary, each neuron and even each synaptic junction would be built into many engrams."

Sixth, there are in the human central nervous system approximately 10^{10} neurons with at least 10^3 connections apiece. Can the details of these connections be important? Specifying, for example, 10^{13} connections at birth would involve the storage of a vast amount of genetic information. It is not likely that the amount of DNA present in the chromosomes could specify so many connections. This suggests that the central nervous system, although globally organized, probably is locally random. Studies by Creutzfeidt and Ito (1968) suggest that neurons in visual cortex receive most of their input from a small number of locally randomly assorted fibers from the lateral geniculate body. Other data suggests the same kind of conclusion.

Seventh, processing in the central nervous system is highly parallel. The visual system, for example, has over a million parallel input channels feeding into cerebral cortex. Other areas of cortex show similar organization and cortical interconnections are dense and highly parallel.

All of the above considerations have been used in the past, in whole or in part, in the construction of various models of nervous system function. Some models have proceeded by analogy with the digital computer. Both the brain and the computer carry on 'processing' in some sense. We believe, however, that the analogy ends there. The modern computer is a very fast digital machine capable of performing serially comparatively simple operations. Faced with a task of biological significance, analyzing a complex or noisy pattern, for example, the computer is a very poor second. Further, considerable evidence now indicates that most neurons in mammals do not behave in binary fashion. Apparently what is important to the nervous system is not the simple presence or absence of a spike, but the average firing frequency of the cell over a brief interval. Perkel and Bullock (1968) list various codes employed in nervous systems. Most of these, and the ones most often found, depend on temporal patterns of spike long in comparison to the duration of a single spike.

Where the computer analogy breaks down most severely is in the storage and retrieval of information. We do not wish to discuss these aspects of computers, other than to say that information is stored locally and retrieval involves looking at a specified memory location. As we shall see, our model stores information globally and is content addressable.

The problem we will now discuss is that of the storage of memory traces (engrams) which we will define as large patterns of individual neuron activities which tend to act as units in operations of the system. We will argue that a simple rule of synaptic plasticity is sufficient for the storage of information and is in fact an optimum way to do so. (Anderson, 1972, p. 203) The idea that synaptic change with use could serve as a mechanism for memory goes back at least as far as the nineteenth century. (Tanzi, 1893). More recently, Eccles has reviewed the evidence for synaptic change with use and considered several specific mechanisms. A great deal of experimental effort has gone into this research with as yet no conclusive results. Various mechanisms have been proposed, and many of these are compatible with our model.

In line with the evidence presented, let us proceed to construct an idealized system and to make certain reasonable assumptions.

1. We consider a system, common in cortex, where one large group of neurons, α, projects to another large group of neurons, β. α and β do not have to be distinct systems. Often a group of neurons projects to itself via recurrent collaterals having a long conduction time. Examples of such projection systems are, the projections of thalamic nuclei to cortex, and the intracortical projections.

2. The trace (or engram of memory) is the simultaneous activities shown by a large group of neurons.

3. Synaptic interactions add linearly.

4. Synaptic weights are coded so that change in synaptic weight is proportional to the product of pre- and post-synaptic activities at a given time.

With these assumptions we shall see that the system is capable of behavior suggestive of a biological memory. Our system can, among other things,

1. Recognize a previously presented (and incorporated) trace.

2. Store associations in the sense that if a trace f is associated with another trace g by making the proper synaptic adjustments according to our rules, then presentation of f gives rise to g plus noise.

In order to have a concrete physiological system in mind we imagine our model to be of cerebral cortex and we identify a trace as the simultaneous pattern of individual activities of cortical cells. As is well known each neuron has a resting rate of firing and upon stimulation this rate can change. At any time then, we can represent the state of a neuron by a number representing the level of its activity. For reasons that we discuss later, we can just as well represent the state of the cell by the algebraic value of the neuron's instantaneous firing rate above or below spontaneous rate. (Thus if a neuron unstimulated has a firing rate of 17 spikes per second, and we observe it firing at 13 spikes per second, we would represent its state as -3). If we have N neurons, we can represent the state of the system by making N entries into a column; in other words, by a vector whose components are the states of the N neurons. We identify the vector with the trace. An input can then be characterized in terms of its effect on the neurons of cerebral cortex by the vector $|f\rangle$, where we have borrowed the Dirac Bra-Ket notation from physics. (In this notation, a vector $\mathbf{f}$ is written as $|f\rangle$ and the inner product of the two vectors $\mathbf{f}$ and $\mathbf{g}$ as $\langle f|g\rangle$. The most useful property of this notation drives from the fact that if we have a complete set of vectors, say three non-coplanar vectors in three dimensional space, we can write $\sum |f\rangle\langle f| = 1$ with proper ORTHO orthonormalization, where the sum is over vectors of the complete set.)

Some of the assumptions we have made require further justification. For one, we have assumed a linear system for the response characteristics of the neurons to stimuli. Certainly we do not expect strictly linear behavior in so complex a system. From a practical point of view, assumption of linearity allows immense simplification of the mathematics. Fortunately, assumption of linearity is quite close to reality as far as the central nervous system is concerned provided we define precisely what system characteristics we are interested in. Mountcastle (1967) has proposed as a general rule that there is a linear relation between the output of first order afferent fibers and the sensory response of the nervous system. He has considerable data on the tactile responses of monkeys indicating preservation of linear transduction of first order afferent output up to units in cortex. In the visual system, Maffei *et al.* (1967) have shown, the firing rate of lateral geniculate body cells follow in a linear fashion the sinusoidal intensity modulation of a light stimulus. Further, Maffei (1968) has shown that LGB cells apparently use spatial averaging in order to preserve linearity of cell response. As in many physical systems, the linearity assumption has a strict domain of validity for small changes in stimulus intensity. Neurons, though having highly nonlinear portions of their individual responses, may respond, on the average in group, in quite linear fashion.

We have implicitly assumed that the behaviour of neurons is a direct reflection of the stimuli applied. In connection with this we would like to mention a very significant set of experiments conducted by Hirsh and Spinelli (1971). Hirsh and Spinelli raised kittens in an environment where they received as their sole visual input three horizontal stripes to one eye and three vertical stripes to the other. They then studied the orginization of receptive fields in visual cortex and found, first, that a given cell in visual cortex was driven by one eye or the other, but not both, in contrast to normal cats where around 80% of cells are binocular. Second, the elongated receptive fields of the visual cortical cells now conform to the direction of the input driving them, that is, receptive field orientations are horizontal or vertical depending on whether they were connected to the eye receiving horizontal or vertical input. These experiment provide clear evidence that cell activity may mirror stimulus form in a very simple way. Let us try to reflect this in our model.

In order to proceed we must make some simplifying assumptions in order to facilitate calculations. The inner product of a trace is taken to be the 'power' of the trace. In some intuitive sense, the power stands for the strength of the trace and for convenience we assume all traces have equal power. We also assume all traces have a mean value of zero.

We can now state the central assumption of our model. To form the memory for a group of traces, we simply form the vector sum of all the traces of the group. Thus, if there are K traces to be stored, then we form the memory vector $|s\rangle$ as

$$|s\rangle = \sum_{k=1}^{k} |f^k\rangle .$$

We further assume that all the traces are statistically independent and the statistics of all elements in the vectors are the same. (That is, neurons are similar statistically to each other.)

As an aside, Noda and Adey (1970) found that when two cells in parietal cortex were recorded simultaneously with the same microelectrode, the two cells although physically very close together were not correlated in their discharge when the cat was awake or in REM sleep. Thus adjacent cells tended to act as 'individuals', as if each cell sampled the environment independently.

By the central limit theorem, the sum of many uncorrelated traces should closely approximate a vector whose elements are the values taken by a normally distributed random variable. (Thus the noise-like atmosphere of the nervous system may be a simple consequence of this kind or organization.) Since we know nothing of the details of the traces in the memory in general we will calculate average values over many sets of allowable traces. In order to get a quantitative measure of how good the system is, we will ask how close is the reconstructed trace to the desired trace. In the language of the electrical engineer we will ask for the signal to noise ratio.

ACKNOWLEDGEMENTS

We have been greatly assisted by many people both at Brown University and at the Rockefeller University. We would like to thank and acknowledge the assistance of Professor C. Elbaum and Professor H. Kucera at Brown University and Bruce W. Knight at the Rockefeller University.

REFERENCES

[1] J. A. Anderson, *Math. Biosci.* **8**, 137 (1970).

[2] J. A. Anderson, *Math. Biosci.* **14**, 197 (1972).

[3] O. D. Creutzfeldt and M. Ito, *Exp. Brain Res.* **6**, 324 (1968).

[4] J. C. Eccles, in *Brain and Human Behavior*, edited by A. G. Karczmar and J. C. Eccles, Springer, Berlin (1972).

[5] H. V. B. Hirsh and D. N. Spinelli, *Exp. Brain Res.* **13**, 509 (1971).

[6] L. Maffei and G. Rizzolatti, *J. Physiol.* **195**, 215 (1968).

[7] V. B. Mountcastle, in *The Neurosciences*, edited by G. C. Ouarton, T. Melnechuk, and F. O. Schmitt, Rockefeller University Press, New York (1967).

[8] H. Noda and W. R. Adey, *J. Neurophysiol.* **33**, 572 (1970).

[9] D. H. Perkel and T. H. Bullock, *Neurosciences Research Program Bulletin* **6**, 221 (1968).

[10] E. Tanzi, *Riv. Sep. Trenia* **19**, 149 (1893).

[11] W. Wickelgren, Multitrace Strength Theory, in Models of Human Memeory, edited by D. A. Norman, Academic Press, New York (1972).

A Possible Organization of Animal Memory and Learning

Proceedings of the Nobel Symposium on Collective Properties of Physical Systems, Sweden, p. 252 (1973)

In the fall of 1972 I delivered a series of lectures to a newly formed interdisciplinary group at Brown; this was to evolve into the Center for Neural Studies, then the Center for Neural Science, and now the Department of Neuroscience and the Institute for Brain and Neural Systems. In these lectures, I presented some ideas about how a distributed memory could be put into an actual nervous system and some of the consequences.

As Jack Cowan remarked to me about that time, "You're learning fast." (I had expected vast praise.) Even today the field is not completely organized. At that time no textbooks existed. No journals were devoted to the subject and prior ideas, if they were printed at all, were scattered throughout the literature. Therefore one spent ones early efforts rediscovering what everyone else had done. Passive learning I so happily introduce in the following paper is very little more than what Donald Hebb had proposed in 1947. The fact that Hebb is not even mentioned suggests that I had not yet heard of him.

My lectures were interrupted by a particularly happy event. It was mid October when I learned that with Bardeen and Schrieffer I had won the Nobel Prize for our work in superconductivity. Since the publication of our 1957 paper and the almost instant acclaim it received, I must say, I had existed in a state of some nervous anticipation when November came around.

It was, however, an October evening when my wife, Kay, greeted me in the lobby of Barus–Holley, the physics building at Brown University. She was to drive me home since my famous convertible Camaro was being repaired — as was not unusual. A mysterious smile was on her face. Why the smile? *"Well"*, she said, *"We had a call from Swedish radio. They wouldn't tell me what they wanted but said they would call back in an hour or so."* My first reaction was, *"It's too early, don't they have any sense of propriety?"* (That year, the Nobel committee changed the timing of the announcements from November to October.) We came home and waited. An hour became two, several glasses of scotch and finally dinner. No telephone. Finally, about ten or eleven, a call came. It was a reporter from Swedish radio who announced the decision and recorded my first reaction: *"I hope this isn't a false alarm."*

For the next few months, amid the incredible flurry of publicity, I had to remind myself what superconductivity was about for my Nobel lecture.

By coincidence, the next term was a sabbatical that we spent in Paris. There, in the winter and spring of 1973, I began to put together my thoughts on the acquisition and storage of distributed memory.

In June of 1973 there was a Nobel symposium on the collective properties of physical systems; and I chose that forum to introduce these thoughts. I can't say that my colleagues weren't a bit puzzled. It would be almost ten years before neural networks became super-high fashion among a subgroup of physicists. As Einstein is reputed to have said, "*Physicists are like horses, when one breaks down the barn door they all rush out into the field*".

In any case, in the article that follows, I tried to lay it all out. Stan Ulam, after reading this paper, said to me, "*Its full of ideas*". He was too polite to add that if rich in themes, it was short on development.[2] In some ways, we have been developing ever since.

---

[2]At about that time at a banquet associated with a series of scientific conferences organized by the Institut de la Vie: From Theoretical Physics to Biology, during an overly extensive predinner locution Stan passed us the following note:

"*I have fame*"

After a bit of puzzling, we decoded this to be

"*J'ai faim*"

He was hungry.

A Possible Organization of Animal Memory and Learning[1]

Reprinted from the Proceedings of the Nobel Symposium on Collective Properties of Physical Systems. © 1973

L. N. Cooper

Brown University, Providence, R.I. 02912, USA

Summary

A brief account is given of some of the properties of a neural model which displays on a primitive level features which suggest some of the mental behavior associated with animal memory and learning. The model as well as the basic passive procedure by which the neural network modifies itself with experience is consistent with known neurophysiology as well as with what information might be available in the neuron network. One must, however, assume that there exists a means of communication (electrical or chemical) between the cell body and dendrite ends—communication in a direction opposite to the flow of electrical signals. This same modification procedure could also lead to the formation of cells of the type observed by Hubel & Wiesel in the cat's visual cortex. It is suggestive that a network modification procedure that could produce such early processing cells might also be responsible, in cortex, for 'higher' mental processes. The explicit mathematics employed here is that of linear transformations on a vector space. However, as only certain topological properties are used, it is possible to construct a more general non-linear theory.

We have been analyzing a class of neural models that display, on a primitive level, features such as recognition, association and generalization, which suggest some of the mental behavior associated with animal memory and learning [1]. The mechanisms employed seem to be biologically plausible and are not inconsistent with known neurophysiology. In addition the networks that result seem to be a reasonable outcome of evolutionary development under the pressure of survival. Some of the ideas discussed are related to or are generalizations of earlier concepts such as perceptrons or similar models [2–4]. In addition non-local memories have been explored previously among others by Longuet-Higgins [5, 6].

[1] This work was supported in part by the US National Science Foundation.

Nobel 24 (1973) Collective properties of physical systems

Quasi-Random Network

Because of the enormous complexity of the neural network, it seems reasonable to assume that an animal's central nervous system is not completely predetermined genetically. This view is reinforced by the observed fact that in mammals a functioning system is relatively resistant to the death or malfunction of individual units. A system is required, therefore, which is either highly redundant, in which the exact placement of individual units is not critical, or in some other way can continue functioning in spite of the failure of individual units.

A quasi-random network is in some ways analogous to a many-body system as treated in statistical mechanics. There the number of degrees of freedom is large compared to the number of parameters such as temperature or pressure, usually specified to characterize any given ensemble. The specification of the few parameters ensures that any actual configuration is a member of a particular ensemble.

We consider the possibility that genetic information determines a small number of overall parameters governing the growth of neurons. These might be, for example, density or types of neurons in various regions, approximate regions of projection, general direction of growth, extent of dendritic arborization, number and type of synaptic connections, etc. There is some evidence that this may be the case, particularly in cerebral cortex. In visual cortex there has long been known to be strong topographic organization, but there also appears to be local randomness of connection [7]. Auditory cortex seems to show even less obvious organization [8], and the olfactory and taste systems show no easily understood organization at all [9]. The model we consider accepts such local randomness as a fundamental principle of organization. Such a system could arise naturally in evolution by allowing progressive modification of previously completely specified neural systems.

Linear Mappings

Neuron behavior is complex and highly nonlinear. However, in spite of the short time nonlinear behavior of neurons, there is presently justification for considering linear or quasilinear models in which the neuron potentials are averaged over short periods of time as surprisingly good approximations of some aspects of neural response [10]. Anderson explored the properties of a simple neural model based on a linear mapping between one set of neurons and another [11, 12]. This mapping contains a non-local memory with some of the properties of the holographic-type memory discussed by Longuet-Higgins [5, 6], but in contrast to the model of Longuet-Higgins, this mapping seems easily realizable with known physiology. Anderson showed that his model would act as a matched filter and be capable of recognition of events to which it is previously exposed. We have developed further the consequences of such systems and have shown that they are capable of recognition, association and a form of generalization which seems possible to relate in a primitive way to some animal mental behavior.

Network Modification

In addition we have been able to introduce a method of network modification in which learning takes place in an effortless or what we call a passive manner. The system is placed in an environment and, without any search procedure, forms an internal representation of its external world—an internal representation which enables it to recognize and associate.

The means of modifying the neural network employed here has a long history. It involves changing the strength of the synaptic junction (the ratio of output to input spiking frequencies) according to products of pre- and postsynaptic activity. This is physiologically possible (as will be discussed in more detail) if there is a means of communicating information between the cell bodies and dendrite ends—communication in a direction opposite to the flow of electrical signals. If such communication exists, alterations in the synaptic junctions could be made according to what information is locally available. (One can easily add to this the possibility of overall (global) controls affecting all of the network.)

Though there is no difficulty imagining a variety of possible mechanisms by which such changes might take place, experimental evidence bearing on which of these might be operative (or whether in fact synaptic modifications occur) is very sparse.

The modification procedure could also account for the formation of cells which respond to particular patterns as observed, for example, by Hubel & Wiesel in visual cortex of the cat [13]. The suggestive possibility emerges that the action of a single adaptive mechanism designed originally, perhaps, for the formation of cells to facilitate processing in the outer portions of the brain, can account also for 'higher' mental processes such as memory.

Space of Events and Representations

The duration and extent of an 'event' should be defined self-consistently by the interaction between the environment and the system itself. We proceed initially, however, as though an event is a well-defined objective happening and envision a space of events, E, labelled $e^1, e^2, e^3 \ldots$ Imagine that these are mapped by the sensory and early processing devices of the system through the mapping P (processing) into signal distributions in the neuron space $f^1, f^2, f^3 \ldots$ The mapping, P, is denoted by the double arrow, fig. 1. For the moment we maintain the fiction that this mapping is not modified by experience. (What seems actually to be the case is that such early processing systems are at least partially constructed in the youth of the animal and become 'hardened' at some relatively early stage in its development [14]).

Although we do not discuss the mapping P in any detail, it can be very complex; it must be rich and detailed enough so that a sufficient amount of information is preserved to be of interest. *We assume that the mapping P from E to F has the fundamental property of preserving in a sense (not yet completely defined) the closeness or separateness of events.*

Two events e^ν and e^μ map into f^ν and f^μ whose separation is related to the separation of the

Nobel 24 (1973) Collective properties of physical systems

original events. In a vector representation we imagine that two events as similar as a white cat and a grey cat map into vectors which are close to parallel while two events as different as the sound of a bell and the sight of food map into vectors which are close to orthogonal to each other.

Given the signal distribution in *F* which is the result of an event in *E*, we imagine that the signal distribution *f* is mapped onto another set of neurons (or onto the same set) by a mapping, *A*, denoted by the single arrow, fig. 1. This latter type of mapping is modifiable, and we propose that it is in such mappings that animal memory is contained.

The cortex of higher animals is of course very complex, and if any such systems of neurons exist they would be expected not only to be complicated but to be interspersed with neuron assemblies which do specific processing. Further one would imagine that large numbers of such systems arranged serially or in parallel would be required to reproduce even a portion of the complexity and variety of mental processes.

As will be seen later, the behavior of such neuron sub-networks is dependent on parameters which could easily be imagined to vary from sub-network to sub-network resulting in quite different characteristics. It is possible that sequences of such sub-networks with different values of the biological parameters could be involved in actual mental behavior.

In what follows we construct an idealized model of a network which incorporates a modi-

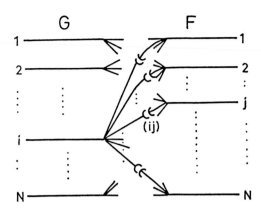

Fig. 2. The ideal associator unit. Each of the *N* incoming neurons in *F* is connected to each of the *N* outgoing neurons in *G* by a single ideal junction. (Only the connections to i are drawn.) We assume that the firing rate of neuron i in *G*, g_i, is mapped from the firing rates of all of the neurons in *F* by: $g_i = \Sigma_j A_{ij} f_j$.

fiable mapping and explore some of its properties.

Consider *N* neurons 1, 2 ... *N*, each of which has some spontaneous firing rate r_{j0}. (This need not be the same for all of the neurons nor need it be constant in time.) We can then define an *N*-tuple whose components are the difference between the actual firing rate r_j of the jth neuron and the spontaneous firing rate r_{j0}.

$$f_j \equiv r_j - r_{j0}$$

By constructing two such banks of neurons connected to one another (or even by the use of a single bank which feeds signals back to itself), we arrive at a simplified model as illustrated in fig. 1.

The actual connections between one neuron and another are generally complex and redundant; we idealize the network by replacing this multiplicity of connections between axon and dendrites by a single ideal junction which summarizes logically the effect of all of the synaptic contacts between the incoming axon branch from neuron j in the *F* bank and the dendrite tree of the outgoing neuron i in the *G* bank (fig. 2). Each of the *N* incoming neurons, in *F*, is connected to each of the *N* outgoing neurons, in *G*, by a single ideal junction. We then assume that: *the firing rate of neuron i in G, g_i, is mapped*

Fig. 1. The *N* neurons in the *F* bank are connected via synaptic junctions to the *N* neurons of the *G* bank.

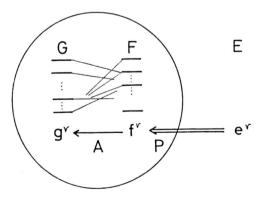

from the firing rates of all of the neurons, f_j, in F by:

$$g_i = \sum_{j=1}^{N} A_{ij} f_j$$

This is the fundamental linear assumption which gives the influence of firing rates in F on those in G. Although most of the results we obtain below do not require so strong an assumption, the simplicity of this hypothesis makes worthwhile an exploration of its consequences. In making this hypothesis, we are focusing our attention on firing rates, on time averages of the instantaneous signals in a neuron (or perhaps a small population of neurons). We are further using the known integrative properties of dendrite branches. Although we have available both excitatory or inhibitory synaptic junctions, such a module could be built if necessary of excitatory neurons alone since a decrease of incoming signal (if $r_j - r_{j0} < 0$) would decrease the output (less excitation).

The Associative Mapping, Memory and Mental Processes

It is in modifiable mappings of the type A that the experience and memory of the system are stored. In contrast with machine memory which is at present local (an event stored in a specific place) and addressable by locality (requiring some equivalent of indices and files) animal memory is likely to be distributed and addressable by content or by association. In addition for animals there need be no clear separation between memory and 'logic'. We show below that the mapping A can have the properties of a memory that is non-local, content addressable and in which 'logic' is a result of association and an outcome of the nature of the memory itself.

The suggestion that animal memory is non-local goes back at least to Lashley [15]; it is implied in Perceptron-like devices [2–4]. The holographic memory of Longuet-Higgins [5, 6] is non-local and content addressable but difficult to realize physiologically. The form we employ here was introduced by Anderson [11, 12].

Anderson's associative mapping is most easily written in the basis of the mapped vectors the

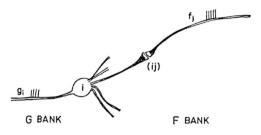

Fig. 3. The ideal junction.

system has experienced:

$$A = \sum_{\mu\nu} c_{\mu\nu} g^{\mu} \times f^{\nu}$$

Although this is a transparent mathematical form, its meaning as a mapping among neurons deserves some discussion. The ij'th element of A gives the strength of the ideal junction between the incoming neuron j in the F bank and the outgoing neuron i in the G bank (fig. 3).

Thus, if only f_j is non-zero

$$g_i = A_{ij} f_j.$$

Since

$$A_{ij} = \sum_{\mu\nu} c_{\mu\nu} g_i^{\mu} f_j^{\nu}$$

the ij'th junction strength is composed of a sum of the entire experience of the system as reflected in firing rates of the neurons connected to this junction. Each experience or association ($\mu\nu$), however, is stored over the entire array of $N \times N$ junctions. This is the essential meaning of a distributed memory: Each event is stored over a large portion of the system, while at any particular local point many events are superimposed.

Recognition and Recollection

The fundamental problem posed by a distributed memory is the address and accuracy of recall of the stored events. Consider first the 'diagonal' portion of A

$$(A)_{\text{diagonal}} \equiv \mathcal{R} \equiv \sum_{\nu} c_{\nu\nu} g^{\nu} \times f^{\nu}$$

An arbitrary event, e, mapped into the signal, f, will generate the response in G

$$g = Af$$

Nobel 24 (1973) Collective properties of physical systems

If we equate recognition with the strength of this response g, say the value of

$$(g, g),$$

then the mapping A will distinguish between those events it contains (the f^ν, $\nu = 1, 2, ..., k$) and other events separated from these.

The word 'separated' in the above context requires definition. In a type of argument given previously [11, 12], the vectors f^ν are thought to be independent of one another and to satisfy the requirements that on the average

$$\sum_{i=1}^{N} f_i^\nu = 0$$

$$\sum_{i=1}^{N} (f_i^\nu)^2 = 1.$$

Any two such vectors have components which are random with respect to one another so that a new vector, f, presented to R above gives a noise-like response since on the average (f^ν, f) is small. The presentation of a vector seen previously, f^λ, however, gives the response

$$Rf^\lambda = c_{\lambda\lambda} f^\lambda + \text{noise}$$

It is then shown that if the number of imprinted events, k, is small compared to N, the signal to noise ratios are reasonable.

If we define separated events as those which map into orthogonal vectors, then clearly a recognition matrix composed of k orthogonal vectors $f^1 f^2, ..., f^k$

$$R = \sum_{\nu=1}^{k} c_{\nu\nu} g^\nu \times f^\nu$$

will distinguish between those vectors contained, $f^1 ... f^k$, and all vectors separated from (perpendicular to) these. Further the response of R to a vector previously recorded is unique and completely accurate

$$Rf^\lambda = c_{\lambda\lambda} g^\lambda$$

In this special situation the distributed memory is as precise as a localized memory.

In addition this type of memory has, as has been pointed out before [5, 6], the interesting property of recalling an entire associated vector g^λ

even if only part of f^λ is presented. Let

$$f^\lambda = f_1^\lambda + f_2^\lambda$$

If only part of f^λ, say f_1^λ, is presented, we obtain

$$Rf_1^\lambda = c_{\lambda\lambda}(f_1^\lambda, f_1^\lambda) g^\lambda + \text{noise}$$

The result is thus the entire response to the full f^λ with a reduced coefficient plus noise.

Association

If we now take the point of view that presentation of the event e^ν which generates the vector f^ν is recognized if

$$Rf^\nu = cg^\nu + \text{noise}$$

Then the off-diagonal terms

$$A \equiv \sum_{\mu \neq \nu} c_{\mu\nu} g^\mu \times f^\nu$$

may be interpreted as leading to association of events initially separated from one another

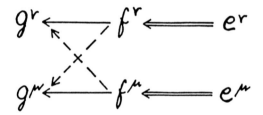

where $(f^\nu, f^\mu) = 0$.

For with such terms the presentation of the event e^ν will generate not only g^ν (which is equivalent to recognition of e^ν) but also (perhaps more weakly) g^μ which should result with the presentation of e^μ. Thus, for example, if g^μ will initiate some response (originally a response to e^μ) the presentation of e^ν when $c_{\mu\nu} \neq 0$ will also initiate this response.

We, therefore, can write the association matrix

$$A = \sum_{\mu\nu} c_{\mu\nu}' g^\mu \times f^\nu = R + A$$

where

$$R = (A)_{\text{diagonal}} \equiv \sum_{\nu} c_{\nu\nu} g^\nu \times f^\nu \quad [\text{recognition}]$$

and

$$A = (A)_{\text{off-diagonal}} \equiv \sum_{\mu \neq \nu} c_{\mu\nu} g^\mu \times f^\nu \quad [\text{association}]$$

The $c_{\mu\nu}$ are then the 'direct' recognition and association coefficients.

"Logic"

In actual experience the events to which the system is exposed are not in general highly separated nor are they independent in a statistical sense. There is no reason, therefore, to expect that all vectors, f^ν, printed into A would be orthogonal or even very far from one another. Rather it seems likely that often large numbers of these vectors would lie close to one another. Under these circumstances a distributed memory of the type contained in A will become confused and make errors. It will 'recognize' and 'associate' events never in fact seen or associated before.

To illustrate, assume that the system has been exposed to a class of non-separated events $\{e^1 ... e^k\} : \{e^\alpha\}$ which map into the k vectors $\{f^1 ... f^k\} : \{f^\alpha\}$.

The closeness of the mapped events can be expressed in a linear space by the concept of community. We define the community of a set of vectors such as $\{f^\alpha\}$ above as the lower bound of the inner products (f^s, f^t) of any two vectors in this set. *The community of the set of vectors $\{f^\alpha\}$ is Γ, $C(f^\alpha) = \Gamma$, if Γ is the lower bound of (f^s, f^t) for all f^s and f^t in $\{f^\alpha\}$.*

If each exposure results in an addition to A (or to $\mathcal{R}$) of an element of the form

$$c_{\nu\nu} g^\nu \times f^\nu,$$

then the response to an event f^s from this class $f^s \in \{f^\alpha\}$ is

$$\mathcal{R}f^s = g = \sum_\nu c_{\nu\nu} g^\nu (f^\nu, f^s) = c_{ss} g^s + \sum_{\nu \neq s} (f^\nu, f^s) c_{\nu\nu} g^\nu$$

where $(f^\nu, f^s) \geq \Gamma$. If Γ is large enough the response to f^s is, therefore, not very clearly distinguished from that of any other f contained in $\{f^\alpha\}$. (In the next section we discuss how such an A might be constructed.)

If a new event, e^{k+1}, not seen before is presented to the system and this new event is close to the others in the class α (for example, suppose that e^{k+1} maps into f^{k+1} which is a member of the community $\{f^\alpha\}$) then $\mathcal{R}f^{k+1}$ will produce a response not too different from that produced for one of the vectors $f^s \in \{f^\alpha\}$. Therefore, the event e^{k+1} will be recognized though not seen before.

This, of course, is potentially a very valuable error. For the associative memory recognizes

and then attributes properties to events which fall into the same class as events already recognized. If in fact the vectors in $\{f^\alpha\}$ have the form

$$f^\nu = f^0 + n^\nu$$

where n^ν varies randomly, f^0 will eventually be recognized more strongly than any of the particular f^ν actually presented.

We have here an explicit realization of what might loosely be called 'animal logic'—which, of course, is not logic at all. Rather what occurs might be described as the result of a built-in directive to 'jump to conclusions'. The associative memory by its nature takes the step

$$f^0 + n^1, f^0 + n^2 ... f^0 + n^k ... \to f^0$$

which one perhaps attempts to describe in language as passing from particulars: cat^1, cat^2, cat^3 ... to the general: cat.

How fast this step is taken depends (as we will see in the next section) on the parameters of the system. By altering these parameters, it is possible to construct mappings which vary from those which retain all particulars to which they are exposed, to those which lose the particulars and retain only common elements—the central vector of any class.

In addition to 'errors' of recognition, the associative memory also makes errors of association. If, for example, all (or many) of the vectors of the class $\{f^\alpha\}$ associate some particular g^β so that the mapping A contains terms of the form

$$\sum_{\nu=1}^k c_{\beta\nu} g^\beta \times f^\nu$$

with $c_{\beta\nu} \neq 0$ over much of $\nu = 1, 2, ..., k$, then the new event e^{k+1} which maps into f^{k+1} as in the previous example will not only be recognized

$$\mathcal{R}f^{k+1}, \mathcal{R}f^{k+1}) \quad \text{large}$$

but will also associate g^β

$$Af^{k+1} = cg^\beta + ...$$

as strongly as any of the vectors in $\{f^\alpha\}$.

If errors of recognition lead to the process described in language as going from particulars to the general, errors of association might be described as going from particulars to a universal: cat^1 meows, cat^2 meows ... $\to$ all cats meow.

There is, of course, no 'justification' for this animal process. It is performed as a consequence of the nature of the system. Whatever efficacy it has will depend on the order of the world in which the animal system finds itself. If the world is properly ordered, an animal system which 'jumps to conclusions' in the sense above may be better able to adapt and react to the hazards of its environment and thus survive. The animal philosopher sophisticated enough to argue 'the tiger ate my friend but that does not allow me to conclude that he might want to eat me' might then be a recent development whose survival depends on other less sophisticated animals who jump to conclusions.

By a sequence of mappings of the form above (or by feeding the output of A back to itself) one obtains a fabric of events and connections

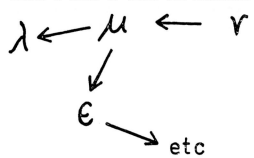

which is rich as well as suggestive. One easily sees the possibility of a flow of electrical activity influenced both by internal mappings of the form A and the external input. This flow is governed not only by direct association coefficients $c_{\mu\nu}$ (which can be explicitly learned as described next) but also by indirect associations due to the overlapping of the mapped events as indicated in fig. 4. In addition one can easily imagine situations arising in which direct access to an event, or a class of events, has been lost ($c_{\gamma\gamma} = 0$ in fig. 4) while the existence of this event or class of events in A influences the flow of electrical activity.

One serious problem in making the identifications suggested above is a direct consequence of the assumption of the linearity of the system. Any state is generally a superposition of various vectors. Thus one has to find a means by which events—or the entities into which they are mapped are distinguished from one another.

There are various possibilities; neurons are so non-linear that it is not at all difficult to imagine

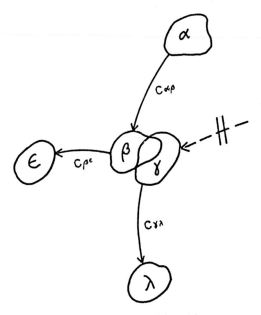

Fig. 4. There will be a flow from $\{\alpha\}$ to $\{\lambda\}$ even though no directly learned coefficient $c_{\alpha\lambda}$ exists, due to the overlap between the $\{\beta\}$ and $\{\gamma\}$ classes. In the particular situation above direct access to $\{\gamma\}$ has been lost.

non-linear or threshold devices that would separate one vector from another. But the occurrence of a vector in the class $\{f^x\}$ in a distributed memory is a set of signals over a large number of neurons each of which is far from threshold. A basic problem, therefore, is how to associate the threshold of a single cell or a group of cells with such a distributed signal. How this might come about will be described in a later section.

In addition to the appearance of such 'pontifical' cells, there will be a certain separation of mapped signals due to actual localization of the areas in which these signals occur. For example, optical and auditory signals are subjected to much processing before they actually meet in cortex. It is possible to imagine that identification of optical or auditory signals (as optical or auditory) occurs first from where they appear and their immediate cluster of associations. Connections between an optical and an auditory event might occur as suggested in fig. 5.

I need hardly mention that even assuming that the physiological assumptions that underlie these constructions are possible (or even correct), there is a distance to be travelled before it is

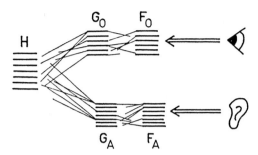

Fig. 5. A Model Optical-Auditory System.

shown that combinations of such elements as introduced above, working together could really reproduce animal mental behavior. The most important step is to make contact between theory and experiment. Some attempts in this direction will be discussed below.

Network Modification, Learning

We now ask how a mapping of the type A might be put into the network. The ij'th element of the associative mapping A

$$A_{ij} = \sum_{\mu\nu} c_{\mu\nu} g_i^\mu f_j^\nu$$

is a weighted sum over the j components of all mapped signals, f^ν, and the i components of the responses, g^μ, appropriate for recollection or association. Such a form could be obtained by additions to the element A_{ij} of the following type:

$$\delta A_{ij} \sim g_i f_j$$

This δA_{ij} is proportional to the product of the differences between the actual and the spontaneous firing rates in the pre and post synaptic neurons i and j[16]. The addition of such changes in A for all associations $g^\mu \times f^\nu$ results finally in a mapping with the properties discussed in the previous section.

For such modifications to occur, there must be a means of communication between the cell body and dendrite ends in order that the necessary information be available at the appropriate junctions; this information must move in a direction opposite to the flow of electrical signals [17]. The junction ij, for example, must have informa-

tion of the firing rate f_j (which is locally available) as well as the firing rate g_i which is somewhat removed (fig. 6). One possibility would be that the integrated electrical signals from the dendrites produce a chemical or electrical response in the cell body which controls the spiking rate of the axon and at the same time communicates to the dendrite ends the information of the integrated slow potential.

There are a variety of means by which the coefficient A_{ij} might be modified, given that the necessary information is available at the ij'th junction. Among these might be growth of additional dendritic spines, adding new synaptic junctions, activation of synaptic junctions previously inactive, changes in membrane resistivity and/or changes in the amount of transmitter in a synapse. Although some structural changes have been observed, there is little evidence yet to choose among the possibilities mentioned above or in fact little evidence that such processes take place at all in the cortex of an adult animal.

To make the modifications

$$\delta A \sim g^\mu \times f^\nu$$

by any of the mechanisms suggested above, the system must have the signal distribution f^ν in its F bank and g^μ in its G bank. It is easy to obtain f^ν since this is mapped in from the event e^ν by P. But to get g^μ in the G bank is more difficult since this in effect is what the system is trying to learn.

In what we denote as active learning (which has been much explored in the past) the system is presented with some f^λ, searches for a response, and is given some indication of when it is coming

Fig. 6. In order that the junction (ij) be modified in proportion to $g_i f_j$, a means is needed for communicating the firing rate g_i which is the result of signals incoming from all the dendrites $g_i = \sum_j A_{ij} f_j$ back to the junction (ij).

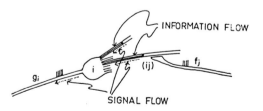

Nobel 24 (1973) Collective properties of physical systems

closer. When (after what could be a long time) by some procedure or another it finds the "right" response, say g^ω, it is "rewarded" and responds to the reward by printing into A the information:

$$\delta A_{ij} = \eta g_i^\omega f_j^\lambda$$

(The information is available at the time of the reward since at that time the system is mapping f^λ, responding g^ω, and thus has just the desired spiking frequencies in the F and G banks of neurons.) Active learning probably describes a type of learning in which a system response to an input is matched against an expected or desired response and judged correct or incorrect.

However, there is a type of animal learning which does not seem from visible external indications to require this type of a search procedure. It is the type of learning in which, as far as can be seen, an animal is placed in an environment and seems to learn to recognize and to recollect in a passive manner.

To arrive at an algorithm which produces passive learning, we utilize a distinction between forming an internal representation of events in the external world as opposed to producing a response to these events which is matched against what is expected or desired in the external world.

The simple but important idea is that *the internal electrical activity which in one mind signals the presence of an external event is not necessarily (or likely to be) the same electrical activity which signals the presence of that same event for another mind.* There is nothing that requires that the same external event be mapped into the same neural patterns by different animals. The event e^ν which for one animal is mapped into the signal distributions f^ν and g^ν, in another animal is mapped into f'^ν and g'^ν. What is required for eventual agreement between animals in their description of the external world is not that electrical signals mapped be identical but rather that the relation of the signals to each other and to events in the external world be the same (fig. 7).

Passive Learning

Call $A^{(t)}$ the A matrix after the presentation of t events. We write

$$A^{(t)} = \gamma A^{(t-1)} + \eta g^t \times f^t$$

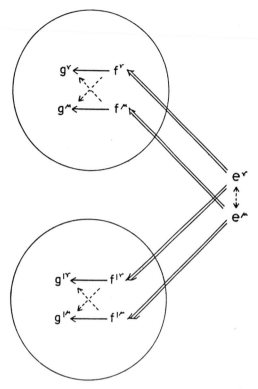

Fig. 7. Representations in two different systems of the same external fabric of events. The two representations are not identical but they each stand in a one-to-one relation to the external fabric and to each other.

In the equation γ is dimensionless and is a measure of the uniform decay of information at every site (a type of forgetting). One would expect that

$$0 \leqslant \gamma \leqslant 1$$

It turns out that values of γ close to one are of most interest. Since f or g are firing rates (spikes/second), η or $\varepsilon = \eta/\gamma$ have the dimensions of s^2. In what follows we normalize all vectors $(f,f) = (g,g) = 1$ so that η and ε become dimensionless.

If we now say that g^t is

$$g^t = A^{(t-1)} f^t + g_R^t + g_A^t$$

We see that the total post-synaptic potentials are composed of three terms: a passive response, $A^{(t-1)} f^t$, an active but random term, g_R^t, and an active response, g_A^t. For purely passive learning we consider only the first term so that

$$\delta A = \eta g^t \times f^t = \eta A^{(t-1)} f^t \times f^t.$$

Here the post synaptic potentials are just those produced by the existing mapping, $A^{(t-1)}$, when the vector f^t in F is mapped into G.

$$g^t = A^{(t-1)} f^t$$

The passive learning algorithm is then

$$A^{(t)} = A^{(t-1)}(\gamma + \eta f^t \times f^t)$$

$$= \gamma A^{(t-1)}(1 + \varepsilon f^t \times f^t)$$

where in general $\varepsilon = n/\gamma$ is presumably much smaller than one. Before any external events have been presented, A has the form $A^{(0)}$ which could be random. This will contain among other things the connectivity of the network.

With this algorithm, after k events A has the form

$$A^{(k)} = \gamma^k A^{(0)} \prod_{\nu=1}^{k} {}_0 (1 + \varepsilon f^\nu \times f^\nu)$$

where $\prod_0$ is an ordered product in which the factors with lower indices stand to the left:

$$\prod_{\nu=1}^{k} {}_0 f(\nu) = f(1) f(2) \dots f(k)$$

This can also be written:

$$A^{(k)} = \gamma^k A^{(0)} \left[1 + \varepsilon \sum_{\nu=1}^{k} |f^\nu\rangle\langle f^\nu| \right.$$

$$+ \varepsilon^2 \sum_{\nu<\mu} |f^\nu\rangle\langle f^\nu|f^\mu\rangle\langle f^\mu| + \dots$$

$$\left. + \varepsilon^k |1\rangle\langle 1|2\rangle\langle 2|3\rangle\langle 3|4\rangle\dots\langle k-1|k\rangle\langle k| \right]$$

It is striking that the passive learning algorithm generates its own response $A^{(0)} f^\nu$ to the incoming vector f^ν, a response that depends on the original configuration of the network through $A^{(0)}$ and on the vector f^ν mapped from the event e^ν. For example if f^ν is the only vector presented, A eventually takes the form

$$A \sim g^\nu \times f^\nu$$

where

$$g^\nu = A^{(0)} f^\nu.$$

Special cases of A

We now display the form of A in several special cases; in all of these ε is assumed to be constant and small.

(1) If the k vectors are orthogonal, A becomes

$$A^{(k)} = \gamma^k A^{(0)} \left(1 + \varepsilon A^{(0)} \sum_{\nu=1}^{k} f^\nu \times f^\nu \right).$$

Letting $A^{(0)} f^\nu = g^\nu$, the second term takes the form of the "diagonal" part of A

$$(A)_{\text{diagonal}} \equiv \mathcal{R} = \sum_{\nu=1}^{k} g^\nu \times f^\nu$$

and will serve for the recognition of the vectors $f^1 \dots, f^k$. (It should be observed that the associated vectors g^ν are not given in advance; they are generated by the network.) If ε is small however this seems inadequate for recognition since the recognition term would be weak. Further one would expect recognition to build up only after repeated exposure to the same event.

(2) The following example demonstrates that the passive learning algorithm does build up recognition coefficients for repeated inputs of the same event. If the same vector f^0 is presented l times, A becomes eventually

$$A^{(l)} \simeq \gamma^l A^{(0)} (1 + e^{l\varepsilon} f^0 \times f^0).$$

If l is large enough so that $e^{l\varepsilon} \gg 1$, the recognition term will eventually dominate. Presumably when $e^{l\varepsilon}$ becomes large enough there should be no further increase in the coefficient. This can be accomplished by making ε a function of the response to the incoming vector in some way so that beyond some maximum value there is no further increase of the coefficient.

(3) The presentation of m orthogonal vectors $l_1, l_2, \dots l_m$ times results in a simple generalization of the second result. When $\gamma = 1$ (for simplicity)

$$A^{(l_1+l_2+\dots+l_m)} = A^{(0)} \left(1 + \sum_{\nu=1}^{m} e^{l_\nu \varepsilon} f^\nu \times f^\nu \right)$$

which is just a separated associative recognition and recall matrix

$$A \simeq \sum_{\nu=1}^{m} c_{\nu\nu} g^\nu \times f^\nu$$

if

$$e^{l_\nu \varepsilon} \equiv c_{\nu\nu} \gg 1.$$

Nobel 24 (1973) Collective properties of physical systems

(4) Some of the effect of non-orthogonality can be displayed by calculating the result of an input consisting of l noisy vectors distributed randomly around a central f^0

$$f^v = f^0 + n^v$$

Here n^v is a stochastic vector whose magnitude is small compared to that of f^0. We obtain

$$A^{(l)} \simeq \gamma^l A^{(0)} \exp\left(l\varepsilon \frac{n^2}{N}\right)(1 + e^{l\varepsilon}f^0 \times f^0)$$

where n is the average magnitude of n^v. We see that the generated $A^{(l)}$, with the additional factor due to the noise, is just of the form for recognition of f^0. Thus the repeated application of a noisy vector of the form above results in an A which recognizes the central vector f^0.

Association Terms

Off-diagonal or associative terms can be generated as follows. Assume that A has attained the form

$$A = \sum_{v=1}^{k} A^{(0)} f^v \times f^v = \sum_{v=1}^{k} g^v \times f^v.$$

Now present the events e^α and e^β so that they are associated so that the vectors f^α and f^β map together. (The precise conditions which result in such a simultaneous mapping of f^α and f^β will depend on the construction of the system. The simplest situation to imagine is that in which $(f^\alpha + f^\beta)/\sqrt{2}$ is mapped if e^α and e^β are presented to the system close enough to each other in time.) We may assume that e^α and e^β are separated so that $(f^\alpha, f^\beta) = 0$. In the F bank of neurons we then have $(f^\alpha + f^\beta)/\sqrt{2}$, where the vector is normalized for convenience.

After one such presentation of e^α and e^β, A becomes:

$$A^{(1)} \simeq \sum_{v=1}^{k} g^v \times f^v + \frac{\eta}{2}(g^\beta \times f^\alpha + g^\alpha \times f^\beta).$$

The second term gives the association between α and β with the coefficient

$$c_{\alpha\beta} = c_{\beta\alpha} = \eta/2$$

which presumably (except in special circumstances) would be small. If f^α and f^β do not occur again

in association, $c_{\alpha\beta}$ or $c_{\beta\alpha}$, although they do grow upon the presentation of f^α or f^β separately, always remain small compared to the respective recognition coefficients $c_{\beta\beta}$ or $c_{\alpha\alpha}$. However if $(f^\alpha + f^\beta)/\sqrt{2}$ is a frequent occurrence (appearing for example l times) the coefficient of the cross term becomes

$$c_{\alpha\beta} \simeq \frac{\gamma^l}{2} e^{l\varepsilon}$$

as large as the recognition coefficient.

We are at present persuing analytical calculations and machine simulations to explore further the development of A in a variety of situations.

Hubel-Wiesel Type Cells

It is most important to make some contact between theory and experiment. We have been working in several directions in such attempts; I will describe one below.

In a classic series of experiments Hubel & Wiesel [13] have shown that cells in the cat's visual cortex respond in a very specific way to external visual patterns to which the cat is exposed. Such patterns can for example be vertical or horizontal lines of specific sizes, in specific positions or with definite motions. Recently Hirsch & Spinelli [18] have shown that depriving the cat of visual experience (such as the experience of horizontal lines) results in the absence of Hubel-Wiesel cells which respond to such lines in the adult. We outline below one means by which such cells might develop, assuming that network modification can occur in a similar manner to that employed previously for the construction of the mapping A.

Since these cells occur in the early processing regions we divide the mapping P so that

$$P = M_2 \times M_1$$

$$c^\alpha <\!\!\sim\!\!\sim d^\alpha <\!\!=\!\!= e^\alpha$$
$$\qquad M_2 \qquad M_1$$

where now M_1 is taken to be prewired but M_2 is modifiable (such modification being possible presumably only in the early development of the animal). As M_2 is to be modified by the same mechanism used previously we write:

$$\delta M_2 \sim g^\mu \times f^\nu.$$

We now assume that the spontaneous firing rate of the cells in the left hand bank of M_2 (other than the random spikes we admit below) is very low or zero and that $M_2^{(0)}$ (the initial value of the mapping M_2) is small, so that the cells in the left hand bank, at least in the development stages, are in general below threshold and do not respond strongly to external signals. These cells therefore are in a pre-threshold and highly non-linear region.

We suppose further (as is in fact the case) that the cells in the left hand bank fire occasional spikes spontaneously and at random. We can then write:

$$M_2^{(t)} = \gamma M_2^{(t-1)} + \eta P_n M_2^{t-1} d^t \times d^t + \eta c^n \times d^t$$

where $P_n M_2$ is the projection of M_2 onto the n'th outgoing cell

$$P_n M_2 = \sum_i \delta_{ni}(M_2)_{ij} \quad \text{for all} \quad j.$$

As before the first term gives the uniform loss of memory, the second the passive response, while the third gives active additions to M_2 due to the simultaneous arrival of the mapped vector d^t and the spontaneous and random firing of the n'th cell in the left hand bank

$$c_i^n = 0 \quad i \neq n$$

$$c_i^n = 1 \quad i = n$$

Since M_2 is initially thought to be small, an incoming vector d^t produces no change in M_2 unless one of the cells in the left hand bank is firing simultaneously. If, however, the entry of the pattern d^α coincides with the firing of the n'th cell we add a term to M_2 of the form

$$\delta M_2 = \eta P_n M_2 d^\alpha \times d^\alpha + \eta c^n \times d^\alpha$$

It is reasonable to assume that a single such entry is not sufficient to fire the n'th cell upon a second presentation of d^α. However, with l coincidences of d^α and the firing of c^n we obtain

$$M_2^{(l)} \simeq \gamma^l e^{l\varepsilon} c^n \times d^\alpha$$

Therefore, there is a build-up of such recognition cells and, depending upon the parameters, they will eventually fire upon the presentation of the pattern with which they have been associated.

There are several problems. One would expect several cells to respond to the same pattern. This is very likely to be the case in fact. However, the addition of a mechanism, such as exists in the visual processing system of Limulus, by which a firing cell suppresses the activity of its neighbours would reduce the number of cells which would respond to the same pattern. In addition it might occur that the same cell would pick up several patterns. How likely this is is being tested by computer simulation with various values of the parameters. Whether this happens in fact is an open question which could be answered experimentally. Such multiple patterns could also eventually be diminished by the addition of a Limulus-like suppression.

All of the prior considerations of picking central vectors out of noisy entries would apply; the details would depend on the parameters chosen. Such cells, linked to F or G type neuron banks, might also be of use for the separation of one vector from another as suggested at the end of the second section.

If these considerations in some measure correspond to facts, it becomes intriguing to speculate that a form of network modification which arose originally in evolutionary development to assist in the formation of special cells for early processing—to increase the flexibility of systems so that they required less genetic pre-programming—might (functioning in a region of cortex which retains for much or all of a lifetime its ability for modification) be responsible also for what we like to call "higher" mental processes.

The author wishes to express his appreciation for the hospitality offered to him by the Laboratoires de Physique Théorique et Hautes Energies, Universités Paris VI and Paris XI (Orsay) where part of this article was written. He would also like to thank those colleagues, in particular Professors Jack Cowen, Charles Elbaum and Bruce Knight, who have offered helpful criticism.

References

1. Anderson, J, Cooper, L, Nass, M, Freiberger, W & Grenander, U, AAAS symposium, theoretical biology and bio-mathematics (1972).
2. Block, H D, Rev mod phys 1962, 34, 123.
3. Block, H D, Knight, B W, Jr & Rosenblatt, F, Rev med phys 1962, 34, 135.

Nobel 24 (1973) Collective properties of physical systems

4. Minsky, M & Papert, S, Perceptrons: An introduction to computational geometry. MIT Press (1969).

5. Longuet-Higgins, H C, Nature, London 1968, 217, 104.

6. — Proc roy roc London B 1968, 171, 327.

7. Creutzfeldt, O D & Ito, M, Exptl brain res 1968, 6, 324.

8. Goldstein, M, Jr, Hall, J L, II & Butterfield, B O, J acoustical society of America 1968, 42, 444.

9. Lettvin, J Y & Gesteland, R C, Cold Spring Harbor symp quant biol 1965, 30.

10. Mountcastle, V B, The neurosciences (ed G C Quarton, T Melnechuk & F O Schmitt) p. 393. Rockefeller University Press, New York, 1967.

11. Anderson, J A, Math biol-sci 1970, 8, 137.

12. — Ibid 1972, 14, 197.

13. Hubel, D H & Wiesel, T N, J physiol 1962, 160, 106.

14. Wiesel, T N & Hubel, D H, J neurophysiol 1965, 28, 1029.

15. Lashley, K S, Arch neurol psychiat, Chicago 1924, 12, 249.

16. Alterations in junction strengths proportional to f_j or the immediate dendrite response to f_j would also seem to be physiologically possible and in some situations might be useful. However, such modifications do not result in the various properties discussed here.

17. Such 'retrograde signalling' has also been postulated by J P Changeux & A Danchin. Private communication.

18. Hirsch, H V & Spinelli, D N, Exptl brain res 1971, 12, 509. See also Blakemore, C & Cooper, G F, Nature 1970, 228, 477.

Discussion

Anderson: Some theories of memory I have seen (not understood!) discuss the question of capacity. Is there corresponding discussion here?

Cooper: Yes, one can discuss the capacity. There seems to be a lot of it. Whether it is enough, I do not know.

Wilson: If you stick to your linear theory the capacity of the brain in your theory seems to be limited, namely there are at most 10^8 linearly independent f's if there is one component of f for each neuron. This would limit the brain to much less than 10^8 distinct remembered patterns.

Cooper: I am not sure whether or not 10^8 distinct patterns (if that is the number) would or would not be enough. One should also recall that each of the "distinct" remembered patterns can be very complex. Further one does not expect a complete absence of non-linearity.

Nobel 24 (1973) Collective properties of physical systems

A Theory for the Development of Feature Detecting Cells in Visual Cortex

(with M. M. Nass)

Biol. Cybern. **19**, 1 (1975)

I was convinced that the ideas presented in the previous paper were at least qualitatively correct. However, the appearance of vaporwave was difficult to avoid. Many ideas, as Stan Ulam said, some possibly interesting, but no real connection with the world in which we happen to live — a limited contribution to a field, if one could call it a field, that was and is plagued with excessive mathematical and philosophical wheel spinning.

It seemed essential to me that theory must be made sufficiently concrete so that it could be confronted by experimental results. At the time theory was a somewhat novel idea for biologists: plausibly so since, for the most part, that specialty, the spinning out of consequences of ideas in long and complex arguments, while accepted (although occasionally the subject of some ridicule) in the community of physicists, had not really been required in biology. There, the connection between idea and experiment was straightforward enough so that every self-respecting experimentalist insisted on doing it himself. This skepticism was also justified, in my opinion, since with a few striking exceptions, many previous so-called theoretical attempts were totally removed from reality. Physicists, in particular, displayed a kind of arrogance in talking to biologists that was not designed to inspire friendly relations. I recall, a presentation at a conference (From Theoretical Physics to Biology sponsored by the Institut de la Vie, a fascinating series of conferences that unfortunately seems to have run out of funding) in which an eminent physicist said, in effect, *"Here is the Schrödinger equation. Here we have 10^{23} electrons and protons subject to electrical forces. One of the consequences is life"*.

With that kind of theory one can appreciate skepticism.

What Menasche and I attempted in the following paper, really the first of a series of such attempts, is to make specific connections between fundamental ideas of synaptic modification and observable and testable experiments in actual animals, to produce a theoretical structure concrete enough so that one would know precisely what the assumptions were, and so that one could see one's way through the arguments and know exactly which conclusions followed from which assumptions. The primary object is not to be right, (although that certainly is one of the hopes) it is to be crystal clear so, to paraphrase Galileo, *"One knows what follows from what one has said before"*. His teachers of mathematics taught him this method.

We chose visual cortex because of the large number of experiments that had been done in that region of the brain. At the time, in addition to the work of Hubel and Wiesel, there was a large body of (very controversial) experimental work suggesting that the response properties of cells of visual cortex could be altered by the visual experience of the animal. This indicated to us that one might be observing experience-dependent cellular changes, analysis of which could reveal the systematics of synaptic modification.

Biol. Cybernetics 19, 1—18 (1975)
© by Springer-Verlag 1975

A Theory for the Development of Feature Detecting Cells in Visual Cortex*

Menasche M. Nass and Leon N. Cooper

Department of Physics and Center for Neural Studies, Brown University, Providence, Rhode Island USA

Received: September 16, 1974

Abstract

Passive modification of the strength of synaptic junctions that results in the construction of internal mappings with some of the properties of memory is shown to lead to the development of Hubel-Wiesel type feature detectors in visual cortex. With such synaptic modification a cortical cell can become committed to an arbitrary but repeated external pattern, and thus fire every time the pattern is presented even if that cell has no genetic pre-disposition to respond to the particular pattern. The additional assumption of lateral inhibition between cortical cells severely limits the number of cells which respond to one pattern as well as the number of patterns that are picked up by a cell. The introduction of a simple neural mapping from the visual field to the lateral geniculate leads to an interaction between patterns which, combined with our assumptions above, seems to lead to a progression of patterns from column to column of the type observed by Hubel and Wiesel in monkey.

Introduction

In a classic series of experiments Hubel and Wiesel [1, 2] demonstrated the existence of what might be called "feature detectors" in visual cortex (area 17) of the adult cat. They determined that single cells in this region of cortex respond in a fairly specific way to elongated shapes of illumination or darkness of various sizes, at various angles and with definite directions of motion. Recent experiments [3–7] seem to indicate that these "feature detectors" will be absent in the adult animal if it is deprived of the visual experience of the feature during a "critical" period.

In this paper we present a theory outlined earlier [8] for the development of such feature detectors. We find that the same passive network modification that leads to the construction of internal mappings with some of the properties of memory also results in the development of Hubel and Wiesel type cells without any pre-programming of specific patterns to specific cells. Malsburg [9] and Perez, Glass, and Shlaer [10] have recently reported results of computer simulation studies using a network modification scheme similar in some respects to ours. They find

* Work supported in part by the National Science Foundation and the Alfred P. Sloan Foundation.

that "cells" come to respond to oriented lines [9], in some cases even in the absence of initial predisposition to respond to these lines [10].

Our results show that a cell can develop a response to an arbitrary, repeated external pattern so that in time the cell will give a maximal response upon presentation of this pattern. The important point is that a cell which "learns" to respond (or commits itself) to a given external pattern need in no way have been genetically pre-programmed to do so.

With the further assumption of interaction (excitation and lateral inhibition) between cortical cells, we show that the number of cells which pick up a single pattern as well as those that pick up more than one pattern is limited. Further, such interaction can produce correlations in single columns as well as between columns.

I. Some Experimental Results

In their experiments Hubel and Wiesel [1, 2] showed that cells in the cat's visual cortex respond optimally to a restricted class of patterned stimuli, namely to elongated regions of illumination or darkness. Their further investigations [2, 11] revealed that these cells are organized according to their receptive field properties. Within an anatomical column all cells have receptive fields in roughly the same region of retina and respond to lines of roughly the same orientation. As an electrode penetrates cortex obliquely a succession of columns are recorded whose optimal orientation is sometimes seen to rotate around a full circle. Hubel and Wiesel concluded that there was a correspondence between functional and anatomical columns in cortex. Within a column all cells respond to a line of fixed orientation (ideally) and neighboring columns respond to lines of "neighboring orientation" on the retina.

Although Talbot and Marshall [12] and others [13] already demonstrated the topographic nature of the retinal-cortical mapping, so regular an organization in so complex a network demanded additional explanation. Hubel and Wiesel [2, 14] hypothesized an ascending hierarchy of cell organization from retina to cortex in which a few cells from the lateral geniculate body would synapse on a single cortical simple cell. The receptive fields would be so matched as to produce the response properties of a simple cell. The implication of this structure is a large degree of anatomical specificity in cortex – raising the issue of whether such an organization could be innate.

The few experiments thus far performed on new born kittens are in conflict on whether orientation specificity is innate. Hubel and Wiesel [15] maintain that such specificity is already present in the new-born kitten; Barlow and Pettigrew [16] claim that while binocularity and direction selectivity are innate, orientation selectivity and disparity are not.

Hirsch and Spinelli's [4] experiments indicate that orientation specificity can be severely altered if the animal is exposed to a restricted set of stimuli in its youth. Blakemore and Mitchell [5] confirmed the existence of a "critical period", similar to the one for binocularity discovered by Wiesel and Hubel [17] during which selective exposure has significant effects on the nervous system. They report that on the 28th day of a cat's life selective exposure for as little as one hour will affect the response properties of cortical cells. Imbert [7] found that not only will a few hours of active visual experience result in "normal" response characteristics of cortical cells, but that the orientation selectivity of a cortical cell can be altered while recording from that cell, by presenting patterned stimulation of a particular orientation.

Both Hirsch and Spinelli [4] and Blakemore and Cooper [3] selectively exposed their animals to straight lines or gratings with fixed orientations. Pettigrew and Freeman [6] examined the effect on the response of cortical cells of raising kittens in a planetarium like environment, where the only stimuli presented to the animals were spots of light, smaller than the receptive field size of geniculate cells. They found that the optimal stimulus for their animals was a spot of light of approximately the same size as the original stimulus.

In short, while it is not clear whether orientation specificity is innate, it does seem clear that the response properties of cells in visual cortex can be significantly altered during the early developmental period of the animal.

III. Application to Visual Cortex

The results of the last section indicate that a network which modifies itself passively so that the synaptic strength of the junction (ij) changes according to

$$\delta A_{ij} \sim g_i f_j$$

will construct by itself an internal mapping which has some of the properties required for memory. Although there is general agreement that neural interconnections can in some circumstances be modified, there is little agreement as to what this modification is or how it comes about. In what follows we take the point of view that the same form of modification which we have assumed in order to construct a distributed memory operates generally in those portions of the nervous system which are capable of change. We then explore whether such a form of modification will account for the development and organization of visual cortex.

Definition of M

From retina to cortex one traverses the neural pathway:

visual cortex ← lateral geniculate cells ← ganglion cells.

[We go from right to left in order to be consistent with the matrix notation employed.] To schematize this we envision a bank of N neurons, D, projecting

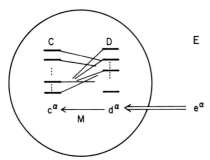

Fig. 4. The N neurons in the D bank are connected via synaptic junctions to the N neurons of the C bank

onto another bank of N neurons, C. We may regard the neurons in D as being either lateral geniculate or retinal ganglion cells, while those in C are neurons of area 17 of visual cortex. Although there have been previous suggestions of much detailed architecture in the neural projections from retina to cortex [2, 14] we begin here with a minimum of pre-programmed circuitry and represent this by connecting, at least in principle, every neuron in D to every neuron in C. [This we call (N, N) connectivity.] We now imagine that a pattern, e^α, in the external world, E, is mapped into activity, d^α, in the D bank of neurons and this is mapped by M into activity c^α in the cortical cells, C, Fig. 4.

Experimental work of Herz and others [20] indicates that spontaneous activity decreases significantly as one proceeds from retina to visual cortex. The mean values given by these workers are 35.5 spikes per second in the optic tract, 14.0 spikes per second in the lateral geniculate and 5.7 spikes per second in visual cortex. Further there is evidence [21] that the spontaneous activity of cells in visual cortex increases from very low values at birth to adult levels with visual experience. We therefore assume that the spontaneous activity in D is non-zero, whereas we take the point of view that, initially at least, the spontaneous firing rate in cortex is very low or zero. Let

$$d_j = r_j - r_{j0} \tag{3.1}$$

be the activity of the jth cell of the D bank which is equal to the actual firing rate minus the spontaneous firing rate, and

$$c_i = s_i - s_{i0} \tag{3.2}$$

be the activity of the ith cell in the C bank which is equal to the actual firing rate minus the spontaneous firing rate (which may be zero).

A non-zero value of d_j (representing activity, different from spontaneous activity, among neurons in

the D bank) may result in non-zero activity in C due to synaptic connections between the neurons in the two regions. This is represented by the mapping M

$$c \xleftarrow{\quad M \quad} d. \tag{3.3}$$

The many to many connectivity is contained in the fact that all $N \times N$ synaptic connections may in principle be unequal to zero. A stimulus in D results in an integrated slow potential at the axon hillock of a cell in C as is discussed in the Appendix. When this potential exceeds the threshold for generation of an action potential, the cell will fire. In the region, far above threshold (as in the prior theory where the signal and response neural banks were both far from threshold) the response firing rates are linearly dependent upon the signal firing rates so that M becomes a matrix and

$$c_i = \sum_{j=1}^{N} M_{ij} d_j. \tag{3.4}$$

Below threshold a non-zero value of d results in a non-zero potential variation in the axon hillock of the C-cell, resulting sometimes in action potentials (due to random potential fluctuations) and sometimes not. The simple relationship between firing rates in the C and D banks no longer holds.

However, using the function $P[x]$ defined in the Appendix (A 14), we can write for the firing rate of the ith cortical cell

$$c_i = P \left[\sum_{j=1}^{N} M_{ij} d_j + s_{i0} + s_{iR} \right]. \tag{3.5}$$

Here $\sum_{j=1}^{N} M_{ij} d_j$ is proportional to the integrated slow potential, produced by the mapped pattern d, at the axon hillock of the ith cortical cell, while s_{i0} and s_{iR} are continuing and random fluctuations in the membrane potential. Following (A21) we then have

$$c_i = (Md)_i + s_{i0} + s_{iR} - \alpha/2 \quad \text{if} \quad (Md)_i + s_{i0} + s_{iR} > \alpha$$
$$= 0 \quad \text{otherwise.} \tag{3.6}$$

The decay time, α, and M_{ij}, the "efficacy" of the $(ij)^{th}$ synaptic junction are defined in the Appendix.

Let c^i or d^j denote unit vectors referring to the ith cortical cell of the C bank or the jth lateral geniculate cell of the D bank. The elements of the vector c^i are all zero except the ith which is equal to one:

$$c_l^i = \delta_{il} \tag{3.7}$$

and this corresponds to unit activity in the ith cortical cell. In the same way

$$d_k^j = \delta_{jk} \tag{3.8}$$

corresponds to unit activity in the jth lateral geniculate cell. In this representation M can be written

$$M(t) = \sum_{ij} M_{ij}(t)\, c^i \times d^j . \qquad (3.9)$$

It will sometimes be convenient to choose a representation using some or all the mapped patterns d^v where the vector d^v denotes an arbitrary pattern in the D space

$$d^v = \sum_{j=1}^{N} d_j^v d^j . \qquad (3.10)$$

In this representation

$$M(t) = \sum_{iv} m_{iv}(t)\, c^i \times d^v . \qquad (3.11)$$

The coefficients $M_{ij}(t)$ or $m_{iv}(t)$ change with the experience of the animal and, for example, $m_{iv}(t)$ is related to the response of the ith cortical cell to the incoming normalized pattern d^v.

For orthogonal patterns, $d^1 \ldots d^K$, with these conventions when

$$m_{iv}(t) < \alpha$$

the ith cell remains below threshold for the incoming pattern d^v, while

$$m_{iv}(t) \geqq \alpha$$

means that the ith cell has reached threshold for the vth pattern. At threshold there is a discontinuity in the response of the cortical cells to the incoming signals. We now say that *a cortical cell, i, is committed to the pattern $e^v \to d^v$, when the coefficient m_{iv} satisfies*

$$m_{iv} \geqq \alpha .$$

When this is the case the ith cell will fire with every presentation of v independent of random activity[1].

Above threshold M maps firing rates in the D space into firing rates in the C space so that in the absence of spontaneous or random fluctuations we obtain:

$$c_i = \sum_{j=1}^{N} M_{ij} d_j - \alpha/2 \qquad (3.12)$$

where now c_i is the firing rate of the ith cortical cell while d_j is the actual minus the spontaneous firing rate of the jth incoming cell in the D bank. We could also write

$$c_i = \sum_{v} m_{iv}(d^v, d) - \alpha/2 . \qquad (3.13)$$

---

[1] If we allow $s_{i0} + s_{iR}$ to take negative as well as positive values so that $|s_{i0} + s_{iR}| < m_R^{\text{max}}$, for all i, we might as an alternative definition of commitment use $m_{iv} > \alpha + m_R^{\text{max}}$.

Modification of M

An arbitrary pattern e^v produces activity d^v in the D bank of neurons; this is mapped by M initially (prior to visual experience) into

$$M(0)\, d^v .$$

If at least one cell is already above threshold for this pattern, say the ith cell, then

$$(M(0)\, d^v)_i \equiv (c^i, M(0)\, d^v) \geqq \alpha \qquad (3.14)$$

and this i^{th} cell will fire in response to e^v and d^v.

As will be seen, if a cell responds initially (above threshold) to some pattern, d^v, then it is likely that this cell will come, in the adult animal, to serve as a "feature detector" for this pattern. Whether or not any cell responds initially to an arbitrary pattern depends on $M(0)$ (in effect the initial connectivity of the lateral geniculate to area 17).

There is some question, whether or not cells in the new-born animal will respond at all to an external pattern. Hubel and Wiesel report that cortical cells in the new-born kitten are sluggish in their response. Recording from 8 and 16 day old kittens they note [21] that

> Perhaps the most marked difference between these experiments and our usual recordings from cortex in adult cats was in the maintained activity and responsiveness of the cells. With steady diffuse background illumination cells tended to be silent or to fire at a very low rate. Perhaps partly because of this paucity of maintained activity the number of cells studied in a penetration was unusually small, a few in each passage from surface to white matter as against 20–30 in an ordinary adult penetration. Cells were not only sluggish in their spontaneous activity but also responded grudgingly to the most effective patterned stimuli. This relative difficulty in eliciting responses reminded one of similar difficulties in driving cells in very deeply anesthetized adult cats ⋯ . Just as in deeply anesthetized adults, some cells did not respond at all to patterned stimuli unless the stimuli were moving.

The important question for us, therefore is whether a cortical cell can become "comitted" (or "respond optimally") to a relatively arbitrary pattern (that is a pattern arbitrary in shape but within the receptive field of that cell) assuming little or no initial response of that cell to this pattern.

How M develops when exposed to repeated incoming patterns depends on the form of network modification. That form used previously, when applied to the present situation would alter the mapping elements, M_{ij}, according to

$$\delta M_{ij} = \eta c_i d_j \qquad (3.15)$$

where c_i and d_j are the activities of the ith cortical cell and the jth lateral geniculate axon respectively. This modification would be proportional to the product of differences between the actual and spontaneous firing rates in the pre- and post-synaptic *axons j* and *i* both of which are assumed to be far above threshold. However in the present situation some, if not all, of the neurons in the C bank are below threshold initially at least. The extension of the prior form of modification to the region close to threshold is not unique and there is little experimental evidence to guide us.

It would be possible to couple pre-synaptic variations in firing rates to all post-synaptic potential fluctuations. When these exceed the level necessary to stimulate the action potential response we regain the form above if $s_{i0} = 0$. Or we might take the point of view that no alterations occur unless the action potential stimulus is exceeded, that a depolarization sufficient to produce the action potential is necessary for modification. In what follows, we take this latter position. (Our qualitative results however do not depend upon this particular assumption.)

A reason for this last assumption is that, as has been discussed before, in order that the required information be available at the synaptic junctions, it is necessary that the firing rate in the axon of the post-synaptic cell be communicated backward to its dendrite junctions. This might just possibly be accomplished chemically, but backward propagation of electrical signals seems easier to achieve. Below threshold no spikes are produced so that any backward propagation to the synaptic junctions by electrical signals would be passive and therefore greatly attenuated. Above threshold, when spiking occurs, the spikes can be propagated backward towards the dendrites as well as forward[2]. This provides a possible mechanism for the required flow of information from axon hillock to dendrite ends and at the same time would give non-zero modification (or at least a much larger modification) only when the action potential was exceeded in the post-synaptic cell.

With these assumptions we arrive at the following algorithm by which the mapping, M, develops:

$$M(t+1) = \gamma M(t) + \eta c(t+1) \times d(t+1). \quad (3.16)$$

Here $d(t+1)$ is the $t+1$ mapped incoming pattern in the D bank, and $c(t+1)$ is the post-synaptic response to that pattern in the C bank. Following (A24) the post-synaptic potential may be written

$$c(t+1) = P[\gamma M(t) d(t+1) + s_0(t+1) + s_R(t+1)]. \quad (3.17)$$

We note that the expression in the brackets contains three terms. The first represents the passive response proportional to the integrated slow potentials in the axon hillocks due to $d(t+1)$, while the second and third terms are the continuing and random membrane potential fluctuations.

The constant, γ, represents a uniform decay at all of the junctions while η determines the rate at which a junction is modified. It should be noted that neither η nor γ need remain constant throughout the entire life of the animal. During the learning phase (that is to say during the period in which the animal acquires information) it is valuable to have η fairly large and γ smaller than one. When the feature detectors have been formed it would be valuable to end the uniform decay in this region of cortex by setting $\gamma = 1$ and $\eta = 0$. If γ is much smaller than one details tend to be lost; this would assist in extracting features from complex inputs.

The junction modification which is assumed here requires that information of the post-synaptic spiking rate be transmitted to all of the dendrite junctions to which the axon is physically connected. The uniform decay could, for example, come about if the transmitter substance decays in proportion to the amount accumulated. No communication is required between dendrite junctions of the same cell as is assumed in the normalization condition used by Malsburg [9] and Perez *et al.* [10].

In our prior considerations the incoming and outgoing signals were divided into events. Such a division implies a time duration for each "event" or for its mapping. In the present situation the time interval for the division into events suggests itself in a natural way since the lateral geniculate cells respond in short bursts upon the presentation of an optical stimulus in the receptive field. Increased times of exposure do not increase spiking frequency. In fact the opposite is more likely to be the case. (It has been suggested that simple cells receive their input from the X or tonic cells of Enroth-Cugell and Robson [22, 23].) If this is the case, the "time interval of exposure" can be taken to mean the initial elevated burst of activity in the lateral geniculate axons. And as Hubel and Wiesel [21] have noted, cortical cells in kitten respond with a brief burst, lasting about 1 sec, followed by a long period of silence. Thus both stimulus to and output of a cortical cell can be considered as bursts of activity over a standard time interval. We therefore define events in our present situation as being mapped into the initial bursts described by the firing rates d and of duration τ

---

[2] For a discussion of this question and references see Eccles, J C. The Physiology of Nerve Cells, Baltimore, The Johns Hopkins Press, 1957, p. 50.

8

(of the order of a second) and the constants η and γ are adjusted accordingly.

IV. Development of Feature Detecting Cells

One Cell, One Pattern

To demonstrate the fundamental point that a cell can become committed to an arbitrary but repeated pattern without any genetic pre-disposition to respond to that pattern, we focus our attention on a single cell in the C bank – say the kth, and imagine that a single pattern – say e^β, which leads to activity in the D bank, d^β, is presented repeatedly. The relevant element of the mapping M is

$$m_{k\beta}(t)\, c^k \times d^\beta \qquad (4.1)$$

and we will be concerned with the growth of $m_{k\beta}(t)$.

With the algorithm (3.16) modification of M proceeds according to

$$M(t+1) = \gamma M(t) + \delta M \qquad (4.2)$$

where

$$\delta M = \eta c(t+1) \times d(t+1) \qquad (4.3)$$

and $c(t+1)$ and $d(t+1)$ represent post- and pre-synaptic neural activity in the C and D banks at the $t+1$ event. Since we are concerned for the moment only with the firing rate of the kth cortical cell and the incoming activity d^β, the relevant part of δM can be written

$$\delta m_{k\beta}\, c^k \times d^\beta \,. \qquad (4.4)$$

[We note that in this representation modification of the N synaptic junctions $k1$, $k2$,... are treated together: $\delta M_{kj} = \eta c_k d_j^\beta$.]

Presentation of the pattern, e^β, in the receptive field stimulates the kth cortical cell by the amount

$$(M d^\beta)_k = \sum_\nu m_{k\nu}(d^\nu, d^\beta) = m_{k\beta}\,. \qquad (4.5)$$

[We can assume either that d^β is the only pattern or that each of the patterns $\nu = 1, 2, \ldots$ map into vectors $d^1, d^2 \ldots$ which are orthogonal to one another.] This produces spiking in the kth cortical cell (see the Appendix) if

$$m_{k\beta} \geqq \alpha \,.$$

We are particularly interested in the situation in which initially, there is little or no disposition for the kth cell to respond to the pattern, β. We therefore consider the case in which $m_{k\beta}(0)$ is either zero, or in any case too small to produce firing in the kth cell:

$$m_{k\beta}(0) < \alpha \,.$$

The kth cortical cell, therefore, does not respond to the pattern β, prior to visual experience.

In addition to stimulation of the kth cortical cell due to an incoming pattern, the cell may also be stimulated by continued activity in the D bank, s_{i0}, and/or by random fluctuations in the membrane potential, s_{iR}, as discussed in the Appendix. Equation (4.2) then becomes:

$$\sum_{i\nu} m_{i\nu}(t+1)\, c^i \times d^\nu = \gamma \sum_{i\nu} m_{i\nu}(t)\, c^i \times d^\nu$$
$$+ \eta P[\gamma m_{k\beta}(t) + s_{k0} + s_{kR}]\, c^k \times d^\beta \,. \qquad (4.6)$$

This yields the recursion relation for $m_{k\beta}$:

$$m_{k\beta}(t+1) = \gamma m_{k\beta}(t) + \eta P[\gamma m_{k\beta}(t) + s_{k0} + s_{kR}] \,. \qquad (4.7)$$

We can now, without confusion, drop some subscripts so that (4.7) becomes

$$m(t+1) = \gamma m(t) + \eta P[\gamma m(t) + s_0 + s_R] \,. \qquad (4.8)$$

According to the discussion of the Appendix

$$P[\gamma m(t) + s_0 + s_R] = \gamma m(t) + s_0 + s_R - \alpha/2$$
$$\text{if} \qquad \gamma m(t) + s_0 + s_R \geqq \alpha \qquad (4.9)$$
$$= 0 \quad \text{otherwise}$$

is the firing rate of the cortical cell for the time interval during which $\gamma m(t) + s_0 + s_R \geqq \alpha$. During the developmental period this will occur when the cell is randomly stimulated to fire in coincidence with the incoming pattern β. Since the interval over which s_R acts is τ_R (Appendix) the effective value of η becomes:

$$\eta \rightarrow \frac{\tau_R}{\tau}\eta \equiv \eta_0 \qquad (4.10)$$

so that during such periods of coincidence

$$m(t+1) = \gamma m(t) + \eta_0(\gamma m(t) + s_0 + s_R - \alpha/2) \,. \qquad (4.11)$$

When there is no post-synaptic spiking [i.e., when $\gamma m(t) + s_0 + s_R < \alpha$], $c_k = 0$ and $\delta m = 0$. It becomes convenient then to define as events occurrences in which the incoming pattern e^β is accompanied by some firing in the kth cortical cell, $c_k \neq 0$. The decay constant, γ, is then interpreted so that γM is the decay of M in the interval between such coincidences.

With these assumptions and definitions we have

$$m(t+1) = \gamma m(t) + \eta_0(\gamma m(t) + s_0 + s_R - \alpha/2) \,. \qquad (4.12)$$

This can be written

$$m(t+1) = \gamma(1 + \eta_0)\, m(t) + \eta_0 m_R(t+1) \qquad (4.13)$$

where we define $m_R(t)$, the term independent of the incoming pattern,

$$m_R(t) \equiv s_0(t) + s_R(t) - \alpha/2 \,. \qquad (4.14)$$

The coefficient, $m(t)$, obtained from the recursion Relation (4.13) is

$$m(t+1) = \gamma^{t+1}(1+\eta_0)^{t+1} m(0) + \gamma^t(1+\eta_0)^t \eta_0 m_R(1)$$
$$+ \gamma^{(t-1)}(1+\eta_0)^{t-1} \eta_0 m_R(2) + \cdots$$
$$+ \gamma^0(1+\eta_0)^0 \eta_0 m_R(t+1) \qquad (4.15)$$

$$= e^{(t+1)\Delta_0} m(0) + \eta_0 \sum_{n=0}^{t} e^{n\Delta_0} m_R(t+1-\eta)$$

where

$$e^{\Delta_0} \equiv \gamma(1+\eta_0) . \qquad (4.16)$$

Since $m_R(t)$ contains a random component, it is most convenient to calculate the expected value of $m(t)$. Let

$$\overline{m_R(t)} = m_R . \qquad (4.17)$$

[Since the continuing or spontaneous component of $m_R(t)$, s_0, probably increases with time, this approximation underestimates the growth of $m(t)$.] We then have

$$\overline{m(t+1)} = m(0) e^{(t+1)\Delta_0} + \eta_0 m_R \sum_{n=0}^{t} e^{n\Delta_0} . \qquad (4.18)$$

The behavior of $\bar{m}$, especially for large values of t, depends on the value of $\gamma(1+\eta_0) = e^{\Delta_0}$. There are three regions of interest, $\gamma(1+\eta_0)$ smaller than, equal to or larger than one, or Δ_0 smaller than, equal to or larger than zero. For $\Delta_0 < 0$

$$\lim_{t\to\infty} \overline{m(t)} = m_R \frac{\eta_0}{1 - e^{\Delta_0}} \qquad (4.19)$$

and therefore approaches a constant value independent of $m(0)$, the initial connectivity. For $\Delta_0 = 0$

$$\overline{m(t)} = m(0) + m_R \eta_0 t , \qquad (4.20)$$

increases linearly with t and for large t becomes independent of $m(0)$. For $\Delta_0 > 0$

$$\lim_{t\to\infty} \overline{m(t)} = \left[m(0) + \frac{\eta}{e^{\Delta_0} - 1} m_R \right] e^{t\Delta_0} . \qquad (4.21)$$

For $\Delta_0 > 0$, both $m(0)$ and m_R contribute almost equally and in the asymptotic region $\overline{m(t)}$ grows exponentially as $e^{\Delta_0 t}$. Figure 5 shows $\overline{m(t)}$ as a function of t in these three cases.

Although values of $\Delta_0 < 0$ produce a natural maximum, it seems more reasonable to assume that $\Delta_0 > 0$ in the development phase. (The maximum value of the m_{iv} would then be related to characteristics of cells or junctions. For example, as we will assume in later sections, a cell firing maximally might shut off further modification of its dendrite junctions.) Then the growth of the coefficient $m_{k\beta}$ is exponentially rapid due to passive modification in combination with random coincidences. It is interesting to note

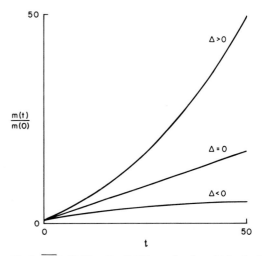

Fig. 5. $\overline{m(t)}/m(0)$ [Equation (4.18)] as a function of t for the three cases discussed in the text. The values of Δ_0 for the three curves are: -0.036, 0, and 0.028. At $t = 0$, $\frac{m(t)}{m(0)} = 1$

that in this region of exponential growth both $m(0)$, the initial predisposition, and m_R, the random term, contribute equally. Therefore a cell could develop a strong response and become committed to a pattern in the absence of any initial predisposition. However with fairly random connectivity, some cells will likely respond more strongly to a given pattern than others and these, as will be discussed further, will be more likely to "acquire" the pattern.

N Cells, One Pattern

In actual experience an animal will see a large variety of patterns and if a given pattern lies within the receptive field of one or more cells, any of those cells may eventually respond to the pattern. Thus many cortical cells can respond in principle to one or many patterns.

To generalize the result of the last section to such situations we consider first the case in which a single pattern is presented but in which N cortical cells can respond. Let the pattern once more be e^β which results in the neural activity d^β in the D bank. Now (4.3) becomes

$$\delta M = \eta c(t+1) \times d(t+1) = \eta c(t+1) \times d^\beta . \qquad (4.22)$$

Since

$$M(t) d^\beta = \sum_i m_{i\beta}(t) c^i \qquad (4.23)$$

we obtain

$$c(t+1) = P\left[\gamma \sum_i m_{i\beta}(t) c^i + s_0(t+1) + s_R(t+1) \right] \qquad (4.24)$$

where s_0 and s_R are vectors. All together this gives

$$M(t+1) \qquad (4.25)$$
$$= \gamma M(t) + \eta P\left[\gamma \sum_i m_{i\beta}(t) c^i + s_0(t+1) + s_R(t+1)\right] \times d^\beta.$$

We assume that initially

$$m_{i\beta} < \alpha$$

so that the passive response by itself is not sufficient to produce spiking; the incoming pattern must therefore be accompanied by random activity sufficient to exceed threshold if spiking is to occur in a post-synaptic cell. Accordingly to make an analytic calculation we now classify as events, t, those instances in which the incoming pattern is accompanied by random activity which produces spiking in at least one cell in the cortical bank.

For any event, t, one can assign a probability, $p_{k\beta}(t)$ that the kth cell fires during this event. (There is an increase in "events" per given time for a constant input of patterns since it becomes more likely that some cell fires as M increases. This is accounted for in the definition.) With (4.10) and (4.14) the recursion relation at the $t+1$ event for $\overline{m}_{k\beta}$ is then

$$\overline{m_{k\beta}(t+1)} = \gamma \overline{m_{k\beta}(t)} + \eta_{k\beta}[\gamma \overline{m_{k\beta}(t)} + m_R] \quad (4.26)$$

where

$$\eta_{k\beta} = \eta_0 p_{k\beta}(t+1) \qquad (4.27)$$

and m_R is defined as in (4.14) and (4.17).

We assume that random activity does not favor one cell or one pattern over another so that initially $p_{k\beta}$ is independent of k and β

$$p_{k\beta}(0) = p. \qquad (4.28)$$

The probability that the kth cell will fire upon the occurrence of d^β will increase as $m_{k\beta}$ increases. (This is clearly seen in our machine simulations.) For the moment we neglect this effect and let the probability remain constant

$$p_{k\beta}(t) = p, \qquad (4.29)$$

so that we can define

$$\eta_1 = \eta_0 p. \qquad (4.30)$$

We thus underestimate the rate at which $m_{k\beta}$ grows; more important, we underestimate probabilistic variations in the values of the m_{iv} at any t. This will be of importance in what we do later.

With these assumptions we obtain

$$\overline{m_{k\beta}(t+1)} = m(0) \, e^{(t+1)\Delta_1} + \eta_1 m_R \sum_{n=0}^{t} e^{n\Delta_1}. \quad (4.31)$$

This is analogous to (4.18) if we replace η_0 and Δ_0 by η_1 and Δ_1. The coefficients, $m_{k\beta}$, grow exponentially

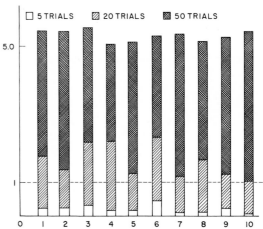

Fig. 6. A computer simulation of the responses of 10 cortical cells to repeated inputs of the same pattern without inhibition. Results are shown after 5, 20, and 50 trials. After 5 trials all cells are below threshold (set equal to 1 in all simulations). After 20 trials all cells have crossed threshold, but have not yet attained the maximum firing rate of 5/sec (Section 4). Above threshold, cell responses grow rapidly. After 50 trials all cells are firing near the maximal level. Having attained or exceeded the maximal firing rate, a cell's synaptic junctions remain unmodified until uniform decay reduces its response below maximum. At this point, the cell is once again subject to modification. This results in oscillation of the cell's response about the maximum. For this simulation $\gamma = 0.98$, $\eta = 0.1$, $\alpha = 1$ and $s_{i0} + s_{iR}$ was taken to be a random number, independent of i and t, uniformly distributed between 0 and 1.75

when $\Delta_1 > 0$. If the probabilities $p_{k\beta}$ are equal, the mean values of all coefficients grow in the same manner so that eventually every cell $k = 1 \ldots N$ will become committed to the pattern β.

Although the mean values of the coefficients are equal the variance is large[3]. Therefore the actual progress of the growth of the coefficients is in general that one of the N cells say the kth leads its neighbors in the response to the pattern β. When this cell becomes committed to β (and therefore fires everytime β is presented) the growth of $m_{k\beta}$ becomes very rapid. However, the neighbors eventually catch up so that finally all N cells become committed to β. This is clearly seen in Fig. 6, a computer simulation of (4.25) with ten cells and one pattern.

If we introduce the assumption that there is lateral inhibition between cortical cells, this, combined with the large variance noted above will insure that all of the cells do not become committed to the same pattern. Although there is little direct evidence of lateral inhibition among cortical cells, there is some experimental suggestion [24–26] that such inhibition

---

[3] Professor U. Grenander has made an exact calculation of the variance which will be published elsewhere.

occurs. In addition, there exists a very complex structure of lateral projections of the kind that produce lateral inhibition in the retina.

We therefore assume that (4.25) is modified so that the firing rate of the kth cortical cell upon the incoming pattern d^β is

$$c_k(t+1) = P[\gamma(M(t)\,d^\beta)_k + s_{k0} + s_{kR} \\ - \sum_{j \neq k} K_j c_j(t+1)]. \quad (4.32)$$

We allow the inhibition constant, K_j, to vary with the distance of the cells from one another. The addition of the lateral inhibition term of course converts (4.25) into a complex set of non-linear equations whose exact solution is difficult to obtain. The qualitative nature of the solutions, however, can be seen easily. Now the lead cell, which first becomes committed to a pattern and thus fires every time this pattern is presented, inhibits the firing of its neighbors (so that their response to the pattern no longer grows) and this with the decay due to γ diminishes their response.

We have approximated (4.32) in our machine simulations in the following way. We first calculate $c_i^{(0)}$ defined by (4.24) and then calculate the inhibition term using these $c_i^{(0)}$. What is used in the calculations is

$$c_k^{(1)}(t+1) = P[\gamma(M(t)d^\beta)_k + s_{k0} + s_{kR} - \sum_{j \neq k} K_j c_j^{(0)}(t+1)]. \quad (4.33)$$

The results are shown in Fig. 7. It is seen that once a cell becomes committed to the pattern not only does it grow rapidly but it suppresses further growth of its neighbors. With the decay constant γ set smaller than one the response of the neighbors to the pattern gradually diminishes until finally the comitted cell is the only one that responds at all.

N Cells, K Orthogonal Patterns

We now consider the situation in which K orthogonal patterns $d^1, d^2 \ldots d^K$ such that

$$(d^\alpha, d^\beta) = \delta_{\alpha\beta} \quad (4.34)$$

are presented to the N cortical cells $c^1 \ldots c^N$.

In general, of course, there is no reason that two arbitrary patterns e^α and e^β would produce activity in D, d^α, and d^β so that

$$(d^\alpha, d^\beta) = 0. \quad (4.35)$$

The precise value of this inner product depends on the "closeness" of the two patterns as well as the neural

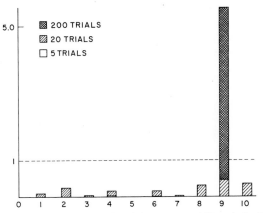

Fig. 7. A computer simulation of the responses of 10 cortical cells to repeated inputs of the same pattern, with intercellular inhibition [Eq. (4.33)]. The inhibition function depends on the firing rate of the cells as well as the "distance" between cells, $K_j = 0.5 e^{-|j-k|^{0.1}}$. The other parameters for this simulation are the same as for Fig. 6. Cell 9, which after 20 trials is "leading" the rest of the cells, crosses threshold first and inhibits the other cells. After 200 trials, cell 9 is firing maximally while the other cells are all below threshold. This configuration is stable, with the firing rate of cell 9 oscillating above the maximal level. The exact form of the inhibition function is not critical

mapping by which visual patterns are mapped into activity in lateral geniculate and retinal cells. The vector in the D space is a vector of very high dimensionality. We might imagine that two patterns separated from one another in the visual space will map into vectors in the many dimensional space which are orthogonal to one another. As we will see in the next section the relation of various vectors in the D space can be more subtle. In any case if patterns are separated (do not overlap) or sometimes even if they do overlap they will generate activity in the D space such that (4.35) is satisfied. We explore such a situation next.

Equation (3.16) can be written

$$\sum_{iv} m_{iv}(t+1)\, c^i \times d^v = \gamma \sum_{iv} m_{iv}(t)\, c^i \times d^v \\ + \eta P[\gamma M(t)\, d(t+1) + s_0(t+1) + s_R(t+1)] \times d(t+1). \quad (4.36)$$

Again, for the purpose of analysis, we define as events those occurrences in which any of the K incoming patterns are accompanied by activity in any of N cortical cells.

If we now define $p_\beta(t+1)$ as the probability that the βth pattern is the one presented at the $t+1$ event and $p_{k\beta}(t+1)$ as the probability that the kth cortical cell responds above threshold to the presentation of this β^{th} pattern on the $t+1$ event and assume,

12

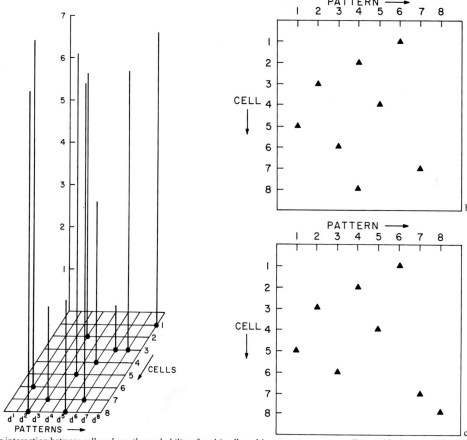

Fig. 8. (a) The interaction between cells reduces the probability of multi-cell, multi-pattern responses, as discussed in the text. This is illustrated for 8 cells and 8 orthogonal patterns with $\gamma = 0.995$, $\eta = 0.5$ and $K_j = \exp[-|j - k|^{0.1}]$. In addition $s_0 + s_R$ is uniformly distributed between 0 and 5. The large values for γ, η, and $s_0 + s_R$ were chosen because the growth of coefficients for orthogonal vectors is slow. The responses of the cells after 1300 trials (each trial consists of the presentation of one of the randomly chosen patterns) are shown. For clarity, many below threshold responses were set equal to zero in the figure. All cells have crossed threshold for some pattern and there is no duplication of committment. A configuration, without duplication, once attained is very stable. Although not shown here, there is no change in configuration after an additional 100 trials. The firing rates, having attained the maximum value, oscillate about that value. (b) and (c). With the same parameters, these figures illustrate the instability of a configuration with duplication. The triangles indicate which cell(s) is (are) responding to which pattern(s). After 1200 trials (this run is independent of the run shown in Fig. 8a) both cells 2 and 8 have crossed threshold for pattern 4 (Fig. 8b). This sets off a competition of inhibitory interactions so that after 200 additional trials, only one survivor is left, in this case cell 2 (Fig. 8c). The actual magnitudes of the responses have not been shown here

following the argument of the last sections, that these are independent of β, k, and t so that

$$p_\beta(t) = p_a$$
$$p_{k\beta}(t) = p_b,$$
(4.37)

(4.36) yields

$$\overline{m_{k\beta}(t + 1)} = \gamma\overline{m_{k\beta}(t)} + \eta_2(\gamma\overline{m_{k\beta}(t)} + m_R) \quad (4.38)$$

where

$$\eta_2 = \eta_0 p_a p_b \quad (4.39)$$

and m_R is defined as before.

We again obtain

$$\overline{m_{k\beta}(t + 1)} = m(0)\, e^{(t + 1)\Delta_2} + \eta_2 m_R \sum_{n=0}^{t} e^{n\Delta_2} \quad (4.40)$$

which is the same result as (4.18) if η_2 and Δ_2 replace η_0 and Δ_0. The analysis of (4.40) follows in the same way as before; in particular, for $\Delta_2 > 0$ all of the coefficients grow exponentially so that eventually all cells would become committed to all patterns.

However the variance is large. If we again introduce lateral inhibition, as in the previous section we find that the action of this single mechanism serves at the same time to limit the number of cells that respond to a single pattern as well as to limit the number of patterns to which a single cell becomes committed. There now arises a remarkable cooperative effect. Considering a particular pattern, we find that a leader arises among the cells in their response to this pattern. When this leader becomes committed to the pattern it suppresses the response to that pattern of its neighbors so that, with the action of the decay constant γ, the response of the neighbors diminishes, until finally the leader will be the only cell to respond at all to the pattern.

At the same time the cells interact to reduce the probability that any particular cell pick up more than one pattern. In order that a particular cell pick up two patterns this cell must be the leader for both of these patterns. The probability that the same cell be the leader for more than one pattern diminishes as the number of cells increases, but grows as the number of patterns increases. The same inhibition mechanism, therefore, serves to prevent more than one cell from picking up a given pattern and at the same time severely inhibits the cases in which a single cell picks up more than one pattern.

In our machine simulations this is clearly seen, Fig. 8. Furthermore, we rarely see a pattern picked up by more than one cell though we occasionally see a single cell pick up more than one pattern. The latter occurrence is virtually eliminated by the following maximum response condition.

When

$$c_k \geq c_k^{\max}, \delta M_{kj} = 0 \quad \text{for all } j, \text{ for } T \text{ trials} . \quad (4.41)$$

For sufficiently large values of T, the probability of a cell picking up two or more patterns becomes vanishingly small. In our simulations this number of trials is

$$T \cong \frac{\ln(c_k^{\max}/c_k)}{\ln \gamma} \quad (4.42)$$

where c_k is the value of response actually attained when the response exceeded maximum. This means that when a cell fires at or above the maximum value, $c_k^{\max}$, no further modification occurs at its dendrite junctions until uniform decay has diminished the response below this maximum value.

V. Organization of Visual Cortex

We have now shown that passive modification with lateral inhibition produces an organization of cortical cells in which on the average a single cell responds to a single pattern. Within the range of inhibition of a cell, when that cell becomes committed to a pattern, it prevents its neighbors from doing so. In addition, the interaction between cells limits the number of patterns to which a single cell becomes committed. With such an organization there is no connection between the pattern a cell picks up and the patterns its neighbors respond to. Therefore as one goes from one region of cortex to another one should find on the average no correlation between the patterns to which cells are committed.

Hubel and Wiesel, however, have occasionally observed striking correlations between neighboring cells in visual cortex. They note that an electrode penetration of this region results in a succession of cells with a highly correlated organization of their receptive fields and the patterns to which they respond. Within an anatomical column all cells have roughly the same receptive field position on the retina and respond optimally to lines of the same orientation. In addition, in monkey [27] at least, the "optimal line" rotates in a regular manner as the electrode, goes from one column to another, the pattern, sometimes going all the way around a circle.

If an anatomical column is pre-wired so that the firing of one cell excites the rest of the column, the uniformity of response of the cells in such a column to external patterns would follow. To explain a progression of patterns from one column to another by any pre-wiring would require a very elaborate genetic programming and would be difficult to make consistent with experimental results on development. Our problem, therefore, becomes to explain such an organization without any assumption of pre-wiring. We attempt to do this using what interaction there is between cells due to lateral inhibition and possible interactions between patterns.

In actual experience an animal would be presented with a great variety of visual situations. These are mapped by the neural mapping into activity in the ganglion and lateral geniculate cells. The experimental work discussed in Section II, leads us to assume that actual feature detectors which an animal constructs are extracted from repeated visual elements in its environment. These presumably occur over and over in the midst of more complex overall patterns. We have attempted to deal with this situation by repeating the features themselves and we have been able to show that cells do become committed to repeated patterns. We imagine that if the rest of the visual image is not repeated sufficiently often it behaves as noise. A more detailed exploration of this question is left to a later publication.

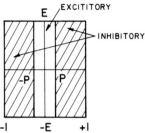

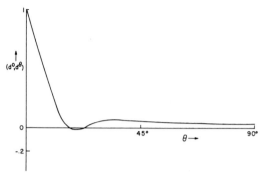

Fig. 9. A simplified representation of cell activities produced in a layer of lateral geniculate cells in response to a bar of light on the retina. The excitatory and inhibitory regions represent areas in which firing rates of cells are increased and decreased respectively when the stimulus is presented. The pattern is constructed from the on-center, off-surround retinal receptive fields given the topographic mapping of retina to lateral geniculate nucleus. The width of the excitatory region is $2P$ while the width of each inhibitory flank is $1 - P$. Both regions are of height $2E$

Fig. 10. The inner products (d^0, d^θ) are obtained by a computer simulation of the rotation of the pattern shown in Fig. 9. With variation of detail, the form of (d^0, d^θ) as a function of θ, as shown in this figure, is preserved over a large range of values of the parameters. Of particular importance here is the existence of a minimum. The value of θ for which the minimum occurs is a decreasing function of E (Fig. 9), the elongation of the pattern, while the value of (d^0, d^θ) at the minimum depends on P as well as on the "strengths" of the response in the different regions. For the figure shown, $P = 0.6$ and $E = 3.1$. The ratio of strengths of the excitatory and inhibitory regions is 1.1. If the regions are of equal strength the inner product goes to zero as θ goes to $\frac{\pi}{2}$

Although the details of the neural mapping from retina to cortex are not known, it is known that lateral geniculate and ganglion cells have approximately circular receptive fields organized into on-centers and off-surrounds[4] [28, 29]. The "neural mapping" of a spot of light (that is the profile of cell activities) is related to the receptive field of a neuron. For example, a spot of light on the retina would map into a region of activity in the lateral geniculate layer composed of an excitatory center and an antagonistic surround. The neural mapping of a bar would then look something as shown in Fig. 9. It is important to note that the bar maps into regions in D space which fire at a higher level than the spontaneous rate and others which fire at a lower level.

A quantity of great interest in calculating the development of the mapping, M, is the inner product between the mapping in the D space of the various patterns. In what follows we consider K non-orthogonal patterns: bars in the visual space oriented at fixed angles to one another about a common center. Each bar maps into a neural pattern in lateral geniculate, as shown in Fig. 9, suitably rotated. The inner products of these patterns with one another (d^0, d^θ) are shown in Fig. 10. It is striking, and important for what follows that (d^0, d^θ) has a minimum and rises again. This minimum is not critically dependent upon the exact size of the bar, nor the neural

mapping. It depends primarily on the fact that the on region is surrounded by an off region and occurs, with differences of detail, for a variety of shapes.

We now consider the development of M when there are N cortical cells and the K non-orthogonal patterns $d^1, d^2 \ldots d^K$. Although the $d^1 \ldots d^K$ are not orthogonal, they can be chosen to be linearly independent. In addition, any pattern in the D space, in which we are interested can be expressed as a linear combination of $d^1 \ldots d^K$. We can therefore use the $d^1 \ldots d^K$ as a non-orthogonal basis and write

$$M(t) = \sum_{iv} m_{iv}(t) \, c^i \times d^v . \qquad (5.1)$$

The developmental algorithm (3.16) now becomes

$$\sum_{iv} m_{iv}(t+1) \, c^i \times d^v = \gamma \sum_{iv} m_{iv}(t) \, c^i \times d^v \qquad (5.2)$$

$$+ \eta P [\gamma M(t) d(t+1) + s_0(t+1) + s_R(t+1)] \times d(t+1) .$$

Suppose that the $t + 1$ pattern is d^β. Then

$$M(t) \, d^\beta = \sum_{i\lambda} m_{i\lambda}(t) \, (d^\lambda, d^\beta) \, c^i . \qquad (5.3)$$

It is convenient to define

$$(d^\lambda, d^\beta) \equiv f(\lambda, \beta) . \qquad (5.4)$$

With the patterns we employ

$$f(\lambda, \beta) = f(\lambda - \beta) \qquad (5.5)$$

---

[4] A spot of light shined on the central region of the receptive field excites the cell, while an annular light stimulus surrounding the central zone inhibits the cell (reduces the response below the spontaneous level). Cells of both the on-center, off-surround as well as off-center, on-surround variety are found in the ganglion cell layer and lateral geniculate nucleus. For simplicity we deal here only with the first cell type.

is a function of the angle between the axis of the two patterns λ and β. Therefore (5.3) can be written:

$$M(t)\,d^\beta = \sum_{i\lambda} m_{i\lambda}(t)\,f(\lambda-\beta)\,c^i. \qquad (5.6)$$

If again $p_\beta(t+1)$ is the probability that the βth pattern is presented at the $t+1$ event and $P_{k\beta}(t+1)$ is the probability that the kth cell fire upon the presentation of β at the $t+1$ event, and these as before are assumed to be independent of β, k, and t, we obtain

$$\sum_{iv}\left[\overline{M_{iv}(t+1)}-\gamma\overline{M_{iv}(t)}\right]c^i \times d^v$$
$$= \eta_2\left[\gamma\overline{M(t)}\,d^\beta + m_R\right]\times d^\beta \qquad (5.7)$$

where η_2 and m_R are defined by (4.39) and (4.17). Using the linear independence of the d^v yields

$$\overline{m_{i\beta}(t+1)}=\gamma\overline{m_{i\beta}(t)}+n_2\left[\gamma\sum_\lambda \overline{m_{i\lambda}(t)}\,f(\lambda-\beta)+m_R\right]. \quad (5.8)$$

We leave the solution of this set of coupled equations to a future publication. However, just as for the orthogonal vectors, the growth of the coefficients can be exponential. Due to the non-orthogonality of the vectors however the growth is more rapid and this growth depends on the inner products of the vectors.

In this situation a cell will respond not only to a pattern to which it is committed but, because of the non-orthogonality, $(d^\alpha, d^\beta)\neq 0$, to other patterns also. Thus lateral inhibition in this situation is even more important to limit the number of patterns to which a cell responds.

We include the effect of lateral inhibition in our simulations according to Eqs. (4.24) and (4.33); if the $t+1$ pattern is d^β, the developmental algorithm becomes

$$m_{iv}(t+1)=\gamma m_{iv}(t)\quad v\neq\beta \qquad (5.9)$$

$$m_{i\beta}(t+1)=\gamma m_{i\beta}(t)+\eta P\left[\gamma\sum_\lambda m_{i\lambda}(t)\,f(\lambda-\beta)+s_{i0}(t+1)\right.$$
$$\left.+s_{iR}(t+1)-\sum_{j\neq i}K_j c_j^{(0)}(t+1)\right].$$

We assume, in addition a maximal response of a cell to any pattern, as discussed at the conclusion of Section IV.

Our computer simulations behave in the following manner. A first cell, call it the ith, commits itself to some pattern, say β. Because of rapid growth above threshold, this cell soon attains its maximum firing level upon presentation of the pattern, β. During this period the other cells also modify their responses, but are prevented by lateral inhibition from picking up the β pattern. Since lateral inhibition falls off with the distance between cells, it is the cells

neighboring cell i that are most strongly inhibited when i fires. Once cell i has acquired the pattern β, the neighboring cells most easily acquire the pattern which excites the ith cell the least.

If for example the ith cell has become committed to the β pattern due to the growth of the coefficient, $m_{i\beta}$, then the ith cell may also respond to the α pattern as

$$(Md^\alpha)_i = \sum_v m_{iv}(d^v, d^\alpha) = m_{i\beta}(d^\beta, d^\alpha)+\cdots \quad (5.10)$$

is not equal to zero if $(d^\beta, d^\alpha)\neq 0$. The probability that the ith cell fire upon the presentation of α is enhanced depending on the value of (d^β, d^α) and this (with lateral inhibition) will suppress the activity of the neighboring cell, thus preventing it from becoming committed to α. The neighboring cells therefore, most easily acquire a pattern which will excite i the least – the pattern d^α for which (d^β, d^α) is a minimum.

From Fig. 10 we see that such a minimum occurs for patterns rotated from one another by the angle θ (about $18°$ for the parameters we have chosen) so that it becomes natural to define such rotated patterns as "neighbors". It then follows that a cell neighboring a committed cell has an enhanced probability of becoming committed to a "neighboring" pattern. [Suceeding cells, by similar arguments, can be shown to have an enhanced probability of acquiring the "next neighboring" pattern.] The result is that the interaction of mapped patterns due to their non-orthogonality, combined with lateral inhibition which diminishes with distance, yields a most probable configuration of committed cells which is the one observed by Hubel and Wiesel in monkey.

This most probable configuration is dependent on the values of the parameters and so could differ from species to species. Even if the parameters are favorable, this configuration need not always emerge. Two "leaders" may arise in different regions, resulting in competition between their neighborhoods reducing the probability of a completely rotated organization. In addition, less probable configurations, can, of course, also emerge.

In our computer simulations, the remarkable rotation of receptive fields around a full circle does occur not infrequently, even in a linear chain of small numbers of cells. When this organization does not occur, there is still in most cases a strong correlation between receptive field orientation of neighboring cells. Since we have so far dealt only with relatively short chains, and since the organization is dependent on the values of the parameters, it is too early to state definite conclusions. We are presently pursuing further simulations for planar arrays of a larger number of cells which we expect will clarify the situation. It seems likely to us that for such larger arrays with a reasonable choice of parameters, the organization should become more striking.

VI. Conclusion

We have shown that passive modification of the strength of synaptic junctions which leads to the construction of neural mappings with some of the properties of memory can also account for the development of feature detectors in visual cortex. With such synaptic modification a cortical cell will become committed to an *arbitrary* but repeated external pattern (as long as that pattern lies within its receptive field) and thus come to fire every time that pattern is presented *even if that cell has no genetic pre-disposition to respond to the particular pattern*.

With the additional assumption of lateral inhibition between cortical cells, the number of cells which respond to a pattern as well as the number of patterns that are picked up by a cell are severely limited.

The introduction of a simple neural mapping from the visual field to the lateral geniculate leads to an interaction between patterns which combined with our previous assumptions seems to lead to a progression of patterns from column to column of the type observed by Hubel and Wiesel in monkey.

Appendix

Relation of the Matrix M to the Firing Rate of Cortical Cells

Let T_{ij} be the total amount of transmitter deposited on all dendrites of cortical cell i due to activity in the lateral geniculate axon j or on the post-synaptic membrane of the ideal junction (ij) due to spiking in the pre-synaptic axon j. According to a standard theory of synaptic transmission [30] transmitter is destroyed in proportion to the amount present and is released at axon j in proportion to the axonal spiking frequency. Therefore we may write

$$\frac{dT_{ij}}{dt} = -\alpha T_{ij} + M'_{ij} d_j \qquad (A1)$$

where αT_{ij} is the amount of transmitter destroyed per unit time and M'_{ij} is the amount of transmitter deposited in the synaptic cleft per spike in axon j. The transmitter deposited on the dendrites of the cell body alters the dendrite membrane conductance to ionic flow, thereby altering the potential difference across the membrane. This produces a current flow along and through the dendrite membrane resulting finally in an altered potential difference across the axon hillock; there the total potential difference due to transmitter deposited on all of the dendrites is integrated.

Let V_{ij} be the potential difference across the axon hillock due to transmitter deposited at the synapse (ij). As a simple representation of a complex sequence of events we write

$$V_{ij} = k_{ij} T_{ij}. \qquad (A2)$$

Here we neglect all time lapses and distortions of the shape of the signals in time; however we do include possible differences of resistance and other non-time dependent effects in the individual dendrites. Putting this into (A1) we have

$$\frac{dV_{ij}}{dt} = -\alpha V_{ij} + k_{ij} M'_{ij} d_j. \qquad (A3)$$

The total potential difference, V_i, across the axon hillock of cell i is given by summing the contributions from all the cell's dendrites

$$V_i = \sum_{j=1}^{N} V_{ij} \qquad (A4)$$

so that

$$\frac{dV_i}{dt} = -\alpha V_i + D_i \qquad (A5)$$

where

$$D_i \equiv \sum_{j=1}^{N} k_{ij} M'_{ij} d_j. \qquad (A6)$$

The solution of the differential Eq. (A5) (which is just Knight's forgetful integrate and fire model of neuronal spiking [31]) with the initial condition $V_i(0) = 0$ is

$$V_i(t) = e^{-\alpha t} \int_0^t e^{\alpha t'} D_i(t') \, dt'. \qquad (A7)$$

Since lateral geniculate cells respond in short bursts of duration τ (about one second) to stimuli in the receptive field we consider the situation in which the incoming spikes d_j continue for the time interval, τ, so that

$$D_i(t') = D_i \qquad 0 < t < \tau$$
$$= 0 \qquad \tau < t. \qquad (A8)$$

Putting this into (A7) gives

$$V_i(t) = \frac{D_i}{\alpha} (1 - e^{-\alpha t}) \qquad t \lessgtr \tau$$
$$= \frac{D_i}{\alpha} (e^{\alpha \tau} - 1) e^{-\alpha t} \qquad t > \tau. \qquad (A9)$$

The axon hillock of cell i is presumably the most excitable region of the cell so that the cell fires if V_i is above threshold, θ, at the axon hillock: $V_i > \theta$. As Knight has shown this will be the case if $D_i / \alpha\theta \gtrless 1$; further (well above the threshold) the spiking frequency c_i will become proportional to D_i / θ so that

$$c_i = \frac{D_i}{\theta} = \Sigma M_{ij} d_j \qquad \left(\frac{D_i}{\theta} \gg 1\right). \qquad (A10)$$

We have defined M_{ij} [the "efficacy" of the $(ij)^{\text{th}}$ synaptic junction] so that

$$M_{ij} \equiv \frac{1}{\theta} k_{ij} M'_{ij}. \qquad (A11)$$

This will contain the "strength" of the $(ij)^{\text{th}}$ synapse as well as such other factors as the resistance of the $(ij)^{\text{th}}$ dendrite.

Above threshold, in general, c_i is related to D_i (in Knight's forgetful integrate and fire model) by

$$c_i = \frac{-\alpha}{\ln\left(1 - \frac{\alpha\theta}{D_i}\right)}. \qquad (A12)$$

Threshold occurs at $D_i / \alpha\theta = 1$; below this value (although there may be irregular spiking due to potential fluctuations) the value of c_i (due to excitation from the dendrites) is zero. In the region

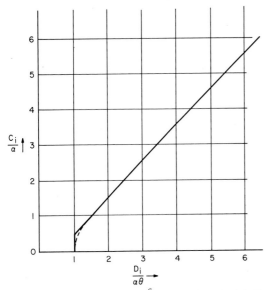

Fig. 11. The dashed curve shows $\dfrac{c_i}{\alpha}$ as a function of $D_i/\alpha\theta$. Above threshold ($D_i/\alpha\theta = 1$) the nonlinear rise is approximated by the projection operator, $P[D_i/\alpha\theta]$, the solid line

interesting to us we can approximate the exact form of c_i as given by (A12) by the first two terms of its expansion

$$c_i \simeq \frac{D_i}{\theta} - \alpha/2 \qquad \frac{D_i}{\alpha\theta} \geqq 1$$
$$= 0 \qquad\qquad \frac{D_i}{\alpha\theta} < 1 . \tag{A13}$$

(In this way we linearize the non-linear rise just above threshold. See Fig. 11). It is therefore convenient to define the function $P[x]$ such that

$$P[x] = x - \alpha/2 \qquad x/\alpha \geqq 1$$
$$= 0 \qquad\qquad x/\alpha < 1 . \tag{A14}$$

Spontaneous and Random Stimulation

In addition to the V_i of Eqs. (A4) and (A9) there can be a stimulus at the axon hillock, i, due either to spontaneous activity in the lateral geniculate cell and/or random fluctuations in its own membrane potential (possible pacemaker activity). Following (A5) the first can be expressed as

$$\sum_{j=1}^{N} k_{ij} M'_{ij} r_{j0} \tag{A15}$$

where r_{j0} is the spontaneous activity (spiking frequency) of lateral geniculate cell j (presumably this should be independent of j). The second contribution, D_{iR}, is the random stimulus fluctuation at the axon hillock of cell i. Equation (A5) becomes

$$\frac{dV_i}{dt} = -\alpha V_i + \sum_{j=1}^{N} k_{ij} M'_{ij} d_j + \sum_{j=1}^{N} k_{ij} M'_{ij} r_{j0} + D_{iR} . \tag{A16}$$

Since r_{j0} does not vary rapidly in time, it is constant over the period of a burst, τ, so that all of the results of (A1) to (A14) apply if

$$D_i \rightarrow D_i + D_{i0} \tag{A17}$$

where

$$D_{i0} \equiv \sum_{j=1}^{N} k_{ij} M'_{ij} r_{j0} \tag{A6a}$$

and again M_{ij} is defined by (A11).

If $M_{ij}(0) = 0$ and there are only excitatory junctions, $M_{ij} > 0$, then D_{i0} will increase with time so that, if the threshold level remains unchanged, spontaneous activity will increase. If however $M_{ij}(0) > 0$ or if there are inhibitory as well as excitatory junctions so that

$$\sum_{j} M_{ij}(t)$$

can remain constant, spontaneous activity due to this term need not increase.

Random fluctuations in the stimulus probably should be regarded as rapid compared to τ. If we assume that D_{iR} is a constant stimulus over a time τ_R

$$\tau_R < \tau .$$

(A9) is modified, for times shorter than τ_R by replacing

$$D_i \rightarrow D_{iT} = D_i + D_{i0} + D_{iR} . \tag{A18}$$

We then have

$$V_{iT}(t) = V_i + V_{i0} + V_{iR}$$
$$= \frac{1}{\alpha}(D_i + D_{i0} + D_{iR})(1 - e^{-\alpha t}) \; t \gtrless \tau_R . \tag{A19}$$

Since random or spontaneous spiking is very low at early stages in development we imagine that such stimuli are separated by times large compared to α^{-1} so that there is no accumulation of potential due to these stimuli and they result in potential fluctuations which reach their maxima at τ_R and which decay exponentially thereafter. Even in the absence of any random or continuing stimuli (when $D_{iT} = D_i$ since $\alpha\tau \gg 1$, the potential (A9) rises rapidly to its maximum value. This is D_i/α, which may be regarded as an "average" value of $V_i(t)$

$$V_i \equiv \frac{1}{\tau} \int_0^\infty V_i(t') \, dt' = D_i/\alpha .$$

In a case of particular interest to us $V_i(t) + V_{i0}(t)$ is below threshold in the absence of random excitation. If $V_i(t) + V_{i0}(t) = \dfrac{D_i + D_{i0}}{\alpha}$ is below threshold, the occurrence of random excitation of the form above will put one above threshold if

$$\frac{D_i + D_{i0} + D_{iR}}{\alpha} \gtrless \theta . \tag{A20}$$

Putting these results together and using the operator defined (A14) we obtain for the spiking frequency of the ith cortical cell during the time interval of spontaneous excitation, τ_R,

$$c_i = P[(Md)_i + s_{i0} + s_{iR}] = (Md)_i + s_{i0} + s_{iR} - \alpha/2$$
$$\text{if} \quad (Md)_i + s_{i0} + s_{iR} \geqq \alpha \tag{A21}$$
$$= 0 \quad \text{otherwise} .$$

Where we have defined

$$s_{i0} = \frac{D_{i0}}{\theta}$$
$$\tag{A22}$$
$$s_{iR} = \frac{D_{iR}}{\theta} .$$

It is also convenient to extend the definition of the function $P[x]$ so that it can operate on a vector. If u is an arbitrary vector $P[u]$ is defined as a vector whose i^{th} component is

$$(P[u])_i = u_i - \alpha/2 \qquad u_i \geqq \alpha$$
$$= 0 \qquad u_i < \alpha . \qquad (A\,23)$$

The vector of post-synaptic spiking frequencies, c may then be written

$$c = P[Md + s_0 + s_R] . \qquad (A\,24)$$

Acknowledgements. We would like to express our appreciation to all of our colleagues at the Center for Neural Studies for their encouragement and advice. We have profited particularly from conversations with Professors Anderson, Elbaum, Grenander, and McIlwain and Mr. Jack Silverstein. In addition, we would like to thank Mr. David Ferster for his help in writing programs.

References

1. Hubel, D. H., Wiesel, T. N.: J. Physiol. **148**, 574 (1959)
2. Hubel, D. H., Wiesel, T. N.: J. Physiol. **160**, 106 (1962)
3. Blakemore, C., Cooper, G. F.: Nature **228**, 477 (1970)
4. Hirsch, H. V. B., Spinelli, D. N.: Exp. Brain Res. **13**, 509 (1971)
5. Blakemore, C., Mitchell, D. E.: Nature **241**, 467 (1973)
6. Pettigrew, J. D., Freeman, R. D.: Science **182**, 599 (1973)
7. Imbert, M.: to be published
8. Cooper, L. N.: In: Lundquist, B., Lundquist, S. (Eds.): Proceedings of the Nobel symposium on collective properties of physical systems. London-New York: Academic Press 1973
9. Malsburg, C. von der: Kybernetik **14**, 85 (1973)
10. Perez, R., Glass, L., Shlaer, R.: To be published
11. Hubel, D. H., Wiesel, T. N.: J. Physiol. **165**, 559 (1963)
12. Talbot, S. A., Marshall, W. H.: Amer. J. Opthamol. **24**, 1255 (1941)
13. Daniel, P. M., Whitteridge, D.: J. Physiol. **159**, 203 (1961)
14. Hubel, D. H., Wiesel, T. N.: J. Neurophysiol. **28**, 229 (1965)
15. Hubel, D. H., Wiesel, T. N.: J. Neurophysiol. **26**, 994 (1963)
16. Barlow, H. B., Pettigrew, J. C.: J. Physiol. **218**, 98P (1971)
17. Wiesel, T. N., Hubel, D. H.: J. Neurophysiol. **28**, 1029 (1965)
18. Anderson, J. A.: Math. Bio-Sci. **8**, 137 (1970) and Anderson, J. A.: Math. Bio-Sci. **14**, 197 (1972)
19. Anderson, J., Cooper, L., Nass, M., Freiberger, W., Grenander, U.: AAAS Symposium, Theoretical Biology and Bio-Mathematics (1972)
20. Herz, A., Creutzfeldt, O., Fuster, J.: Kybernetik **2**, 61 (1964)
21. Hubel, D. H., Wiesel, T. N.: J. Neurophysiol. **26**, 994 (1963)
22. Enroth-Cugell, C., Robson, J. G.: J. Physiol. **187**, 517 (1966)
23. Hoffmann, K. P., Stone, J.: Brain Research **32**, 460 (1971)
24. Scheibel, M. E., Scheibel, A. B.: In: Schmitt, F. O. (Ed.): The neurosciences, Vol. II. New York: Rockefeller University Press, N. Y. 1970
25. Pettigrew, J. D., Nikara, T., Bishop, P. O.: Exp. Brain Research **6**, 391 (1968)
26. Creutzfeldt, O., Ito, M.: Exp. Brain Research **6**, 324 (1968)
27. Hubel, D. H., Wiesel, T. N.: J. Physiol. **195**, 215 (1968)
28. Kuffler, S. W.: J. Neurophysiol. **16**, 37 (1953)
29. Hubel, D. H., Wiesel, T. N.: J. Physiol. **155**, 385 (1961)
30. Stevens, C. F.: Neurophysiology: A Primer. New York: John Wiley and Sons 1966
31. Knight, B. W.: J. Gen. Physiol. **59**, 734 (1972)

Prof. Dr. L. N. Cooper
Dept. of Physics
Brown University
Providence, Rhode Island 02912, USA

A Theory for the Acquisition and Loss of Neuron Specificity in Visual Cortex

(with F. Liberman and E. Oja)

Biol. Cybern. **33**, 9 (1979)

In the previous paper, Nass and I focused on the environmental driven development of Hubel–Wiesel-type feature detectors, showing that with Hebbian-type modification, exposure of the cells to repeated patterns would result in cells strongly responsive to those patterns. We encountered the usual, now well-known, problems with Hebbian mechanisms. A cutoff was required in order that the system not run away. In addition, we had to introduce the idea of lateral inhibition between the cells so that they didn't all acquire the same pattern.

Similar attempts were made at approximately the same time by von der Malsburg and Peres *et al.* They ran into the same problems. (As early as this one sees the introduction of various ideas to prevent the Hebbian growth of synapses from running away.) von de Malsburg and Peres introduced the idea that, somehow, the sum of synaptic strengths would remain constant. Nass and I used the device of ending the modification of synaptic strengths when the cell response reached a maximum level. (Depending on the precise formulation, these are not unrelated.)

In the fall of 1976 I began what was to be become a seven-year stint as Professor of the Fondation de France. In this arrangement, I spent a few months a year in Paris coordinating the work of our laboratory at Brown with the laboratories of Jean-Pierre Changeux and Michel Imbert at the Institut Pasteur and the College de France. As part of the arrangement, I was to give a short series of lectures each December followed by a seminar series in the month of May. This, coupled with our acquiring a truly exquisite rental on the Place Furstenberg available for just those months we were to be in Paris, resulted in a not unpleasant situation in which Kay and I spent several months a year living and working in Paris. Although it was really only about two or three months each year, myths grew and it was rumored that we were mostly in Paris and only occasionally in the United States.

For years Howard Swearer, then President of Brown, would greet me when he saw me on campus saying, "*Leon, how come you're not in Paris?*" The best I could ever do for a response was something like, "*I'm leaving tomorrow*", "*I came back yesterday*", or "*How come you're not fishing?*"

My lectures in the fall of '76 concerned general properties of neural networks. I interacted strongly with the people in the laboratory, particularly Michel Imbert and Pierre Buisseret, and was very much struck with their experimental results showing that the tuning properties of visual cortical cells depended strongly on the environment in which the young kittens were reared. If the kittens were reared in a normal environment the cells were sharply tuned. Dark reared, the cells were broadly tuned; very few of the normal Hubel–Weisel cells appeared. And, most striking, the situation could be reversed by very short exposure to a normal visual environment at the height of the critical period. This environmental dependence of selectivity cried to be explained.

While I was lecturing in December of 1977 on various approaches to neural networks and visual cortex, it occurred to me that one could generalize the rule that Nass and I had employed to another in which synapses increased if the post-synaptic cell response was above what came to be known the modification threshold and decreased if the response was below. This would lead to the results that Nass and I had obtained and, in addition, could produce selectivity in a patterned environment and no selectivity in a noisy environment. The combination seemed ideal to explain both the results of Imbert and Buisseret along with the prior results.

Erkki Oja, from Tuovo Kohonen's laboratory in Finland, was visiting us when I returned to Brown and we decided to try to put these ideas together. This resulted in the following paper. There passive modification picked up from the prior two papers, now would be called threshold-passive modification. The general result of the paper is that, with a variety of somewhat artificial assumptions, we can in fact obtain selectivity in patterned environments and lack of selectivity in a noise-like environment.

Biol. Cybernetics 33, 9–28 (1979)

Biological
Cybernetics
© by Springer-Verlag 1979

A Theory for the Acquisition
and Loss of Neuron Specificity in Visual Cortex*

Leon N Cooper, Fishel Liberman, and Erkki Oja[1]

Center for Neural Science and Department of Physics, Brown University, Providence, R.I., USA

Abstract. We assume that between lateral geniculate and visual cortical cells there exist labile synapses that modify themselves in a new fashion called threshold passive modification and in addition, non-labile synapses that contain permanent information. In the theory which results there is an increase in the specificity of response of a cortical cell when it is exposed to stimuli due to normal patterned visual experience. Non-patterned input, such as might be expected when an animal is dark-reared or raised with eyelids sutured, results in a loss of specificity, with details depending on whether noise to labile and non-labile junctions is correlated. Specificity can sometimes be regained, however, with a return of input due to patterned vision. We propose that this provides a possible explanation of experimental results obtained by Imbert and Buisseret (1975); Blakemore and Van Sluyters (1975); Buisseret and Imbert (1976); and Frégnac and Imbert (1977, 1978).

Introduction

Experimental work of the last generation, beginning with the pathbreaking work of Hubel and Wiesel (1959, 1962), has shown that there exist cells in visual cortex (areas 17, 18 and 19) of the adult cat that respond in a precise and highly tuned fashion to external patterns – in particular bars or edges of given orientation and moving in a given direction. Much further work (Blakemore and Cooper, 1970; Blakemore and Mitchell, 1973; Hirsch and Spinelli, 1971; Pettigrew and Freeman, 1973) has been taken to indicate that the number and response characteristics

---

* This work was supported in part by a grant from the Ittleson Foundation, Inc.

1 *Permanent address*: Department of Technical Physics, Helsinki University of Technology, SF-02150 Espoo 15, Finland

of such cortical cells can be modified. It has been observed in particular by Imbert and Buisseret (1975); Blakemore and Van Sluyters (1975); Buisseret and Imbert (1976); and Frégnac and Imbert (1977, 1978), that the relative number of cortical cells that are highly specific in their response to visual patterns varies in a very striking way with the visual experience of the animal during the critical period.

These results provide strong evidence for the modification by experience of the response characteristics of individual cortical neurons. A question of great interest is just what the form of such modification is. In this paper, we show that a proposed new form of neuron (synaptic) modification, closely related to prior forms that have been shown to lead to distributed memories in neural networks, can also account for many of the experimental results.

Imbert and Buisseret have classified cortical cells that respond to visual stimuli into three groups – aspecific, immature and specific. They and Frégnac and Imbert have measured the relative proportions of these groups depending on the visual experience of the animal.

The classification of cells used was as follows:

Non-specific Units

The receptive fields of these cells are usually circular and often very large (5–15°). They are activated equally well by bars or spots of light moving in any direction over their receptive field.

Immature Units

The receptive fields of these cells tend to be rectangular ($10° \times 8°$) and the response to rectilinear stimuli is greater than to spots. When moving stimuli are used, their selectivity as a function of the direction of movement is greater when bars or slits are used than with spots. They exhibit a degree of orientation selec-

0340-1200/79/0033/0009/$04.00

10

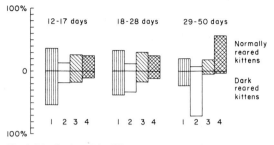

Fig. 1. Distribution of the different types of cells in three age groups in the normally reared kittens (upper part) and in the dark-reared kittens (lower part), from Frégnac and Imbert (1977, 1978). We have normalized the ordinate so that the heights are the percentages of cells in the various functional groups. Type 1, non-activatable (▥); 2, non-specific (☐); 3, immature (▨); 4, specific (▩)

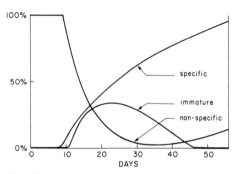

Fig. 2. Evolution of the development of the various specificity groups in cats raised normally. The curves taken from Y. Frégnac (1978) are the results of his unweighted regression analysis of experimental data on 1050 cells

tivity although responding to a range of orientation up to 45° either side of the optimal orientation. In all cases there is a clear null orientation, orthogonal to the optimal, which has no effect.

Specific Units

These cells are orientation-specific and exhibit all the characteristics of simple or complex cells in the adult cat (Hubel and Wiesel, 1962; Henry et al., 1974; Pettigrew, 1974). Such cells exhibit sharp tuning curves and do not respond to orientations more than 30° on either side of the optimal orientation. In addition, their receptive fields are smaller in size (2° × 4°) than those of immature cells.

The distribution of the different cell types in three age groups is shown in Figure 1.

Examination of these results, which were obtained from the study of 1050 cells, confirms that cells having some of the highly specific response properties of adult visual cortical neurons, especially concerning orientation selectivity are present in the earliest stages of post-natal development independent of visual experience (Frégnac and Imbert, 1977, 1978). However, visual experience between 17 and 70 days is critical in determining the evolution of these cells. Animals reared normally showed a marked increase in the number of specific cells as compared with aspecific. (The period between 17 and 28 days is usually sufficient to reach the normal adult level of specificity.) The reverse is true for animals reared in the dark. A statistical analysis of this evolution, performed by Frégnac, (1978) shows clearly the striking dependence of the ratio of sharply tuned to broadly tuned cells depending on the experience of the animal (Figs. 2 and 3).

In addition as has been shown by Imbert and Buisseret (1975); Buisseret and Imbert (1976); and Buisseret et al. (1978) as little as six hours of normal visual experience at about 42 days of age can alter in a striking fashion the ratio of specific or immature to aspecific cells (Fig. 4). That such a short visual experience can change the tuning ratios so markedly is clear evidence of the great plasticity of these cortical cells at the height of the critical period.

These results seem to us to provide direct evidence for the modifiability of the response of single cells in the cortex of a higher mammal according to its visual experience. Depending on whether or not patterned visual information is part of the animal's experience, the specificity of the response of cortical neurons varies widely. With normal patterned experience specificity, developing normally, increases. Deprived of normal patterned information (dark-reared or lid-sutured at birth, for example) specificity decreases. Further, even a short exposure to patterned information after six weeks of dark-rearing can reverse the loss of specificity and produce an almost normal distribution of cells.

The data also indicate that for some cells, at least, some orientation preference is built-in and develops independent of very early visual experience. This initial preference seems to be quite stable in acute experiments (Bienenstock and Frégnac, to be published) and further may be stable under limited dark-rearing conditions since orientation preference can be retrieved with visual experience. There is some evidence however that the orientation columns that result after initial deprivation of patterned experience are not necessarily the same as those that result with normal experience (Blakemore and Van Sluyters, 1974; Movshon, 1976).

In what follows we present a theory (which generalizes the theory of Nass and Cooper (NC), 1975) in which we attempt to account for many of these results.

I. Mapping from the Visual Field to Visual Cortex

The organization and receptive fields of retinal ganglion cells as well as the relay through the lateral geniculate nucleus (LGN) have been intensively studied and the neural pathway from retina to visual cortex is one of the best known signal transmitting systems in a brain. For our present purpose, however, the precise details of this mapping are not as important as the requirement that there exist a correspondence that is sufficiently one-to-one, between external stimuli in the visual environment and the actual sensory input to the cortical neuron. This is a basic assumption in what follows.

In a visually active animal, the stimuli converging on a cortical neuron at a given time, are predominantly determined by (or have as a dominating component) the visual pattern falling on the retina. We generalize the mapping of Nass and Cooper (1975) in which a pattern e^k in the external world is mapped into activity in cortical cells to include two classes of synaptic junctions between cortical and lateral geniculate cells: those which are modifiable and those which are not at all (or substantially less) modifiable. (A similar hypothesis has been made by Blakemore and Van Sluyters (1975).) We do this to include the possibility that innate instructions are contained in non-modifiable junctions. This results, under circumstances that will become clear later, in the ability of a cortical cell to retrieve specificity to its original orientation preference even if it has lost this specificity due to lack of patterned visual input during some portion of the critical period. (The possibility of having innate information in non-modifiable junctions that already existed in NC is also retained.)

We thus divide stimuli carried by input fibers to a cortical cell into two groups, depending on the modifiability of the junctions through which they are transmitted to the post-synaptic membrane. These stimuli are represented as two simultaneous spatial pattern vectors, one comprised of those signals that are input to modifiable junctions, the other comprised of those signals that are input to weakly modifiable, or non-modifiable junctions as shown in Fig. 5. A pattern e^k in the external world impinges on the retina and is transmitted through the retinal-LGN pathway, evoking a set of responses at the outputs of LGN cells to visual cortex.

These responses are the elements of two signal vectors b^k and d^k which then are input into layer IV C of visual cortex to cortical neurons sharing approximately the same receptive field, the one containing the pattern e^k.

It should be stressed that the vectors b^k and d^k are labelled by definition, according to the modifiability of their respective synaptic junctions with the cortical cell. The elements of both vectors are just the responses in different LGN axons to one and the same external stimulus e^k. It is therefore possible that these two

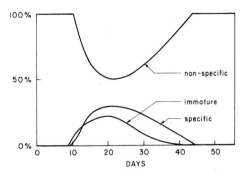

Fig. 3. Evolution of the development of the various specificity groups in cats raised in total darkness. The curves taken from Y. Frégnac (1978) are the results of his unweighted regression analysis of experimental data on 1050 cells

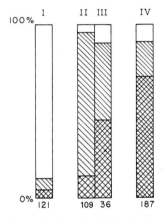

Fig. 4. Distribution in % of the three types of visual cortical units (area 17) recorded after visual exposure in six week old dark-reared kittens, from Buisseret et al. (1978). Visual stimuli, slits or spots, were displayed on a tangent screen. Columns: I, dark-reared, IV, normally reared kittens. During 6 h of exposure, conditions were: II and III, freely moving; in III, 12 h in the dark followed the 6 h of exposure. Numbers of visual cells recorded are given under each column. Specific cells (▨) are activated by orientated stimuli within a sharp angle (<60°). Immature cells (◩) are activated by orientated stimuli within a larger angle (<150°). Non-specific cells (□) are activated by non-orientated stimuli moving in any direction. A statistical analysis reveals no significant difference in the percentages of immature and specific units between columns III and IV. Therefore it may be that a six hour exposure to visual input followed by twelve hours in the dark is sufficient to produce a distribution of cortical cells similar to that of normally reared animals

vectors are equal (which would be the case, for example, if the two types of synapses share the same input fibers) or that one or the other is zero, corresponding to a situation in which one or the other type of synapses is missing. It is also possible that modifiable or non-modifiable synaptic junctions lie close together and have similar basic properties (excitatory, inhibitory) or it might be that they are separated and have

12

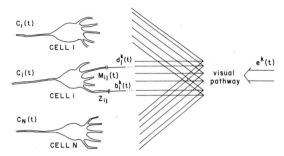

Fig. 5. A visual pattern $e^k(t)$ (e.g., a bar or an edge of a given orientation and moving in a given direction) falls on the retina at time t. This pattern evokes a set of parallel responses in the axons of the LGN cells, and these are presynaptic inputs to a layer of modifiable neurons in visual cortex. The inputs are divided in two parts in the model: vector $d^k(t)$ contains $d_j^k(t)$ that fall on modifiable junctions, vector $b^k(t)$ contains $b_j^k(t)$ that fall on non-modifiable junctions. The strengths at time t of the modifiable junctions, $M_{ij}(t)$, comprise a matrix $M(t)$, while the strengths of the non-modifiable junctions, Z_{ij}, comprise a matrix Z. The responses of cortical neurons, making up a vector $c(t)$, are obtained from weighted sums of the different inputs with weights given by the corresponding junction strengths $M_{ij}(t)$ and Z_{ij}

different properties. (For example, the modifiable junctions might be purely excitatory.)

The response of a given cortical neuron depends on the combined potentials produced by both b^k and d^k. Assuming linearity, the effects of these individual stimuli can be expressed using two junctions matrices, here called M and Z, instead of just one as in NC. The elements M_{ij} of the ith row of M denote the strengths of the modifiable junctions of the ith cortical neuron, and the elements Z_{ij} of the ith row of Z denote in a similar fashion, the non-modifiable junction strengths of that same neuron. The basic component of the post-synaptic potential, produced by the external stimulus e^k, on the soma of the ith cortical cell then is:

$$\sum_{j=1}^{L_d} M_{ij}d_j^k + \sum_{j=1}^{L_b} Z_{ij}b_j^k.$$

Here, using the same notational rules as NC, d_j^k and b_j^k are the jth element of d^k and b^k, respectively. L_d and L_b are the dimensions of d^k and b^k or the numbers of columns in the matrices M and Z, and need not be equal. It is also likely that the number of the cortical neurons, N, (the number of rows in both matrices M and Z) is different from either L_d or L_b. The mutual ratios of N, L_d, and L_b depend on the amount of convergence in the mapping from LGN to visual cortex, and on the relative numbers of modifiable vs. non-modifiable junctions on the cortical neurons.

The external stimulus e^k is not the only factor producing activity in cortical cells. Even in the absence of patterned external stimuli, as is the case, for exam-

ple, when the eyes are covered, there is neural activity. This is due in part to variations in spontaneous activity in the presynaptic pathways, dark discharges from the retina, input from the reticular formation or other such inputs that are non-specific with respect to visual stimuli (Batini and Buisseret, 1974; Buisseret et al., 1978). For simplicity, these will all be put together to form an additional component x, that can be characterized as channel noise which, in this paper, will be taken to be a random fluctuation about the post-synaptic passive potential. We then have for the post-synaptic potential on the soma of the ith cortical cell, denoted by c_i^0:

$$c_i^0 = \sum_{j=1}^{L_d} M_{ij}d_j^k + \sum_{j=1}^{L_b} Z_{ij}b_j^k + x \qquad (1.1)$$

Channel noise should be distinguished from another possible random component in the input, whose sources are variations in the actual input pattern e^k. In a normal visual environment, optimal patterns will not usually appear alone on the retina: they will generally be immersed in a background that is likely to vary from one realization to another.

If e^k is given a time index $e^k(t)$, where it is understood that $e^k(t)$ stands for the external pattern actually occurring at time t, in which the optimal pattern e^k is a prominent component, then d^k and b^k will also become dependent on time and will consist of an optimal pattern and a component (that will be shown to be noise-like in the next section) as follows:

$$d^k(t) = d^k + r(t)$$
$$b^k(t) = b^k + s(t) \qquad (1.2)$$

The constant patterned parts of $d^k(t)$ and $b^k(t)$ continue to be written d^k and b^k and, as will be shown, the noise vectors $r(t)$ and $s(t)$ can be taken to have zero-mean.

It is generally agreed that the input-output function by which a cortical neuron maps dendritic post-synaptic potentials into axon firing rates is approximately linear in the central part of the frequency scale, and cutoff at upper and lower ends. At the lower end, where the combined potential from all input channels is small, there is a threshold due to the inability of the input to raise the membrane potential to a firing level. At the upper end, a natural saturation limit in the output frequency is imposed by the refractoriness or "dead time" of the neuron membrane.

Measuring the post-synaptic potential from that required to produce the mean spontaneous firing rate, let θ_F be the post-synaptic potential at the firing threshold and let μ be that corresponding to the saturation limit. We let all proportionality constants be absorbed in θ_F and μ in such a way that it is meaningful to compare the input frequencies d_j^k and b_j^k

weighted by the junction strengths M_{ij} and Z_{ij}, to θ_F and μ. Now changing the conventions of NC somewhat define P (essentially the input-output function of a cortical cell) as follows:

$$P(u) = \begin{cases} = \mu & \text{if} \quad u > \mu \\ = u & \text{if} \quad \theta_F \leqq u \leqq \mu \\ = \theta_F & \text{if} \quad u < \theta_F \end{cases} \quad (1.3)$$

Collecting the above yields for the actual instantaneous firing rate c_i of the ith cortical neuron (in the absence of lateral inhibition or excitation from other neurons)

$$c_i(t) = P[c_i^0(t)] =$$

$$P\left[\sum_{j=1}^{L_d} M_{ij}(t-1)d_j^k(t) + \sum_{j=1}^{L_b} Z_{ij}b_j^k(t) + x(t) \right] \quad (1.4)$$

at a time t when an image of the external pattern $e^k(t)$ appears on the retina.[1]

In (1.3) and (1.4) both the input and the output are given on a scale in which *zero level means the mean spontaneous firing frequency*. So inputs $d_j^k(t)$ and $b_j^k(t)$, as well as output $c_i(t)$, may be negative or positive. The random component $x(t)$ represents fluctuations in the spontaneous frequency and gives a measure to the amount of variance in it. If spontaneous activity is larger than zero, the firing threshold, θ_F, is negative. In the above, one difference as compared to previous models, is the appearance of the term containing nonmodifiable junction strengths Z_{ij}. This makes possible a reversible behavior that shows a close resemblance to experimental results.

II. The External World as Seen by a Cortical Cell

Some visual cortical cells are known to respond preferentially to edges or bars of given orientation in the earliest stages of post-natal development, independent of visual experience. In addition, cortical cells are

organized in orientation columns so that cells in a single column will respond preferentially to a bar or an edge of given orientation in the visual field (possibly exerting a mutually excitatory effect on one another).

Although it is likely that all orientation preferences occur among cortical cells, for a normally reared animal, orientation preferences shift discontinuously by about 15° from one orientation column to the next. These shifts are seen in electrode penetrations in the plane perpendicular to the column direction (Hubel and Wiesel, 1959). It thus seems reasonable to suppose that interacting orientation columns, those corresponding to a single mapping (M, Z), correspond to discrete orientations separated by about 15°.

We therefore assume that in the orientation preference space there are K basic patterns $e^1, e^2, ..., e^K$, corresponding, for example, to bars of differing fixed orientations moving across the receptive field of the cortical cell. A typical value used in our simulations is $K = 7$. (This could correspond to a situation in which the orientation directions differ from one another by 15° angles and cover the full circle if we assume that the tuning curves of cortical neurons are symmetrical around their peak orientation and have no directional preference.)

One way in which innate orientation preferences can be built into the mapping from lateral geniculate to cortical cells is as follows: Suppose that initially the contribution of the modifiable junctions can be neglected and further that Z has the form

$$Z = \theta_M \sum_{k=1}^{K} c^k x b^k. \quad (2.1)$$

here $b^1 ... b^K$ are the signal vectors in the LGN pathway, input to the stable junctions, corresponding to $e^1 ... e^K$, and c^k denotes a unit vector referring to the kth cortical neuron

$$c_j^k = \delta_{jk}.$$

θ_M is the modification threshold, whose meaning will be explained in Section III. Z has the form of a widely tuned correlation matrix memory as has been employed, for example, by Anderson (1970, 1972) and Kohonen (1972). In the absence of noise the passive potential on the body of the ith cell due to the input pattern pair (d^i, b^i) corresponding to e^i is:

$$(c^i, Md^i + Zb^i) \simeq \theta_M \quad (2.2)$$

We assume that the vectors b^i and d^i are normalized. Any other pattern pair (d^l, b^l) would give

$$(c^i, Md^l + Zb^l) \simeq \theta_M(b^i, b^l) \quad (2.3)$$

which in general would be smaller than θ_M if the patterns do not overlap too much so that $(b^i, b^l) < 1$.

Thus in the absence of noise, cell i will respond preferentially to pattern e^i. This preferential response could be enhanced by excitation from other cells in the same orientation column and by inhibition from cells in neighboring columns of different orientation preference[2].

In normal visual experience optimal patterns such as straight lines or edges seldom appear alone; they are generally immersed in a highly variable environment that is extremely difficult to characterize in any satisfactory fashion. A normally reared animal encounters an immense variety of patterns and shapes in a vast number of combinations. In addition, the animal for reasons of its own selects those objects and shapes which are of interest upon which to fix its gaze.

A cortical cell biased to some pattern (according to (2.2) and (2.3), for example) will, on the average, respond most strongly when its preferred pattern is in its receptive field. It seems reasonable to suppose that for a normal visual environment, the rest of the visual image is uncorrelated with the preferred pattern so that, averaged over the times the cortical cell is firing most strongly, *the visual image seen by the cortical cell, other than the preferred pattern might be regarded as noise.* (This now becomes a definition of a "normal" environment. Many deprivation experiments (cats raised with goggles, in planetaria, etc.) are designed precisely to alter this "normal" environmental situation.)

We therefore conclude that from the point of view of a cortical cell biased for one reason or another to respond preferentially to some specific pattern, the "normal" environment, excluding the preferred pattern, might be treated as noise; thus a visual pattern firing the cell, where the preferred e^k is one of the components, can be represented at a certain moment as

$$e^k(t) = e^k + \text{noise} \tag{2.4}$$

where noise is essentially unbiased and in the first approximation at least, can be assumed to be "white" or non-correlated from one time to another.

It now becomes meaningful to speak of K pattern classes, with the kth class consisting of all the possible representations of pattern e^k with its noise-like visual accompaniment, i.e., of visual patterns like (2.4). In the mapping from retina to visual cortex, this pattern gives rise to a pair of parallel signal patterns $(d^k(t), b^k(t)) = (d^k + r(t), b^k + s(t))$. Since the noise components

---

2   The dependence of the inner product (b^i, b^i) or (d^i, d^i) with increasing angle has been discussed in Nass and Cooper, p. 14 (1975). It is seen that a rather sharp fall off is expected. In our simulations we have used such a set of inner products. However, the form of $Md + Zb$ at time zero is chosen to be more complicated than (2.1) to allow for broader initial tuning curves

$r(t)$ and $s(t)$ are in fact representations of the random component $e^k(t) - e^k$, the best way to describe them is to use the mathematical correlation between $r(t)$ and $s(t)$, and assume that this correlation can have different values. On the other hand, since by definition $E(e^k(t)) = e^k$, or the random environment in $e^k(t)$ has no orientation bias, it is reasonable to assume that both $r(t)$ and $s(t)$ have zero means. (This is a matter of definition; if $r(t)$ and $s(t)$ do not have zero means, it suffices to redefine the noisy vs. patterned components in the external pattern $e^k(t)$ in such a way that, after the constant mapping of the visual pathway, the noise components have zero mean.)

The visual stimuli received by a dark-reared animal (or an animal deprived of patterned visual input by some procedure such as suturing the eyelids – though eyelid suturing and dark-rearing are not equivalent (Spear et al., 1978)) can probably be characterized as time-varying white noise. Such stimuli arrive on a cortical cell via modifiable or non-modifiable synaptic junctions so that we again denote them by $r(t)$ and $s(t)$. The signal patterns in this case are

$$(d^k(t), b^k(t)) = (r(t), s(t)) \tag{2.5}$$

The noise inputs, $r(t)$ and $s(t)$ may be mutually correlated at fixed times, but from one step, t, to another they both are zero-mean, independent, and at least weakly stationary. Also, the vector $r(t)$ will be assumed to have non-singular covariance structure.

III. Modifiability of Cortical Synapses

Synaptic modification dependent on inputs alone, of the type already directly observed in *Aplysia* (Kandel, 1976), is sufficient to construct a simple memory – one that distinguishes what has been seen from what has not, but does not easily separate one input from another (Anderson and Cooper, 1978). To distinguish between inputs as well requires synaptic modification dependent on information that exists at different places on the neuron membrane, what we call two (or higher) point modification. In order that such modification take place, information must be communicated from, for example, the axon hillock to the synaptic junction to be modified. This implies the possibility of internal communication of information within the neuron. One might guess that once the physiological mechanism for such communication was available, different types of two (or higher) point modification evolved in various ways. It is tempting to conjecture that a liberating evolutionary step was just the development of this means of internal communication which, coupled with the ability of synapses to modify, created the possibility for a new organization principle.

A number of related articles on the development of cortical neurons have appeared recently (von der Malsburg, 1973; Nass and Cooper, 1975; Perez et al., 1975; Kohonen and Oja, 1976; Kohonen et al., 1977 and Anderson et al., 1977). In these, the modification hypothesis may be regarded as different realizations of the "conjunction rule" of Hebb (1949) or of the two-point modification mentioned above. Cooper (1973) and Nass and Cooper (1975) explored some of the consequences of a passive two-point modification of synaptic junctions. In Kohonen's models (1977) the concept of optimal mappings has been introduced. The error-correcting and noise-attenuating properties of the optimal associative mappings make such mappings better able to extract a basic pattern immersed in a noisy environment, and also to separate between the different pattern classes.

In this paper, we employ a variation of the two-point modification which captures some of the properties of passive modification (in that a cell can learn to respond or increases its response to a repeated external pattern) while at the same time modifying its response so that it responds to no more than one pattern. In this way, the tuning curve sharpens and the mapping from input to output becomes optimal.

Threshold Passive Modification

The general modification algorithm for the mapping, M may be written[3]

$$M(t) = \gamma M(t-1) + \delta M \tag{3.1}$$

In passive modification (Cooper, 1973) the change in strength of the labile junction M_{ij} of the efferent cell i is given by

$$\delta M_{ij} \sim (\text{output})_i (\text{input})_j \tag{3.2}$$

when the output is below a maximum (saturation) level, μ; there the output (excluding spontaneous fluctuations) is just the result of the input applied to already existing synapses. When the output is equal to or above the maximum level μ,

$$\delta M_{ij} = 0 \tag{3.3}$$

so that above this maximum level, all of the synapses $M_{i1}, M_{i2}, \ldots$ associated with the cell, i, stop modifying. (The only further change in synaptic strengths is then due to the uniform decay of all of the junctions if $\gamma < 1$ (3.1).) This results in a maximum firing rate (which in any case is physiologically necessary) in such a way as to preserve the information in the synaptic junctions.

Applied to visual cortex, where the spontaneous firing rate is low (Herz et al., 1964), occasionally the incoming pattern is unable to make a particular cell fire. (The cell remains below threshold.) In such a case the modification of that cell's synapses is zero.

In what follows this passive modification algorithm is altered so that below a threshold called the modification threshold, θ_M, (which might, for example, be the actual threshold, θ_F, for the firing of the cell, the level of spontaneous activity, or some higher level) the synapses modify according to

$$\delta M_{ij} \sim -(\text{"output"})_i (\text{input})_j \tag{3.4}$$

Here the "output" is the actual output of the cell (on a scale where the mean spontaneous firing level is denoted by zero according to 1.3) if the cell is above the firing threshold, θ_F, or is just the integrated passive potential if the "output" is below the firing threshold θ_F. In this latter case we call this a force-no-fire (FNF) situation which results due to the integrated potential forcing the cell but not succeeding in firing it.

The modification threshold has the effect of increasing the response of a cell to an input to which it responds sufficiently strongly while decreasing its response (negative or positive) toward the level of spontaneous activity to inputs to which it responds too weakly. It is this which drives the system to an optimal mapping.

We thus assume that the modifiability of a synaptic junction is dependent on events that occur at different parts of the same cell and on the rate at which the cell responds: below θ_M, above θ_M but below the maximum rate μ, and above μ.

Suppose now that at time t the external pattern $e^k(t)$, $k \in (1, 2, \ldots K)$, appears and maps into the pattern pair $(d^k(t), b^k(t))$. Adding channel noise, including the effect of lateral inhibition[4] and using (1.4) we can write for the output of the ith cell

$$c_i(t) = P[c_i^\kappa(t)] \tag{3.5a}$$

where

$$c_i^\kappa(t) = c_i^0(t) - \sum_{j \neq i} \kappa_{ij} c_j^0(t) \tag{3.5b}$$

and

$$c_i^0(t) = \sum_{j=1}^{L_d} M_{ij}(t-1) d_j^k(t) + \sum_{j=1}^{L_b} Z_{ij} b_j^k(t) + x(t) \tag{3.5c}$$

Here the κ_{ij} are the coefficients of lateral inhibition. It is convenient to write the modification algorithm from

---

3   In this paper we use discrete time, $t = 0, 1, 2 \ldots$ where one unit of time can be thought to correspond approximately to the duration of a burst of activity in the LGN and cortical cell axons. For further details, see NC

4   We use here a form of "forward" lateral inhibition that leads to linear equations and is somewhat simpler than the form employed by NC. Our results are relatively independent of the precise form used

the point of view of this i^{th} cell. What we are primarily interested in then is the ith row of the matrix M. Denoting this by m^i and the corresponding row of Z by z^i we can rewrite (3.5c) as

$$c_i^0(t) = (m^i(t-1), d^k(t)) + (z^i, b^k(t)) + x(t) \qquad (3.5d)$$

With this the threshold passive modification algorithm becomes[1]

$$m^i(t) = \gamma m^i(t-1) + \eta^+ c_i^\kappa(t) d^k(t)$$
$$\text{if} \quad \mu > c_i^\kappa(t) \geqq \theta_M;$$
$$m^i(t) = \gamma m^i(t-1)$$
$$\text{if} \quad c_i^\kappa(t) \geqq \mu;$$
$$m^i(t) = \gamma m^i(t-1) - \eta^- c_i^\kappa(t) d^k(t)$$
$$\text{if} \quad c_i^\kappa(t) < \theta_M \qquad (3.6)$$

In this form the modification algorithm is discontinuous at μ. A more convenient form for analysis results if the change in synaptic modification at $c_i^\kappa = \mu$ does not take place abruptly, but rather weakens when output frequency approaches its maximum level. Therefore, the proportionality coefficient in front of (output) × (input) will be assumed to depend on output frequency in such a way that learning is fast when output is far from saturation, but tends to zero as output tends to saturation limit. A suitable functional relationship for this proportionality or gain factor would be

$$(\text{constant}) \times \left[\frac{\mu}{(\text{output})} - 1 \right] \qquad (3.7)$$

which clearly decreases to zero as output grows to the saturation limit μ. As will be seen presently, the above functional form is especially suitable for mathematical treatment.[5] To show the relation between this modified algorithm and some algorithms previously studied (notably stochastic approximation type algorithms in regression problems and pattern recognition (Kohonen, 1977)) we substitute

$$\eta^+(t) = \eta^+ \left(\frac{\mu}{c_i^\kappa(t)} - 1 \right)$$

for η^+ in (3.6). This now gives

$$m^i(t) = \gamma m^i(t-1) + \eta^+ (\mu - c_i^\kappa(t)) d^k(t)$$
$$\text{if} \quad \mu > c_i^\kappa(t) \geqq \theta_M$$
$$m^i(t) = \gamma m^i(t-1)$$
$$\text{if} \quad c_i^\kappa(t) \geqq \mu$$
$$m^i(t) = \gamma m^i(t-1) - \eta^- c_i^\kappa(t) d^k(t)$$
$$\text{if} \quad c_i^\kappa(t) < \theta_M \qquad (3.8)$$

---

5 A somewhat more complicated form makes the algorithm continuous at θ_M as well as μ without substantially altering our results

Now, if threshold is surpassed, an *additive* term μ appears in (3.8).

This makes the algorithm considerably easier to handle from a mathematical point of view.

IV. Acquisition and Loss of Specificity

An Ideal Case:
Input of Patterns Without Noise

To show as clearly as possible how threshold passive modification can lead to an optimal mapping for which each cell responds only to a single orientation pattern and thus becomes specific (Imbert and Buisseret, 1975; Blakemore and Van Sluyters, 1975; Buisseret and Imbert, 1976; Frégnac and Imbert, 1977, 1978) or "sharply tuned" we first consider an ideal case. In this the cortical neurons receive only stimuli due to pure patterned inputs – the fixed pattern pairs (d^k, b^k) without noise and without specific inclusion of effects of other cells in the same orientation column (excitation) or cells in nearly orientation columns (lateral inhibition). Such other effects are discussed later.

As a starting-point of the analytical approach, assume that in the initial situation $(t=0)$ the leading pattern pair (d^1, b^1) produces an above threshold response, $c_1 > \theta_M$, in neuron one while the response of this first neuron to the rest of the pattern pairs remains below modification threshold:

$$(m^1(0), d^1) + (z, b^1) > \theta_M$$
$$(m^1(0), d^j) + (z, b^j) < \theta_M \quad j \neq 1. \qquad (4.1)$$

In what follows $m(t)$ and z are understood to mean $m^1(t)$ and z^1 respectively.

The actual stimuli received by a cortical cell contain at least two distinct "noise-like" components: one due to fluctuations in the firing rates of the neurons, variations in spontaneous activity, the other due to the visual background in which the optimal pattern (e.g., a bar or an edge) is immersed. The addition of such noise components leads to responses above θ_M for patterns other than the optimal one (d^1, b^1).

For $\theta_M = \theta_F$ (the force-no-fire situation) noise is required to make the cell fire at all for non-optimal input so that in the ideal case the cell is in a sense already "sharply tuned" since it does not respond to any pattern other than the leading pattern pair (d^1, b^1). If θ_M is at the level of spontaneous activity, noise is required to give response levels above the level of spontaneous activity for non-optimal input. If θ_M is above the level of spontaneous activity one would obtain such responses to non-optimal inputs in the absence of noise. (Experimental observation of the variance in the level of spontaneous activity (Bienenstock and Frégnac, Private Communication) indicates that normal variations are large enough to produce responses to non-optimal input. In our simulations we have used noise levels consistent with such observed variance.)

Equation (3.8) can be regarded as yielding a Markov-process with continuous states and discrete time, where the probability measures are in principle derivable from the statistical properties of the input process (the fixed pattern pairs enter in a random order) and the initial state. No mean-square convergence or convergence in probability of $m(t)$ to a fixed value is possible for the algorithm of (3.8). There will be fluctuations which in the real world might be reflected in a certain percentage of neurons failing to achieve sharp tuning and remaining aspecific or losing their ability to respond entirely. It should be stressed that while, for simplicity, only one of the modifiable cells is considered here, the overall effect of orientation specificity is due to the entire network of neurons.

To obtain asymptotic values at large times analytically, we concentrate our attention on the "successful" neurons. A linearization of the algorithm then offers a fairly good approximation; this is confirmed by simulations. Here linearization means the linearization of the recursion for $E[m(t)]$, the expectation of $m(t)$. While in reality the choice of strengthening vs. weakening of the response, according to the threshold modification hypothesis, is statistically dependent on $m(t)$ itself, we now make the approximation that this choice depends only on the input. This leads to a linear algorithm for $E[m(t)]$. Otherwise the ensueing non-linearities prevent us from obtaining any closed-form recursive relationship from which the asymptotics might be derived with reasonable ease.

Theorem 1. Let

$$m(t+1) = \gamma m(t) + \eta^+ [\mu - (m(t), d^1) - (z, b^1)] d^1$$

if the pattern pair (d^1, b^1) enters (4.2)

and

$$m(t+1) = \gamma m(t) - \eta^- [(m(t), d^j) + (z, b^j)] d^j$$

if the pattern pair (d^j, b^j), $j \neq 1$, enters. (4.3)

Also let $\eta^+ > 0$, $\eta^- > 0$, $1 \geq \gamma > 0$. Let d^i-vectors be linearly independent and let each pair (d^i, b^i) appear with equal probability at step t, independent of the input at previous step $t-1$.

Then if η^+ and η^- are small enough, the vector $\sigma(t)$ with elements

$$E[(m(t), d^k)] + (z, b^k) \quad (k = 1, 2, ..., K),$$

giving the average responses of the cell to pattern pairs (d^k, b^k), tends asymptotically to

$$\bar{\sigma} = -[(1-\gamma)I + H]^{-1}(1-\gamma)y + \mu c^1, \quad (4.4)$$

where H is a square matrix with

$$H_{ij} = \frac{\eta^-}{K}(d^i, d^j) \quad (j > 1),$$

$$H_{i1} = \frac{\eta^+}{K}(d^i, d^1), \quad (4.5)$$

y is a K-dimensional vector with

$$y_j = -(z, b^j) \quad (j > 1),$$

$$y_1 = \mu - (z, b^1), \quad (4.6)$$

and c^1 is the unit vector $(1 \ 0 \ 0...0)^T$.

The proof of this theorem is given in the Appendix.

The meaning of the asymptotic result is not easily visualized, if $\gamma < 1$. However, since the limits of the average responses to the K different pattern classes are continuous functions of γ as $\gamma \to 1$, a good approximation for γ close to one (which is the realistic situation) is given by the vector μc^1, whose elements are $\mu, 0, 0...0$. Thus the response of the cell remains on average at saturation, μ, for the leading pattern pair, but tends to zero (or, in terms of actual firing frequencies, to the level of spontaneous activity) for all the other pattern pairs. This is an optimal state. It should be observed that if there are patterns (d^j, b^j) that are geometrically close to the leading pair (d^1, b^1), the response for them will nevertheless tend to zero which means that only a very narrow part of the visual environment manages to elicit responses that deviate from the spontaneous activity level. So the tuning is indeed sharp.

To obtain a better picture of the asymptotic result of (4.4) when γ is close to but not quite equal to one, a simple corollary to Theorem 1 is given in the following.

Corollary 1. In Theorem 1, let $1 - \gamma = \varepsilon$ be small. Then in (4.4)

$$\bar{\sigma} = \mu c^1 - \varepsilon H^{-1} y + 0(\varepsilon^2), \quad (4.7)$$

where H, y, and c^1 are defined in Theorem 1.

The proof is given in the Appendix.

In the simulation of Figs. 6a and b, the initial situation was such that the response of the cell under study was slightly above the modification threshold for one of the pattern pairs (say d^1, b^1) and below this threshold for the others.

The modifiable part of the memory mapping was now changed in the simulation according to (3.8). The simulation shows that the response to the leading pattern pair (d^1, b^1), to which the neuron was sensitive from the very start, increases towards saturation level causing strong firing; at the same time, the other responses decrease toward the level of spontaneous activity. This sharpening results in an optimal state in

18

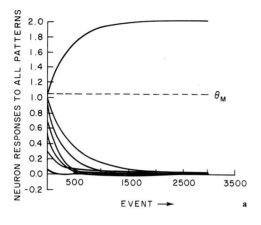

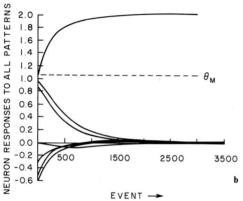

Fig. 6a and b. The responses of a neuron to 7 different noiseless patterns as functions of time. In these simulations, the parameters have the following values: $\gamma=1.0$, $\eta^+=0.032$, $\eta^-=0.017$, $\mu=2.0$, $\theta_M=1.05$, and $\kappa=0$. All noise levels were zero. The individual patterns entered in a pseudorandom order such that within each set of 7 steps in the algorithm, each pattern enters once but otherwise in a random sequence. z is such that it alone gives response 1 to the leading pattern and 0.5 to all the other patterns. Upper curves: the response to the leading pattern to which the initial response was higher than θ_M. Lower curves: responses to the other 6 patterns. In **a** all of the initial responses of the cell to the input patterns were positive; in **b** some of the initial responses of the cell were negative. The only difference between **a** and **b** is in the value of $m(0)$

which the leading response is close to saturation while the other responses are close to zero. Because of the finite length of simulation runs, the exact asymptotic values and especially their dependence on the system parameters is not always evident; however, they are given by Theorem 1 above.

In all of our simulations the vectors d were set equal to the vectors b. The dependence of the inner products of these vectors (b^i, b^l) or (d^i, d^l) with increasing angle was chosen to duplicate the sharp fall-off discussed in NC and assumes no directional pre-

ference.[2] The inner products used were $(b^i, b^l)=(d^i, d^l)=f(i-l)$; $f(0)=1$, $f(1)=0.4$, $f(2)=0.3$, $f(3)=0.2$, $f(4)=0.2$, $f(5)=0.3$, $f(6)=0.4$.

In Fig. 7, the synaptic strengths were "frozen" at given times between inputs, showing what the postsynaptic potentials of the cell would be if the noiseless ideal patterns were input. In the simulation there are 7 cells and 7 input classes. These classes might be thought of as modeling single vectors produced by an illuminated bar on the retina, with 7 different orientations at 15° intervals, as explained in Sect. II. The cell shown is typical in its response. The patterns to the right of the leading one correspond to increments of 15° in the angle of the bar; the response of the cell to those to the left, corresponding to decrements of 15°, has been set symmetric to those on the right. The 13 points then show the postsynaptic potentials produced by the pure patterns, i.e., the function

$$(m, d^k) + (z, b^k)$$

as k is varied but m and z are the fixed "frozen" values at that step. These points have been connected by straight lines.

Input of Patterns with Noise

With the inclusion of noise the consequences of threshold passive modification are more difficult to analyze. There are several problems. It is hard to estimate the relative strengths of the actual external stimulus $(d(t), b(t))$ and channel noise $x(t)$. It is also difficult to estimate in the external stimulus the relative strengths of the fixed standard inputs (d^i, b^i) and the rest of the image field $(r(t), s(t))$, reflecting the changing background in which the primary patterns always appear in normal visual experience. A safe method seems to be to choose the noise components to be relatively large; if the system works in a satisfactory way with a great deal of noise, it is likely to give even better results in less noisy situations.

Even when there is an initial tendency for a cell to respond more strongly to a single pattern class than to the others, noise of large magnitude can cause many misfirings and the cells might pick up more than one pattern class, especially if the inner products between different patterns d^i and d^j are large. A similar situation was analyzed by Nass and Cooper (1975). There the introduction of lateral inhibition between cortical cells limited the number of cells that respond to a single pattern, and the combination of lateral inhibition with the upper limit beyond which no modification occurs limits the number of patterns to which a single cell responds. We employ a similar lateral inhibition here. Thus the innate tendency for a single cell to respond most strongly to a single pattern is enhanced by lateral inhibition between cells in different orientation columns. At the same time it is expected that excitation within an orientation column will enhance the response of a cell in that column to the "preferred" pattern.

In our simulations, in spite of the fact that there are a considerable number of misfirings the neurons very seldom become asymptotically sensitive to other pattern classes than their own. The simulations show that the lateral inhibition term (which in any case was small) has a considerable effect on modification only in the beginning, tending to disappear rapidly as learning continues. Then the contribution of lateral inhibition is small compared to the threshold passive modification effect, which alone is sufficient to drive the neurons towards a limiting state that is a distorted version of the optimal mapping obtained in the noiseless case.

Let $m^i(t)$ and z^i be the modifiable and non-modifiable synaptic vectors of the ith cell at time t. Then including lateral inhibition the response before thresholding, proportional to the post-synaptic potential of the cell, is

$$c_i^\kappa(t+1)=[(m^i(t),d(t+1))$$
$$+(z^i,b(t+1))+x_i(t+1)]$$
$$-\sum_{j\ne i}\kappa_{ij}[(m^j(t),d(t+1))$$
$$+(z^j,b(t+1))+x_j(t+1)] \tag{4.8}$$

where $(d(t+1), b(t+1))$ is the pair of input vectors, $x_i(t+1)$ is the channel noise input to the ith cell and κ_{ij} are lateral inhibition coefficients.

We now write again

$$d(t+1)=d^k+r(t+1)$$
$$b(t+1)=b^k+s(t+1)$$

where k denotes the pattern class input at $(t+1)^{st}$ step. Neglecting the term due to lateral inhibition we have for $c_i^\kappa(t+1)$

$$(m^i(t),d^k)+(z^i,b^k)+(m^i(t),r(t+1))$$
$$+(z^i,s(t+1))+x_i(t+1). \tag{4.9}$$

There $(m^i(t),d^k)+(z^i,b^k)$ would be the post-synaptic potential if no noise were present, and the three other components in the sum represent noise.

In case all elements of $r(t+1)$ and $s(t+1)$ and $x_i(t+1)$ are non-correlated and zero-mean we would obtain for the variance of the noise component

$$\text{var(noise)}=\sum_{j=1}^{L_d} m_j^i(t)^2\,\text{var}(r_j(t+1))$$
$$+\sum_{j=1}^{L_b}(z_j^i)^2\,\text{var}(s_j(t+1))$$
$$+\text{var}(x(t+1)). \tag{4.10}$$

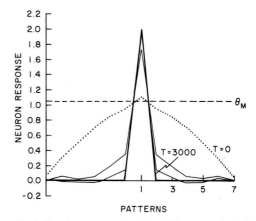

Fig. 7. The responses of a neuron to 7 different noiseless patterns at intervals from $t=0$ to $t=3000$. (This is the same simulation as that shown in Fig. 6a.) The responses of the neuron to patterns on the left side of pattern 1 were set equal to the responses to the corresponding patterns on the right. This give a symmetrical tuning curve. (In general we would obtain this result in an actual simulation. However, machine time available did not permit running simulations with thirteen patterns.)

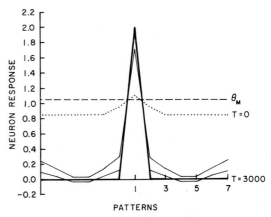

Fig. 8. The responses of a neuron to seven different noiseless patterns. This simulation is identical to that of figure seven except that $m(0)$ is chosen to make the initial responses of the neuron relatively flat (as is the case for some aspecific cells)

Furthermore, if $\text{var}(r_j(t+1))=V_1=\text{constant}$

$$\text{var}(s_j(t+1))=V_2=\text{constant}$$

and

$$\text{var}(x(t+1))=V_3=\text{constant}$$

we obtain

$$\text{var(noise)}=\|m^i(t)\|^2 V_1+\|z^i\|^2 V_2+V_3. \tag{4.11}$$

If there is correlation between $r(t)$ and $s(t)$, then (4.11) will not hold exactly. To take a simple special case, assume that the cross-covariance of r and s is diagonal, i.e., $\text{cov}[r_i(t+1), s_j(t+1)] = \delta_{ij} V_4$, with δ_{ij} the Kroenecker delta. Then it is easy to show that we have

$$\text{var(noise)} = \|m^i(t)\|^2 V_1 + \|z^i\|^2 V_2 + V_3 + 2(m^i(t), z^i) V_4$$

(4.12)

So the passive potential on the cell can be divided into two parts: the noiseless part due to pure input patterns, and a noise component with zero mean and variance given by the above equations. Since noise is the sum of several independent components, its distribution is close to the normal distribution.

The dependence of the asymptotic limits on the different parameters can again be approximated by a mathematical result of a linearized algorithm, where the starting point is a sub-optimal situation with low probabilities of misfirings. A modification of Theorem 1 leads to the following result:

Theorem 2. *Let $m(t)$ be the vector of junction strengths of one of the cells, say, the one for which (d^1, b^1) are the leading patterns. Let $m(t)$ change according to*

$$m(t) = \gamma m(t-1) + \eta^+ [\mu - (m(t-1), d^1(t))$$
$$- (z, b^1(t)) - x(t)] d^1(t)$$

if $(d^1(t), b^1(t))$ enters;

$$m(t) = \gamma m(t-1) + \eta^- [-(m(t-1), d^i(t))$$
$$- (z, b^i(t)) - x(t)] d^i(t)$$

if $(d^i(t), b^i(t))(i \neq 1)$ enters. Let the parameters satisfy $0 < \gamma \leq 1$, $\eta^+ > 0$, $\eta^- > 0$. Assume the following for the inputs: for each i, $d^i(t) = d^i + r(t)$, $b^i(t) = b^i + s(t)$, where $E[r(t)] = 0$, $E[s(t)] = 0$, $E[r(t) \times r(t)] = R$ is positive definite and bounded, $E[s(t) \times s(t)]$ is bounded, and the probability of occurrence at step t of a pattern from the ith class is uniform, i.e., $\frac{1}{K}$, and independent of the previous step. Also, assume that channel noise $x(t)$ satisfies $E[x(t)] = 0$ and is independent of $r(t)$ and $s(t)$. Let d^i be linearly independent and let $m(0)$ have finite mean.

Then for η^- and η^+ sufficiently small

$$\lim_{t \to \infty} E(m(t)) = U^{-1} a, \quad (4.13)$$

where U is the matrix

$$U = R + d^1 \times d^1 + \frac{\eta^-}{\eta^+} \sum_{j=2}^{K} (R + d^j \times d^j) + \frac{K}{\eta^+}(1-\gamma)I,$$

(4.14)

and a is the vector

$$a = [\mu - (z, b^1)] d^1 - p - \frac{\eta^-}{\eta^+} \sum_{j=2}^{K} [(z, b^j)d^j + p], \quad (4.15)$$

with

$$p = E[r(t) \times s(t)] z.$$

Proof. See Appendix.

By the above the limit of $m(t)$ in mean sense is close to the optimal vector; it becomes optimal, as can be expected, if γ tends to one (no forgetting) and the noise in d-patterns is zero, as is shown by Theorem 1 and Corollary 1. In the general case, noise tends to distort even the mean value from the ideal case obtained with noiseless learning and given by Theorem 1. However, the sharpening of the tuning curve is evident. Small residual responses of a cell to non-optimal patterns are likely related to spontaneous activity.

In the following simulations, the maximal intensity of the channel noise $x(t)$ at cortical cell inputs has been approximately one half of the input; this is consistent with experimental data (Bienenstock and Frégnac, Private Communication). If smaller channel noise is used the convergence is better. (In an experimental situation the visual inputs are usually adjusted to include an optimal input and to be relatively free of variations. The reported standard deviations then give a rough picture of the fluctuations of the firing frequency in response to noise-less input patterns, i.e., what has been termed here as channel noise.)

The relative intensities of the patterns vs. background were varied; the maximum background-to-pattern ratios $\|r(t)\|/\|d^j\|$ and $\|s(t)\|/\|b^j\|$ were well over one. When this situation is considered geometrically in the L_a- and L_b-dimensional pattern spaces, it becomes clear that this kind of "noise" is already very high. Assume, for example, that the inner product of two unit length pattern vectors d^i and d^j is 0.4, which is a typical value and corresponds to an angle of $66.5°$ in the pattern space. (This is not the same as the angle between the corresponding visual images e^i and e^j on the retina, due to the complicated way in which the two-dimensional visual patterns are mapped to the frequency – coded signal vectors d^i and d^j.) If noise vectors with larger than unit length and completely random directions are now added to the patterns d^i and d^j, a very considerable overlapping of the pattern regions takes place. In fact d^j could even be considered as a noisy version of d^i and vice versa.

In Fig. 9 simulation results in the noisy case are shown. This plot was constructed along the same principles as Fig. 7. The noiseless patterns (d^k, b^k) have been used in computing the plot so that a comparison with Fig. 7 would be possible; however, when m^i has been modified in this run, then the noisy patterns have been used. At some points in Fig. 9, the noise limits have been plotted. Each individual realization of a "tuning curve" would be inside these limits.

Input of Noise Without Patterns

A picture of the functioning of threshold modification now emerges: the neurons are driven towards the optimal feature extraction state by patterned input information immersed in a noise-like sensory environment and carried by noisy neural pathways. The final state is one in which there is sharp tuning; each cell is very sensitive to one pattern class and almost completely insensitive to the other classes. The post-synaptic potential caused by the leading pattern is large, leading to almost maximal firing frequency, while the post-synaptic potentials to other patterns tend to be such that even large variations do not cause significantly intense firings.

We now consider what happens if the flow of patterned information ceases. There is some loss of information due to the factor γ. However, the experimental situation suggests that the effective γ should be very close to one (if not equal to one) since cortical cells are known experimentally to continue to respond to visual stimuli even if the animal is dark reared. There is another effect, however, which follows from threshold passive modification that drives the cells from sharp to broad tuning. The asymptotic state, however, is not simply zero but is more subtle, depending on correlations among the noise inputs.

If the flow of normal patterned visual stimuli ceases (for example due to dark-rearing or eyelid suturing) input to cortical cells continues but is changed in nature. When the eyelids are sutured diffuse light falls on the retina and produces occasional firings of the retinal ganglion cells. When the animal is dark-reared there still occur dark discharges of retinal cells. The resulting stimuli, however, are totally different from those assumed previously: instead of certain fixed patterns in a noise-like environment, the input now can be considered to be totally noise-like. We can denote the signal patterns caused by these noise-like visual stimuli in the two fixed parallel pathways as $r(t)$ and $s(t)$. There is also channel noise present emanating from the signal-carrying pathway itself, although the intensity of this channel noise may be different from that connected with strong visual stimuli. Let channel noise be, as before, $x(t)$.

Once more omitting the lateral inhibition term the adaptation algorithm for one of the cells is, according to (3.8)

$$m(t) = \gamma m(t-1) + \eta^+ \left[\mu - (m(t-1), r(t)) \right.$$

$$\left. - (z, s(t)) - x(t) \right] r(t)$$

if $\quad \mu > (m(t-1), r(t)) + (z, s(t)) + x(t) > \theta_M$

$$m(t) = \gamma m(t-1)$$

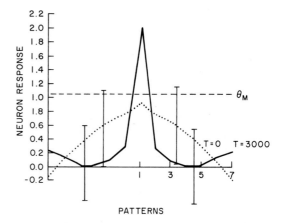

Fig. 9. The responses of a neuron to the 7 patterns at times $t=0$ and $t=3000$. In this simulation, the input vectors were the patterns with noise. Channel noise was also present. All elements of noise vectors $r(t)$ and $s(t)$ were uncorrelated and uniformly distributed over $[-0.3, +0.3]$. The input vectors used were of the form $d^k + r(t)$, $b^k + s(t)$. Channel noise $x(t)$ was uncorrelated with signal noise and uniformly distributed over $[-0.5, +0.5]$. Different input classes ($k = 1, 2, \ldots, 7$) entered in a pseudorandom order. Dotted curve: responses at time $t = 0$. Solid curve: responses at time $t = 3000$. The vertical bars represent regions, inside which the responses would be when channel noise is added. Therefore each individual realization of a "tuning curve" would be inside these regions, with the curves giving the average tuning at the respective times $t = 0$ or $t = 3000$. The parameters used in this simulation were: $\gamma = 0.9999$, $\eta^+ = 0.035$, $\eta^- = 0.017$, $\kappa = 0.3$, $\mu = 2$, $\theta_M = 1.05$

if $\quad (m(t-1), r(t)) + (z, s(t)) + x(t) > \mu$

$$m(t) = \gamma m(t-1) - \eta^- \left[(m(t-1), r(t)) \right.$$

$$\left. + (z, s(t)) + x(t) \right] r(t)$$

if $\quad (m(t-1), r(t)) + (z, s(t)) + x(t) < \theta_M \qquad (4.16)$

To get an analytical picture of the situation, the behavior is again approximated by a linear algorithm. Assume in the above equation that during one step of the recursion, corresponding to an integration period of about 1 second, the noisy input is changing so rapidly around its mean value zero compared to the rate of change of synaptic strengths, that the integrated or averaged input activity alone is not sufficient to fire the cell with a frequency higher than the modification threshold and cause synaptic growth. So in effect we concentrate on the third of (4.16), omitting the two low-probability cases.

The following theorem allows a variety of values for the possible correlation of $x(t)$ with $r(t)$ and $s(t)$, as well as the correlation between the two inputs $r(t)$ and

22

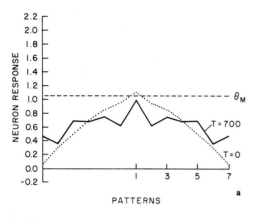

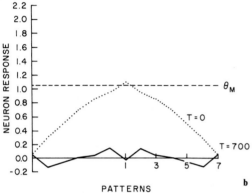

Fig. 10a and b. The response of a neuron to signal and channel noise. The channel noise was uncorrelated with the signal noise and was uniformly distributed over $[-0.5, +0.5]$. The signal noise ($r(t), s(t)$) was uniformly distributed over $[-0.3, +0.3]$. In **a** the two signal noise inputs, $r(t)$ and $s(t)$ were uncorrelated, while in **b** they were completely correlated ($r(t) = s(t)$). The parameters in this simulation were: $\gamma = 1$, $\eta^+ = 0.035$, $\eta^- = 0.5$, $\mu = 2$, $\theta_M = 1.05$, $\kappa = 0$. η^- was increased in order to decrease simulation time. Dotted upper curve: responses at time $= 0$. Solid curve: responses at time $= 700$

$s(t)$. It is shown there how the asymptotic mean value of vector $m(t)$ is related to the various parameters and statistical correlations, and how choosing the gain factor η^- small improves the asymptotic mean square error.

Theorem 3. *Consider the algorithm*

$$m(t) = \gamma m(t-1) - \eta^- [(m(t-1), r(t))$$
$$+ (z, s(t)) + x(t)] r(t) \quad (4.17)$$

There $r(t)$ and $s(t)$ are sequences of independent random vectors and $x(t)$ is a sequence of independent random scalars satisfying: $E[r(t)] \equiv 0$, $E[s(t)] \equiv 0$, $E[r(t) \times r(t)] \equiv R$ is positive definite and $r(t)$, $s(t)$, and $x(t)$ are bounded for each t. Let $E[r(t) \times s(t)] z \equiv p$, and $E[x(t) \times r(t)] \equiv q$.

Let $m(0)$ have finite mean and covariance and let $0 < \gamma \leqq 1$.

Then there exists a number $\delta > 0$ such that if $0 < \eta^- < \delta$,

$$\lim_{t \to \infty} E[m(t)] = \bar{m} = -\eta^- [(1-\gamma)I + \eta^- R]^{-1}(p+q) \quad (4.18)$$

and

$$E[\|m(t) - \bar{m}\|^2] < \omega \quad (4.19)$$

for all t with ω a finite constant. Furthermore, $\lim_{t \to \infty} \sup E[\|m(t) - \bar{m}\|^2]$ as a function of η^- tends to zero as $\eta^- \to 0$.

Proof. See Appendix.

Several conclusions can be drawn from the asymptotic result given in the theorem. If it is desirable to have the limit as close to zero as possible, this would imply that the gain should be small, the cross-correlations of the random processes weak and forgetting strong, as is of course clear from physical considerations. On the other hand, strong cross-correlations and large values of the two system parameters lead to a "noisy" limit with mean value away from zero, possibly closer to the negative of the non-modifiable part, with the effect that after a period with only noise stimulating the cells the "tuning" tends to be even less sharp than that given by the nonplastic part of the mapping only.

Figure 10a shows a simulation run using the original algorithm in (4.16). $s(t)$, $r(t)$ and $x(t)$ had zero mean. There was no correlation between noise values at different times.

Although no sensory patterns were now used as input, the plots in Figs. 10a and b were constructed along the same principles as in ones in Figs. 7 and 9. The plot at a given step reveals what the response of the cell would be if the values were again "frozen" and patterned input were used to determine how the cell is responding. It is seen that the sharpening, shown in Figs. 7 and 9 and obtained using patterned input, is now practically reversed in Fig. 10a; the tuning curve seems to settle to a wide tuning with the non-modifiable z-part of the mapping predominant. Simulations showed that now there will be firings occasionally, but they are not selective and do not manage to maintain the tuning.

In Fig. 10b, another simulation of the non-patterned situation is shown. The difference from the simulation of Fig. 10a is that in b there is correlation between $r(t)$ and $s(t)$; to emphasize the effect $r(t)$ and $s(t)$ were set equal for each t. Theorem 3 predicts that the total response of the cell will tend on the average to zero, i.e., $m(t)$ tends now to $-z$ in the mean sense. This is what in fact happens: however η^- had to be chosen to be large in this run so that the fluctuations in $m(t)$ are rather large.

Reversibility of Gain and Loss of Specificity

We now ask whether a return of patterned stimuli after a period of darkness would result in regaining specificity of response to these stimuli (see the experimental result of Fig. 4). This is indeed the case in the present model.

The simulation in Figs. 11 and 12 exhibit a complete run, where a model neuron is shown to gain sharp tuning with noiseless patterned input, then lose it again under noise-like input, but regain it when the noiseless patterned stimuli are once more allowed to appear in the input. For the first 3000 time units (the actual length of this period depends on the values of various parameters) the neuron receives input consisting of patterns without noise. A sharpening takes place. For the next 500 steps the patterns are absent in the input and only uncorrelated noise is presented. Partial loss of specificity occurs. Then, for another period of 2500 steps, noiseless patterns appear again and the modification is seen to be very similar to the original sharpening effect in the beginning of the run. This is what might be expected, since the modification scheme and the inputs are the same and there are only small differences in starting values in these two sharpening periods in the plot.

V. Conclusion and Discussion

We have assumed that between lateral geniculate and visual cortical cells there exist labile synapses which are modified according to the threshold form of passive modification as well as non-labile synapses which contain permanent information. These latter give the cortical cells a weak initial tendency to fire more strongly and readily to some orientations in the visual field than to others.

In the theory that results, there is an increase in the specificity of the response of a cortical cell to visual input (sharpening of its tuning curve) when that cell has been exposed to stimuli that are the results of normal patterned visual experience. When exposed to noise-like input, such as might be expected when an animal is dark-reared or raised with eyelids sutured, there is a loss of specificity. Specificity can be regained, however, with a return of input due to patterned vision. This seems to us to provide a possible explanation of the experimental results obtained by Imbert and Buisseret (1975); Blakemore and Van Sluyters

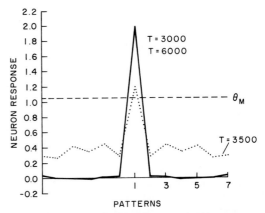

Fig. 11. Loss and retrieval of specificity. At $T=3000$ the neuron is sharply tuned (Heavy line). Noise was presented from $T=3001$ to $T=3500$. The channel noise was uniformly distributed over $[-0.5, +0.5]$. The signal noise was uniformly distributed over $[-0.3, +0.3]$ and was uncorrelated. Parameters used were, $\gamma=1$, $\eta^+=0.035$, $\eta^-=0.1$, $\theta_M=1.05$, $\mu=2$, $\kappa=0$. At $T=3500$ the cell was broadly tuned (Dotted line). Then (noiseless) patterns were presented from $T=3501$ to $T=6000$; η^- was changed back to 0.017 to make the result comparable with Fig. 7. At $T=6000$ (Solid line) the neuron has regained all of its previous sharp tuning. [The parameter η^- is varied to decrease simulation time. For the same η^- loss of specificity is slower than increase of specificity.]

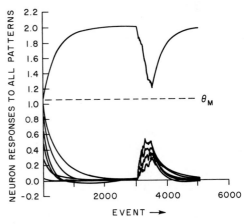

Fig. 12. Loss and retrieval of specificity. This figure is plotted in a manner similar to Fig. 6a and b. The parameters used are those of Fig. 11. From $T=0$ to $T=3000$ noiseless patterns were presented and the neuron acquired a sharp tuning. Uncorrelated noise was presented to the neuron from $T=3001$ to $T=3500$ and a broadening (i.e., loss of specificity) occurred. Noiseless patterns were presented from $T=3501$ to $T=6000$ and the neuron regained its former sharp tuning (i.e., retrieval of specificity)

(1975); Buisseret and Imbert (1976); and Frégnac and Imbert (1977, 1978).

In addition to this basic behaviour, simulations and mathematical results on the asymptotic states of the neural network show some more subtle phenomena that depend upon values of system parameters, notably the amount of decay (forgetting per unit time), the strength of selective modification of synaptic junctions, and the different statistical properties of noise factors. In the following discussion, some of these effects are illuminated by making simplifying approximations.

Sharpening of Tuning with Learning

The theoretical outcome of a long period of learning with patterned stimuli is given in Theorems 1 and 2. Sharpest tuning is achieved if forgetting is very slow, i.e., if memory traces tend to be rather permanent over short periods at least, and if the patterns occurring in the course of learning are as noiseless representations of a set of prototypes as possible. Then the modifiable part, M, of the memory mappings tends to become such that it practically cancels out the effect of the non-modifiable part Z, except for the optimal stimuli which give a sharp and high peak in the tuning curve. The response to these optimal patterns is close to the maximum firing level.

With stimuli corresponding to normal visual input, the modification of the tuning curve of a cell is monotone, going from broadly tuned to sharply tuned. The neuron changes from a non-specific unit, showing some response to all input patterns, to an immature one, for which some orientations have already lost the ability to elicit any kind of response above the spontaneous activity of the cell. Then the range of orientations for which the neuron is sensitive diminishes further until only a high and narrow peak, centered at the leading pattern for that cell, is visible in the tuning curve: the neuron has become a specific unit. This is in good agreement with experimental observation of the progression from a-specific to immature to specific as reported by Frégnac and Imbert (1977, 1978). In addition, this suggests that the classifications (Imbert and Buisseret, 1975; Buisseret and Imbert, 1976) of a-specific, immature and specific are relatively arbitrary divisions in what is really a continuum of response.

In learning with patterned stimuli, the outcome is not very sensitive to system parameters; the gross behavior of sharpening of tuning close to the optimum was achieved in simulations under a considerable range of parameter values. It is especially notable that even with relatively high noise levels, with signal-to-noise ratios considerably smaller than one, we still obtain qualitatively very similar behaviour to the noiseless learning cases. This confirms the good averaging properties of the modification algorithm.

Loss of Specificity

With noise-like stimuli, corresponding to non-patterned visual input, widening of the cortical cell tuning curve takes place, practically independent of how sharp the tuning was originally. The main effect is a rapid decay of the central peak of the tuning curve; the change in response for non-optimal patterns is smaller. The overall time constant of the loss of specificity was larger than that of gain of specificity.

This seems to be in agreement with experimental results, where giving patterned stimuli to dark-reared animals produces remarkably fast gains of specificity (see Fig. 4), while the loss of specificity is a more gradual process.

In this situation the effect of parameters is more important. In the absence of the overwhelming influence of strong patterned stimuli, some rather subtle phenomena, due to correlation properties among different noise components, now become visible. To illustrate these effects the following simplifications are made in Theorem 3:

We let the two parallel noise-like input vectors, $r(t)$ and $s(t)$, have equal numbers of elements ($L_a = L_b$) and for all i let these elements satisfy

$$\text{var}(r_i(t)) = \text{var}(s_i(t)) = V_1,$$
$$\text{cov}(r_i(t), s_i(t)) = V_2, \tag{5.1}$$

but all the other correlations between elements are zero. This might correspond to a structural situation in which each modifiable synapse on the cortical neuron has its non-modifiable counterpart, and the inputs $r_i(t)$ and $s_i(t)$ entering these two synapses are correlated. In fact if $V_1 = V_2$ in (5.1), then this correlation is complete (essentially $r_i(t) = s_i(t)$) and we see a situation where these two synapses, one modifiable and the other non-modifiable, share the same input signal.

We denote possible correlation between channel noise x and input r_i (for all i) by

$$\text{cov}(x(t), r_i(t)) = Q.$$

Then Theorem 3 gives the limit of $E(m(t))$ as

$$\bar{m} = \frac{-\eta^-}{(1-\gamma) + \eta^- V_1} (V_2 z + q) \tag{5.2}$$

where q is a vector with each element equal to Q.

Equation (5.2) has several implications. If $1 - \gamma$ is very small and Q can be neglected, we have approximately

$$\bar{m} \approx -\frac{V_2}{V_1} z. \tag{5.3}$$

This shows that strong correlation ($V_2 \approx V_1$) leads to a limit $\bar{m}$ such that the *total* response of the cell will be close to zero, while weak correlation ($V_2 \approx 0$) drives $E(m(t))$ close to zero, leaving the non-modifiable part z alone to determine the tuning of the cell. If, however, $1 - \gamma$ is not small and Q is not negligible, then $\bar{m}$ may be close to zero even in the correlated case.

We see then that if there is strong correlation between r and s, the noise inputs to labile and non-labile junctions, the initial bias of the cell can be entirely lost. A return of specificity would then not necessarily be to the same orientation as that preferred originally. (If excitation is assumed between cells in an orientation column, other cells which retained their original bias might still guide a cell that has lost its bias to its original orientation preference. However, an entire column could shift its orientation preference (Blakemore and Van Sluyters, 1974; Movshon, 1976)).

It should be stressed that the theorems give only weak convergence, so $\bar{m}$ in (5.2) would be only an average limit in a large number of similar cells. Although with η^- small the actual limits tend to be close to $\bar{m}$ according to Theorem 3, the non-zero variance in $\bar{m}$ would cause variations in asymptotic tuning in individual cells. In some cases our model neuron might even change its preferred pattern. It is very important which pattern happens to give the highest response when learning with patterned input is again commenced, since it is this initial situation alone which determines the optimal pattern for the given cell. It should also be mentioned that we have not taken into account effects due to binocular interactions. Such effects are at present being investigated and will be discussed in a future publication.

If the theory presented here bears any relation to the facts, then the interesting relation of the amount of loss of specificity to correlations between different noisy inputs might provide information on the connectivity of the neural network as well as providing a variety of opportunities for further experimental verification.

VI. Appendix

Proof of Theorems 1 and 2

Since the algorithm in Theorem 1 is just a special case of that in Theorem 2 (when $s(t)$, $r(t)$ and $x(t)$ have zero variance), the proofs can be combined. Take the algorithm of Theorem 2:

$$m(t) = \gamma m(t-1) + \eta^+ [\mu - (m(t-1), d^1(t))$$

$$- (z, b^1(t)) - x(t)] d^1(t)$$

if $(d^1(t), b^1(t))$ enters;

$$m(t) = \gamma m(t-1) + \eta^- [-(m(t-1), d^i(t))$$

$$- (z, b^i(t)) - x(t)] d^i(t)$$

if $(d^i(t), b^i(t))$ ($i \neq 1$) enters.

Calculate first the conditional expectation value

$$E\{[(m(t-1), d^i(t)) + (z, b^i(t)) + x(t)] d^i(t) | m(t-1)\}$$

using the assumptions made on $b^i(t)$ and $d^i(t)$. We obtain

$$E\{[(m(t-1), d^i) + (m(t-1), r(t)) + (z, b^i)$$

$$+ (z, s(t)) + x(t)] (d^i + r(t)) | m(t-1)\}$$

$$= (d^i \times d^i) m(t-1) + R m(t-1) + (z, b^i) d^i + p.$$

When these are multiplied by the class probabilities $\dfrac{1}{K}$ and summed, noting the difference when $i = 1$, we obtain

$$E[m(t)|m(t-1)] = \frac{1}{K} \gamma m(t-1)$$

$$+ \frac{1}{K} \eta^+ [\mu d^1 - (d^1 \times d^1) m(t-1) - R m(t-1)$$

$$- (z, b^1) d^1 - p]$$

$$+ \sum_{j=2}^{K} \left\{ \frac{1}{K} \gamma m(t-1) - \frac{1}{K} \eta^- [(d^j \times d^j) m(t-1)) \right.$$

$$\left. + R m(t-1) + (z, b^j) d^j + p] \right\},$$

yielding

$$E[m(t)|m(t-1)] = \gamma m(t-1) +$$

$$\frac{\eta^+}{K} \cdot [\mu d^1 - (d^1 \times d^1) m(t-1)$$

$$- R m(t-1) - (z, b^1) d^1 - p]$$

$$- \frac{\eta^-}{K} \sum_{j=2}^{K} [(d^j \times d^j) m(t-1)$$

$$+ R m(t-1) + (z, b^j) d^j + p]$$

$$= \bar{U} m(t-1) + \bar{a},$$

with

$$\bar{U} = \gamma I - \frac{\eta^+}{K} [R + (d^1 \times d^1)]$$

$$- \frac{\eta^-}{K} \sum_{j=2}^{K} [R + (d^j \times d^j)] \tag{A.1}$$

and

$$\bar{a} = \frac{\eta^+}{K} [\mu d^1 - (z, b^1) d^1 - p]$$

$$- \frac{\eta^-}{K} \sum_{j=2}^{K} [(z, b^j) d^j + p]. \tag{A.2}$$

Taking expectations over $m(t-1)$ yields $E(m(t)) = \bar{U} E(m(t-1)) + \bar{a}$. Since $\gamma \leq 1$, $\eta^+ > 0$, $\eta^- > 0$, R positive definite and finite, it is obvious that the norm of $\bar{U}$ will be smaller than one if η^+ and η^- are not too large. Then the equation is stable and the solution tends to the fixed point $\lim\limits_{t \to \infty} E(m(t)) = (I - \bar{U})^{-1} \bar{a}$, which is seen to be equal to $U^{-1} a$ when both $I - \bar{U}$ and $\bar{a}$ are multiplied with $\dfrac{K}{\eta^+}$.

26

To derive from this Theorem 1, set $R=0$, $p=0$ in $\bar{U}$ and $\bar{a}$. Then $\lim_{t\to\infty} E(m(t))=\bar{m}$ is a solution of the equation

$$(I-\bar{U})\bar{m}=\bar{a}$$

or

$$\left[I-\gamma I+\frac{\eta^+}{K}(d^1\times d^1)+\frac{\eta^-}{K}\sum_{j=2}^{K}(d^j\times d^j)\right]\bar{m}$$
$$=\left[\frac{\eta^+}{K}(\mu-(z,b^1))d^1-\frac{\eta^-}{K}\sum_{j=2}^{K}(z,b^j)d^j\right].$$

That $\bar{m}$ indeed satisfies (4.4) of Theorem 1, is seen if we denote by D the matrix with columns $d^1, d^2, ..., d^K$, by Λ the diagonal matrix with diagonal elements $\frac{\eta^+}{K}$, $\frac{\eta^-}{K}$, $\frac{\eta^-}{K}, ..., \frac{\eta^-}{K}$, and by y the vector with elements $\mu-(z,b^1)$, $-(z,b^2), ..., -(z,b^K)$. With this notation the above equation reads $[(1-\gamma)I+D\Lambda D^T]\bar{m}=D\Lambda y$. Multiplying by D^T from the left yields $[(1-\gamma)I+D^TD\Lambda]D^T\bar{m}=D^TD\Lambda y$. This in turn implies

$$[(1-\gamma)I+D^TD\Lambda]D^T\bar{m}=[(1-\gamma)I+D^TD\Lambda]y-(1-\gamma)y,$$

yielding

$$D^T\bar{m}-y=-[(1-\gamma)I+D^TD\Lambda]^{-1}(1-\gamma)y,$$
$$=-[(1-\gamma)I+H]^{-1}(1-\gamma)y, \qquad (A.3)$$

where $H=D^TD\Lambda$ is easily seen to have elements H_{ij} given by (4.5) in Theorem 1.

The nonsingularity of matrix $(1-\gamma)I+H$ follows from the fact that matrix $(1-\gamma)D^TD+D^TD\Lambda D^TD$ $=[(1-\gamma)I+H]D^TD$ is non-singular because of the matrix inversion lemma (see Kohonen, 1977). Since also D^TD is nonsingular because D has linearly independent columns, then $(1-\gamma)I+H$ must be nonsingular.

Since the first element of vector $D^T\bar{m}-y$ is $\lim_{t\to\infty} E[(m(t),d^1)]+(z,b^1)-\mu$ (from (4.6)), and the ith element $(i>1)$ is $\lim_{t\to\infty} E[(m(t),d^i)]+(z,b^i)$, we see that

$$\bar{\sigma}=D^T\bar{m}-y+\mu c^1. \qquad (A.4)$$

Substituting this in (A.3) establishes Theorem 1.

Proof of Corollary 1

With $1-\gamma=\varepsilon$ small, (4.4) yields

$$\bar{\sigma}=-(\varepsilon I+H)^{-1}\varepsilon y+\mu c^1$$
$$=-\varepsilon[(\varepsilon H^{-1}+I)H]^{-1}y+\mu c^1$$
$$=-\varepsilon H^{-1}(I+\varepsilon H^{-1})^{-1}y+\mu c^1$$
$$=-\varepsilon H^{-1}(I-\varepsilon H^{-1}+\varepsilon^2 H^{-2}-...)y+\mu c^1$$
$$=-\varepsilon H^{-1}y+0(\varepsilon^2)+\mu c^1,$$

as was to be shown.

Proof of Theorem 3

Denote $v(t)=m(t)-\bar{m}$, where $\bar{m}$ is the vector $-\eta^-[(1-\gamma)I+\eta^-R]^{-1}(p+q)$. Then $\bar{m}$ satisfies $\bar{m}=\gamma\bar{m}-\eta^-R\bar{m}-\eta^-(p+q)$. Substituting $v(t)$ in the recursion yields $v(t)=m(t)-\bar{m}=\gamma m(t-1)-\gamma\bar{m}$

$$-\eta^-[(m(t-1)-\bar{m},r(t))+(\bar{m},r(t))+(z,s(t))$$
$$+x(t)]r(t)+\eta^-[R\bar{m}+p+q]$$
$$=\gamma v(t-1)-\eta^-[(r(t)\times r(t))v(t-1)$$
$$+(r(t)\times r(t))\bar{m}-R\bar{m}$$
$$+(r(t)\times s(t))z-p+x(t)r(t)-q].$$

Now taking expectations over $r(t)$, $s(t)$, $x(t)$, but keeping $v(t-1)$ fixed yields

$$E[v(t)|v(t-1)]=\gamma v(t-1)-\eta^-[Rv(t-1)$$
$$+R\bar{m}-R\bar{m}+p-p+q-q]=(\gamma I-\eta^-R)v(t-1),$$

since $v(t-1)$ depends only on $s(k)$, $r(k)$, $x(k)$ for $k\leq t-1$, hence is independent of $r(t)$.

Finally taking expectations over $v(t-1)$ yields

$$E[v(t)]=(\gamma I-\eta^-R)E[v(t-1)].$$

Now, $E[v(t)]$ converges to zero if and only if $\|\gamma I-\eta^-R\|<1$. This is true if $\gamma-\eta^-\lambda_0<1$ and $\gamma-\eta^-\lambda_1>-1$, where λ_0 and λ_1 are the smallest and largest eigenvalues of R. Since $\gamma\leq 1$, $\lambda_0>0$, it is sufficient for the first inequality that $\eta^->0$; for the second inequality it is sufficient that $\eta^-<\frac{1+\gamma}{\lambda_1}$ which is positive since λ_1 is finite, by the assumption on boundedness of $r(t)$.

To show uniform boundedness of $E[\|m(t)-\bar{m}\|^2]=E[\|v(t)\|^2]$ we write

$$\|v(t)\|^2=\gamma^2\|v(t-1)\|^2$$
$$+(\eta^-)^2\|u\|^2-2\gamma\eta^-(u,v(t-1)),$$

where

$$u=(r(t)\times r(t))v(t-1)+(r(t)\times r(t))\bar{m}$$
$$-R\bar{m}+(r(t)\times s(t))z-p+x(t)r(t)-q.$$

Now it would be possible to limit the variance explicitly. However, since this limit would not be a tight one anyway, let us use a briefer method: the form of u and the boundedness of $r(t)$, $s(t)$ and $x(t)$ reveal that there exist finite non-negative constants ω_0, ω_1, ω_2 such that

$$E[\|u\|^2|v(t-1)]\leq\omega_0\|v(t-1)\|^2$$
$$+\omega_1\|v(t-1)\|+\omega_2.$$

The expectation of $(u, v(t-1))$ is easier to handle: we obtain

$$E[(u, v(t-1))|v(t-1)]$$
$$= (v(t-1), Rv(t-1)) + (v(t-1), R\bar{m})$$
$$- (v(t-1), R\bar{m}) + (v(t-1), p)$$
$$- (v(t-1), p) + (v(t-1), q)$$
$$- (v(t-1), q) \geqq \|v(t-1)\|^2 \lambda_0.$$

Combining these yields

$$E[\|v(t)\|^2|v(t-1)] \leqq \gamma^2 \|v(t-1)\|^2$$
$$+ (\eta^-)^2 [\omega_0 \|v(t-1)\|^2$$
$$+ \omega_1 \|v(t-1)\| + \omega_2] - 2\gamma(\eta^-)\lambda_0 \|v(t-1)\|^2$$
$$= [\gamma^2 + (\eta^-)^2 \omega_0 - 2\gamma\lambda_0\eta^-]\|v(t-1)\|^2$$
$$+ (\eta^-)^2 [\omega_1 \|v(t-1)\| + \omega_2].$$

It is easily established that $(\eta^-)^2 \omega_0 - 2\gamma\lambda_0\eta^- + \gamma^2 - 1 = 0$ has two real roots in η^-, one non-positive, one positive, between which the expression is negative. Also, because of continuity, there is a number $\delta > 0$ such that $(\eta^-)^2 \omega_0 - 2\gamma\lambda_0\eta^- + \gamma^2 \in (0, 1)$ as long as $0 < \eta^- < \delta$. Let us pick η^- from this interval and denote then

$$\alpha = (\eta^-)^2 \omega_0 - 2\gamma(\eta^-)\lambda_0 + \gamma^2, \quad 0 < \alpha < 1.$$

Now taking expectations with respect to $v(t-1)$ yields

$$E[\|v(t)\|^2] \leqq \alpha E[\|v(t-1)\|^2]$$
$$+ \omega_1(\eta^-)^2 E[\|v(t-1)\|] + \omega_2(\eta^-)^2.$$

By Jensen's inequality,

$$E[\|v(t-1)\|] \leqq (E[\|v(t-1)\|^2])^{1/2}.$$

The boundedness of $E[\|v(t)\|^2]$ follows from the fact that

$$\alpha E[\|v(t-1)\|^2] + \omega_1(\eta^-)^2$$
$$\cdot (E[\|v(t-1)\|^2])^{1/2} + \omega_2(\eta^-)^2$$

with

$$\alpha \in (0, 1), \eta^- > 0, \omega_1 > 0, \omega_2 > 0,$$

becomes smaller than $E[\|v(t-1)\|^2]$ when the latter is large enough, and in this region $E[\|v(t)\|^2] < E[\|v(t-1)\|^2]$. So it is clear that $E[\|v(t)\|^2]$ will always be bounded by some number ω.

Finally, consider the recursion for $E[\|v(t)\|^2]$ as η^- tends to zero. Since ω cannot increase as η^- decreases, we can write

$$E[\|v(t)\|^2] \leqq \alpha E[\|v(t-1)\|^2] + \omega_1(\eta^-)^2 \omega^{1/2}$$
$$+ \omega_2(\eta^-)^2 = \alpha E[\|v(t-1)\|^2] + (\eta^-)^2 \omega_3$$

for all η^- small enough. There $\omega_3 = \omega_1 \omega^{1/2} + \omega_2$ is a non-negative constant. The solution satisfies

$$E[\|v(t)\|^2] \leqq \alpha^t E[\|v(0)\|^2]$$
$$+ (\eta^-)^2 \omega_3 \sum_{i=1}^{t} \alpha^{t-i} = \alpha^t E[\|v(0)\|^2]$$
$$+ (\eta^-)^2 \omega_3 \frac{1 - \alpha^t}{1 - \alpha},$$

tending to $\dfrac{(\eta^-)^2 \omega_3}{1 - \alpha}$ as $t \to \infty$ for all sufficiently small $\eta^- > 0$.

Letting $\eta^- \to 0$ establishes the last part of the theorem, since

$$\frac{(\eta^-)^2 \omega_3}{1 - \alpha} = \frac{(\eta^-)^2 \omega_3}{1 - \gamma^2 + 2\gamma\lambda_0\eta^- - \omega_0(\eta^-)^2} \to 0 \text{ as } \eta^- \to 0$$

(for $\gamma = 1$, notice that $\lambda_0 > 0$, R being positive definite).

Acknowledgements. We would like to thank our colleagues in the Center for Neural Science for their encouragement and advice. In particular, we thank Mr. Elie Bienenstock for carefully reading the manuscript and suggesting several improvements. One of the authors (LNC) would like to express his appreciation to the Fondation de France and the CNRS for their support during his stays in France. He would also like to express his appreciation to his colleagues in the laboratory of Michel Imbert at the College de France for their hospitality and for the patience with which they explained the intricacies of the experimental results. In addition, financial support to one of the authors (EO) from the Academy of Finland and the Emil Aaltonen Foundation is gratefully acknowledged.

References

Anderson, J.A.: Two models for memory organization using interacting traces. Math. Biosci. **8**, 137–160 (1970)

Anderson, J.A.: A simple neural network generating an interactive memory. Math. Biosci. **14**, 197–220 (1972)

Anderson, J.A., Silverstein, J.W., Ritz, S.A., Jones, R.S.: Distinctive features, categorical perception, and probability learning. Some applications of a neural model. Psychoanal. Rev. **84**, 413–451 (1977)

Anderson, J.A., Cooper, L.N.: Les modeles mathematiques de l'organization biologique de la memoire. Plurisci. 168–175 (1978)

Batini, C., Buisseret, P.: Sensory peripheral pathway from extrinsic eye muscles. Arch. Ital. Biol. **112**, 18–32 (1974)

Bienenstock, E., Frégnac, Y.: Stability of response of single cells in kittens visual cortex. (to be published)

Blakemore, C., Cooper, G.F.: Development of the brain depends on the visual environment. Nature **228**, 477–478 (1970)

Blakemore, C., Mitchell, D.E.: Environmental modification of the visual cortex and the neural basis of learning and memory. Nature **241**, 467–468 (1973)

Blakemore, C., Van Sluyters, R.C.: Reversal of the physiological effects of monocular deprivation in kittens. Further evidence for a sensitive period. J. Physiol. (London) **237**, 195–216 (1974)

Blakemore, C., Van Sluyters, R.C.: Innate and environmental factors in the development of the kitten's visual cortex. J. Physiol. (London) **248**, 663–716 (1975)

Buisseret, P., Imbert, M.: Visual cortical cells. Their developmental properties in normal and dark reared kittens. J. Physiol. (London) **255**, 511–525 (1976)

Buisseret, P., Gary-Bobo, E., Imbert, M.: Ocular motility and recovery of orientational properties of visual cortical neurons in dark-reared kittens. Nature **272**, 816–817 (1978)

Cooper, L.N.: A possible organization of animal memory and learning. In: Proceedings of the Nobel Symposium on Collective Properties of Physical Systems. Lundquist, B., Lundquist, S., eds. London, New York **24**, 252–264 (1973)

Frégnac, Y., Imbert, M.: Cinetique de developement des cellules du cortex visuel. J. Physiol. (Paris) **6**, T.73 (1977)

Frégnac, Y., Imbert, M.: Early development of visual cortical cells in normal and dark-reared kittens. Relationship between orientation selectivity and ocular dominance. J. Physiol. (London) **278**, 27–44 (1978)

Frégnac, Y.: Cinetique de development du cortex visuel primaire chez le chat. Effets de la privation visuelle binoculaire et modele de maturation de la selective a l'orientation. Doctoral thesis, Université René Descartes (1978)

Hebb, D.O.: The organization of behavior. New York: Wiley 1949

Henry, G.H., Dreher, B., Bishop, P.O.: Orientation specificity of cells in cat striate cortex. J. Neurophysiol. **137**, 1394–1409 (1974)

Herz, A., Creutzfeldt, O., Fuster, J.: Statistische Eigenschaften der Neuronaktivität im ascendierenden visuellen System. Kybernetik **2**, 61–71 (1964)

Hirsch, H.V.B., Spinelli, D.N.: Modification of the distribution of receptive field orientation in cats by selective visual exposure during development. Exp. Brain Res. **12**, 509–527 (1971)

Hubel, D.H., Wiesel, T.N.: Receptive fields of single neurons in the cat striate cortex. J. Physiol. (London) **148**, 574–591 (1959)

Hubel, D.H., Wiesel, T.N.: Receptive fields binocular interaction and functional architecture in the cat's visual cortex. J. Physiol. (London) **160**, 106–154 (1962)

Imbert, M., Buisseret, P.: Receptive field characteristics and plastic properties of visual cortical cells in kittens reared with or without visual experience. Exp. Brain Res. **22**, 2–36 (1975)

Kandel, E.R.: Cellular basis of behavior. San Francisco: Freeman 1976

Kohonen, T.: Correlation matrix memories. IEEE Trans. Comput. C-**21**, 353–359 (1972)

Kohonen, T.: Associative memory – a system – theoretical approach. Berlin, Heidelberg, New York: Springer 1977

Kohonen, T., Oja, E.: Fast adaptive formation of orthogonalizing filters and associative memory in recurrent networks of neuron like elements. Biol. Cybernetics **21**, 85–95 (1976)

Kohonen, T., Lethiö, P., Rovamo, J., Hyvärinen, J., Bry, K., Vainio, L.: A principle of neural associative memory. Neuroscience **2**, 1065–1076 (1977)

Movshon, J.A.: Reversal of the physiological effects of monocular deprivation in the kittens visual cortex. J. Physiol. (London) **261**, 125–174 (1976)

Nass, M., Cooper, L.N.: A theory for the development of feature detecting cells in visual cortex. Biol. Cybernetics **19**, 1–18 (1975)

Perez, R., Glass, L., Shlaer, R.J.: Development of specificity in the cat visual cortex. J. Math. Biol. **1**, 275–288 (1975)

Pettigrew, J.D., Freeman, R.D.: Visual experience without lines. Effects on developing cortical neurons. Science **182**, 599–601 (1973)

Pettigrew, J.D.: The effect of visual experience on the development of stimulus specificity by kitten cortical neurons. J. Physiol. **237**, 49–74 (1974)

Spear, P.D., Tong, L., Langsetmo, A.: Striate cortex neurons of binocularly deprived kittens respond to visual stimuli through the closed eyelids. Brain Res. **155**, 141–146 (1978)

von der Malsburg, C.: Self-organization of orientation sensitive cells in the striate cortex. Kybernetic **14**, 85–100 (1973)

Received: December 1, 1978

Prof. Dr. L. N. Cooper
Center for Neural Science
Brown University
Providence, R.I. 02912, USA

Seat of Memory: Brain Theory Meets Experiment in Visual Cortex

(with M. Imbert)

The Sciences, pp. 10–13, pp. 28–29 (February 1981)

Experimentalists have their problems too. The deprivation experiments that seemed to indicate modification of the response properties of visual cortical cells were often not well received. In addition to problems of interpretation, the feeling among those most influential in the field was that even if effects were there, they were not significant. *"Why would you worry about plasticity in visual cortex when it is at best a small effect? Genetics determines visual cortex architecture almost entirely?"* The idea that we were not primarily interested in the architecture of the visual cortex but in learning and memory storage and that whatever plasticity there was could be studied to reveal the systematics as well as the molecular basis for learning and memory storage was not generally appreciated. In the face of this skepticism, Michel Imbert and I took the opportunity to present our point of view in the following somewhat popularized short article.

The Sciences

Seat of Memory

by Leon N Cooper and
Michel Imbert

Reprinted from *The Sciences*, February 1981

Although the properties of individual neurons are relatively well understood, the manner in which large interacting networks of these nerve cells produce mental activity remains almost a complete mystery. This is due, in part, to the complexity of the central nervous systems of higher animals, as well as to the great difficulty of observing these systems without destroying them. It also seems likely that higher central nervous system properties such as storage and retrieval of memory are of unusual subtlety, involving small changes in the activities of large numbers of neurons.

Many ways to store and retrieve information exist: filing cabinets, libraries and computers. But the fact that an animal's memory is held in a living structure, and is successfully utilized even though the animal may have no idea of where his memories are stored or how they are ordered, places special requirements on theory. Current computer memories, for example, are made of elements in which yes/no information is recorded and which can be recalled by addressing the location of an element. These computers perform sequences of elementary operations with incredible speed and accuracy, completely beyond the capability of living cells. A basic problem in understanding the organization of memory in a biological system is to understand how a vast quantity of information can be stored and recalled by a system composed of vulnerable and relatively unreliable elements and with no knowledge of how or where the information has been filed.

In 1950, Karl Lashley proposed that central nervous system memory is distributed: "It is not possible to demonstrate the isolated localization of a memory trace anywhere within the nervous system. . . . The same neurons which retain the memory traces of one experience must also participate in countless other activities. . . . Recall involves the synergic action or some sort of resonance among a very large number of neurons."

This point of view was first regarded as unpromising because it seemed as if individual memories would interfere with one another if there were no privileged sites in the brain for the storage of memory items in isolation from one another. But it has now become theoretically acceptable to say that the storage of memory in an animal central nervous system *is* distributed rather than

Leon N Cooper is Thomas J. Watson, Sr. Professor of Science and Co-Director of the Center for Neural Science at Brown University. In 1972, he received the Nobel Prize in Physics. Michel Imbert is Deputy Director of the Laboratory of Neurophysiology at the College de France, Paris. A more complete account of the current status of this collaboration between theorists and experimentalists interested in central nervous system organization will be given in the proceedings of a workshop recently held at Brown University under the auspices of the Alfred P. Sloan Foundation: SYNAPTIC MODIFICATION, NEURON SELECTIVITY AND NERVOUS SYSTEM ORGANIZATION.

Seat o

Brain theory meets experiment
in visual cortex

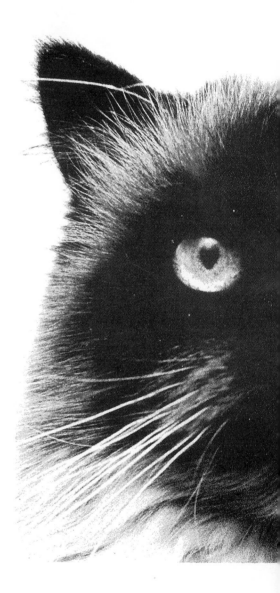

Memory

by Leon N Cooper and
Michel Imbert

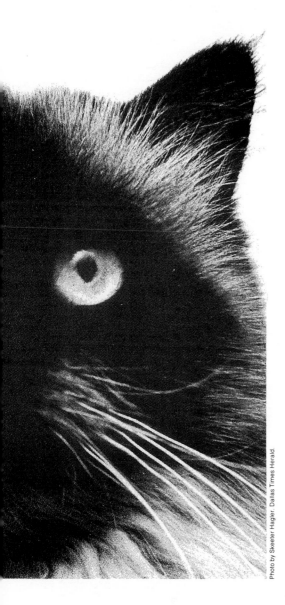

Photo by Skeeter Hagler, Dallas Times Herald

local (more like a hologram than a photograph), and that such distributed memory is stored over large regions of the neural network by small but coherent modifications of large numbers of synaptic junctions (the gaps between axons and dendrites, signal-sending and signal-receiving branch-like structures of adjoining neurons), thus altering the relation between presynaptic and postsynaptic electrical potentials.

Many conjectures have been made concerning whether and how such synaptic modification occurs, what precise form it takes, and what the physiological basis of this modification is. There is direct experimental evidence that at least some modification of synaptic strength occurs in invertebrates and there are various indications that synaptic modification is a rather general phenomenon. It has been suggested that the existence of a fundamental biological mechanism could lead to an entire class of interesting modifications among which are those that could produce distributed memories with very attractive properties.

Modeling Memory

For several years, we and our colleagues have been constructing and analyzing neural models that can organize themselves so that they acquire and store distributed memories. These models display, on a primitive level, features such as recognition, association and generalization, that suggest some of the mental behavior associated with animal memory and learning. The mechanisms employed in these constructions seem to be plausible biologically and are not inconsistent with known neurophysiology. In addition, the networks modeled seem to be a reasonable outcome of evolutionary development under the pressure of survival. Some of these ideas are related to or are generalizations of earlier neural model concepts.

In a distributed memory it is the simultaneous or near-simultaneous activities of many different neurons (the result of external or internal stimuli) that is of interest. The brain is composed of vast numbers of neurons (10^{11} is one estimate commonly given for humans) held together, fed and cleansed by various supporting structures, blood vessels and glial (meaning glue) cells. It is thought that the information processing and storage functions of the brain are accomplished by the neurons and that the other tissue is occupied primarily with housekeeping.

A single neuron collects information in the form of electrical potentials in its dendrite system from the axon branches of other neurons. These potentials are passively transmitted to the cell body, where they are integrated; the integrated potential then determines the firing rate of the cell. The electrochemical spikes that result propagate with minimal degradation over sometimes long distances along the axon trunk to all the axon branches. The information, contained in the frequency of spiking or the

THE SCIENCES FEBRUARY 1981 11

number of spikes in a burst, is then communicated, usually chemically, across synaptic junctions from the axon terminals to the dendrite branches of other neurons producing potentials in the dendrites; thus, the information flow continues.

A large spatially distributed pattern of neuron discharges or firings could contain important, if hard to detect, information. To illustrate some of the important features of such systems, consider the behavior of an idealized neural network that might be regarded as a model component of a nervous system.

Think of two large banks of neurons, the first receiving incoming signals, the second receiving the outgoing signals of the first. The actual synaptic connections between one neuron and another are generally complex and redundant. For utter simplicity, imagine that each of the incoming neurons is connected to each of the outgoing neurons by a single ideal junction that summarizes the effect of all the actual synaptic contacts between them.

We may then regard the synaptic strengths of the ideal junctions as a transformation that takes us from the incoming activities of the first bank of neurons to the outgoing activities of the second. It is in such modifiable sets of synaptic junctions that the experience and memory of the system are proposed to be stored. In contrast

with machine memory that is, at present, local (an event stored in a specific place and addressable by locality, requiring some equivalent of indices and files), animal memory is likely to be distributed and addressable by content or by association. In addition, for animals, there need be no clear separation between memory and "logic."

Each synaptic junction stores some portion of the entire experience of the system, as reflected in the firing rates of the neurons connected to this junction. Each experience or association, however, is stored over a large array of junctions. This is the essential meaning of a distributed memory: Each event is stored over a large portion of the system, while, at any particular local point, many events are superimposed.

How could such a distributed network be constructed in an actual nervous system? We have been able to show mathematically that such a network is a possible consequence of behavior of individual cells, assuming that these cells are able to modify their output behavior in certain ways. We have made mathematical formulations of possible kinds of synaptic modifications and explored various characteristics of distributed memory (concerning recognition, recollection, association and so on) that result from using these constructs.

It seems to us that the central issue has become to

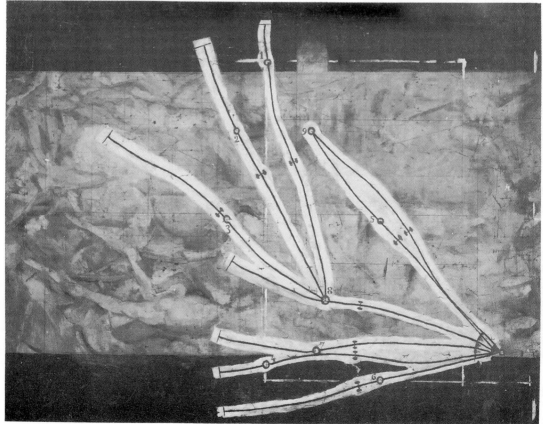

Network of Stoppages (1914) by Marcel Duchamp. Oil and pencil on canvas, 58⅝″ x 77⅝″. Collection, The Museum of Modern Art, New York. Abby Aldrich Rockefeller Fund and gift of Mrs. William Sisler.

confront the various theoretical ideas with hard experimental results. We believe that this is possible at present and indeed may have already proceeded further than is generally realized. In what follows we present some recent experimental and theoretical results relating to the development and modification of neurons in cat visual cortex which we feel provide just such a meeting of experiment and theory. This is not intended to be a review of all work done on visual cortex; rather it gives an account of a particular collaboration between experimentalists and theoreticians.

Effects of Visual Experience

Experimental work of the last generation, beginning with the pathbreaking work of David Hubel and Torstin Wiesel at Harvard University Medical School, has shown that neurons exist in the visual cortex of the adult cat that respond in a precise and highly tuned fashion to visual patterns—in particular, they fire electrical impulses when the animal is shown bars or edges of light of a given orientation (vertical, horizontal or slanted position) moving in a given direction.

Further work indicates that both the number of orientation selective cortical neurons and their response properties can be modified. It has been observed that the relative number of cortical neurons that are highly specific in their response to visual patterns varies in a very striking way with the visual experience of the animal during the critical period, the first seven to ten weeks after its birth.

For experimental purposes, these responsive cells have been classified by Michel Imbert and his collaborators into three groups—aspecific, immature and specific—according to their response properties and receptive field arrangements. Aspecific cells, the most indiscriminately responsive of the three, are usually activated by a circular stimulus moving in any direction across their receptive field, which is characteristically circular and rather large in size. The immature cells, more discriminating or highly tuned in their response, are generally activated by a moving rectilinear stimulus correctly oriented within their receptive field, which is still large but rectangular in shape. The orientation of the stimulus is rather imprecise; but in all cases there exists a direction in which the oriented stimulus is ineffective. The specific cells have receptive fields that are also rectangular but smaller in size than that of immature cells; they are the most highly tuned cells of the three groups.

The relative distribution of these different cell types depends on visual experience and has been compared in two groups of kittens of different ages. The first group was normally reared, that is to say, reared in a normal visual environment with natural alternation of day and night. The second group was reared in complete darkness from the first or second day of age.

The results of these experiments show that cells having some of the highly specific response properties of adult visual cortical neurons are present in the earliest stages of postnatal development, independent of visual experience. However, visual experience between 17 and 70 days is critical in determining the evolution of these cells. Animals reared normally show a marked increase in the number of specific cells as compared with aspecific. The reverse is true for animals reared in the dark.

It has also been shown that as little as six hours of normal visual experience at about 42 days of age can alter in a striking fashion the ratio of specific or immature or aspecific cells. That such a short visual experience can change the tuning ratios so markedly is clear evidence of the great plasticity of these cortical cells at the height of the critical period.

In addition, ocular dominance (the eye which is most influential in firing a cell) does not seem to be constant during the first stages of postnatal development. A significant increase in binocularly driven neurons (neurons which are fired by stimuli from both eyes) occurs with age. For kittens younger than three weeks, whatever their rearing conditions, most of the oriented cells (immature and specific cells) are activated by the eye opposite the hemisphere in which they are located. With or without visual experience, for kittens under three weeks old, ocular dominance and orientation preference appear as two linked parameters characterizing the intrinsic organization of the visual cortex. Beyond four weeks, ocular dominance is independent of orientation preference, for aspecific, immature and specific cells and for both normal and dark rearing conditions.

It has been proposed by Yves Fregnac of the Laboratory of Neurophysiology at the College de France and Michel Imbert "that binocularly driven cells are the modifiable cells which may be specified or despecified in terms of orientation preference under the control of visual experience. The monocular horizontal or vertical detector cells form a special subpopulation and are stable and resistant to the absence of visual input or to the selective exposure to an orientation to which they do not respond as long as they are influenced only by one eye." If no binocular imbalance is produced by experimental manipulations, the level of binocularity increases with age, and the horizontal and vertical cells which were still monocularly driven at three weeks become mainly binocularly activated after four weeks and can be affected by visual experiences. Thus, it is proposed that the absence of visual stimulation caused by rearing kittens in complete darkness leads to a complete loss of orientation preference because the development of binocularity remains unaffected. Such experiments demonstrate that there exist two distinct periods in the functional development of the visual cortex. During the first period, which ends at the end of the third week, there exists a maturation process independent of visual experience; during the

(Continued on page 28)

second period, starting at the end of the third week, visual experience is necessary for specificity to be maintained and to develop. They demonstrate further that in the early and maturing visual cortex there exist different types of cells. Certain cells, monocularly activated by the eye opposite the hemisphere in which they are located, respond primarily to vertical or horizontal orientations. These are relatively stable, being more resistant to deprivations as long as they remain monocularly driven. Other cells, mostly binocularly activated and responding to slanted orientations, can be made more or less specific, depending upon visual experience. After three weeks old, visual experience plays a very strong role in the development of specificity.

Such results seem to us to provide direct evidence for the modifiability of the response of single cells in the cortex of a higher mammal according to its visual experience. Depending on whether or not patterned visual information is part of the animal's experience, the specificity of the response of cortical neurons varies widely. With normal patterned visual experience, cell specificity increases. Deprived of normal patterned information (dark-reared or lid-sutured at birth, for example) specificity decreases. Further, even a short exposure to patterned information after six weeks of dark-rearing can reverse the loss of specificity and produce an almost normal distribution of cells.

We do not claim and it is not necessary to our theory that all neurons in the visual cortex be so modifiable. Nor is it necessary that modifiable neurons are especially important in producing the architecture of visual cortex. It is our hope that the general form of modifiability we require to construct distributed memory mappings manifests itself for at least some cells of visual cortex that are accessible to experiment. We thus make the conservative assumption that biological mechanisms, once established, will manifest themselves in more or less similar forms in different regions. If this is the case, modifiable individual neurons in visual cortex can provide evidence for such modifications more generally. In what follows, we present some of the results of a theory recently developed by Leon Cooper, Fishel Liberman and Erkki Oja which accounts for some of the experimental data mentioned above.

Comparison with Theory

In this theory, it is assumed that between the cells of the lateral geniculate (a relay station between the eyes and the visual cortex) and visual cortical cells, there exist labile or changeable synapses that are modified according to a form of passive modification called threshold passive modification, as well as non-labile or unchangeable synapses that contain permanent information. The latter give the cortical cells a weak initial tendency to fire more strongly and readily to some orientations in the visual field than to others. Threshold passive modification is a variation of two-point modification (synaptic modification which depends on information that exists at different places on the neuron membrane) for which a cell can learn to respond or increases its response to a repeated external pattern, while at the same time changing its response so that it responds to no more than one pattern.

We assume that the modifiability of a synaptic junction is dependent on events that occur at different parts of the same cell and on the rate at which the cell responds, and have proven several theorems which show that, with this form of passive modification, there is an increase in the specificity of the response of a cortical cell to visual input (sharpening of its tuning curve) when that cell is exposed to stimuli that are the result of normal patterned visual experience, and a loss of specificity when that cell is exposed to noise-like input, such as might be expected when an animal is dark-reared. Specificity can be regained, however, with a return of input due to patterned vision.

Computer simulation and mathematical results on the asymptotic states of the neural network indicate some more subtle phenomena that depend upon values of system parameters, notably the amount of decay (forgetting per unit time), the strength of selective modification of synaptic junctions, and the different statistical properties of noise factors. In the following discussion, some of these effects are illuminated by using the results of these simulations or by making simplifying approximations.

With stimuli corresponding to normal visual input, the tuning curve of cells sharpens, going from broadly tuned to sharply tuned. The neuron changes from a non-specific unit, showing some response to all input patterns, to an immature one which responds more selectively. Then the range of orientations to which the neuron is sensitive diminishes further until only a high and narrow peak is visible in the tuning curve: The neuron has become a specific unit. This theoretical result is in good agreement with experimental observation of the progression from aspecific to immature to specific. Additionally, this suggests that the classification of aspecific, immature and specific are relatively arbitrary divisions in what is really a continuum of response.

In simulations using patterned stimuli, the outcome is not very sensitive to system parameters; specificity of response close to the optimum was achieved in simulations under a considerable range of parameter values. It is especially notable that even with relatively high noise levels (interfering unpatterned stimuli), with signal-to-noise ratios considerably smaller than one, we still obtain qualitatively very similar behavior to the noiseless (no interfering stimuli) learning cases.

In simulations using noise-like stimuli (such as one might expect when an animal is dark-reared), a decrease of specificity takes place, practically independent of how specific the tuning of the cell was originally. The main effect is a rapid decay of the central peak of the tuning curve; the change in response for non-optimal patterns is smaller. The overall time constant for loss of specificity is larger than that for gain of specificity. This seems to be in agreement with the experimental results, showing that exposing dark-reared animals to patterned stimuli produces remarkably fast gains of specificity; loss of specificity is a more gradual process.

For the noise-input situation the effect of parameters is important. In the absence of the overwhelming influence of strong patterned visual stimuli, some rather subtle phenomena, due to correlation properties among different noise components, become visible. Strong correlation between the noise inputs to modifiable and non-modifiable junctions leads to a limit such that the total response of the cell is close to zero. Weak correlation drives the modifiable part of the synapse junction mapping close to zero, leaving the non-modifiable part alone to determine the tuning of the cell. A return of patterned stimuli (after noisy stimuli have broadened the response of a cell) can then restore the sharp tuning.

We see that if there is strong correlation between the noise inputs to labile and non-labile junctions, the initial bias of the cell can be entirely lost. A return of specificity would then not necessarily be to the same orientation as that preferred originally. If excitation is assumed between cells in an orientation column, other cells that retained their original bias might still guide a cell that had lost its bias to its original orientation preference. However, an entire column could shift its orientation preference.

These results seem to us to be, at least qualitatively, in agreement with the experimental data cited previously. An extension of the theory to include the effects of binocular interactions has been developed with Elie Bienenstock and Paul Munro. With this extension, we have a fairly explicit theory, applicable to a wide variety of situations in visual cortex, in agreement with present observation and with a number of new and hopefully testable consequences. □

The theoretical work on which this article is based was supported in part by a grant from the Ittleson Foundation, Inc. and the Fondation de France. The experimental work was supported by grants from the French CNRS.

Theory for the Development of Neuron Selectivity: Orientation Specificity and Binocular Interaction in Visual Cortex

(with E. L. Bienenstock and P. W. Munro)

J. Neurosci. **2**, 32 (1982)

Although the previous work with Oja and Liberman seemed promising, there were several obvious flaws. First, we dealt with a single cortical cell, neglecting the fact that visual cortex is an extraordinary rich and complex mélange of many cells receiving input from LGN and from each other. Second, our modeling of the external visual environment, while plausible, was artificial. Third, the modification algorithm itself had what seemed to be a serious problem. The threshold had to be set very carefully: cell response to one pattern above threshold, response to the others below. In many ways our work in this area since that time has been designed to correct these flaws.

The first problem we addressed was that of the artificiality of setting the modification threshold. If it was set too high, the response of all patterns would be below threshold and the cell would lose its responsiveness to all patterns. If it was set too low and more than one pattern responded above threshold, the cell would become responsive to more than one pattern and thus would not be selective.

Fishel Liberman, who was the youngest member of CLO, a graduate student at that time, and who was assigned the task of running the simulations, would often appear in my office complaining that he was losing cells. My response was that he should set the threshold more carefully. Upon some deep thinking we decided that it was unlikely that the cells in visual cortex each would have assigned to them their own Fishel Liberman to set thresholds; for the algorithm to be viable there had to be some automated mechanism by which the threshold found its proper place.

Thus we began to toy with the idea of a threshold that moved depending on a cell property. A likely property seemed to be some average of the post-synaptic cells responsiveness, say its firing rate or the post-synaptic depolarization (designated henceforth as c). Since, if the cell were not responsive, the threshold would be low and the cell's response to inputs would increase, while if the cell's response was large, the threshold would be high and responsiveness (to at least some patterns) would decrease. We thus began to experiment with a threshold movingin some fashion, depending on some average value of c; the results immediately appeared promising.

At about that time there was a changing of the guard. Fishel Liberman graduated; Paul Munro took his place as youngest team member and Elie Bienenstock joined our group. I had met Elie, a young student in Michel Imbert's laboratory, in the College de France; he attended my lectures and expressed an interest in doing work in this area for his Ph. D. He joined our laboratory in the fall of '78. Although he didn't take immediately to New England cuisine (I recall that he was particularly impressed by a type of hamburger prepared with blue cheese known as a "blueburger") he has since become thoroughly acclimated to life in New England and now visits us frequently. It appears that his modification threshold has adjusted significantly.

Elie's mathematical skills were ideal for working out the consequences of the nonlinear modification threshold introduced about that time. In his thesis he did a very thorough exploration of the stability of the fixed points that would result in a two-dimensional situation. The following paper is the basic paper in which the algorithm is explored in its almost completely developed form, discussing the fixed points and their stability. It is shown that, in a patterned environment, the cell becomes selective; in a noisy environment the cell loses selectivity, all in a very robust fashion. In addition, under various deprivation conditions, in particular monocular deprivation, the response of the cells is in agreement with what is observed experimentally. This has come to be known as the BCM algorithm. In later papers in this collection we demonstrate that this algorithm has powerful statistical properties. We are thus tempted to conjecture that the algorithm evolved in such a way as to enable cortical neurons to process early information in a manner that was valuable statistically.

0270-6474/82/0201-0032$02.00/0
Copyright © Society for Neuroscience
Printed in U.S.A.

The Journal of Neuroscience
Vol. 2, No. 1, pp. 32–48
January 1982

THEORY FOR THE DEVELOPMENT OF NEURON SELECTIVITY: ORIENTATION SPECIFICITY AND BINOCULAR INTERACTION IN VISUAL CORTEX[1]

ELIE L. BIENENSTOCK,[2] LEON N COOPER,[3] AND PAUL W. MUNRO

Center for Neural Science, Department of Physics, and Division of Applied Mathematics, Brown University, Providence, Rhode Island 02912

Received June 5, 1981; Revised August 27, 1981; Accepted September 1, 1981

Abstract

The development of stimulus selectivity in the primary sensory cortex of higher vertebrates is considered in a general mathematical framework. A synaptic evolution scheme of a new kind is proposed in which incoming patterns rather than converging afferents compete. The change in the efficacy of a given synapse depends not only on instantaneous pre- and postsynaptic activities but also on a slowly varying time-averaged value of the postsynaptic activity. Assuming an appropriate nonlinear form for this dependence, development of selectivity is obtained under quite general conditions on the sensory environment. One does not require nonlinearity of the neuron's integrative power nor does one need to assume any particular form for intracortical circuitry. This is first illustrated in simple cases, e.g., when the environment consists of only two different stimuli presented alternately in a random manner. The following formal statement then holds: the state of the system converges with probability 1 to points of maximum selectivity in the state space. We next consider the problem of early development of orientation selectivity and binocular interaction in primary visual cortex. Giving the environment an appropriate form, we obtain orientation tuning curves and ocular dominance comparable to what is observed in normally reared adult cats or monkeys. Simulations with binocular input and various types of normal or altered environments show good agreement with the relevant experimental data. Experiments are suggested that could test our theory further.

It has been known for some time that sensory neurons at practically all levels display various forms of stimulus selectivity. They may respond preferentially to a tone of a given frequency, a light spot of a given color, a light bar of a certain length, retinal disparity, orientation, etc. We might, therefore, regard stimulus selectivity as a general property of sensory neurons and conjecture that the development of such selectivity obeys some general rule. Most attractive is the idea that some of the mechanisms by which selectivity develops in embryonic or early postnatal life are sufficiently general to allow a unifying theoretical treatment.

In the present paper, we attempt to construct such a mathematical theory of the development of stimulus selectivity in cortex. It is based on (1) an elementary definition of a general index of selectivity and (2) stochastic differential equations proposed as a description of the evolution of the strengths of all synaptic junctions onto a given cortical neuron.

The ontogenetic development of the visual system, particularly of higher vertebrates, has been studied very extensively. Since the work of Hubel and Wiesel (1959, 1962), it has been known that almost all neurons in the primary visual cortex (area 17) of the normally reared adult cat are selective; they respond in a precise and sometimes highly tuned fashion to a variety of features—in particular, to bars or edges of a given orientation and/or those moving in a given direction through their receptive fields. Further work has shown that the response characteristics of these cortical cells strongly depend on the visual environment experienced by the animal during a *critical period* extending roughly from the 3rd to the 15th week of postnatal life (see, for example, Hubel and Wiesel, 1965; Blakemore and Van Sluyters, 1975; Buis-

[1] This work was supported in part by United States Office of Naval Research Contract N00014-81-K-0136, the Fondation de France, and the Ittleson Foundation, Inc. We would like to express our appreciation to our colleagues at the Brown University Center for Neural Science for their interest and helpful advice. In particular, we thank Professor Stuart Geman for several useful discussions.

[2] Present address: Laboratoire de statistique appliquée, Batiment 425, Université de Paris Sud, 91405 Orsay, France.

[3] To whom correspondence should be addressed at Center for Neural Science, Brown University, Providence, RI 02912.

seret and Imbert, 1976; Frégnac and Imbert, 1978; Frégnac, 1979). Although these experiments show that visual experience plays a determining role in the development of selectivity, the precise nature of this role is still a matter of controversy.

Applying our general ideas to the development of *orientation selectivity* and *binocular interaction* in area 17 of the cat visual cortex, we obtain a theory based on a single mechanism of synaptic modification that accounts for the great variety of experimental results on monocular and binocular experience in normal and various altered visual environments. In addition, we obtain some new predictions.

It is known that various algorithms related to Hebb's principle of synaptic modification (Hebb, 1949) can account for the formation of associative and distributed memories (see, for example, Marr, 1969; Brindley, 1969; Anderson, 1970, 1972; Cooper, 1973; Kohonen, 1977). We therefore suggest that it may be the same fundamental mechanism, accessible to detailed experimental investigation in primary sensory areas of the nervous system, which is also responsible for some of the higher forms of central nervous system organization.

In sections I to III, our ideas are presented in general form, section IV is devoted to the development of orientation selectivity primarily in a normal visual environment, whereas in section V, it is shown that our assumptions also account for normal or partial development of orientation selectivity and binocularity in various normal or altered visual environments.

I. Preliminary Remarks and Definitions

Notation. We simplify the description of the dynamics of a neuron by choosing as variables not the instantaneous incoming time sequence of spikes in each afferent fiber, the instantaneous membrane potential of the neuron, or the time sequence of outgoing spikes but rather the pre- and postsynaptic firing frequencies. These may be thought of as moving time averages of the actual instantaneous variables,[4] where the length of the averaging interval is of the order of magnitude of the membrane time constant, τ. Throughout this paper, these firing frequencies are used as instantaneous variables. This formal neuron is thus a device that performs spatial integration (it integrates the signals impinging all over the soma and dendrites) rather than spatiotemporal integration: the output at time t is a function of the input and synaptic efficacies at t, independent of past history.

A synaptic efficacy m_j characterizes the *net effect* of the presynaptic neuron j on the postsynaptic neuron (in most of the paper, only one postsynaptic neuron is considered). This effect may be mediated through a complex system including perhaps several interneurons, some of which are excitatory and others inhibitory. The resulting "ideal synapse" (Nass and Cooper, 1975) thus may be of either sign, depending on whether the net effect is excitatory or inhibitory; it also may change sign during development.

A further simplification is to assume that the *integrative power* of the neuron is a linear function, that is:

$$c(t) = \Sigma_j m_j(t) d_j(t) \qquad (1)$$

where $c(t)$ is the output at time t, $m_j(t)$ is the efficacy of the jth synapse at time t, $d_j(t)$ is the jth component of the input at time t (the firing frequency of the jth presynaptic neuron), and Σ_j denotes summation over j (i.e., over all presynaptic neurons). We can then write:

$$m(t) = (m_1(t), m_2(t), \ldots, m_N(t))$$
$$d(t) = (d_1(t), d_2(t), \ldots, d_N(t))$$
$$c(t) = m(t) \cdot d(t) \qquad (2)$$

$m(t)$ and $d(t)$ are real-valued vectors, of the same dimension, N (i.e., the number of ideal synapses onto the neuron), and $c(t)$ is the inner product (or "dot product") of $m(t)$ and $d(t)$. The vector $m(t)$ (i.e., the array of synaptic efficacies at time t) is called the *state* of the neuron at time t. (Note that $c(t)$ as well as all components of $d(t)$ represent firing frequencies that are measured from the level of average spontaneous activity; thus, they might take negative as well as positive values; $m_j(t)$ is dimensionless.)

The precise form of the integrative power is not essential: our results remain unchanged if, for instance, $c(t) = S(m(t) \cdot d(t))$, with S being a positive-valued sigmoid-shaped function (see Bienenstock, 1980). This is in contrast to other work (e.g., von der Malsburg, 1973) that does require nonlinear integrative power (see "Appendix B").

Selectivity. It is common usage to estimate the orientation selectivity of a single visual cortical neuron by measuring the half-width at half-height—or an equivalent quantity—of its orientation tuning curve. The selectivity is then measured with respect to a parameter of the stimulation, namely the orientation, which takes on values over an interval of 180°. In the present study, various kinds of inputs are considered, e.g., formal inputs with a parameter taking values on a finite set of points rather than a continuous interval. It will be useful then to have a convenient general index of selectivity, defined in all cases. We propose the following:

$$\mathrm{Sel}_d(\mathcal{N}) = 1 - \frac{\text{mean response of } \mathcal{N} \text{ with respect to } d}{\text{maximum response of } \mathcal{N} \text{ with respect to } d} \qquad (3)$$

With this definition, selectivity is estimated *with respect to* or *in* an *environment for the neuron*, that is, a random variable d that takes on values in the space of inputs to the neuron $\mathcal{N}$. The variable d represents a random input to the neuron; it is characterized by its probability distribution that may be discrete or continuous. (During normal development, the input to the neuron (or neuronal network) is presumably distributed uniformly over all orientations. In abnormal rearing conditions (e.g., dark reared), the input during development could be different from the input for measuring selectivity. How this should be translated in the formal space R^N will be discussed in section IV.) This distribution defines an environment, mathematically a random variable d. Selectivity is estimated (before or after develop-

---

[4] The precise form of the averaging integral (i.e., of the convolution kernel) is not essential. Exponential kernels $K(t) = \exp(-t/\tau)$ often are used in this context (see, e.g., Nass and Cooper, 1975; Uttley, 1976).

ment) with respect to this same environment.[5] Obviously, $\mathrm{Sel}_d(\mathcal{N})$ always falls between 0 and 1 and the higher the selectivity of $\mathcal{N}$ in $\boldsymbol{d}$, the closer $\mathrm{Sel}_d(\mathcal{N})$ is to 1.

When applied to the formal neuron in state m, definition 3 gives:

$$\mathrm{Sel}_d(m) = 1 - \frac{E[m \cdot \boldsymbol{d}]}{\mathrm{ess\ sup}(m \cdot \boldsymbol{d})}$$

where $\boldsymbol{d}$ is any R^N-valued random variable (the formal environment for the neuron). The symbol $E[\ldots]$ stands for "expected value of $\ldots$" (i.e., the mean value with respect to the distribution of $\boldsymbol{d}$) and "ess sup of $\ldots$" (essential supremum) is equivalent to "maximum of $\ldots$" in most common applications. This is illustrated in Figure 1.

II. Modification of Cortical Synapses

The various factors that influence synaptic modification may be divided broadly into two classes—those dependent on global and those dependent on local information. Global information in the form of chemical or electrical signaling presumably influences in the same way most (or all) modifiable junctions of a given type in a given area. Evidence for the existence of global factors that affect development may be found, for instance, in the work of Kasamatsu and Pettigrew (1976, 1979), Singer (1979, 1980), and Buisseret et al. (1978). On the other hand, local information available at each modifiable synapse can influence each junction in a different manner. In this paper, we are interested primarily in the effect of local information on the development of selectivity.

An early proposal as to how local information could affect synaptic modification was made by Hebb (1949). His, now classical, principle was suggested as a possible neurophysiological basis for operant conditioning: "when an axon of cell A is near enough to excite a cell B and repeatedly or persistently takes part in firing it, some growth process or metabolic change takes place in one or both cells such that A's efficiency, as one of the cells firing B, is increased." Thus, the increase of the synaptic strength connecting A to B is dependent upon the correlated firing of A and B. Such a correlation principle has inspired the work of many theoreticians on various topics related to learning, associative memory, pattern recognition, the organization of neural mappings (retinotopic projections), and the development of selectivity of cortical neurons.

It is fairly clear that, in order to actually use Hebb's principle, one must state conditions for synaptic decrease as specific as those for synaptic increase: if synapses are allowed only to increase, all synapses will eventually saturate; no information will be stored and no selectivity

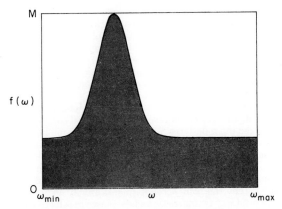

Figure 1. Computing the selectivity with respect to an environment uniformly distributed between $\omega_{\min}$ and $\omega_{\max}$. The abscissa displays a parameter of the stimulus (e.g., orientation $(\omega_{\max} - \omega_{\min} = 180°)$) and on the ordinate, the neuron's response 0 is the level of the average spontaneous activity; M is the maximum response. The selectivity of the neuron then is given by

$$\mathrm{Sel}_d(\mathcal{N}) = 1 - \frac{1}{M(\omega_{\max} - \omega_{\min})} \int_{\omega_{\min}}^{\omega_{\max}} f(\omega)\, d\omega$$

$$= \frac{\text{light area}}{\text{total box area}}$$

This is a simple measure of the breadth of the peak: curves of same selectivity have approximately the same half-width at half-height. (Think, for instance, of triangularly shaped tuning curves.) Typical values for orientation selectivity of adult cortical cells vary between 0.7 and 0.85 ("specific" cells). Selectivity of broadly tuned but still unimodal cells (e.g., those termed "immature" by Buisseret and Imbert (1976) and Frégnac and Imbert (1978)) lies between 0.5 and 0.7. Obviously, 0 is the selectivity of an absolutely flat curve, whereas 1 is the selectivity of a Dirac δ function.

will develop (see, for example, Sejnowski, 1977a, b). What is required is thus a complementary statement to Hebb's principle giving conditions for synaptic decrease.[6]

Such statements usually have resulted in a form of synaptic competition. Consider, for example, one that was proposed by Stent (1973): "when the presynaptic axon of cell A repeatedly and persistently fails to excite the postsynaptic cell B while cell B is firing under the influence of other presynaptic axons, metabolic changes take place in one or both cells such that A's efficiency, as one of the cells firing B, is decreased." According to Stent's principle, the increase of the strength of certain synapses onto neuron B is accompanied by simultaneous decrease of the strength of other synapses onto the same

---

[5] The mathematical concept that is needed in order to represent the environment, $\boldsymbol{d}$, during the development period is that of a *stationary stochastic process*, $\boldsymbol{d}(t)$, that is (roughly), a time-dependent random variable whose distribution is invariant in time. For example, $\boldsymbol{d}$ could represent an elongated bar in the receptive field of the neuron, rotating in some random manner around its center. At each instant, the probability of finding the bar in any given orientation is the same as at any other: the distribution of $\boldsymbol{d}(t)$ is time invariant, uniform over the interval $(0, 180°)$.

---

[6] Nonspecific conditions for synaptic decrease, such as uniform exponential decay, are clearly insufficient too: in Nass and Cooper (1975) for instance, no selectivity is achieved without lateral intracortical inhibition. Other models (von der Malsburg, 1973; Perez et al., 1975) use a normalization rule in conjunction with a hebbian scheme for synaptic increase, which actually results in decrease as well as increase. This normalization rule is discussed in "Appendix B."

neuron B. There thus occurs a *spatial competition between convergent afferents*. A competition mechanism of this kind provides a qualitative explanation of some experimental results on cortical development (e.g., monocularly deprived animals (Stent, 1973)) as well as some aspects of certain more complex deprivation paradigms such as those recently reported by Rauschecker and Singer (1981).

In the present work, we present a mechanism of synaptic modification that results in a temporal competition between input patterns rather than a spatial competition between different synapses. With this mechanism, whether synaptic strength increases or decreases depends upon the magnitude of the postsynaptic response as compared with a variable modification threshold. We show that this can account quantitatively in a more powerful way for increases and decreases in selectivity as well as for a great variety of other experimental results in diverse rearing conditions.

We propose that the change of the jth synapse's strength at the time t obeys the following rule:

$$\dot{m}_j(t) = \phi(c(t))d_j(t) - \epsilon m_j(t) \tag{4}$$

where $\phi(c)$ is a scalar function of the postsynaptic activity, $c(t)$, that changes sign at a value, θ_M, of the output called the modification threshold:

$$\phi(c) < 0 \quad \text{for} \quad c < \theta_M; \qquad \phi(c) > 0 \quad \text{for} \quad c > \theta_M$$

The term, $-\epsilon m(t)$, produces a uniform decay of all junctions; this, in most cases, does not affect the behavior of the system if ϵ is small enough. However, as will be seen later, it is important in some situations. Other than this uniform decay, the vector m is driven in the direction of the input d if the output is large (above θ_M) or opposite to the direction of the input if the output is small (below θ_M). As required by Hebb's principle, when $d_j > 0$ and c is large enough, m_j increases. However, when $d_j > 0$ and c is *not large enough*, m_j decreases. We may regard this as a form of *temporal competition between incoming patterns*.

The idea of such a modification scheme was introduced by Cooper et al. (1979). Their use of a constant threshold θ_M, however, resulted in a certain lack of robustness of the system: the response to all patterns could slip below θ_M and then decrease to zero. In the absence of lateral inhibition between neurons, the response might increase to more than one pattern, leading to stable states with a maximal response to more than one pattern.

In this paper, we will see that making an appropriate choice for $\theta_M(t)$ allows correct functioning under quite general conditions and provides remarkable noise tolerance properties.

In our threshold modification scheme, the change of the jth synapse's strength is written as a product of two terms, the presynaptic activity, $d_j(t)$, and a function, $\phi(c(t), \bar{c}(t))$, of the postsynaptic variables, the output, $c(t)$, and the average output, $\bar{c}(t)$. Making use of $\bar{c}(t)$ in the evolutive power of the neuron is a new and essential feature of this work. It is necessary in order to allow both boundedness of the state and efficient threshold modification.

Neglecting the uniform decay term, for the moment

($\epsilon = 0$), in vector notation, we have

$$\dot{m}(t) = \phi(c(t), \bar{c}(t)) \, d(t) \tag{5}$$

This, together with equation 2, yields:

$$\dot{m}(t) = \phi(m(t) \cdot d(t), m(t) \cdot \bar{d}) \, d(t) \tag{6}$$

The crucial point in the choice of the function $\phi(c, \bar{c})$ is the determination of the threshold $\theta_M(t)$ (i.e., the value of c at which $\phi(c, \bar{c})$ changes sign). A candidate for $\theta_M(t)$ is the average value of the postsynaptic firing rate, $\bar{c}(t)$. The time average is meant to be taken over a period T preceding t much longer than the membrane time constant τ so that $\bar{c}(t)$ evolves on a much slower time scale than $c(t)$. This usually can be approximated[7] by averaging over the distribution of inputs for a given state $m(t)$

$$\bar{c}(t) = m(t) \cdot \bar{d}$$

This results in an essential feature, the *instability of low selectivity points*. (This can be most easily seen at 0 selectivity equilibrium points, where, with any perturbation, the state is driven away from this equilibrium, whatever the input.)

Therefore, if stable equilibrium points exist in the state space, they are of high selectivity. However, do such points exist at all? The answer is generally yes provided that the state is *bounded from the origin and from infinity*. These conditions, instability of low selectivity equilibria as well as boundedness, are fulfilled by a single function $\phi(c, \bar{c})$ if we define $\theta_M(t)$ to be a *nonlinear function of $c(t)$* (for example, a power with an exponent larger than 1). The final requirement on $\phi(c, \bar{c})$ thus reads:

$$\text{sign} \, \phi(c, \bar{c}) = \text{sign}\left(c - \left(\frac{\bar{c}}{c_0}\right)^p \bar{c}\right) \quad \text{for} \quad c > 0 \tag{7}$$

$$\phi(0, \bar{c}) = 0 \quad \text{for all} \quad \bar{c}$$

where c_0 and p are two fixed positive constants.[8] The

---

[7] Replacing the time average by an average over the distribution of d is allowed provided that (1) the process $d(t)$ is stationary, (2) the interval, T, of time integration is short with respect to the process of synaptic evolution (i.e., $m(t)$ changes very little during an interval of length T), (3) T is long compared to the mixing rate of the process d (i.e., during a period of length T, the relative time spent by the process $d(t)$ at any point d in the input space is nearly proportional to the weight of the distribution of d at d). Now, synaptic modification of the type involved in changes of selectivity is probably a slow process, requiring minutes or hours (if not days) to be significant, whereas elementary sensory patterns (e.g., oriented stimuli in the receptive field of a given cortical neuron) are normally all experienced in an interval of the order of 1 min or less. Thus, we are able to choose T so that a good estimate of $\bar{c}(t)$ can be available to the neuron. In some experimental situations in which the environment is altered, there are subtle dependences of the sequence by which the final state is reached depending on how rapidly $\bar{c}$ adjusts to the changed environment.

[8] The sign of $\phi(c, \bar{c})$ for $c < 0$ is not crucial since c is essentially a positive quantity: cortical cells in general have low spontaneous activity and, at any rate, are rarely inhibited much below their spontaneous activity level. For the sake of mathematical completeness, one may, however, wish to define $\phi(c, \bar{c})$ for negative c; $\phi(c, \bar{c}) > 0$ is then the most convenient for it allows us to state theorems 1 to 3 below under the most general initial conditions. In addition, the form of ϕ for $c < 0$ can affect calculations such as those of "Appendix C."

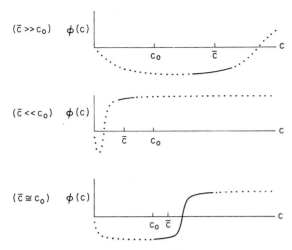

Figure 2. A function satisfying condition 7. The three diagrams show the behavior of $\phi(c, \bar{c})$ as a function of c for three different constant values of $\bar{c}$. In each diagram, the *solid part* of the *curve* represents $\phi(c, \bar{c})$ in the vicinity of $\bar{c}$, which of course is the relevant part of this function. In the *upper diagram* ($\bar{c} \gg c_0$), although $\phi(c, \bar{c})$ is not negative for all c values as was formally required (see text), the probability that $\phi(c, \bar{c}) > 0$ is small and gets even smaller as $\bar{c}$ increases. The important point in the definition of ϕ is the nonlinearity of $\theta_M(\bar{c})$ which makes it increase or decrease faster than $\bar{c}$, while $\theta_M(\bar{c})$ is of the same order as $\bar{c}$, if $\bar{c}$ itself is of the same order as c_0.

threshold $\theta_M(\bar{c}) = (\bar{c}/c_0)^p \bar{c}$ thus serves two purposes: allowing threshold modification when $\bar{c} \simeq c_0$ as well as driving the state from regions such that $\bar{c} \ll c_0$ or $\bar{c} \gg c_0$. Equation 7 is illustrated in Figure 2.

The process of synaptic growth, starting near zero to eventually end in a stable selective state, may be described as follows. Initially, $\bar{c} \ll c_0$; hence, $\phi(c, \bar{c}) > 0$ for all inputs in the environment: the responses to all inputs grow. With this growth, $\bar{c}$ increases, thus increasing θ_M. Now some inputs result in postsynaptic responses that exceed θ_M, while others—those whose direction is far away from (close to orthogonal to) the favored inputs—give a response less than θ_M. The response to the former continues to grow, while the response to the latter decays. This results in a form of *competition between incoming patterns* rather than competition between synapses. The response to unfavored patterns decays until it reaches zero, where it stabilizes for $\phi(0, \bar{c}) = 0$ for any $\bar{c}$ (equation 7). The response to favored patterns grows until the mean response $\bar{c}$ is high enough, and the state stabilizes. This occurs in spite of the fact that many complicated geometrical relationships may exist between different patterns (i.e., that they are not orthogonal since different patterns may and certainly do share common synapses).

Any function, ϕ, that satisfies equation 7 will give the results that we describe below. The precise form of this function (e.g., the numerical values of p and c_0) will affect the detailed behavior of the system, such as rate of convergence, the height of the maximum response for a selective cell, etc., and would have to be determined by experiment.

III. Mathematical Results

The behavior of system 6 depends critically on the environment, that is, on the distribution of the stationary stochastic process, d. Two classes of distributions may be considered—discrete distributions and continuous distributions. Discrete distributions include K possible inputs $d^1, \ldots, d^K$. These will generally be assumed to occur with the same probability $1/K$. The process d is then a jump process which randomly assumes new values at each time increment. The vector m is (roughly) a Markov process. In the present work, the only continuous distribution that will be considered is a uniform distribution d over a closed one-parameter curve in the input space R^N (section IV).

Although the principles underlying the convergence to selective states are intuitively fairly simple (see the preceding section), mathematical analysis of the system is not entirely straightforward, even for the simplest d. Mathematical results, obtained only for certain discrete distributions, are of two types: (1) equilibrium points are locally stable if and only if they are of the highest available selectivity with respect to the given distribution of d and (2) given any initial value of m in the state space, the probability that $m(t)$ converges to one of the maximum selectivity fixed points as t goes to infinity is 1. Results of the second type are much stronger and require a tedious geometrical analysis. Results are stated here in a somewhat simplified form (obvious requirements of a very mathematical character are omitted). For exact statements and proofs, the reader is referred to Bienenstock (1980).

We first study the simple case where d takes on values on only two possible input vectors, d^1 and d^2, that occur with the same probability:

$$P[d = d^1] = P[d = d^2] = \tfrac{1}{2}$$

Whatever the actual dimension N of the system, it reduces to two dimensions. (Any component of m outside of the linear subspace spanned by d^1 and d^2 will eventually decay to zero due to the uniform decay term.)

It follows immediately from the definition that the maximum value of $\text{Sel}_d(m)$ in the state space is $\tfrac{1}{2}$. It is reached for states m which give a null response when d^1 comes in (i.e., are orthogonal to d^1) but a positive response for d^2—or vice versa. Minimum selectivity, namely zero, is obtained for states m such that $m \cdot d^1 = m \cdot d^2$. Equilibrium states of both kinds indeed exist.

Lemma 1. Let d^1 and d^2 be linearly independent and d satisfy $P[d = d^1] = P[d = d^2] = \tfrac{1}{2}$. Then for any value of ϕ satisfying equation 7, equation 6 admits exactly four fixed points, $m^0, m^1, m^2,$ and $m^{1, 2}$ with: $\text{Sel}_d(m^0) = \text{Sel}_d(m^{1, 2}) = 0$ and $\text{Sel}_d(m^1) = \text{Sel}_d(m^2) = \tfrac{1}{2}$. (Here the superscripts indicate which of the d^i are *not* orthogonal to m. (m^0 is the origin.) Thus, for instance, $m^1 \cdot d^1 > 0, m^1 \cdot d^2 = 0$.)

The behavior of equation 6 depends on the geometry of the inputs, in the present case, on $\cos(d^1, d^2)$. The crucial assumption needed here is that $\cos(d^1, d^2) \geq 0$. This is a reasonable assumption which is obviously satisfied if all components of the inputs are positive, as is assumed in some models (von der Malsburg, 1973; Perez et al., 1975). We then may state the following.

Theorem 1. Assume that, in addition to the conditions of

lemma 1, $\cos(d^1, d^2) \geq 0$. Then m^0 and $m^{1,2}$ are unstable, m^1 and m^2 are stable, and whatever its initial value, the state of the system converges almost surely (i.e., with probability 1) either to m^1 or to m^2.

Theorem 1 is the basic result in the two-dimensional setting: it characterizes evolution schemes based on *competition between patterns* and states that the state eventually reaches maximal selectivity even when the two input vectors are very close to one another. Obviously this requires that some of the synaptic strengths be negative since the neuron has linear integrative power. Inhibitory connections are thus necessary to obtain selectivity (see also section IV below). Some selectivity is also realizable with no inhibitory connections—not even "intracortical" ones—if the integrative power is appropriately nonlinear. However, whatever the nonlinearity of the integrative power, theorem 1 could not hold for evolution equations based on *competition between converging afferents* (see "Appendix B").

In theorem 1, we have a discrete sensory environment which consists of exactly two different stimuli—a situation, although simple mathematically, not often encountered in nature. It may, however, very well correspond to a visual environment restricted to only horizontally and vertically oriented contours present with equal probability. Theorem 1 then predicts that cortical cells will develop a selective response to one of the two orientations, with no preference for either (other than what may result from initial connectivity). Thus, on a large sample of cortical cells, one should expect as many cells tuned to the horizontal orientation as to the vertical one. (So far, no assumption is made on intracortical circuitry. See "Appendix D.")

The proof of theorem 1 is based on the existence of *trap regions* around each of the selective fixed points.

Theorem 2. Under the same conditions as in theorem 1, there exists around $m^1(m^2)$ a region $F^1(F^2)$ such that, once the state enters $F^1(F^2)$, it converges almost surely to $m^1(m^2)$.

The meaning of theorem 2 is the following: once $\boldsymbol{m}(t)$ has reached a certain selectivity, it cannot "switch" to another selective region. Applied to cortical cells in a patterned visual environment, this means that, once they become sufficiently committed to certain orientations, they will remain committed to those orientations (provided that the visual environment does not change), becoming more selective as they stabilize to some maximal selectivity. Theorems 1 and 2 are illustrated in Figure 3.

It is worth mentioning that, when $\cos(d^1, d^2) < 0$, the situation is much more complicated: trap regions do not necessarily exist and periodic asymptotic behavior (i.e., limit cycles) may occur, bifurcating from the stable fixed points when $\cos(d^1, d^2)$ becomes too negative (see Bienenstock, 1980).

We now turn to the case where $\boldsymbol{d}$ takes on K values. The following is easily obtained.

Lemma 2. Let $d^1, d^2, \ldots, d^K$ be linearly independent and $\boldsymbol{d}$ satisfy $P[\boldsymbol{d} = d^1] = \ldots = P[\boldsymbol{d} = d^K] = 1/K$. Then, for any function ϕ satisfying equation 7, equation 6 admits exactly 2^K fixed points with selectivities $0, 1/K, 2/K, \ldots, (K-1)/K$. There are K fixed points $m^1, \ldots, m^K$ of selectivity $(K-1)/K$.

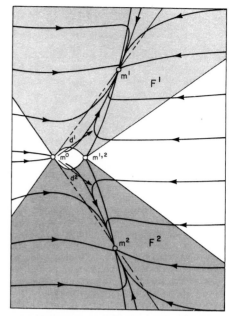

Figure 3. The phase portrait of equation 6 in an environment consisting of two inputs, d^1 and d^2 (theorems 1 and 2). The diagram shows the trajectories of the state of the system, starting from different initial points. This is a computer simulation performed with one given function ϕ satisfying condition 7. Using a different function may slightly change the shape of the trajectories without any essential change in the behavior. The unstable fixed points are $m^{1,2}$ and m^0; the stable ones are m^1 and m^2. The system is a *stochastic* one, which means that the trajectories depend, in fact, on the precise sequence of inputs. As long as the state is in the *unshaded region*, it is not yet known whether it will eventually be attracted to m^1 or m^2. This is determined as the state enters one of the trap (*shaded*) regions, F^1 or F^2. The trajectories shown here are deterministic ones, obtained by alternating $\boldsymbol{d}$ regularly between d^1 and d^2. They are, in fact, the averaged trajectories of the state and are much more regular and smooth than the actual stochastic ones.

Obviously, $(K-1)/K$ is also the maximum possible selectivity with respect to $\boldsymbol{d}$. It means a positive response for one and only one of the inputs. The situation is now much more complicated than what it was with only two inputs: it is not obvious whether, in all cases, the assumption that all of the cosines between inputs are positive is sufficient to yield stability of the maximum selectivity fixed points. However, we may state the following.

Theorem 3. Assume, in addition to the conditions of lemma 2, that $d^1, \ldots, d^K$ are all mutually orthogonal or close to orthogonal. Then the K fixed points of maximum selectivity are stable, and whatever its initial value, the state of the system converges almost surely to one of them.

The proof of theorem 3 also involves trap regions around the K maximally selective fixed points, and the analog of theorem 2 is true here.

Although the general case has not yet been solved analytically, as will be seen in the next section, computer simulations suggest that, for a fairly broad range of environments, if $d^i \cdot d^j \geqslant 0$, even if $d^1, \ldots, d^K$ are far from being mutually orthogonal, the K fixed points of maximum selectivity are stable.

Simulations suggest further (see, for instance, Fig. 4b) that, even if the $d^1, \ldots, d^K$ are *not* linearly independent and are very far from being mutually orthogonal, the asymptotic selectivity is close to its maximum value with respect to d.

IV. Orientation Selectivity and Binocular Interaction in Visual Cortex

We now apply what has been done to a concrete example, orientation selectivity and binocular interaction in the primary visual cortex. The ordinary development of these properties in mammals depends to a large extent on normal functioning of the visual system (i.e., normal visual experience) during the first few weeks or months of postnatal life. This has been demonstrated many times by various experiments, based mainly on the paradigm of rearing the animal in a restricted sensory environment. In the next two sections, it is shown how equations 4 to 7 account for both normal development as well as development in restricted visual environments.

Consider first a classical test environment used to construct the tuning curve of cortical neurons. This environment consists of an elongated light bar successively presented or moved in all orientations—preferably in a random sequence—in the neuron's receptive field. Thus, all of the parameters of the stimulus are constant except one, the orientation, which is distributed uniformly on a circularly symmetric closed path. We assume that the retinocortical pathway maps this family of stimuli to the cortical neuron's space of inputs in such a way as to preserve the circular symmetry (as defined below). Thus, the typical theoretical environment that will be used for constructing the formal neuron's tuning curve is a random variable d uniformly distributed on a circularly symmetric closed one-parameter family of points in the space R^N. The parameter coding orientation in the receptive field is, in principle, continuous. However, for the purpose of numerical simulations, the distribution is made discrete. Thus, d takes on values on the points d^1, $\ldots, d^K$.

The requirement of circular symmetry is expressed mathematically as follows: the matrix of inner products of the vectors $d^1, \ldots, d^K$ is circular (i.e., each row is obtained from its nearest upper neighbor by shifting it one column to the right) and the rows of the matrix are unimodal. A random variable, d, uniformly distributed on such a set of points will be, hereafter, called a *circular environment*. Such a d may be roughly characterized by three parameters: N, K, and a measure of the mutual geometrical closeness of the d^i vectors, for instance, $\min \cos(d^1, d^i)$.

Now we are faced with the difficult problem of specifying the stationary stochastic process that represents the time sequence of inputs to the neuron during development. In a first analysis, there is no choice but to oversimplify the problem by giving the stochastic process exactly the same distribution as the circular d defined above. In doing so, we assume that development of

orientation selectivity is to a large extent independent of other parameters of the stimulus (e.g., contrast, shape, position in the receptive field, retinal disparity for binocular neurons, etc.). The elementary stimulus for a cortical neuron is a rectilinear contrast edge or bar. Any additional pattern present at the same time in the receptive field is regarded as random noise. (A discussion of this point is given in Cooper et al. (1979).)

IVa. Normal Monocular Input

The behavior of a monocular system in circular environments is investigated by numerically simulating equation 6 with a variety of circular environments, d, and functions ϕ satisfying equation 7. In the simulations presented here, the dimension of the input and state space is generally $N = 37$; the number K of input vectors varies from 12 to 60. (Various kinds of functions ϕ were used: some were stepwise constant; others were smooth, bounded, or unbounded.) One may reasonably expect the system's behavior to be fairly independent of N and K if these are high enough. However, the geometry of d may be determining: if the inputs, d^i, are closely packed together in the state space (i.e., if $\min \cos(d^1, d^i)$ is close to 1), convergence to selective states may presumably be difficult to achieve or even impossible.

Simulations show the following behavior:

1. The state converges rapidly to a fixed point or *attractor*.
2. Various such attractors exist. For a given d and ϕ, they all have the same selectivity, which is close to its maximum value in d.
3. The asymptotic tuning curve is always unimodal. Thus, one may talk of the preferred orientation of an attractor.
4. There exists an attractor for each possible orientation.
5. If there is no initial preference, all orientations have equal probability of attracting the state. (Which one will become favored depends on the exact sequence of inputs.) This does not hold for environments which are not perfectly circular, at least for a single neuron system such as the one studied here.

In Figure 4, a and b show, respectively, the progressive buildup of selectivity and the tuning curve when the state has virtually stabilized.

In summary then, the system behaves in circular environments exactly as we might have expected from the results of the preceding section. However, one should note one important difference: the maximum selectivity for a continuous environment cannot be calculated as simply as it was before. It is only when d is distributed uniformly on K linearly independent vectors that we know that $\max \mathrm{Sel}_d(m) = (K - 1)/K$ (lemma 2). Theorem 3 indicates that, if, in addition, the vectors are nearly orthogonal to one another, this selectivity is indeed asymptotically reached. We could not prove that this is also true when the vectors are arranged circularly but are not mutually orthogonal. However, it could not be disproved by any numerical simulation; therefore, we *conjecture* that this is indeed true. (Reasonable selectivity is attained even in most unfavorable environments. As an

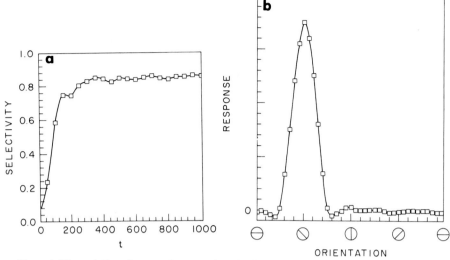

Figure 4. The evolution of a synaptic system in a circular environment. Here, $K = 40$ and $N = 37$ so that the vectors are linearly dependent. The value of the maximum selectivity with respect to d is therefore not precisely calculable. The asymptotic selectivity is approximately 0.9, perhaps the maximum selectivity. *a* demonstrates the progressive buildup of the selectivity in a circular environment d, while *b* shows the resulting tuning curve at $t = 1000$.

example, in a circular d such that all cosines fall between 0.94 and 1, a selectivity of 0.68 was reached after 12,000 iterations.) Notice that, in the present context, this question is only of theoretical interest, since naturally occurring environments are continuous rather than discrete. The behavior of our system in such an environment is very well approximated by a discrete circular d, provided that K is large enough. K is then presumably larger than N, the K inputs are linearly dependent, and we have no explicit formula for max $\text{Sel}_d(m)$.

The system thus functions well in a large class of environments. It should be stressed that the numerical value of the parameters that appear explicitly in the evolution equation, namely, c_0 and the exponent p, are not at all critical. Simulations performed with a constant d, with p being varied from 0.01 to 10, yield the same asymptotic limit for the selectivity; the height of the asymptotic tuning curve (i.e., max$(m \cdot d)$) is, however, highly dependent on p. This invariance property validates in a sense the definition of $\text{Sel}_d(x)$.

Inhibitory synapses are essential here exactly as they are in the two-dimensional case. One way to show this is to substitute 0 for all negative components in the state once it has become selective. This typically results in a drastic drop of selectivity (e.g., from 0.81 to 0.55) although a slight preference generally remains for the original orientation. This may be related to the experimental finding that local pharmacological deactivation of inhibitory connections strongly impairs orientation selectivity by rendering all orientations effective in triggering the cell's response (Sillito, 1975).

Finally, it should be mentioned that the system displays a good noise tolerance, particularly when the state has already reached a selective region. The system then resists presynaptic additive noise with a signal-to-noise

ratio of the order of 1 and postsynaptic noise with a signal-to-noise ratio as small as $\frac{1}{4}$.

IVb. Restricted Monocular Input

To discuss this situation, we now must include the exponential decay term, $-\epsilon_m(t)$, previously neglected (equation 4). It is clear that the results stated above will be preserved if ϵ is sufficiently smaller than the average of $|\phi(c, \bar{c})|$ (i.e., competition mechanisms are faster than decay). However, exponential decay does become crucial in some situations. One of these is the response of the cell to patterns that were not represented in the environment during development.

Consider, for instance, an environment consisting of a single stimulus d^1. It is then easily shown that system 6 with condition 7 admits one attractor m^1 that satisfies $m^1 \cdot d^1 \simeq c_0$ for small ϵ ($m^1 \cdot d^1 = c_0$ for $\epsilon = 0$). Obviously, for $\epsilon > 0$, m^1 will satisfy $m^1 \cdot d = 0$ for any d orthogonal to d^1. However, the response to a pattern d not orthogonal to d^1 will depend both on ϵ and on $\cos(d, d^1)$. One may for instance find that $m^1 \cdot d \simeq \frac{1}{2}(m^1 \cdot d^1)$ for $\cos(d, d^1) = 0.5$. The selectivity of the neuron in state m^1 with respect to a circular environment $(d^1, \ldots, d^K)$, such that min $\cos(d^1, d^i) = 0.5$, is then lower than 0.5. This should be contrasted with the high selectivity reached by a neuron exposed to all inputs, $d^1 \ldots d^K$.

The one-stimulus environment may be regarded as a case corresponding to rearing the animal in a visual world where only one orientation is present. No controversy remains at present that rearing in such a visual environment results in a cortex in which all visually responsive cells are tuned to the experienced (or nearby) orientations (Blakemore and Cooper, 1970; Hirsch and Spinelli, 1970, 1971; see also Stryker et al., 1978). We see that our theory is in agreement with these findings; moreover, we predict that, in such a cortex, the average selectivity of these cells should be lower than normal. Although there is so far no detailed quantitative study on this point, in a recent study, there is some indication that this may indeed be true: "more neurons with normal orientation tuning were found in the kittens that could see all orientations, or at least horizontal and vertical, than in the kittens that had experienced only one orientation" (Rauschecker and Singer, 1981).

IVc. Binocular Input

We now consider a binocularly driven cell. The firing rate of the neuron at time t becomes

$$c(t) = m_r(t) \cdot d_r(t) + m_l(t) \cdot d_l(t) \tag{8}$$

with evolution schemes for "right" and "left" states m_r and m_l straightforward generalizations of equation 4. Various possibilities now exist for the input (d_r, d_l): one may wish to simulate normal rearing (both d_r and d_l circular and presumably highly correlated), monocular deprivation, binocular deprivation, etc.

Detailed discussion of the results of simulations under various conditions is given in the next section. The main results are summarized here:

1. In an environment simulating normal binocular rearing, the cell becomes orientation selective and binocular, preferring the same orientation through both eyes.
2. In an environment simulating monocular deprivation, the cell becomes monocular and orientation selective, whatever its initial state.
3. In an environment simulating binocular deprivation, the cell loses whatever orientation selectivity it had but does not lose its responsiveness and, in general, remains driven by both eyes.

V. Development under Different Rearing Conditions: Comparison of Theory with Classical Experimental Data

Related Experimental Data

This brief summary is restricted to area 17 of kitten's cortex. Most kittens first open their eyes at the end of the 1st week after birth. It is not easy to assess whether orientation-selective cells exist at that time in the striate cortex: few cells are visually responsive, and the response's main characteristics are generally "sluggishness" and fatigability. However, it is agreed quite generally that, as soon as cortical cells are reliably visually stimulated (e.g., at 2 weeks), some are orientation selective, whatever the previous visual experience of the animal (cf., Hubel and Wiesel, 1963; Blakemore and Van Sluyters, 1975; Buisseret and Imbert, 1976; Frégnac and Imbert, 1978).

Orientation selectivity develops and extends to all visual cells in area 17 if the animal is reared, and behaves freely, in a normal visual environment (NR): complete "specification" and normal binocularity (about 80% of responsive cells) are reached at about 6 weeks of age (Frégnac and Imbert, 1978). However, if the animal is reared in total darkness from birth to the age of 6 weeks (DR), then none or few orientation-selective cells are recorded (from 0 to 15%, depending on the authors and the classification criteria); however, the distribution of ocular dominance seems unaffected (Blakemore and Mitchell, 1973; Imbert and Buisseret, 1975; Blakemore and Van Sluyters, 1975; Buisseret and Imbert, 1976; Leventhal and Hirsch, 1980; Frégnac and Imbert, 1978). In animals whose eyelids have been sutured at birth and which are thus binocularly deprived of pattern vision (BD), a somewhat higher proportion (from 12 to 50%) of the visually excitable cells are still orientation selective at 6 weeks (and even beyond 24 months of age) and the proportion of binocular cells is less than normal (Wiesel and Hubel, 1965; Blakemore and Van Sluyters, 1975; Kratz and Spear, 1976; Leventhal and Hirsch, 1977; Watkins et al., 1978).

Of all visual deprivation paradigms, putting one eye in a competitive advantage over the other has probably the most striking consequences: monocular lid suture (MD), if it is performed during a "critical" period (ranging from about 3 to about 12 weeks), results in a rapid loss of binocularity, to the profit of the open eye (Wiesel and Hubel, 1963, 1965); then, opening the closed eye and closing the experienced one may result in a complete reversal of ocular dominance (Blakemore and Van Sluyters, 1974). A disruption of binocularity that does not favor one of the eyes may be obtained, for example, by provoking an artificial strabismus (Hubel and Wiesel, 1965) or by an alternating monocular occlusion, which gives both eyes an equal amount of visual stimulation (Blakemore, 1976). In what follows, we call this uncorrelated rearing (UR).

Theoretical Results

The aim of this section is to show that the experimental results briefly reviewed above follow from our assumptions if one chooses the appropriate distribution for d. The model system now consists of a single binocular neuron. The firing rate of the neuron at time t is given by

$$c(t) = m_r(t) \cdot d_r(t) + m_l(t) \cdot d_l(t) \tag{8}$$

where the indices r and l refer to right and left eyes, respectively. m_r (or m_l) obeys the evolution scheme described by equations 4 to 6, where d_r (or d_l) is substituted for d. The two equations are, of course, coupled, since $c(t)$ depends at each t on both $m_r(t)$ and $m_l(t)$.

The vector (d_r, d_l) is a stationary stochastic process, whose distribution is one of the following, depending on the experimental situation one wishes to simulate.

Normal rearing (NR)

$d_r(t) = d_l(t)$ for all t, and d_r is circular. (Noise terms that may be added to the inputs may or may not be stochastically independent.)

Uncorrelated rearing (UR)

d_r and d_l are i.i.d. (independent identically distributed): they have the same circular distribution, but no statistical relationship exists between them.

Binocular deprivation

Total light deprivation (DR). The $2N$ components of (d_r, d_l) are i.i.d.: d_r and d_l are uncorrelated noise terms, $(d_r, d_l) = (n_r, n_l)$.

Binocular pattern deprivation (BD). $d_r(t) = \lambda_r(t)e$, $d_l(t) = \lambda_l(t)e$, where e is an arbitrary normalized fixed vector with positive components, and λ_r and λ_l are scalar positive valued and i.i.d.

Monocular deprivation (MD)

d_r is circular; d_l is a noise term: $d_l = n$.

In the NR case, the inputs from the two eyes to a binocular cell are probably well correlated. We therefore assume that they are equal, which is mathematically equivalent. The DR distribution represents dark discharge. The BD distribution deserves a more detailed explanation. In this distribution, it is only the *length* λ_r and λ_l of the vectors d_r and d_l that varies in time. This length is thought to correspond to the intensity of light coming through each closed eyelid, whereas the direction of the vector in the input space is determined by the constant "unpatterned" vector e (e.g., $e = (1/\sqrt{N}) \times (1, 1, .., 1)$). One may indeed assume that, when light falls on the retina through the closed lids, there is, at any instant of time, high correlation between the firing rates of all retinal ganglion cells on a relatively large region of the retina. Inputs from the two eyes, however, are probably to some extent asynchronous (cf., Kratz and Spear, 1976); hence the BD distribution.

Simulations of the behavior of the system in these different environments give the following.

NR (Fig. 5a). All asymptotic states are selective and binocular, with matching preferred orientations for stimulation through each eye.

DR (Fig. 5b). The motion of the state (m_r, m_l) resembles a random walk. (The small exponential decay term is necessary here, too, in order to prevent large fluctuations.) The two tuning curves[9] therefore undergo random fluctuations that are essentially determined by the second order statistics of the input d. As can be seen from the figure, these fluctuations may result sometimes in a weak orientation preference or unbalanced ocular dominance. However, the system never stays in such states very long; its average state on the long run is perfectly binocular and non-oriented. Moreover, whatever the second order statistics of d and the circular environment in which tuning curves are assessed, a regular unimodal orientation tuning curve is rarely observed, and selectivity has never exceeded 0.6. Thus, we may conclude that orientation selectivity as observed in the NR case (both experimental and theoretical) cannot be obtained from purely random synaptic weights. It is worth mentioning here that prolonged dark rearing has been reported to increase response variability (Leventhal and Hirsch, 1980); a similar observation was made by Frégnac and Bienenstock (1981).[10]

BD (Fig. 5c). Unlike the DR case, the state *converges*

(as may easily be proved mathematically). Although there exist both monocular and binocular stable equilibrium points, the asymptotic state is generally monocular if the initial state is taken as 0. The orientation tuning curve then is determined essentially by the relative geometry of the fixed arbitrary vector e and the arbitrary circular environment which serves to assess the tuning curve. Fine unimodal tuning, therefore, is not to be expected.

MD (Fig. 5d). The only stable equilibrium points are monocular and selective. The system converges to such states whatever the initial conditions. In particular, this accounts for reverse suture experiments (Blakemore and Van Sluyters, 1974; Movshon, 1976).

UR (Fig. 5e). This situation is, in a sense, similar to the BD one: the state converges, but monocular as well as binocular equilibria exist. As in the BD case, the asymptotic state generally observed with $m_r(0) = m_l(0) = 0$ is monocular. (This should be attributed to the mismatched inputs from the two eyes, as is done by most authors.) In this case, however, asymptotic states are selective, and when they are binocular, preferred orientations through each eye do not necessarily coincide. It should be mentioned here that Blakemore and Van Sluyters (1974) report that, after a period of alternating monocular occlusion, the remaining binocular cells may differ in their preferred orientations for stimulation through each eye.

These results are in agreement with the classical experimental data in the domain of visual cortex development. Most of them can be obtained fairly easily, with no need of further simulations, as a consequence of the convergence to selective states in the case of a monocularly driven neuron in a circular environment (section IVa).

Some intriguing properties of our theory are more subtle, however, and, in addition to contributing to the results above, provide the opportunity for applications to more complicated experimental paradigms and for new tests. As an example, it is shown in "Appendix C" that, in the MD case, the degree of monocularity of the cortical cell is correlated with its orientation selectivity as well as the diversity of inputs to the open eye. These unexpected predictions agree well with the observation by Cynader and Mitchell (1980) and Trotter et al. (1981) that, after a brief period of monocular exposure, oriented cells are more monocular than non-oriented ones as well as the observation of Rauschecker and Singer (1981) that an open eye with restricted inputs leads to cells oriented to the restricted input that are driven less monocularly than usual. A summary of theoretical results is given below.

VI. Discussion

We have proposed a new mathematical form for synaptic modification and have investigated its consequences on the development of selectivity in cortical neurons. In addition, we have provided a definition of the notion of selectivity with respect to a random variable that might be applied in many different situations (in the domain of development of sensory systems, for example, selectivity of binocular neurons to retinal disparity, etc.). In its application to visual cortex, our theory is in agree-

---

[9] The circular environment which serves to assess the orientation tuning curves is now, in a sense, arbitrary, since it is not at all used in the development period. The same remark applies, of course, to the BD case.

[10] In Figure 1B of Frégnac and Bienenstock (1981), which shows averaged orientation tuning curves of a cell recorded in an 86-day-old DR cat, the selectivity is 0.58 at the beginning of the recording session and 0.28 at the end.

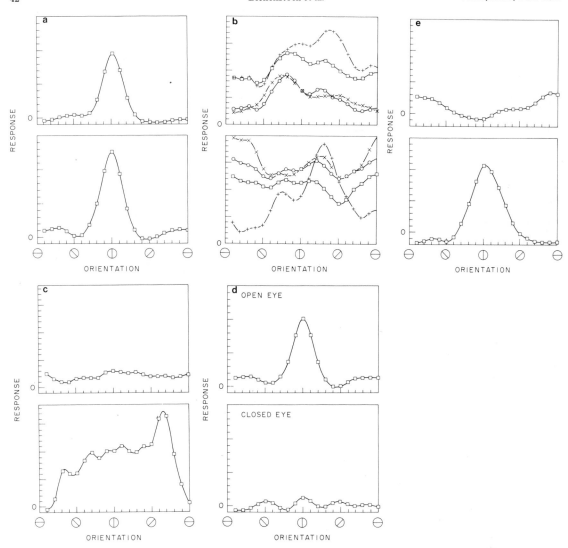

Figure 5. Results of computer simulations corresponding to various rearing conditions. In these simulations, the *upper* and *lower panels* show cell responses to stimuli from the two eyes. *a*, Normal (NR). The cell's response is binocular and selective. *b*, Dark rearing (DR). There is no stable selectivity in the cell's response. The response curve fluctuates randomly. The cell is, on the average, driven binocularly. *c*, Binocular deprivation (BD). The cell reaches a final state corresponding to the arbitrary vector which corresponds to a diffuse input to the retina. The cells sometimes are driven monocularly. This is somewhat analogous to *e* below. *d*, Monocular deprivation (MD). The cell's response is monocular and selective. *e*, Uncorrelated rearing (UR). Both binocular and monocular selective final states are observed.

ment with the classical experimental results obtained over the last generation and offers a number of new predictions, some of which can be tested experimentally. This may lead to the identification of the parameters of the theory and provide indications as to the biochemical mechanisms underlying cortical plasticity.

In a broader context, we may regard our form of synaptic modification as a specific correlation modifica-

tion of a hebbian type. The great majority of models on a synaptic level in domains such as pattern recognition, task learning, or associative memory[11] (which are less accessible to direct neurophysiological experimentation)

---

[11] Notice, for instance, the analogy between states of maximum selectivity as defined here and the optimal associative mappings of Kohonen (1977).

use schemes of a hebbian type with some success. Thus, we are led to conjecture that some form of correlation modification is a very general organizational principle that manifests itself in visual cortex in a manner that is accessible to experiment.

Although synaptic competition is a natural consequence of Hebb's principle, it may be given various mathematical forms. A distinction was made in section II between spatial competition—the form commonly accepted by theoreticians as well as experimentalists—and temporal competition—a new form proposed in this work. Competition is said to be *purely spatial*, or to take place between converging afferents, if the sign of $\dot{m}_j(t)$ is determined by a comparison of the firing rate $d_j(t)$ with firing rates $d_k(t)$ in the other afferents to the neuron at the same time, t. In some schemes (e.g., von der Malsburg, 1973), the sign of $\dot{m}_j(t)$ also depends on the value of the synaptic efficacy $m_j(t)$ relative to $m_k(t)$, $1 \leq k \leq N$. Formally then, we might characterize competition as purely spatial if

$$\text{sign } (\dot{m}_j(t)) = F\left(\frac{d_j(t)}{d_1(t)}, \dots, \frac{d_j(t)}{d_N(t)}, \frac{m_j(t)}{m_1(t)}, \dots, \frac{m_j(t)}{m_N(t)}\right) \quad (9)$$

On the other hand, we say that competition is *purely temporal*, or takes place between incoming patterns, if the sign of $\dot{m}_j(t)/d_j(t)$ is independent of j and is determined by a relationship between the postsynaptic neuron's firing rates, $c(t)$ and $c(t')$, $t' < t$:

$$\text{sign}\left(\frac{\dot{m}_j(t)}{d_j(t)}\right) = F(c(t); c(t')), \quad t' < t, \quad j = 1 \dots N \quad (10)$$

In this work, the modification threshold, θ_M, is given as a function of $\bar{c}$, an average of $c(t')$ over a relatively long time period preceding t, thus satisfying equation 10.

One may, of course, imagine hebbian schemes of a mixed type, involving both spatial and temporal competition.[12] However, the distinction is useful since the performance of the scheme seems to be highly dependent on which of the two classes it is in. This is most clearly seen in the development of selectivity.

In the temporal version used here, asymptotic states are of maximum selectivity with respect to the experienced environment d, independent of the geometry of d. This was rigorously proven analytically in some cases (theorem 1; section III) and conjectured on the basis of numerical results in other cases (circular environments; section IV). In contrast with this, we conjecture that, in any model using pure spatial competition, behavior will usually depend on the geometry of the environment. In any case, maximum selectivity is not reached if the patterns in the environment are not sufficiently separated from one another. This is illustrated in "Appendix B" for one particular model using spatial competition between converging afferents.

We further note that selectivity, as was shown in section V, does not develop in a "pure noise" environment (the distribution termed DR). Some kind of patterned input is required.[13] It follows that, at this level of organization of connectivity, information is being transferred from the environment to the system. This may shed some light on what has been known for a long time as the innate/learned controversy in visual cortex. Our results suggest that this dichotomy is, at best, misleading. The system's potential developmental ability—its evolutive power—may indeed be determined genetically; yet selectivity has no meaning if it does not refer to a given structured environment that determines the final organization of the system.[14]

The present work, however, makes no assumption concerning the initial state of cortex (e.g., the presence or absence of selectivity at eye opening). This question, still a subject of controversy (see Pettigrew, 1978), must be settled experimentally. Further, although we here assume that all synapses are equally modifiable, it could easily be the case that there is variation in modifiability—even one that is time dependent—and that, for example, some of the initial state information including some orientation selectivity is contained in a skeleton of synapses that is less modifiable.[15] Such assumptions can easily be incorporated in fairly obvious extensions of our theory and would, of course, result in the modification of some details of our results. The principal results of our theory, applied to visual cortical neurons and assuming that they are all equally modifiable according to equations 4 to 7, are summarized now. These are either in agreement with existing experimental data or are new and somewhat unexpected consequences of our theory.

Summary of Theoretical Results—New Predictions

Monocularly driven neurons

(*1*) A monocularly driven neuron in a "normal" (patterned) environment becomes selective. The precise pattern to which it becomes selective is determined at random if the initial selectivity is 0 or may be biased toward a particular pattern if there is a built-in preference for this pattern.

(*2*) This same neuron in various deprived environments evolves as follows.

Pure noise. The neuron becomes less selective but

---

[12] More complicated temporal or mixed spatiotemporal schemes are possible and some such have been proposed. For example, Sejnowski (1977a) has suggested a form of modification in which the change of the *j*th synaptic strength involves the co-variance between the *j*th fiber and postsynaptic activities. In addition, interaction between neurons (such as lateral inhibition) can increase selectivity (see, for example, Nass and Cooper, 1975 and "Appendix D").

[13] Pure noise and circular environments may be regarded as two extreme cases: the first totally lacks structure, whereas the second is highly organized. Intermediate cases (i.e., environments consisting of the sum of a noise process and of a circular process) also have been investigated (see, for example, Bienenstock, 1980). There it is shown that the asymptotic selectivity directly depends on a parameter that measures the degree of structure of the environment.

[14] We note, further, that the mechanism of synaptic modification that we have proposed leads both to what are sometimes called "selective" and "instructive" effects (depending on the structure of the environment and the genetic initial state). Thus, as is already suggested by Rauschecker and Singer (1981), this dichotomy is obscured, or does not appear at all, at the synaptic level.

[15] This skeleton might consist primarily of the contralateral pathway and favor the development of the orientation preference for horizontally and vertically oriented stimuli (see, for example, Frégnac, 1979).

continues to be (somewhat) responsive. It may show an orientation preference, but this is relatively unstable.

Exposure to a single pattern (such as vertical lines). The neuron comes to respond preferentially to the single pattern *but with less selectivity (less sharply tuned) than if all orientations were present in the environment.* This last is a natural consequence of temporal competition between incoming patterns and can provide a good test of our theory.[16]

(*3*) Inhibitory synapses are required to produce maximum selectivity. If such inhibitory connections are arbitrarily set equal to 0, selectivity diminishes.

Binocularly driven neurons

(*1*) A binocularly driven neuron in a "normal" (patterned) environment becomes selective and binocular. It is driven selectively by the same pattern from both eyes.

(*2*) This same binocularly driven neuron in various deprived environments evolves as follows.

Uncorrelated patterned inputs to both eyes. The neuron becomes selective, often monocularly driven; if the neuron is binocular, sometimes it is driven by different patterns from the two eyes.

Patterned input to one eye, noise to the other (monocular deprivation). The neuron becomes selective and generally driven only by the open eye. *There is a correlation between selectivity and binocularity. The more selective the neuron becomes, the more it is driven only by the open eye. A non-selective neuron tends to remain binocularly driven.* This correlation is due, in part, to the fact that it is the same mechanism of synaptic change that serves to increase both the selectivity and ocular dominance of the open eye. However (as shown in "Appendix C"), there is also a subtler connection: it is the non-preferred inputs from the open eye accompanied by noise from the closed eye that drive the neuron's response to the closed eye to 0. Thus, for example, if the visual environment were such that there were mostly preferred inputs to the open eye, *even a selective cell would remain less monocular.* (It should prefer the open eye but remain somewhat driven by the closed eye.) As another example, a kitten dark-reared to the age of about 42 days (when there remain few or no specific cells) and then given monocular exposure to nonpatterned input would retain more binocularly driven cells than a similar animal given patterned input.[17]

Noise input to both eyes (dark rearing or binocular deprivation). The neuron remains non-selective (or loses its selectivity) and diminishes its responsiveness but remains binocularly driven (in contrast to the situation in monocular deprivation).

These theoretical conclusions are consistent with experimental data on increases and decreases in selectivity, data concerning changes in ocular dominance in various rearing conditions, as well as data from more complicated paradigms. Although there are indications in recent work that some of the new predictions are in agreement with experimental results, they provide the opportunity for tests of subtler aspects of the theory.

In conclusion, we note that a precise application of our theory to certain complicated experimental situations would probably require inclusion of some anatomical details, interneuronal interactions, as well as a statement of what information is innate and which synapses are modifiable.[18]

Appendix A: Biochemical Mechanism for Temporal Competition

It is probably premature to propose a detailed physiological mechanism for a mathematical synaptic modification algorithm: too many possibilities exist with no present experimental test to decide among them. However, we propose the following as a possible example.

The dependence of our modification threshold upon the mean postsynaptic activity, which regulates the individual neuron modification in an overall manner, might be the result of a physiological mechanism within the framework proposed by Changeux et al. (1973). Their basic hypothesis is that receptor protein on the postsynaptic membrane exists in two states, one *labile* and the other *stable*; *selective stabilization* of the receptor takes place during development in an activity-dependent fashion. The quantity of labile receptor available for stabilization is determined by the neuron's average activity; that is, labile receptor is not synthesized anymore when the neuron's activity is high for a relatively long period of time ($\bar{c} \gg c_0$) (cf., Changeux and Danchin (1976): "The activity of the postsynaptic cell is expected to regulate the synthesis of receptor.").

Our hypothesis that, during the period when competition really takes place (i.e., when $\bar{c}$ is of the order of c_0 in equation 7), the sign of the modification is determined by the instantaneous activity, c, relative to its mean, $\bar{c}$, requires that a single message, the instantaneous activity, be fed back from the site of integration of the incoming message to the individual synaptic sites, on a rapid time scale (i.e., much faster than the one involved in the overall regulation mechanism). This might be contrasted with the assumption implicit in most spatial competition models, namely, that a chemical substance is redistributed between all subsynaptic sites (cf., the principle of conservation of total synaptic strength (von der Malsburg, 1973)).

---

[16] In addition, the principle of temporal competition suggests an experimental paradigm that could be used to increase the selectivity of a cortical neuron while recording from the same neuron. The paradigm consists of controlling the postsynaptic activity of the neuron while presenting sequentially in its receptive field two stimuli, A and B. Stimulus A (or B) should be associated with a high (or low) instantaneous firing rate in such a way as to keep the cell's mean firing rate at its original value. We predict that the cell will prefer stimulus A eventually (i.e., exhibit selectivity with respect to the discrete environment consisting of A and B). Moreover, we predict that presentation of stimulus A alone will lead to less selectivity. An experiment based on this paradigm is currently being undertaken by one of us (E. L. B.) in collaboration with Yves Frégnac.

[17] In this situation, one might have to distinguish between short and long monocular exposures. In very long monocular exposures, the decay term of equation 4 ($-\epsilon m(t)$) eventually could produce decay of junctions from the closed eye independent of the effect discussed above.

[18] This last might be treated as, for example, in the work of Cooper et al. (1979).

Appendix B: von der Malsburg's Model of Development of Orientation Selectivity

A model of development of orientation selectivity using an evolution scheme of the spatial type may be found in the work of von der Malsburg (1973). We present here a brief analysis of this model in view of the definition given in section II. We first show that the type of competition implied by this model is indeed, formally, the spatial one. Next, we investigate the behavior of the system in the simple situation of theorem 1 in section III (i.e., for a two-pattern environment, with the dimension of the system being $N = 2$). We will show why the assumption that is made of nonlinearity of the integrative power is a necessary one. Finally, we prove that the class of two-pattern environments d in which the system behaves nicely (i.e., the state is asymptotically selective with respect to d) is defined by a condition of the type $0 < \cos(d^1, d^2) < a$, where d^1 and d^2 are the two patterns in d, and a is a constant strictly less than 1, which actually depends on the nonlinearity of the integrative power (i.e., on its threshold Θ).

For the purpose of our analysis, we consider a single "cortical" neuron whose integrative and evolutive power are, in our notation, the following:

$$c(t) = (m(t) \cdot d(t))^* \tag{B1}$$

with

$$u^* = \begin{cases} u - \Theta & \text{if } u > \Theta \\ 0 & \text{if } u < \Theta \end{cases} \tag{B2}$$

$$m_j(t + 1) = \gamma(t + 1)(m_j(t) + hc(t)d_j(t)) \quad j = 1, \ldots, N \tag{B3}$$

with h a small positive constant and $\gamma(t + 1)$ such that:

$$\sum_{j=1}^{N} m_j(t + 1) = \sum_{j=1}^{N} m_j(t) = s \tag{B4}$$

The integrative power is thus nonlinear with threshold Θ. The normalizing factor $\gamma(t + 1)$ in the evolution equation B3 keeps the sum of synaptic weights constant and equal to s. All variables are positive.

Our analysis will be carried out on this *reduced version* of von der Malsburg's model: we simply ignore the fixed intracortical connections assumed there, for these are clearly not sufficient to tune the system to a selective state if individual neurons do not display this property already. As is clearly stated by the author himself, the ability to develop selectivity is an intrinsic property of individual neurons, the intracortical connections being there to organize orientation preference in a coherent way in cortex. (This is also the viewpoint in the present work: see "Appendix D.") Notice that this is by no means a contradiction to the fact that, *in the final state*, intracortical connections, particularly the inhibitory ones, significantly contribute to the selectivity of each neuron.

A straightforward calculation shows that equations B3 and B4 are equivalent to the following.

$$\begin{cases} m_j(t + 1) - m_j(t) = K(t)(d_j(t)/d(t) - m_j(t)/s) & j = 1, \ldots, N \\ K(t) = shc(t)d(t)/(s + hc(t)d(t)) \end{cases} \tag{B5}$$

where

$$d(t) = \sum_{j=1}^{N} d_j(t)$$

(In the simulations, $d(t)$ is actually a constant.)

Thus, according to equation B5, the sign of the change of m_j at time t does not depend on the postsynaptic activity $c(t)$ but on the jth fiber activity $d_j(t)$. This is clearly spatial competition as is suggested by the conservation law (equation B4).

We now investigate the behavior of system B5 in a two-pattern environment: $P[d = d^1] = P[d = d^2] = 0.5$. For this purpose, we slightly modify the original setup: there, the dimension is relatively high ($N = 19$), but the firing frequencies in the afferent fibers are discretely valued (i.e., $d_j = 0$ or 1, $j = 1, \ldots, N$). Here, we take $N = 2$, with $d_{1,2}$ allowed to take any value between 0 and 1. By doing so, we still get a broad range of environments ($\cos(d^1, d^2)$ may assume any value between 0 and 1), but the analysis is made considerably easier. To further simplify, we characterize d by a single parameter $0 < \delta < 1$ by writing $d^1 = (1, \delta)$, $d^2 = (\delta, 1)$. Thus, $\cos(d^1, d^2) = 2\delta/(1 + \delta^2)$. We also set $s = 1$.

Under these circumstances, averaging the evolution equation B5 with respect to d leads to the following:

$$E[m_j(t + 1) - m_j(t)] = \tag{B6}$$

$$\tfrac{1}{2}h(2m_j(t) - 1)(\Theta(1 + \delta) - 2\delta), \quad j = 1, 2$$

To obtain equation B6, it has been assumed that both inputs yield above threshold responses (i.e., $m \cdot d^1$ and $m \cdot d^2 > \Theta$). Higher order terms in h have been ignored.

We see that the behavior of the system is determined by the sign of the quantity $\Theta(1 + \delta) - 2\delta$. Notice that, since $s = 1$, Θ cannot be arbitrarily high: in order that states m exist such that $m \cdot d^1$ and $m \cdot d^2 > \Theta$, one has to assume that $\Theta < (1 + \delta)/2$.

It follows from equation B6 that, for δ such that $\Theta(1 + \delta) - 2\delta < 0$, there is one attractor of selectivity 0, namely $(0.5, 0.5)$. When δ gets smaller and $\Theta(1 + \delta) - 2\delta$ becomes positive, the solution bifurcates into two attractors of maximum selectivity. We thus conclude that:

1. If the neuron's integrative power is linear (i.e., $\Theta = 0$), the asymptotic state is non-selective. (When $\Theta = 0$ and d^1 and d^2 are orthogonal (i.e., $\delta = 0$), the first order term in h vanishes, yet the second order term also leads to the non-selective fixed point.)

2. Given a fixed $0 < \Theta < 1$, the environments d that are acceptable to the system are those which satisfy $\delta < \Theta/(2 - \Theta)$, which is equivalent to a condition of the type $\cos(d^1, d^2) < a$ with a strictly less than 1. (Notice that, in the actual simulations, d consists of nine stimuli that are indeed well separated from one another, since $\min_i \cos(d^1, d^i) = \frac{1}{2}$.)

Appendix C: Correlation between Ocular Dominance and Selectivity in the Monocular Deprived Environment

Consider the MD environment in section V: it is defined by (d_r, n), where d_r is circular and n is a "pure noise" vector. We will prove that the state $(m_r^*, 0)$ is stable in this environment provided that m_r^* is a stable selective state in the environment d_r.

Let (x_r, x_l) be a small perturbation from equilibrium. The motion at point $(m_r^* + x_r, x_l)$ is given by:

$$\dot{x}_r = \phi(m_r^* \cdot d_r + x_r \cdot d_r + x_l \cdot n, \, m_r^* \cdot \bar{d}_r + x_r \cdot \bar{d}_r) \, d_r \qquad \text{(C1r)}$$

$$\dot{x}_l = \phi(m_r^* \cdot d_r + x_r \cdot d_r + x_l \cdot n, \, m_r^* \cdot \bar{d}_r + x_r \cdot \bar{d}_r) n \qquad \text{(C1l)}$$

where we assume that the noise has 0 mean.

We analyze separately, somewhat informally, the behavior of the two equations. The stability of equation C1r is immediate from the stability of the selective state m_r^* in the circular environment d_r. To analyze equation C1l, we divide the range of the right eye input d_r into three classes:

1. d_r is such that $m_r \cdot d_r$ is either far above threshold, θ_M, and therefore $\phi(m_r \cdot d_r, m_r \cdot \bar{d}_r) > 0$, or far below threshold, θ_M (but still positive), and therefore $\phi(m_r \cdot d_r, m_r \cdot \bar{d}_r) < 0$. This case might occur before m_r has reached a stable selective state, m_r^*.
2. d_r is such that $m_r^* \cdot d_r$ is near threshold, θ_M, and therefore, $\phi(m_r^* \cdot d_r, m_r^* \cdot \bar{d}_r) \simeq 0$.
3. d_r is such that $m_r^* \cdot d_r \simeq 0$, again resulting in $\phi(m_r^* \cdot d_r, m_r^* \cdot \bar{d}_r) \simeq 0$.

For the first class of inputs, the sign of ϕ is determined by d_r alone, hence equation C1l is the equation of a random walk. To investigate the behavior of equation C1l in the two other cases, we neglect the term x_r and linearize ϕ around the relevant one of its two zeros. It is easy to see that case 2 yields

$$\dot{x}_l \simeq \epsilon_1 (x_l \cdot n) n \qquad \text{(C2)}$$

whereas, in case 3, one obtains

$$\dot{x}_l \simeq -\epsilon_2 (x_l \cdot n) n \qquad \text{(C3)}$$

where ϵ_1 and ϵ_2 are *positive* constants, measuring, respectively, the absolute value of the slope of ϕ at the modification threshold and at zero.[8]

Since n is a noise-like term, its distribution is presumably symmetric with respect to x_l so that averaging equations C2 and C3 yields, respectively

$$\dot{x}_l \simeq \epsilon_1 \overline{n_0^2} x_l \qquad \text{(C4)}$$

$$\dot{x}_l \simeq -\epsilon_2 \overline{n_0^2} \, x_l \qquad \text{(C5)}$$

where $\overline{n_0^2}$ is the average squared magnitude of the noise input to a single synaptic junction from the closed eye.

We thus see that input vectors from the first class move x_l randomly, inputs from the second class drive it away from 0, whereas inputs from the third drive it toward 0. In the case where the range of d_r is a set of K linearly independent vectors and m_r^* is of maximum selectivity, $(K - 1)/K$, case 1 does not occur at all. (The random contribution occurs only before the synaptic strengths from the open eye have settled to one of their fixed points). Case 2 occurs only for one input, e.g., d_r^1, with $m_r^* \cdot d_r^1$ exactly equal to threshold, θ_M, and case 3 occurs for the other $K - 1$ vectors which are all orthogonal to m_r^*. In the general case (d_r any circular environment), the more selective m_r^* with respect to d_r, the higher the proportion of inputs belonging to class 3, the class that yields equation C5 (i.e., that brings x_l back to 0).

The stability of the global system still depends on the ratio of the quantities ϵ_1 and ϵ_2 as well as on the statistics of the noise term n (e.g., its mean square norm). We may, however, formulate two general conclusions. First, under reasonable assumptions (ϵ_1 of the order of ϵ_2 and the mean square norm of n of the same order as that of d_r), $x_l = 0$ is stable on the average for a selective m_r^*. Second, the residual fluctuation of x_l around zero, essentially due to inputs d_r in classes 1 and 2, is smaller for highly selective m_r^* values than it is for mildly selective ones.

Thus, one should expect that, in a monocularly deprived environment, non-selective neurons tend to remain binocularly driven. In addition, since it is the non-preferred inputs from the open eye accompanied by noise from the closed eye (case 3) that drive the response to the closed eye to 0, if inputs to the open eye were restricted to preferred inputs (case 2), even a selective cell would remain less monocular.

Appendix D: Many-neuron Systems

It is very likely that interactions between cortical neurons play an important role in overall cortical function as well, perhaps, as in selectivity of individual cortical cells (Creutzfeldt et al., 1974; Sillito, 1975). The development of selectivity then might be regarded as a many-neuron problem. Since the underlying principles put forward in this work are stated most clearly at the single unit level (where a more complete analysis is also possible), we have chosen this description. However, the methods employed are also applicable to a system of many cortical neurons interacting with one another. Most important, the result that stable equilibria in a stationary environment are selective with respect to their environment can be taken over to the many-neuron system.

Consider such a system in a stationary external environment. The state of each cortical neuron now has two parts: one relative to the geniculocortical synapses, the other to the cortico-cortical ones. The environment of the neuron is no longer stationary, for the states of all other cortical neurons in the system evolve. Yet, when the system reaches global equilibrium, which will occur under reasonable assumptions, each individual environment becomes stationary. The single unit study then allows us to state that, at least in principle (we do not know *a priori* that each environment is circular), the state of each neuron is selective with respect to its own individual environment.

In practice, formulation of the many-neuron problem poses two questions. First, the integrative power of the system should be specified. Since the system includes cortico-cortical loops, it is not obvious what the response to a given afferent message should be. The two major alternatives are: (a) stationary cortical activity is reached rapidly (i.e., before the afferent message changes) and (b) relevant cortical activity is transitory. The second question concerns the evolution of cortico-cortical synaptic strengths: should these synapses be regarded as modifiable at all, and if yes, how? von der Malsburg (1973) assumes *alternative a* above and proposes fixed connectivity patterns, short range excitatory and longer range inhibitory.

We have performed a simulation of a many-neuron system using the much simpler (and probably more nat-

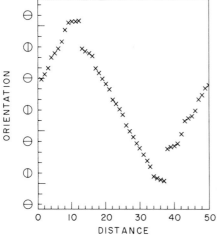

Figure 6. A regular distribution of preferred orientation in a one-dimensional cortex. The system is an array of 50 "cortical" cells arranged in a cyclic way (cell 1 and cell 50 are neighbors) and interconnected according to a fixed short-range-excitation-long-range-inhibition pattern. (Coefficients of interactions are, as a function of increasing intercell distance: 0.4, 0.4, −0.2, −0.4, 0, . . .). The environment d of the system is the usual circular one. Integrative and evolutive powers are described in the text. When the system reaches equilibrium, one has $0.73 \leq \text{Sel}_d(m_i) \leq 0.77$ for all i values between 1 and 50. The diagram shows preferred orientation as a function of cortical coordinate.

ural) *assumption b* above. (*1*) Only monosynaptically and disynaptically mediated components of the afferent message are taken into account for the computation of each cortical neuron's activity before modification is performed and a new stimulus is presented. (*2*) Intracortical connections are fixed and spatially organized as in the work of von der Malsburg (1973). (*3*) The state of each neuron evolves according to equations 6 and 7 of the present work. The results are the following. (*1*) The system's state converges. (*2*) At equilibrium, each neuron stands in a selective state with respect to the environment. (*3*) Preferred orientation—when the environment is a circular one—is a piecewise continuous function of cortical distance (Fig. 6). (*4*) In the final equilibrium state, the intracortical synapses contribute along with the geniculocortical ones to produce the selectivity.

References

Anderson, J. A. (1970) Two models for memory organization using interacting traces. Math. Biosci. *8:* 137–160.

Anderson, J. A. (1972) A simple neural network generating an interactive memory. Math. Biosci. *14:* 197–200.

Bienenstock, E. (1980) A theory of development of neuronal selectivity. Doctoral thesis, Brown University, Providence, RI.

Blakemore, C. (1976) The conditions required for the maintenance of binocularity in the kitten's visual cortex. J. Physiol. (Lond.) *261:* 423–444.

Blakemore, C., and G. F. Cooper (1970) Development of the brain depends on the visual environment. Nature *228:* 477–478.

Blakemore, C., and D. E. Mitchell (1973) Environmental modification of the visual cortex and the neural basis of learning and memory. Nature *241:* 467–468.

Blakemore, C., and R. C. Van Sluyters (1974) Reversal of the physiological effects of monocular deprivation in kittens. Further evidence for a sensitive period. J. Physiol. (Lond.) *237:* 195–216.

Blakemore, C., and R. C. Van Sluyters (1975) Innate and environmental factors in the development of the kitten's visual cortex. J. Physiol. (Lond.) *248:* 663–716.

Brindley, G. S. (1969) Nerve net models of plausible size that perform many simple learning tasks. Proc. R. Soc. Lond. (Biol.) *174:* 173–191.

Buisseret, P., and M. Imbert (1976) Visual cortical cells. Their developmental properties in normal and dark reared kittens. J. Physiol. (Lond.) *255:* 511–525.

Buisseret, P., E. Gary-Bobo, and M. Imbert (1978) Ocular motility and recovery of orientational properties of visual cortical neurones in dark-reared kittens. Nature *272:* 816–817.

Changeux, J. P., and A. Danchin (1976) Selective stabilization of developing synapses as a mechanism for the specification of neuronal networks. Nature *264:* 705–712.

Changeux, J. P., P. Courrège, and A. Danchin (1973) A theory of the epigenesis of neuronal networks by selective stabilization of synapses. Proc. Natl. Acad. Sci. U. S. A. *70:* 2974–2978.

Cooper, L. N. (1973) A possible organization of animal memory and learning. In *Proceedings of the Nobel Symposium on Collective Properties of Physical Systems*, B. Lindquist and S. Lindquist, eds., Vol., 24, pp. 252–264, Academic Press, New York.

Cooper, L. N., F. Lieberman, and E. Oja (1979) A theory for the acquisition and loss of neuron specificity in visual cortex. Biol. Cybern. *33:* 9–28.

Creutzfeldt, O. D., U. Kuhnt, and L. A. Benevento (1974) An intracellular analysis of visual cortical neurones to moving stimuli: Responses in a cooperative neuronal network. Exp. Brain Res. *21:* 251–274.

Cynader, M., and D. E. Mitchell (1980) Prolonged sensitivity to monocular deprivation in dark-reared cats. J. Neurophysiol. *43:* 1026–1040.

Frégnac, Y. (1979) Development of orientation selectivity in the primary visual cortex of normally and dark reared kittens. Biol. Cybern. *34:* 187–204.

Frégnac, Y., and E. Bienenstock (1981) Specific functional modifications of individual cortical neurons, triggered by vision and passive eye movement in immobilized kittens. In *Pathophysiology of the Visual System: Documenta Ophthalmologica*, L. Maffei, ed., Vol. 30, pp. 100–108, Dr. W. Junk, The Hague.

Frégnac, Y., and M. Imbert (1978) Early development of visual cortical cells in normal and dark-reared kittens. Relationship between orientation selectivity and ocular dominance. J. Physiol. (Lond.) *278:* 27–44.

Hebb, D. O. (1949) *Organization of Behavior*, John Wiley and Sons, New York.

Hirsch, H. V. B., and D. N. Spinelli (1970) Visual experience modifies distribution of horizontally and vertically oriented receptive fields in cats. Science *168:* 869–871.

Hirsch, H. V. B., and D. N. Spinelli (1971) Modification of the distribution of receptive field orientation in cats by selective visual exposure during development. Exp. Brain Res. *13:* 509–527.

Hubel, D. H., and T. N. Wiesel (1959) Receptive fields of single neurons in the cat striate cortex. J. Physiol. (Lond.) *148:* 574–591.

Hubel, D. H., and T. N. Wiesel (1962) Receptive fields, binocular interaction and functional architecture in the cat's visual cortex. J. Physiol. (Lond.) *160:* 106–154.

Hubel, D. H., and T. N. Wiesel (1963) Receptive fields of cells in striate cortex of very young, visually inexperienced kittens. J. Neurophysiol. *26:* 994–1002.

Hubel, D. H., and T. N. Wiesel (1965) Binocular interaction in striate cortex of kittens with artificial squint. J. Neurophysiol. *28:* 1041–1059.

Imbert, M., and P. Buisseret (1975) Receptive field characteristics and plastic properties of visual cortical cells in kittens reared with or without visual experience. Exp. Brain Res. *22:* 2–36.

Kasamatsu, T., and J. D. Pettigrew (1976) Depletion of brain catecholamines: Failure of ocular dominance shift after monocular occlusion in kittens. Science *194:* 206–209.

Kasamatsu, T., and J. D. Pettigrew (1979) Preservation of binocularity after monocular deprivation in the striate cortex of kittens treated with 6-hydroxydopamine. J. Comp. Neurol. *185:* 139–181.

Kohonen, T. (1977) *Associative Memory: A System Theoretical Approach*, Springer, Berlin.

Kratz, K. E., and P. D. Spear (1976) Effects of visual deprivation and alterations in binocular competition on responses of striate cortex neurons in the cat. J. Comp. Neurol. *170:* 141–152.

Leventhal, A. G., and H. V. B. Hirsch (1977) Effects of early experience upon orientation selectivity and binocularity of neurons in visual cortex of cats. Proc. Natl. Acad. Sci. U. S. A. *74:* 1272–1276.

Leventhal, A. G., and H. V. B. Hirsch (1980) Receptive field properties of different classes of neurons in visual cortex of normal and dark-reared cats. J. Neurophysiol. *43:* 1111–1132.

Marr, D. (1969) A theory of cerebellar cortex. J. Physiol. (Lond.) *202:* 437–470.

Movshon, J. A. (1976) Reversal of the physiological effects of monocular deprivation in the kitten's visual cortex. J. Physiol. (Lond.) *261:* 125–174.

Nass, M. M., and L. N. Cooper (1975) A theory for the development of feature detecting cells in visual cortex. Biol. Cybern. *19:* 1–18.

Perez, R., L. Glass, and R. J. Shlaer (1975) Development of specificity in the cat visual cortex. J. Math. Biol. *1:* 275–288.

Pettigrew, J. D. (1978) The paradox of the critical period for striate cortex. In *Neuronal Plasticity*, C. W. Cotman, ed., pp. 311–330, Raven Press, New York.

Rauschecker, J. P., and W. Singer (1981) The effects of early visual experience on the cat's visual cortex and their possible explanation by Hebb synapses. J. Physiol. (Lond.) *310:* 215–240.

Sejnowski, T. J. (1977a) Storing covariance with nonlinearly interacting neurons. J. Math. Biol. *4:* 303–321.

Sejnowski, T. J. (1977b) Statistical constraints on synaptic plasticity. J. Theor. Biol. *69:* 385–389.

Sillito, A. M. (1975) The contribution of inhibitory mechanisms to the receptive field properties of neurons in the striate cortex of the cat. J. Physiol. (Lond.) *250:* 305–329.

Singer, W. (1979) Central-core control of visual functions. In *Neuroscience Fourth Study Program*, F. Schmitt and F. Worden, eds., pp. 1093–1109, MIT Press, Cambridge.

Singer, W. (1980) Central gating of developmental plasticity in the cat striate cortex. Verh. Dtsch. Zool. Ges. 268–274.

Stent, G. S. (1973) A physiological mechanism for Hebb's postulate of learning. Proc. Natl. Acad. Sci. U. S. A. *70:* 997–1001.

Stryker, M. P., H. Sherk, A. G. Leventhal, and H. V. Hirsch (1978) Physiological consequences for the cat's visual cortex of effectively restricting early visual experience with oriented contours. J. Neurophysiol. *41:* 896–909.

Trotter, Y., Y. Frégnac, P. Buisseret (1981) Gating control of developmental plasticity by extraocular proprioception in kitten area 17. In *Fourth European Conference on Visual Perception, Gouzieux, France.*

Uttley, A. M. (1976) A two pathway theory of conditioning and adaptive pattern recognition. Brain Res. *102:* 23–35.

von der Malsburg, C. (1973) Self-organization of orientation sensitive cells in the striate cortex. Kybernetik *14:* 85–100.

Watkins, D. W., J. R. Wilson, and S. M. Sherman (1978) Receptive field properties of neurons in binocular and monocular segments of striate cortex in cats raised with binocular lid suture. J. Neurophysiol. *41:* 322–337.

Wiesel, T. N., and D. H. Hubel (1963) Single-cell responses in striate cortex of kittens deprived of vision in one eye. J. Neurophysiol. *26:* 1003–1017.

Wiesel, T. N., and D. H. Hubel (1965) Extent of recovery from the effects of visual deprivation in kittens. J. Neurophysiol. *28:* 1060–1072.

Mean-Field Theory of a Neural Network
(with C. L. Scofield)
Proc. Natl. Acad. Sci. USA **85**, 1973 (1988)

Appendix D of the previous paper sketches a preliminary attempt to deal with the many-neuron problem. There it is suggested that under reasonable assumptions a cell can become selective to its environment consisting of input from LGN and other cortical neurons.

When Chris Scofield joined our group, his assignment was to put together the rudiments of a cortical architecture so that we could study the evolution of large numbers of neurons linked to one another in a simple but reasonable approximation of the architecture of various layers of visual cortex. This effort has continued until the present. Hopefully, we are approaching a reasonably realistic model that can also be analyzed.

One can divide the inputs to a cell into those that come from LGN and thus convey detailed information from the outside world, those that come from other cortical cells, which themselves may receive input from LGN and a variety of feedbacks from cells further along in the network.

In Chris' simulations we soon found that we could duplicate the selectivity achieved by BCM. In addition, by introducing the proper inhibition in the network, there was some chance that we could reproduce the sequence of orientation selectivity, progressing gradually from column to column as reported by Hubel and Wiesel.

Our general conclusion, supported since by many other simulations, was that although such sequence results could be reproduced, they provided no real test of any particular set of assumptions. My own suspicion is that certain orientation preferences are seeded in some of the columns, possibly verticals and horizontals, and that with lateral inhibition, the non-orthogonal orientation preferences fill themselves in between most of the time. Some of our very recent work suggests, however, that some distinctions can be made between various learning rules.

The most important result of this work is that, in many cases, the full-network effects can be approximated by what we came to call a mean field. We could regard a cortical cell as receiving inputs from LGN as well as lateral inputs from other cortical cells. Since the collaterals are fairly long, the individual lateral inputs might, in some situations, be replaced by an average. The cortical cell is thus regarded in a manner similar to an individual magnet in a solid in the presence of an external field interacting with all of its neighbors. An extremely useful approximation is to replace the detailed interactions with all of the neighbors by a mean field. This gives rise to consistency conditions that have been very successfully employed in the theory of magnetism. Here, similar ideas are applied to the neural network. These result in dramatic simplifications. Computer simulations have shown that, in many circumstances, the mean field is a very good approximation to the full network. The stability and positions of the fixed points of a BCM neuron in such a mean-field network are discussed in the following paper.

Proc. Natl. Acad. Sci. USA
Vol. 85, pp. 1973–1977, March 1988
Neurobiology

Mean-field theory of a neural network

(visual cortex/synaptic modification)

LEON N COOPER AND CHRISTOPHER L. SCOFIELD[†]

Center for Neural Science and Physics Department, Brown University, Providence, RI 02912

Contributed by Leon N Cooper, November 9, 1987

ABSTRACT A single-cell theory for the development of selectivity and ocular dominance in visual cortex has been generalized to incorporate more realistic neural networks that approximate the actual anatomy of small regions of cortex. In particular, we have analyzed a network consisting of excitatory and inhibitory cells, both of which may receive information from the lateral geniculate nucleus (LGN) and then interact through cortico–cortical synapses in a mean-field approximation. Our investigation of the evolution of a cell in this mean-field network indicates that many of the results on existence and stability of fixed points that have been obtained previously in the single-cell theory can be successfully generalized here. We can, in addition, make explicit further statements concerning the independent effects of excitatory and inhibitory neurons on selectivity and ocular dominance. For example, shutting off inhibitory cells lessens selectivity and alters ocular dominance (masked synapses). These inhibitory cells may be selective, but there is no theoretical necessity that they be so. Further, the intercortical inhibitory synapses do not have to be very responsive to visual experience. Most of the learning process can occur among the excitatory LGN–cortical synapses.

A single cell theory for the development of selectivity and ocular dominance in visual cortex has been presented previously by Bienenstock, Cooper, and Munro (1). This has been extended to a network applicable to layer 4 of visual cortex (2). In this paper we present a mean field approximation that captures in a fairly transparent manner the qualitative, and many of the quantitative, results of the network theory.

Visual cortex has been extensively investigated (3, 4). We summarize some of the dominating experimental facts very briefly. Neurons in the primary visual cortex of normal adult cats are sharply tuned for the orientation of an elongated slit of light and most are activated by stimulation of either eye. Both of these properties—orientation selectivity and binocularity—depend on the type of visual environment experienced during a critical period of early postnatal development extending from approximately 3 weeks to 3 months. For example, deprivation of patterned input during this critical period leads to loss of orientation selectivity, while monocular deprivation (MD) results in a dramatic shift in the ocular dominance of cortical neurons such that most will be responsive exclusively to the open eye. The ocular dominance shift after MD is the best known and most intensively studied type of visual cortical plasticity. The consequences of binocular deprivation (BD) on visual cortex stand in striking contrast to those observed after MD. While 7 days of MD during the second postnatal month leave few neurons in striate cortex responsive to stimulation of the deprived eye, most cells remain responsive to visual stimulation through either eye after a comparable period of BD. However, prolonged periods of BD lead to a loss of orientation selectivity, an effect

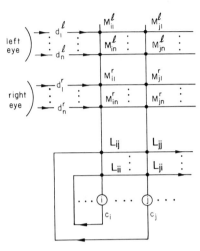

FIG. 1. Network with inputs from left and right eyes, with LGN–cortical and cortico–cortical synapses.

not observed in the response to the open eye after comparable periods of MD. The theory we discuss is concerned primarily with the explanation of these and related facts.

Definitions and Notation

We focus attention on the input from the lateral geniculate nucleus (LGN) and intercortical interactions. Inputs from other regions of cortex are considered part of a background excitation or inhibition contributing to the spontaneous activity of the cell. In addition, the various time delays that result in structure in the poststimulus time histogram are assumed to be integrated over periods of the order of a second for purposes of synaptic modification. This leads to a circuit as shown in Fig. 1.

The output of the cells of the full network can be written

$$c = c^*(Md + Lc),\qquad [1]$$

where c^* is a sigmoidal response function,

$$c = (c_1 \ldots c_N)^T,\qquad [2a]$$

c_i is the output firing rate of the ith cortical cell, and

$$M = (M_{is}^l, M_{is}^r),\qquad [2b]$$

Abbreviations: MD, monocular deprivation; BD, binocular deprivation; LGN, lateral geniculate nucleus; GABA, γ-aminobutyric acid.
[†]Present address: Nestor, Inc., 1 Richmond Square, Providence, RI 02906.

where M_{is}^l and M_{is}^r are the sth LGN "synapses" from the left and right eye to the ith cortical cell.

$$d = (d^l, d^r)^T \text{ and } d^{l,r} = (d_1^{l(r)} \ldots d_n^{l(r)})^T \quad [2c]$$

are the time-averaged inputs from the left and right eye as described in ref. 1,

$$L = (L_{ij}) \quad [2d]$$

is the matrix of cortico–cortical synapses, and L_{ij} is the synapse from the jth cell to the ith cell. (Notice that italicized symbols always contain left and right eye components.)

In the monotonically increasing region above threshold and below saturation, in a linear approximation,

$$c = Md + Lc. \quad [3]$$

We consider a region of cortex for which the neural mapping of the input from the visual field is constant (all of the cells, in effect, look at a given region of the visual field). Under these conditions, for an input, d, constant in time, the equilibrium state of the network would be

$$c = (1 - L)^{-1}Md. \quad [4]$$

Equilibrium as well as nonequilibrium information may be critical to the evolution of the network as well as of primary importance in information processing.[‡]

Mean-Field Approximation

For a given LGN–cortical vector of synapses, m_i (the ith row of M), and for a given input from both eyes, d, Eq. 3 for the firing rate of the ith cortical cell becomes

$$c_i = m_id + \sum_j L_{ij}c_j, \quad [5]$$

where the first term is due to the input from LGN and the second due to input from other cortical cells. We define $\bar{c}$ as the spatially averaged firing rate of all of the cortical cells in the region defined above:

$$\bar{c} = (1/N) \sum_i c_i. \quad [6]$$

The mean-field approximation is obtained by replacing c_j in the sum in Eq. 5 by its average value so that c_i becomes

$$c_i = m_id + \bar{c} \sum_j L_{ij}. \quad [7]$$

Here, in a manner similar to that in the theory of magnetism, we have replaced the effect of individual cortical cells by their average effect (as though all other cortical cells can be replaced by an "effective" cell). It follows that

$$\bar{c} = \bar{m}d + \bar{c}L_0 = (1 - L_0)^{-1}\bar{m}d, \quad [7a]$$

[‡]If we expand $(1 - L)^{-1}$, we obtain $(1 - L)^{-1} = 1 + L + L^2 + \ldots$, an expansion in mono-, di-, tri-, . . . synaptic events. How many synaptic events one includes depends on the time interval of importance. For synaptic modification we assume that time intervals of the order of a half second are appropriate. Thus c represents the average over about ½ sec of the number of spikes that are the result of an external presentation (5). The poststimulus time histogram can be broken into much smaller time intervals, thus separating mono-, di-, trisynaptic events and excitatory and inhibitory synapses.

where

$$\bar{m} = (1/N) \sum_i m_i \quad [8]$$

and

$$L_0 = (1/N) \sum_{ij} L_{ij},$$

so that

$$c_i = \left(m_i + \left(\sum_j L_{ij}\right)(1 - L_0)^{-1}\bar{m}\right)d.$$

If we assume that the lateral connection strengths are a function only of i–j (not dependent on the absolute position of a cell in the network, therefore dependent only on the distance of two cells from one another), L_{ij} becomes a circular matrix so that

$$\sum_i L_{ij} = \sum_j L_{ij} = L_0 = \text{constant} \quad [9]$$

and

$$c_i = (m_i + L_0(1 - L_0)^{-1}\bar{m})d. \quad [10]$$

In the mean-field approximation we can therefore write

$$c_i(\alpha) = (m_i - \alpha)d = (m_i^l - \alpha^l) \cdot d^l + (m_i^r - \alpha^r) \cdot d^r, \quad [11]$$

where the mean field

$$\alpha = (\alpha^l, \alpha^r) = +a(\bar{m}^l, \bar{m}^r) \quad [12]$$

with

$$a = |L_0|(1 + |L_0|)^{-1}, \quad [12a]$$

and we assume that $L_0 < 0$ (the network is, on average, inhibitory).[§]

The Cortical Network

The behavior of visual cortical cells in animals reared in various conditions suggests that some cells respond more rapidly to environmental changes than others. In MD, for example, some cells remain responsive to the closed eye in spite of the very large shift of most cells to the open eye. Singer (6) found, using intracellular recording, that geniculo–cortical synapses on inhibitory interneurons are more resistant to MD than are synapses on pyramidal cell dendrites. In dark rearing some cells become nonresponsive to visual stimuli, while most cells retain some responsiveness (3, 4). Recent work suggests that the density of inhibitory GABAergic synapses (GABA, γ-aminobutyric acid) in kitten striate cortex is also unaffected by MD during the critical period (7, 8).

These results suggest that some LGN–cortical synapses modify rapidly, while others modify relatively slowly, with slow modification of some cortico–cortical synapses. Excitatory LGN–cortical synapses onto excitatory cells may be those that modify primarily. (Since these synapses are formed exclusively on dendritic spines, this raises the possi-

---

[§]The average magnitude of cortico–cortical inhibition, L_0, must be smaller than 1. Otherwise $\bar{c}$ would be smaller than 0 (Eq. 7a). There would be so much inhibition that on average no cells would fire.

Neurobiology: Cooper and Scofield

Proc. Natl. Acad. Sci. USA 85 (1988) 1975

bility that the mechanisms underlying synaptic modification exist primarily in axo–spinous synapses.) To embody these facts we introduce two types of LGN–cortical synapses: those (m_i) that modify (according to the modification rule discussed in ref. 1) and those (z_k) that remain relatively constant. In a simple limit we have

$$\dot{m}_i = \phi(c_i, \bar{\bar{c}}_i)d$$

and [13]

$$\dot{z}_k = 0.$$

(In what follows $\bar{c}$ denotes the spatial average over cortical cells, while $\bar{\bar{c}}_i$ denotes the time-averaged activity of the ith cortical cell.) The function ϕ is discussed in ref. 1. We assume for simplicity, and consistent with the above physiological interpretation, that these two types of synapses are confined to two different classes of cells and that both left and right eye have similar synapses (both m_i or both z_k) on a given cell. We therefore can write

$$c_i = m_i d + \sum_j L_{ij} c_j$$

and [14]

$$c_k = z_k d + \sum_j L_{kj} c_j.$$

Further, in what follows, we assume for maximum simplicity that there is no modification of cortico–cortical synapses, although what experimental results there are suggest only that modification of inhibitory cortico–cortical synapses is slow (7, 8). The consequences of a theory including cortico-cortical synapse modification for the full network were briefly discussed in ref. 1 and will be discussed more fully in the mean-field approximation elsewhere.

In a cortical network with modifiable and nonmodifiable LGN–cortical synapses and nonmodifiable cortico–cortical synapses, the synaptic evolution equations become

$$\dot{m}_i = \phi(c_i, \bar{\bar{c}}_i)d,$$
$$\dot{z}_k = 0,$$

and [15]

$$\dot{L}_{ij} = 0.$$

This leads to a very complex set of coupled nonlinear stochastic evolution equations that have been simulated and partially analyzed elsewhere (2). The mean-field approximation permits dramatic simplification of these equations, leading to analytic results and a fairly transparent understanding of their consequences in various conditions. In this approximation Eqs. 14 become

$$c_i = m_i d + L_0 \bar{c}$$

and [16]

$$c_k = z_k d + L_0 \bar{c},$$

so that we can now write

$$c_i(\alpha) = (m_i - \alpha)d = (m_i^l - \alpha^l)\cdot d^l + (m_i^r - \alpha^r)\cdot d^r$$

and [17]

$$c_k(\alpha) = (z_k - \alpha)d = (z_k^l - \alpha^l)\cdot d^l + (z_k^r - \alpha^r)\cdot d^r,$$

where now $\alpha^{l(r)}$ contains terms from modifiable and nonmodifiable synapses:

$$\alpha^{l(r)} = a(\bar{m}^{l(r)} + \bar{z}^{l(r)}),$$

$$\bar{m}^{l(r)} = N^{-1} \sum_{i=1}^{N_m} m_i^{l(r)},$$ [18]

$$\bar{z}^{l(r)} = N^{-1} \sum_{k=1}^{N_{nm}} z_k^{l(r)}$$

and a is defined in Eq. 12a. ($N = N_m + N_{nm}$, where N_m is the number of cells with modifiable synapses and N_{nm} is the number of cells with nonmodifiable synapses.) Since it is assumed that neither L nor z changes as the network evolves, only $\bar{m}^{l(r)}$ is time dependent.

Position and Stability of Fixed Points of LGN–Cortical Synapses in the Mean-Field Network

We now generalize the arguments given in refs. 1 and 9 for the position and stability of the fixed points of the stochastic nonlinear synaptic modification equations. In the mean-field network

$$\dot{m}_i(\alpha) = \phi(c_i(\alpha), \bar{\bar{c}}_i(\alpha))d = \phi[m_i(\alpha) - \alpha]d, \quad [19]$$

where $c_i(\alpha)$ is defined by Eq. 17 and $\bar{\bar{c}}_i(\alpha)$ is an average of the form

$$\bar{\bar{c}}_i(\alpha) = \tau \int_{-\infty}^{t} \exp[(t' - t)\tau^{-1}]c_i(\alpha, t') \, dt'. \quad [20]$$

The mean field, $\alpha^{l(r)}$ as given by Eq. 12, has a time-dependent component $\bar{m}^{l(r)}$. This varies as the average over all of the network modifiable synapses and, in most environmental situations, should change slowly compared to the change of the modifiable synapses to a single cell:

$$|\dot{\bar{m}}^{l(r)}| << |\dot{m}_i^{l(r)}|. \quad [21]$$

We, therefore, define an adiabatic approximation in which we assume that α is slowly varying and determine the trajectory of m_i for fixed α. (We imagine that m_i reaches its fixed point before α varies. The nonadiabatic situation is analyzed in the *Appendix*. It is shown there that, in any case, the position and stability of the fixed points are unaltered.) In the adiabatic approximation we can write

$$(m_i(\alpha) - \alpha) = \phi[m_i(\alpha) - \alpha]d. \quad [22]$$

We see that there is a mapping

$$m_i' \leftrightarrow m_i(\alpha) - \alpha \quad [23]$$

such that for every $m_i(\alpha)$ there exists a corresponding (mapped) point m_i' that satisfies

$$\dot{m}_i' = \phi[m_i']d, \quad [24]$$

the original equation for the mean-field zero theory. Therefore, if we start from the corresponding initial point

$$m_i'(t_o) = m_i(\alpha, t_o) - \alpha, \quad [25]$$

the m_i' trajectory viewed from the m_i coordinate system is the trajectory of $m_i(\alpha = 0)$. Thus, we can compute $m_i(\alpha)$ from the $\alpha = 0$ trajectory, using

$$m_i(\alpha) = m_i' + \alpha = m_i(\alpha = 0) + \alpha. \qquad [26]$$

The transformation

$$m_i'' = m_i + \alpha \qquad [27]$$

gives a coordinate system whose origin is displaced from the mean-field zero coordinates by α. The trajectory of a solution of the $\alpha = 0$ theory measured from this coordinate system gives a solution of the $\alpha \neq 0$ theory for the corresponding point:

$$m_i''(\alpha) = m_i'(\alpha) + \alpha = m_i(0) + \alpha = m_i(\alpha). \qquad [28]$$

It follows that at corresponding points

$$c_i(\alpha) = c_i(0)$$

and $\qquad\qquad\qquad\qquad\qquad\qquad\qquad\qquad\qquad\qquad$ [29]

$$\bar{\bar{c}}_i(\alpha) = c_i(0),$$

so that the modification threshold, θ_M, is unaltered in this mapping.¶ When this argument is applied to the fixed points, we conclude that for every fixed point of $m_i(\alpha = 0)$ there exists a corresponding fixed point for $m_i(\alpha)$ with *the same selectivity and stability properties*. Therefore, just as for the $\alpha = 0$ theory, for arbitrary α *only selective fixed points are stable*. Further, at corresponding fixed points we obtain the *same* cell output.

From this we see that if the background inhibition is changed (e.g., by long-term application of bicuculine or a GABA agonist) and the LGN–cortical synapses are allowed to evolve to the new fixed points in the same visual environment, the outputs of the cortical cells will evolve to what they were before the background inhibition was altered. (It is presumed that a cortical cell does not jump from one stable fixed point to another in this process.)

The above is limited as follows:

(*i*) The LGN–cortical synapses are restricted to be positive (excitatory). Therefore, if α is too small (insufficient background inhibition), $m_i(\alpha)$ will not be able to reach its fixed points with only positive components.

(*ii*) The LGN–cortical synapses cannot increase beyond some physiological and/or molecular limit. Therefore, if α is too large, the cell will never fire, thus restricting the evolution of $m_i(\alpha)$.

Evolution of the Mean-Field Network Under Various Rearing Conditions

We are now in a position to calculate the evolution of cortical cells under various rearing conditions. In what follows we give as one example the evolution of cortical cells in the mean-field network under MD conditions. The argument is similar to that given in ref. 1. A more detailed analysis including comparisons with experiment will be presented elsewhere (E. E. Clothiaux, M. F. Bear, and L.NC., unpublished).

¶For simplicity we often compute $\bar{\bar{c}}_i$ as an average over the environment $\{d^1 \ldots d^K\}$. Thus, for example, for the monocular case, at a selective fixed point $(m_i - \alpha) \cdot d^1 = \theta_M$ (preferred input), $(m_i - \alpha) \cdot d^j = 0$ ($j > 1$, nonpreferred inputs), so that $\bar{\bar{c}}_i = (1/K) \Sigma_j (m_i - \alpha) \cdot d^j = \theta_M / K$. With the definition $\theta_M = (\bar{\bar{c}}_i)^2$ we obtain $\bar{\bar{c}}_i = K$ so that $\theta_M = K^2$ independent of the mean field, α.

Under MD conditions, the animal is reared with one eye closed. For the sake of analysis assume that the right eye is closed and that only noiselike signals arrive at cortex from the right eye. Then the environment of the cortical cells is

$$d = (d^j, n)^T. \qquad [30]$$

Further, assume that the left eye synapses have reached their selective fixed point, selective to pattern d^1. Then $(m_i^l, m_i^r) = (m_i^{l*}, x_i)$ with $|x_i| << |m_i^{l*}|$. For the preferred open eye pattern (d^1, n) we have $c_i(\alpha) = \theta_M + (x_i - \alpha^r) \cdot n$, while for the nonpreferred open eye patterns (d^j, n), $j > 1$, $c_i(\alpha) = (x_i - \alpha^r) \cdot n$. Following the argument of ref. 1, a time average over the full pattern environment gives

$$\dot{x}_i = -\kappa(x_i - \alpha^r) \text{ with } \kappa \text{ a positive number.} \qquad [31]$$

For a constant or slowly varying mean field this leads to an asymptotic solution for the fixed point:

$$x_i = \alpha^r = a(\bar{x} + \bar{z}^r). \qquad [32]$$

We see that

$$x_i^*(\alpha^r) = x_i^*(0) + \alpha^r = \alpha^r, \qquad [33]$$

as expected from the above general argument. If we now include the self-consistency condition that is a consequence of the variation of the mean field and use

$$\bar{x} = (1/N) \sum_{i=1}^{N_m} x_i, \qquad [34]$$

we obtain

$$\bar{x} = \lambda a(1 - \lambda a)^{-1} \bar{z}^r, \qquad [35]$$

where $\lambda = N_m/N$ is the ratio of the number of modifiable cells to the total number of cells in the network. This yields

$$x_i^* = a(1 - \lambda a)^{-1} \bar{z}^r. \qquad [36]$$

That is, the asymptotic state of the closed eye synapses is a scaled function of the mean field due to nonmodifiable cortical cells. The scale of this state is set not only by the proportion of nonmodifiable cells but in addition by the averaged intracortical synaptic strength L.

Thus, contrasted with the mean-field zero theory, the deprived eye LGN–cortical synapses do not go to zero. Rather they approach the constant value dependent on the average inhibition produced by the nonmodifiable cells in such a way that the asymptotic output of the cortical cell is zero (it cannot be driven by the deprived eye). However, lessening the effect of inhibitory synapses (e.g., by application of bicuculine) reduces the magnitude of α so that one could once more obtain a response from the deprived eye.

Discussion

Having defined a mean-field approximation that greatly simplifies the equations for the response and evolution of cortical cells, we have obtained a fundamental result: the stability and position of the fixed points in this network are related to the fixed points in the absence of mean field ($\alpha^{l(r)} = 0$) by

$$m_i^*(\alpha) = m_i^*(0) + \alpha, \qquad [37]$$

where $m_i^*(\alpha)$ is a fixed point of Eq. 22 in the mean field α,

Neurobiology: Cooper and Scofield

Proc. Natl. Acad. Sci. USA 85 (1988)   1977

while $m_i^*(0)$ is a fixed point of this equation of zero mean field.

Thus if $m_i^*(\alpha)$ is restricted to the first quadrant (positive values for all of its components due to the excitatory nature of LGN–cortical synapses), as long as α is large enough and nonspecific (there is sufficient inhibition for all pattern inputs), $m_i(\alpha)$ can still reach all of the fixed points that would have been reached by $m_i(0)$ (not restricted to the first quadrant). This means that if network inhibition is sufficient, the selective stable fixed points can be reached even though LGN–cortical synapses are excitatory. Once reached, the fixed points, $m_i^*(\alpha)$, have the same stability characteristics as the corresponding $m_i^*(0)$.

We find, consistent with previous theory and with experiment, that most learning can occur in the LGN–cortical synapses; inhibitory (cortico–cortical) synapses need not modify. Some nonmodifiable LGN–cortical synapses are required. It becomes interesting to ask whether these could be associated with some anatomical feature (e.g., might these be synapses into shafts rather than spines).

As in the zero mean-field theory, zero cell output is an unstable fixed point. Thus in BD the cell output could be on average above or below spontaneous activity (depending on the level of inhibition). Some "nonvisual" cells would reappear if excitation were enhanced or inhibition diminished.

In MD the closed eye response goes to

$$c = (x - \alpha) \cdot d \to 0. \qquad [38]$$

Therefore, LGN–cortical synapses do not go to 0. Rather

$$x \to \alpha. \qquad [39]$$

Thus if inhibition is suppressed one would expect some response from the closed eye. This is in agreement with experiment.

Various models for memory storage and retrieval have been suggested. These differ in several ways. One of the most important from the point of view of computational complexity as well as for realization in silicon is the degree of connectivity of each unit. What is suggested here is that much that is significant in at least one layer of visual cortex can be obtained in a primarily feed-forward network of very simplified lateral connectivity. The original connectivity in which each of the N neurons in this layer of cortex is connected to every other [N^2 connectivity] can be replaced by a mean-field network in which a neuron receives n LGN inputs and a single (mean-field) input [(n + 1) connectivity].

Appendix: Asymptotic Behavior of Mean-Field Equations with Time-Dependent Mean Field

From Eqs. 19 and 23, the trajectory of the corresponding point is

$$\dot{m}_i' = \phi[m_i']d - \dot{\alpha}. \qquad [A1]$$

Using Eq. 18, we have

$$\dot{\alpha} = a\,\overline{\dot{m}(\alpha)} = a(1 - a)^{-1}\overline{m}', \qquad [A2]$$

so that

$$\dot{m}_i' = \phi[m_i']d - a(1 - a)^{-1}\overline{m}'. \qquad [A3]$$

At the fixed points $\dot{m}_i' = 0$. When all of the cells of the network have reached their respective fixed points $\dot{m}_i' = 0$ for each cell. Therefore, $\overline{m}' = 0$. It follows that when the network has stabilized at a global fixed point

$$\phi[m_i']d = 0 \qquad [A4]$$

for all inputs. This is the same condition as the $\dot{\alpha} = 0$ (adiabatic) case. Thus the position and stability of the fixed points are the same as those in the adiabatic theory. However, since $0 < (1 - a) < 1$, the absolute value of average movement of the entire network towards the fixed points

$$|\overline{m}'| = (1 - a)\,\overline{\phi[m_i']d} \qquad [A5]$$

is slower than in the adiabatic theory.

This work was supported by the Office of Naval Research and the Army Research Office under Contracts N00014-86-K-0041 and DAAG-29-84-K-0202.

1. Bienenstock, E. L., Cooper, L. N & Munro, P. W. (1982) *J. Neurosci.* **2,** 32–48.
2. Scofield, C. L. & Cooper, L. N (1985) *Cont. Phys.* **26,** 125–145.
3. Sherman, S. M. & Spear, P. D. (1982) *Physiol. Rev.* **62,** 738–855.
4. Fregnac, Y. & Imbert, M. (1984) *Physiol. Rev.* **64,** 325–434.
5. Altmann, L., Luhmann, H. J., Singer, W. & Greuel, J. (1985) *Neuroscience Letters*, Abstracts of the Ninth European Neuroscience Congress (Oxford), Suppl. 22, S353.
6. Singer, W. (1977) *Exp. Brain Res.* **30,** 25–41.
7. Bear, M. F., Schmechel, D. M. & Ebner, F. F. (1985) *J. Neurosci.* **5,** 1262–1275.
8. Mower, G. D., White, W. F. & Rustad, R. (1986) *Brain Res.* **380,** 253–260.
9. Cooper, L. N, Munro, P. W. & Scofield, C. L. (1985) in *Synaptic Modification, Neuron Selectivity and Nervous System Organization*, eds. Levy, W. B., Anderson, J. A. & Lehmkuhle, S. (Erlbaum Assoc., Hillsdale, NJ), pp. 175–192.

Cortical Plasticity: Theoretical Analysis, Experimental Results

Cortical Plasticity, eds. J. P. Rauschecker and P. Marler
(John Wiley & Sons, New York, 1987) p. 177

A key component of our work is the effort to make theory interact with experiment in a serious way. From the beginning it was clear that this would necessitate an experimental additional to our mostly theoretical Center for Neural Studies. This was accomplished in 1977 with the formation of the Center for Neural Science — intended as a merger of the experimental assets of the Section for Neuroscience and the theoretical assets of the Center for Neural Studies. This new Center created the Brown Undergraduate Neuroscience program and has evolved into the Department of Neuroscience as well as the Institute for Brain and Neural Systems. In the early eighties, in addition to Ford Ebner (for many years Co-Director with me of the Center for Neural Science) some of the people involved included Jerry Daniels, Allan Saul, Mike Paradiso, and Mark Bear. Early efforts to confirm or to test various ideas have grown progressively until today, we have developed a rich interaction between experiment and theory. Some of the questions that we attempted to address: local and global controllers, the role of inhibitory neurons are reviewed in the following book chapter. It is included in its entirely since it gives a summary of our thinking at that time. The experiments referred to concerning the correlation between selectivity and ocular dominance were not as conclusive as we would have liked. It seems that population studies of such effects are difficult to interpret. We expect that the needed chronic experiments will be done soon.

9

CORTICAL PLASTICITY: THEORETICAL ANALYSIS, EXPERIMENTAL RESULTS

Leon N. Cooper

Department of Physics and Center for Neural Science, Brown University, Providence, Rhode Island

177

Do all animals learn? And do they learn in the same way? A theme that runs throughout this volume suggests how ubiquitous learning may be and further, possibly more exciting, that common factors may be involved in such seemingly diverse phenomena as the development of selectivity of cells in visual cortex and birdsong learning. We would all agree that some learning must depend on the details of information entering from the outside world. How else would the memory of a new face or the particular sequence that makes a learned birdsong be stored? In addition, there must exist some "global controller" or "enabling factor" that determines whether or not particular learning will take place.

For many years we have heard talk about possible modification of synapses between neurons as the physiological basis of learning and memory storage. These relatively vague ideas are becoming more precise. Insight into the molecular basis of synaptic modification is beginning to appear; the role of possible global controllers such as norepinephrine and acetylcholine is being clarified; and a mathematical structure for the network of neurons is rapidly evolving.

Presented here is a brief summary of some recent theoretical ideas as well as some experimental results related to plasticity in visual cortex, presumably related to the changes that take place when memory is stored and possibly also related to what is known about birdsong learning (see Konishi, Marler, and Margoliash, all this volume). More important than the details, what I hope to convince you is that what is being presented provides us with a language in which these questions can be discussed with clarity and precision.

WHY VISUAL CORTEX?

Experimental work of the last generation, from the path-breaking work of Hubel and Wiesel to some of the most recent results contained in this volume (see Frégnac, Hirsch et al., Rauschecker, Singer, and Timney, all this volume), has shown that there exist cells in the visual cortex of the adult cat that respond in a precise and highly tuned fashion to external patterns—in particular bars or edges of a given orientation and moving in a given direction—and that the number and response characteristics of such cortical cells can be modified. For example, the relative number of cortical cells that are highly specific in their response to visual patterns varies in a very striking way with the visual experience of the animal during the critical period.

Such results seem to provide evidence for the modifiability of the response of single cells in the cortex of a higher mammal according to its visual experience. Depending on whether or not patterned visual information is part of the animal's experience, the specificity of the response of cortical neurons varies widely. Specificity increases with normal patterned experience. Deprived of normal patterned information (e.g., dark reared or lid sutured at birth) specificity decreases. Further, even a short exposure to patterned information (during the critical period) after weeks of dark rearing can reverse the loss of specificity and produce an almost normal distribution of cells.

We do not claim and it is not necessary that all neurons in the visual cortex be so modifiable. Nor is it necessary that modifiable neurons are especially important in producing the architecture of the visual cortex. It is our hope that the general form of modifiability we require to construct interacting neural networks manifests itself for at least some cells of the visual cortex that are accessible to experiment. We thus make the conservative assumption that biological mechanisms, once established, will function in a more or less similar manner in different regions. If this is the case, modifiable individual neurons in the visual cortex can provide evidence for such modification more generally.

SUMMARY OF SINGLE-CELL THEORY

Cortical neurons receive afferents from many sources. In the visual cortex (e.g., layer IV) the principal afferents are those from the lateral geniculate nucleus and from other cortical neurons. This leads to a complex network that we have analyzed in several stages.

In the first stage we consider a single neuron with inputs from both eyes (Fig. 1). Here d^l, d^r, m^l, and m^r are inputs and synaptic junctions from left and right eyes. The output of this neuron (in the linear region) can be written as an inner product:

$$c = m^l{\cdot}d^l + m^r{\cdot}d^r \qquad (1)$$

This means that the neuron firing rate (in the linear region) is the sum of the inputs from the left eye multiplied by the appropriate left-eye

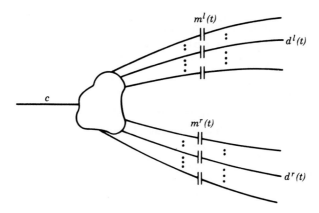

FIGURE 1. A model neuron.

synaptic weights plus the inputs from the right eye multiplied by the appropriate right-eye synaptic weights. Thus, the neuron integrates signals from the left and right eyes.

According to the theory presented by Bienenstock, Cooper, and Munro (1982) (BCM), these synaptic weights modify as a function of local and global variables. To illustrate, we consider the synaptic weight m_j (Fig. 2).

Its change in time, $\dot{m}_j$, is given by

$$\dot{m}_j = F(d_j \ldots m_j; d_k \ldots c; \bar{c} \ldots ; X, Y, Z) \tag{2}$$

Here variables such as $d_j \ldots m_j$ are designated local. These represent information (such as the incoming signal d_j and the strength of the synaptic junction m_j) available locally at the synaptic junction m_j. Variables

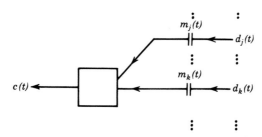

FIGURE 2. Local and quasi-local variables.

such as $d_k \ldots c$ are designated quasi-local. These represent information (such as c, the firing rate of the cell, or d_k, the incoming signal to another synaptic junction) that is not locally available to the junction m_j but is physically connected to the junction by the cell body itself—thus necessitating some form of internal communication between various parts of the cell and its synaptic junctions. Variables such as $\bar{c}$ (the time-averaged output of the cell) are averaged local or quasi-local variables. Global variables are designated X, Y, Z, and so on. These latter variables represent information (e.g., the presence or absence of neurotransmitters such as norepinephrine or the average activity of large numbers of cortical cells) that is present in a similar fashion for all or a large number of cortical neurons (distinguished from local or quasi-local variables presumably carrying detailed information that varies from synapse to synapse). These global variables are candidates for "enabling factors" discussed frequently in this volume (see chapters by Marler, Frégnac, Rauschecker).

In a form relevant to this discussion, BCM modification can be written as

$$\dot{m}_j = \phi(c, \bar{c}; X, Y, Z, \ldots)d_j, \tag{3}$$

so that the jth synaptic junction, m_j, changes its value in time as a function of quasi-local and time-averaged quasi-local variables c and $\bar{c}$ as well as global variables X, Y, and Z through the function ϕ and a function of the local variable d_j. The crucial function, ϕ, is shown in Figure 3.

What is of particular significance is the change of sign of ϕ at the

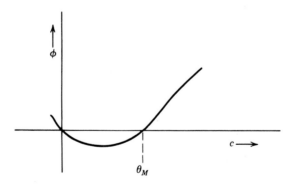

FIGURE 3. The BCM modification function.

modification threshold, θ_M, and the nonlinear variation of θ_M with the average output of the cell c. In a simple situation

$$\theta_M = (\bar{c})^2. \tag{4}$$

The occurrence of negative and positive regions for ϕ drives the cell to selectivity in a "normal" environment. This is so because the response of the cell is diminished to those patterns for which the output c is below threshold (ϕ negative) while the response is enhanced to those patterns for which the output c is above threshold (ϕ positive). The nonlinear variation of the threshold with the average output of the cell, $\bar{c}$, places the threshold so that it eventually separates one pattern from all of the rest. Furthermore, it provides the stability properties of the system.

A detailed analysis of the consequences of this form of modification is given in BCM. The results (as modified by the network analysis outlined next) are in general agreement with what we might call classical experiments of the last generation. Neurons in normal (patterned) environments become selective and binocular. In various deprived environments (MD, BD, etc.) the theoretical behavior follows the experimental results.

An unexpected consequence of this theory is a connection between selectivity and ocular dominance. The analysis given in BCM shows that in monocular deprivation, nonpreferred inputs presented to the open eye are a necessary part of the suppression of deprived eye responses. It follows that the more selective the cell is to the open eye (increasing the probability of nonpreferred inputs), the more the closed eye will be driven to zero, thus increasing the dominance of the open eye.

For an experimental test of these ideas it is important to determine what happens *during* the ocular dominance shift produced by monocular deprivation. Consider the experimental situation in which monocular experience follows a period of dark rearing. [Such experiments are presently being performed by Saul and Daniels (private communication)]. During dark rearing it is known that most area 17 cells become less responsive (sluggish) and lose their selectivity and that some (perhaps 20%) become visually nonresponsive.

Our theoretical analysis indicates that in the course of monocular experience, those cells that have become visually nonresponsive during the dark rearing will first show an increase in responsiveness to the open eye followed by the development of selectivity. Those cells that

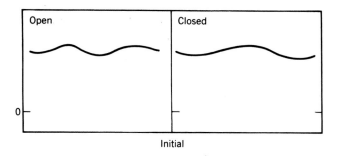

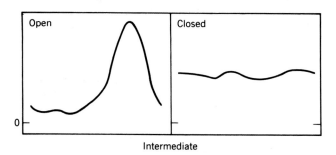

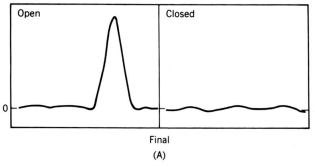

(A)

FIGURE 4. (A) Progression of development of selectivity and ocular dominance. Note that selectivity develops for responsive binocular aspecific cells for the open eye *before* the response to the closed eye is driven to zero. (B) Progression of development of responsiveness and selectivity. Note that responsiveness to the open eye develops before, or along with, selectivity.

183

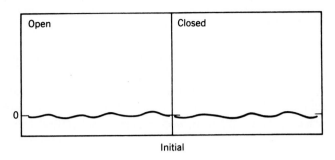

Initial

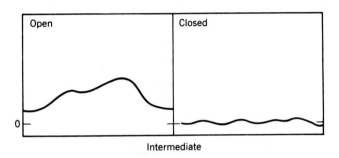

Intermediate

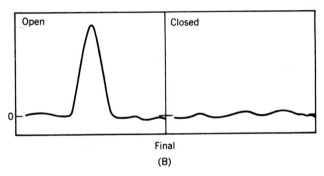

Final

(B)

FIGURE 4. (*continued*)

have survived the period of dark rearing as binocular and aspecific will
exhibit a progression in which their selectivity to the open eye is increased while maintaining their response (often nonselective) to the
closed eye. This should result in the two sequences shown in Figures
4A and 4B.

EXTENSION TO NETWORKS

To better confront these ideas with experiment, the single neuron discussed above must be placed in a network with the anatomical features of the region of interest. For visual cortex this suggests a network in which inhibitory and excitatory cells receive input from the lateral geniculate nucleus (LGN) and from each other. A simplified form of such a network, a first-order representation of the anatomy and physiology of layer IV of cat visual cortex (Fig. 5), has been studied by Scofield and Cooper (1985).

In a network generalization of Equation (1), we write

$$c_i = m_i^l \cdot d^l + m_i^r \cdot d^r + \sum_j L_{ij} c_j \tag{5}$$

where L_{ij} are the intracortical connections.

Analysis by Scofield and Cooper (1985) of the network along lines similar to that of the single-cell analysis described above shows that under proper conditions on the intracortical synapses, the cells converge to states of maximum selectivity with respect to the environment formed by the geniculate signals. Their conclusions are therefore similar to those of BCM with explicit further statements concerning the independent effects of excitatory and inhibitory neurons on selectivity and ocular dominance. For example, shutting off inhibitory cells lessens selectivity and alters ocular dominance. The inhibitory cells may be selective, but there is no theoretical necessity that they be so.

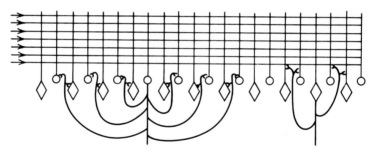

FIGURE 5. A simplified neural network. Shown are the two cell types: inhibitory, represented by circles, and excitatory, by diamonds. Geniculate afferents enter at the top of the figure and synapse with all cells in the network at the intersection of the horizontal and vertical fibers. Also shown are the intracortical fibers for each cell type. The exact ratio of inhibitory to excitatory cells is not important.

A mean field approximation to the above network (Cooper and Scofield, to be published) shows that if the average effect of intracortical connections results in inhibition of individual cells, then in monocular deprivation, the geniculocortical synapses to the cell will converge to nonzero states that give, as the result of stimulation of the closed eye, total responses that are zero. However, the fact that the geniculocortical states are nonzero means that the removal of cortical inhibition through the chemical blocking of inhibitory synapses would uncover responses from previously nonresponsive cells. This result is in accord with the experimental observation of "masked synapses" after the removal of the inhibitory effects of GABA with the blocking agent bicuculline (cf. Duffy et al., 1976; Sillito et al., 1980).

These results are obtained without assuming that the intracortical inhibitory synapses are very responsive to visual experience. Learning can occur entirely among the excitatory LGN–cortical synapses. Another point of view is espoused, for example, by Rauschecker and Singer (1982). They suggest that since cells lose their orientation specificity when binocularly deprived of pattern vision, and since it has been shown that inhibitory connections play a major role in determining orientation selectivity (Sillito, 1975), the cortical inhibitory synapses may suffer more than excitatory ones.

These conflicting ideas led us to perform experiments on changes in inhibitory activity due to visual experience (Bear et al., 1985) (described in the next section). They indicate that one measure of cortical inhibition is relatively constant even during dramatic manipulations of the visual environment.

EXPERIMENTAL TEST OF CHANGES IN INHIBITORY ACTIVITY DUE TO VISUAL EXPERIENCE

One of the consequences of the network theory discussed in the previous section is that experimental results that have been obtained in the visual cortex over the last generation can be explained primarily by modification of the LGN-to-cortex synapses with minimum changes among intracortical synapses. Thus, the possibility is opened that most learning takes place in the geniculocortical synapses. This somewhat surprising result has as one consequence the possibility of great simplification in the analysis of network modification.

An alternate hypothesis that has been considered for some time is that intracortical synapses bear heavy responsibility for modification in cortical circuitry during learning. In particular, it has been suggested that ocular dominance shifts in monocular deprivation are due to increased activity of GABAergic neurons, the open eye suppressing the closed. Sillito (1975) documented, in normal cats, that visually unresponsive cells may be "unmasked" by iontophoretic bicuculline. Thus, it is not unreasonable to speculate that many of the unresponsive cells in visually deprived kittens are being suppressed. Together, these data suggest as a possible hypothesis that *in kitten striate cortex the GABAergic neurons respond to sensory deprivation by forming new synapses.* This hypothesis implies that the density or strength of GABAergic synapses will increase in zones of cortex that are deprived of a normal thalamic input; in the case of monocular deprivation, these zones correspond to the closed-eye ocular dominance columns and to the monocular segment contralateral to the deprived eye. On the other hand, the theory described above suggests that there will be minimal response of GABAergic neurons to sensory deprivation. This hypothesis has been put to the test in a recent series of experiments in our laboratory (Bear et al., 1985) (described below).

To examine the distribution of GABAergic synapses, Bear et al. immunocytochemically localized GAD in sections of striate cortex. While immunocytochemistry is not a quantitative measure, they reasoned that changes restricted to deprived ocular dominance zones should be readily detected with this method. As a quantitative estimate of GABAergic synapse density, they biochemically measured GAD activity in homogenates of striate cortex.

They found no evidence for a change in the distribution of GAD-positive puncta in 12 unilaterally enucleated kittens. The band of layer IV puncta remained uniform even though the periods of monocular deprivation examined would all be sufficient to cause a physiological ocular dominance shift in the striate cortex (Hubel and Wiesel, 1970). GAD immunoreactivity was unchanged even under conditions that produced alterations in the level of the metabolic enzyme, cytochrome oxidase. Measurements of GAD activity showed no consistent or significant difference between either the binocular segments of enucleated and control kittens or the monocular segments of enucleated animals.

This conclusion is in striking agreement with network analysis, which, as mentioned above, suggests that inhibitory synapses are much less modified by experience than excitatory synapses. In addition to its implications for the "site of learning," such a hypothesis leads to important simplifications in the analysis of complex networks.

GLOBAL CONTROLS: THE ELUSIVE "ENABLING FACTOR"

Any theory of learning generally requires global as well as local controls. As discussed previously, local controls are those that determine detailed changes of individual synaptic junctions while global information would be expected to influence all or large numbers of synaptic junctions in the same way. From the point of view of a learning theory, there must be some way to distinguish more interesting from less interesting input. Experimentally it is known that certain areas of cortex (visual, auditory, and somatic sensory) exhibit plasticity during early critical periods but not during adulthood.

For example, most neurons in the visual cortex of newborn kittens (and normal adult cats as well) are activated equally well by both eyes. During the critical period (3 weeks to 3 months in cats), monocular lid suture or misalignment of the eyes (called strabismus) leads to the domination of cortical cells by one eye. In contrast, adult cats that are given monocular visual experience when they are *older* than 6 months of age remain unaffected by the imbalance of visual inputs; adult cortical neurons remain binocularly activated.

A major question has been formulated as follows: What are the global factors acting singly or in combination that affect the development of synapses in cortex?

In recent years it has been suggested that catecholamines (CA) are required for neuronal plasticity in the neocortex. One test of this hypothesis has been made in a series of experiments performed by Kasamatsu and Pettigrew (1976, 1979), who used the monocular deprivation paradigm with kitten visual cortex as a test system. In their control kittens they found the usual effect of monocular deprivation during the critical period—within a week the majority of cells lost their normal binocular responsiveness and could be driven only by stimulation of the nondeprived eye. But in animals given the neurotoxin 6-hydroxydopamine (6-OHDA) to deplete cortical CAs, the ocular dominance shift failed to occur and cells remained binocularly driven.

In later experiments Kasamatsu et al. (1979, 1981) pioneered the use of miniature osmotic pumps to infuse 6-OHDA continuously to local regions of cortex in one hemisphere while they used the other hemisphere as a control by perfusing only the vehicle for the 6-OHDA. Following monocular deprivation, normal plasticity was again disrupted in CA-depleted cortex as indicated by the lack of a shift in ocular dominance of visual cells. Because both noradrenergic and dopaminergic fibers project to the visual cortex, Kasamatsu et al. also used the minipumps to add norepinephrine (NE) to the cortex previously depleted of CAs to demonstrate that NE itself can restore plasticity. This evidence suggests that catecholamines, especially NE, are necessary for the cortical changes observed in kittens that have restricted vision during the critical period.

However, the relationship between static NE levels and plasticity is not simple. In one experiment (Bear and Daniels, 1983; Bear et al., 1983) cortical catecholamines were permanently depleted in newborn kittens by intraperitoneal injections of 6-OHDA. Biochemical analysis demonstrated severe reduction of NE levels; but in this experiment plasticity, as evidenced by ocular dominance shift, remained intact. In the neonatal experiments kittens received 6-OHDA *at birth* and were monocularly deprived for 7 days at about 5 weeks of age. The minipump kittens received 6-OHDA continuously for the 7 days of monocular deprivation. Comparison of the two paradigms suggests the possibility that loss of plasticity is not caused by depletion of NE alone.

Thus, despite Kasamatsu's success in demonstrating support for the NE hypothesis, questions have arisen. The above experiments show that early depletion of cortical NE does not, by itself, prevent the later ocular dominance shift after monocular deprivation. Other experiments confirm this conclusion. Daw et al. (1984) depleted cortical NE by section of the locus coeruleus fiber bundle near lateral hypothalamus and found no diminution of the ocular dominance shift after monocular deprivation. Videen et al. (1984) recorded no difference in the reaction of kitten and adult cat visual cortex neurons to iontophoretically applied NE. Adrien et al. (1982) observed no lack of shift after lesion of the locus coeruleus itself; and that group was unable to reproduce Kasamatsu and Pettigrew's (1979) finding that intraventricular injection of 6-OHDA prevents ocular dominance shift. All of these results reinforce the idea that NE is not the only factor in the global control of learning.

In addition to the NE system, several lines of evidence suggest that the cholinergic (ACh) system may also serve as a global modulator of cortical function. Similar to the locus coeruleus–NE system, the basal forebrain cholinergic (ACh) system has a widespread input to cortex that stands in marked contrast to the highly organized thalamocortical systems that provide specific sensory input to the cortex. In addition, several findings link both the NE and ACh systems to acquisition and storage processes related to learning and memory.

While these global cortical inputs have been related to memory and learning, the specific cellular mechanisms of ACh and NE function are unclear. Present evidence indicates that both systems may modulate the response of cortical neurons to specific sensory inputs. NE appears to improve the signal-to-noise ratio of sensory responses both in somatic sensory and in the primary visual cortex, and NE may potentiate the action of both excitatory and inhibitory transmitters (Waterhouse et al., 1980). These effects of NE could be mediated through α-adrenergic receptors that appear to be concentrated in the deeper layers (IV–VI) of the cortex.

The cholinergic system may function in a manner similar to NE. Application of low levels of ACh enhances the excitatory response of cortical neurons to glutamate and modifies the task-related discharge of cortical neurons during behavior. ACh has also been shown to modify the membrane input resistance of cortical neurons. The slow onset, long duration of action, and sensitizing effects of ACh are all consistent with the conclusion that the cholinergic system acts as a modulator of cortical activity. All of this suggests that both NE and ACh may play a modulatory role in cortex.

This view has been reinforced by recent work of Bear and Singer (1986), which indicates that 6-OHDA, in addition to destroying NE neurons, also interferes with ACh effects on cortical neurons. In addition, Bear and Singer have shown that the simultaneous diminution of both ACh and NE appears to prevent the ocular dominance shift in monocular deprivation while the diminution of either ACh or NE (and not both) does not prevent this shift.

These results may enable a resolution of the apparent contradictions in previous NE experiments discussed above and, in addition, suggest the fascinating possibility that ACh and NE act together to provide a global modulator for plasticity.

ACKNOWLEDGMENT

This work was supported by the Office of Naval Research and the U.S. Army Research Office.

REFERENCES

Adrien, J., P. Buisseret, Y. Frégnac, E. Gary-Bobo, M. Imbert, J. Tassin, and Y. Trotter. (1982). Noradrenaline et plasticité du cortex visuel du chaton: Un reexamen. *C. R. Acad. Sci. Paris Serie III* **295**:745–750.

Bear, M. F., and J. D. Daniels (1983). The plastic response to monocular deprivation persists in kitten visual cortex after chronic depletion of norepinephrine. *J. Neurosci.* **3**:407–416.

Bear, M. F., D. E. Schmechel, and F. F. Ebner (1985). Glutamic acid decarboxylase in the striate cortex of normal and monocularly deprived kittens. *J. Neurosci.* **5**:1262–1275.

Bear, M. F., M. A. Paradiso, M. Schwartz, S. B. Nelson, K. M. Carnes, and J. D. Daniels (1983). Two methods of catecholamine depletion in kitten visual cortex yield different effects on plasticity. *Nature* **302**:245–247.

Bear, M. F. and W. Singer (1986). Modulation of visual cortical plasticity by acetylcholine and noradrenaline. *Nature* **320**:172–176.

Bienenstock, E. L., L. N. Cooper, and P. W. Munro (1982). Theory for the development of neuron selectivity: Orientation specificity and binocular interaction in visual cortex. *J. Neurosci.* **2**:32–48.

Cooper, L. N. and C. L. Scofield (in press). Mean field approximation for a neural network.

Daw, N. W., T. W. Robertson, R. K. Rader, T. O. Videen, and C. J. Coscia (1984). Substantial reduction of cortical noradrenaline by lesions of adrenergic pathway does not prevent effects of monocular deprivation. *J. Neurosci.* **4**:1354–1360.

Duffy, F. H., S. R. Snodgrass, J. L. Burchfiel, and J. L. Conway (1976). Bicuculline reversal of deprivation amblyopis in the cat. *Nature* **260**:256–257.

Hubel, D. H. and T. N. Wiesel (1970). The period of susceptibility to the physiological effects of unilateral eye closure in kittens. *J. Physiol. (Lond.)* **206**:419–436.

Kasamatsu, T. and J. D. Pettigrew (1976). Depletion of brain catecholamines: Failure of ocular dominance shift after monocular occlusion in kittens. *Science* **194**:206–209.

Kasamatsu, T. and J. D. Pettigrew (1979). Preservation of binocularity after monocular deprivation in the striate cortex of kittens treated with 6-hydroxydopamine. *J. Comp. Neurol.* **185**:139–162.

Kasamatsu, T., J. D. Pettigrew, and M. Ary (1979). Restoration of visual cortical plasticity by local microperfusion of norepinephrine. *J. Comp. Neurol.* **185**:163–182.

Kasamatsu, T., J. D. Pettigrew, and M. Ary (1981). Cortical recovery from effects of monocular deprivation: Acceleration with norepinephrine and suppression with 6-hydroxydopamine. *J. Neurophysiol.* **45**:254–266.

Rauschecker, J. P. and W. Singer (1982). Binocular deprivation can erase the effects of preceding monocular or binocular vision in kitten cortex. *Dev. Brain Res.* **4**:495–498.

Sillito, A. M. (1975). The contribution of inhibitory mechanisms to the receptive field properties of neurons in the cat's striate cortex. *J. Physiol.* **250**:304–330.

Sillito, A. M., J. A. Kemp and H. Patel (1980). Inhibitory interactions contributing to the ocular dominance of monocularly dominated cells in the normal cat striate cortex. *Exp. Brain Res.* **41**:1–10.

Scofield, C. L. and L. N. Cooper (1985). Development and properties of neural networks. *Cont. Phys.* **26**(2):125–145.

Videen, T. O., N. W. Daw, and R. K. Rader (1984). The effect of norepinephrine on visual cortical neurons in kitten and adult cats. *J. Neurosci.* **4**:1607–1617.

Waterhouse, B. D., H. C. Moises, and D. J. Woodward (1980). Noradrenergic modulation of somatosensory cortical neuronal responses to iontophoretically applied putative neurotransmitters. *Exp. Neurol.* **69**:30–49.

A Physiological Basis for a Theory of Synapse Modification

(with M. F. Bear, F. F. Ebner)

Science **237**, 42 (1987)

Mark Bear joined Ford Ebner's laboratory in 1979 as a graduate student, having done his undergraduate work with Irving Diamond at Duke. From the beginning Mark displayed an interest in pursuing ideas and a willingness to learn new techniques, when necessary, to do a particularly relevant experiment. With his talent, open personality, and great modesty he interacted readily with the theory group, and gradually we began to shape both theory and experiment to enrich each other. This has resulted in a continuing and very fruitful interaction.

One of the longer term goals of our research was to find the molecular basis for learning and memory storage. Initially this seemed unrealistically far in the future. However, as the years progress, it becomes more and more a possibility. The following paper describes one of our early attempts to provide a physiological basis for the synaptic modification postulated in the BCM theory. We see here initial attempts to link this modification, introduced to explain events in the visual cortex, with the phenomena of long-term potentiation (LTP) first observed in the hippocampus which, because of the rapidly induced long-term changes in neuron response, seemed a likely candidate for a substrate of memory. Although the ideas presented in this paper are speculative and probably not correct in detail, focusing on NMDA receptors and calcium influx have proven to be extremely important in shaping current thinking.

Reprint Series
3 July 1987, Volume 237, pp. 42–48

SCIENCE

A Physiological Basis for a Theory of Synapse Modification

MARK F. BEAR, LEON N. COOPER, AND FORD F. EBNER

A Physiological Basis for a Theory of Synapse Modification

MARK F. BEAR, LEON N COOPER, FORD F. EBNER

The functional organization of the cerebral cortex is modified dramatically by sensory experience during early postnatal life. The basis for these modifications is a type of synaptic plasticity that may also contribute to some forms of adult learning. The question of how synapses modify according to experience has been approached by determining theoretically what is required of a modification mechanism to account for the available experimental data in the developing visual cortex. The resulting theory states precisely how certain variables might influence synaptic modifications. This insight has led to the development of a biologically plausible molecular model for synapse modification in the cerebral cortex.

ALTHOUGH ARISTOTLE IDENTIFIED HEART AS THE SEAT OF intellect, reserving for brain the function of cooling the head, it is now generally believed that it is brain that is the source of thought, the location of memory, the physical basis of mind, consciousness, and self-awareness: all that make us distinct and human. In recent years it has become increasingly fashionable to treat this complex system as a neural network: an assembly of neurons connected to one another by synaptic junctions that serve to transmit information and possibly to store memory.

Since the contents of memory must depend to some extent on experience, the neural network and, in particular, the synapses between neurons cannot be completely determined genetically. This evident reasoning has led to much discussion about possible modification of synapses between neurons as the physiological basis of learning and memory storage. To properly function, neural network models require that vast arrays of synapses have the proper strengths. A basic problem becomes how these synapses adjust their

weights so that the resulting neural network shows the desired properties of memory storage and cognitive behavior.

The problem can be divided into two parts. First, what type of modification is required so that in the course of actual experience the neural network arrives at the desired state? The answer to this question can be illuminated by mathematical analysis of the evolution of neural networks by means of various learning hypotheses. The second part of this problem is to find experimental justification for any proposed modification algorithm. A question of extraordinary interest is: What are the biological mechanisms that underlie the nervous system modification that results in learning, memory storage, and eventually cognitive behavior?

One experimental model that appears to be well suited for the purpose of determining how neural networks modify is the visual cortex of the cat. The modification of visual cortical organization by sensory experience is recognized to be an important component of early postnatal development (1). Although much modifiability disappears after the first few months of life, some of the underlying mechanisms are likely to be conserved in adulthood to provide a basis for learning and memory. We have approached the problem of experience-dependent synaptic modification by determining theoretically what is required of a mechanism in order to account for the experimental observations in visual cortex. This process has led to the formulation of hypotheses, many of which are testable with currently available techniques. In this article we illustrate how the interaction between theory and experiment has suggested a possible

M. F. Bear is an Alfred P. Sloan Research Fellow and assistant professor of neural science in the Center for Neural Science at Brown University; L. N Cooper is Thomas J. Watson, Sr., Professor of Science in the Department of Physics at Brown University; and F. F. Ebner is professor of medical science in the Division of Biology and Medicine at Brown University. L. N Cooper and F. F. Ebner are codirectors of the Center for Neural Science at Brown University, Providence, RI 02912.

molecular mechanism for the experience-dependent modifications of functional circuitry in the mammalian visual cortex.

The Experimental Model

Neurons in the primary visual cortex, area 17, of normal adult cats are sharply tuned for the orientation of an elongated slit of light and most are activated by stimulation of either eye (2). Both of these properties—orientation selectivity and binocularity—depend on the type of visual environment experienced during a critical period of early postnatal development. We believe that the mechanisms underlying the experience-dependent modification of both receptive field properties are likely to be identical. However, for the sake of clarity, we concentrate primarily on the modification of binocular connections in striate cortex.

The majority of binocular neurons in the striate cortex of a normal adult cat do not respond with equal vigor to stimulation of either eye; instead they typically display an eye preference. To quantify this impression Hubel and Wiesel (2) originally separated the population of recorded neurons into seven ocular dominance (OD) categories (Fig. 1). The OD distribution in a normal kitten or adult cat shows a broad peak at group 4, which reflects a high percentage of binocular neurons in area 17 (Fig. 1A). This physiological assay of OD has proved to be an effective measure of the state of functional binocularity in the visual cortex.

Monocular deprivation (MD) during the critical period [extending from approximately 3 weeks to 3 months of age in the cat (3)] has profound and reproducible effects on the functional connectivity of striate cortex. Brief periods of MD will result in a dramatic shift in the OD of cortical neurons such that most will be responsive exclusively to the open eye (Fig. 1B). The OD shift after MD is the best known and most intensively studied type of visual cortical plasticity.

When the MD is begun early in the critical period, the OD shift can be correlated with anatomically demonstrable differences in the geniculocortical axonal arbors of the two eyes (4, 5). However, MD initiated late in the critical period (6) or after a period of rearing in the dark (7) will induce clear changes in cortical OD without a corresponding anatomic change in the geniculocortical projection. Long-term recordings from awake animals also indicate that OD changes can be detected within a few hours of monocular experience (8), which seems too rapid to be explained by the formation or elimination of axon terminals. Moreover, deprived-eye responses in visual cortex may be restored within minutes under some conditions

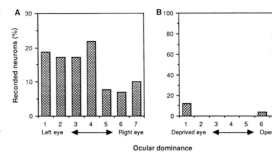

Fig. 1. Representative histograms of OD data (5) obtained from the striate cortex of (**A**) normal cats and (**B**) cats that were monocularly deprived early in life. The bars show the percentage of neurons in each of the seven OD categories. Cells in groups 1 and 7 are activated by stimulation of either the contralateral eye or the ipsilateral eye, respectively, but not both. Cells in group 4 are activated equally well by either eye. Cells in groups 2 and 3, and 5 and 6, are binocularly activated, but show a preference for either the contralateral eye or the ipsilateral eye, respectively. The histogram in (A) reveals that the majority of neurons in the visual cortex of a normal animal are driven binocularly. The histogram in (B) shows that a period of monocular deprivation leaves few neurons responsive to the deprived eye. Fewer than 1 percent of the recorded neurons were classified in groups 2, 3, 4, and 5.)

[such as during intracortical bicuculline administration (9)], which suggests that synapses deemed functionally "disconnected" are nonetheless physically present. Therefore it is reasonable to assume that changes in functional binocularity are explained not only by adjustments of the terminal arbors of geniculocortical axons but also by changes in the efficacy of individual cortical synapses.

The consequences of binocular deprivation (BD) on visual cortex stand in striking contrast to those observed after MD. Although 7 days of MD during the second postnatal month leave few neurons in striate cortex responsive to stimulation of the deprived eye, most cells remain responsive to visual stimulation through either eye after a comparable period of BD (10). Thus, it is not merely the absence of patterned activity in the deprived geniculocortical projection that causes the decrease in synaptic efficacy after MD.

Stent (11) suggested that the crucial difference between MD and BD is that only in the former instance are cortical neurons active. This idea was put to the test in an ingenious series of experiments by Singer and colleagues in which kittens were presented with visual stimuli that created an imbalance in the presynaptic geniculocortical fiber activity from the two eyes, but that were believed to be ineffective in driving cortical neurons (12, 13). Under these conditions, there was no shift in cortical OD. Thus, on a purely descriptive level, it appeared that postsynaptic activation was a necessary condition for synaptic modifications to occur in the visual cortex. This simple rule resembles Hebb's postulate of learning (14), which states that synaptic efficacy should increase only when the pre- and postsynaptic elements are concurrently active. To account for the effects of MD in striate cortex, Stent added the complementary statement that postsynaptic activity is also a necessary condition for the weakening of inactive synapses. According to this idea, postsynaptic activation is necessary for all synaptic modifications and the sign of the change (positive or negative) is dependent on the amount of presynaptic activity.

Subsequent work has suggested, however, that the generation of action potentials in a cortical neuron does not ensure that OD modifications will occur after MD (15). This has led to the idea that there is a critical level of postsynaptic activation that must be reached before experience-dependent modifications will occur, and that this threshold is higher than the depolarization required for soma-spikes (16). Singer (17) has recently proposed a tentative mechanism that

Table 1. Parameters identified as crucial variables for synapse modification.

Parameter	Possible measures	Symbolic notation
Presynaptic activity of the jth synapse*	Firing rate; transmitter release (millisecond time base†)	d_j
Postsynaptic activity	Firing rate; dendritic depolarization (millisecond time base†)	c
Time-averaged postsynaptic activity	Firing rate; dendritic depolarization (minute to hour time base†)	$\bar{c}$
Synaptic transfer function of the jth synapse*	$\Delta c/\Delta d_j$	m_j
"Global" modulation	Dendritic field potentials; second messenger activity	X, Y, Z

*The notation we use for the input activity of a single LGN fiber and its synaptic weight is d_j and m_j, respectively. When we refer to the total input activity and synaptic weight of an array of fibers, we use vector notation, **d** and **m**.   †Time bases can be inferred from experimental results. In this article, d and c are averages over approximately 500 msec, $\bar{c}$ is average over several hours.

could account for this type of modification scheme. Experience-dependent modifications do not normally occur in the visual cortex of anesthetized kittens (18). However, shifts of OD can be induced under anesthesia when cortical excitability is raised by pairing monocular visual stimulation with electrical stimulation of the midbrain reticular formation (19). Only under these conditions can visual stimulation evoke decreases in the extracellular calcium ion concentration as measured with ion-sensitive electrodes (20). These findings led Singer to suggest that the threshold level of postsynaptic activation required for synaptic modification is related to voltage-dependent calcium entry into cortical dendrites. According to this hypothesis, free calcium in the dendrite acts as a second messenger to trigger the molecular changes required for a modification of synaptic efficacy to occur.

This idea that synaptic change is dependent on some suprathreshold output of the postsynaptic neuron can account for many of the observed results in striate cortex after different types of visual deprivation. However, there are several examples of synapse modification in visual cortex that will occur with little or no evoked activity in cortical neurons. For instance, if after a brief period of MD the deprived eye is allowed to see and the experienced eye is sutured closed (known as a reverse suture experiment), then there is a robust OD shift back to the newly opened eye (4, 13, 21). This shift occurs despite the fact that, at the time of the reversal, the unsutured eye was functionally disconnected from the striate cortex. Moreover, if impulse activity from one eye is completely silenced with intraocular tetrodotoxin and the other eye is sutured closed, then an OD shift can sometimes occur in favor of the sutured eye even though this eye is deprived of the visual patterns required to drive cortical neurons

(22). Finally, the OD shift produced by a period of MD will disappear, as will orientation selectivity, if the animal is subsequently binocularly deprived (23). This change in selectivity occurs under conditions where visual cortical neurons are presumably inactive. Hence, it is clear that relative postsynaptic inactivity does not preclude synapse modifications under all conditions. Thus, the Hebb-Stent hypothesis cannot account for the observed data unless further assumptions are made.

Work over the past several years has led to an alternative theoretical solution to the problem of visual cortical plasticity (24). According to Cooper *et al.* (25), when a cortical neuron is depolarized beyond a "modification threshold," θ_M, then synaptic efficacies change along lines envisaged by Hebb (14). However, when the level of postsynaptic activity falls below θ_M, then synaptic strengths decrease. Thus, in this model the sign of a synaptic change is a function primarily of the level of postsynaptic activity. Analysis by Cooper *et al.* confirmed that such a modification scheme could lead to the development of selectivity that is appropriate for the input environment. However, these researchers also noted that a fixed modification threshold leads to certain technical problems. For instance, if the postsynaptic response to all patterns of input activity slipped below θ_M (as might occur during binocular deprivation), then the efficacy of all synapses would decrease to zero. Bienenstock *et al.* (26) solved this problem by allowing the modification threshold to float as a function of the averaged activity of the cell. With this feature, the theory can successfully account for virtually all the types of modification that have been observed in kitten striate cortex over the past 20 years. This theory is outlined in more detail in the next section; then we shall return to the question of possible mechanisms.

Theoretical Analysis

Cortical neurons receive synaptic inputs from many sources. In layer IV of visual cortex the principal afferents are those from the lateral geniculate nucleus (LGN) and from other cortical neurons. This leads to a complex network that has been analyzed in several stages. In the first stage, consider a single neuron with inputs from both eyes (Fig. 2A). Here **d** represents the level of presynaptic geniculocortical axon activity, **m** the synaptic transfer function ("synaptic strength" or "weight"), and c the level of postsynaptic activity of the cortical neuron. These parameters, their symbolic notation, and their possible physiological measures are shown in Table 1. The output of this neuron (in the linear region) can be written:

$$c = \mathbf{m}^\ell \cdot \mathbf{d}^\ell + \mathbf{m}^r \cdot \mathbf{d}^r \qquad (1)$$

which means that the neuron firing rate (or dendritic depolarization) is the sum of the inputs from the left eye multiplied by the left-eye synaptic weights plus the inputs from the right eye multiplied by the right-eye synaptic weights. Thus, the signals from the left and right eyes are integrated by the cortical neuron and determine its level of depolarization (output) at any instant.

The crucial question becomes: How does **m** change in time according to experience? According to the theory of Bienenstock *et al.* (26), **m** modifies as a function of local, quasi-local, and global variables. Consider the synaptic weight of the jth synapse on a neuron, m_j (Fig. 2B). This synapse is affected by local variables in the form of information available only through the jth synapse, such as the presynaptic activity levels (d_j) and the efficacy of the synapse at a given instant in time [$m_j(t)$]. Quasi-local variables represent information that is available to the jth synapse through intracellular communication within the same cell. These include the instantaneous firing rate (or dendritic depolarization) of the cell (c), the

Fig. 2. Illustrated schematically are pyramidal-shaped cortical neurons and the proximal segments of their apical dendrites. The shaded circles attached to the dendrites represent dendritic spines. In the first stage of the theoretical analysis we consider only the inputs to the cell from the LGN (**A**). The signals conveyed along these afferents arise either from the left retina ($\mathbf{d}^\ell$) or the right retina ($\mathbf{d}^r$) and are transferred to the cortical neuron by the synaptic junctions $\mathbf{m}^\ell$ and $\mathbf{m}^r$. The output of the cortical neuron, as measured by the firing rate or the dendritic depolarization, is represented as c, which is the sum of $\mathbf{m}^\ell \cdot \mathbf{d}^\ell$ and $\mathbf{m}^r \cdot \mathbf{d}^r$. The central question is how one of these afferent synapses, m_j, modifies in time as a function of both its level of presynaptic activity d_j and the level of postsynaptic depolarization (**B**).

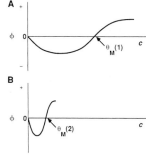

Fig. 3. The φ function at two values of the modification threshold, θ_M. According to the theory of Bienenstock *et al.* (26), active synapses ($d > 0$) are strengthened when φ is positive and are weakened when φ is negative. The φ function is positive when c, the postsynaptic depolarization, is greater than θ_M. The modification threshold, where φ changes sign, is a nonlinear function of the average activity of the postsynaptic neuron ($\bar{c}$). Hence, in this example, θ_M (1) would be expected when cortical neurons have experienced a normal visual environment (**A**) whereas θ_M (2) would result from a prolonged period of binocular deprivation (**B**).

time-averaged firing rate ($\bar{c}$), and the potentials generated at neighboring synaptic junctions [$(dm)_{k,l,m}$ . . .]. Finally, global variables (designated X, Y, and Z in Table 1) represent information that is available to a large number of cortical neurons, including the neuron receiving the *j*th synapse. These variables might include the presence or absence of modulatory neurotransmitters such as acetylcholine and norepinephrine (15) or the averaged activity of large numbers of cortical cells (16).

We can delay consideration of the global variables by assuming that they act to render cortical synapses modifiable or nonmodifiable by experience. In the "plastic state," the Bienenstock *et al.* (26) algorithm for synaptic modification is written:

$$dm_j/dt = \phi(c, \bar{c})d_j \tag{2}$$

so that the strength of the *j*th synaptic junction, m_j, changes its value in time as a function, φ, of the quasi-local states, c and $\bar{c}$, and as a linear function of the local variable d_j. The crucial function, φ, is shown in Fig. 3.

One significant feature of this model is the change of sign of φ at the modification threshold, θ_M. When the input activity of the *j*th synapse (d_j) and φ are both concurrently greater than zero (27), then the sign of the synaptic modification is positive and the strength of the synapse increases. The φ function is positive when the output of the cell exceeds the modification threshold (this type of synaptic modification is "Hebbian"). When d_j is positive and φ is less than zero, then the synaptic efficacy weakens; $\phi < 0$ when $c < \theta_M$. Thus, "effective" synapses will be strengthened and "ineffective" synapses will be weakened, where synaptic effectiveness is determined by whether or not the presynaptic pattern of activity is accompanied by the simultaneous depolarization of the target dendrite beyond θ_M. Since the depolarization of the target cell beyond θ_M normally requires the synchronous activation of converging excitatory synapses, this type of modification will "associate" those synapses that are concurrently active by increasing their effectiveness together.

Another significant feature of this model is that the value of θ_M is not fixed but instead varies as a nonlinear function of the average output of the cell ($\bar{c}$). In a simple situation

$$\theta_M = \bar{c}^2 \tag{3}$$

By allowing θ_M to vary with the average response in a faster-than-linear fashion, the response characteristics of a neuron evolve to maximum selectivity starting at any level within the range of the input environment. It is also this feature that provides the stability properties of the model so that, for instance, simultaneous pre- and postsynaptic activity at a continued high level do not continue to increase the synaptic strength.

Now consider the situation of reverse suture where the right, formerly open eye is closed and the left, formerly sutured eye is reopened. The output of a cortical neuron in area 17 approaches zero just after the reversal since its only source of patterned input is through the eye whose synapses had been functionally disconnected as a consequence of the prior MD. However, as $\bar{c}$ diminishes, so does the value of θ_M. Eventually, the modification threshold attains a value below the small output that is evoked by the stimulation of the synapses of the weak left eye. Now the efficacy of these "functionally disconnected" synapses will begin to increase, because even their low response values exceed θ_M. As these synapses strengthen and the average output of the cell increases, θ_M again slides out until it overtakes the new left-eye response values. At the same time, the efficacy of the right-eye synapses continually decreases because their response values remain below the modification threshold. In its final stable state, the neuron is responsive only to the newly opened eye, and the maximum output to stimulation of this eye equals θ_M.

So far, the discussion has been limited to an idealized single neuron whose inputs arise only from the LGN. The second stage of the theoretical analysis requires that relevant intracortical connections be incorporated into the model. Consider a simple network, illustrated in Fig. 4, in which inhibitory and excitatory cortical neurons receive input from the LGN and from each other. In a network generalization of Eq. 1, the integrated output of the *i*th neuron may be written:

$$c_i = \mathbf{m}_i^{\ell} \cdot \mathbf{d}^{\ell} + \mathbf{m}_i^{r} \cdot \mathbf{d}^{r} + \Sigma L_{ij}c_j \tag{4}$$

where the term $\Sigma L_{ij}c_j$ is the sum of the output from other cells in the network multiplied by the weight of their synapses on the *i*th cell.

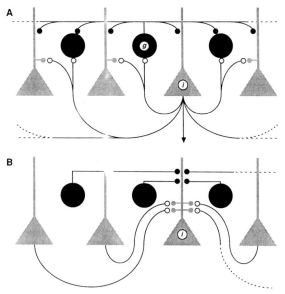

Fig. 4. In the second stage of the theoretical analysis, the neurons of Fig. 2 are placed in a cortical network in which the inhibitory and excitatory cells receive input from the LGN and from each other. (**A**) The efferent intracortical connections of two neurons in the network. The *i*th neuron is excitatory, the *g*th is inhibitory, and both synapse upon every other cell in the network. (**B**) The intracortical inputs to the *i*th neuron. Thus, in addition to the geniculate afferents ($\mathbf{d}^{\ell}$ and $\mathbf{d}^{r}$, shown in Fig. 2A), each neuron in the network receives excitatory and inhibitory intracortical inputs. In a network generalization of Eq. 1, the integrated output of the *i*th neuron may be written as Eq. 4.

The influence of this network on the synaptic modifications of the ith neuron may be analyzed by using a mean field approximation (28). Assuming only that the net influence of the intracortical connections is inhibitory $[(\Sigma L_{ij}c_j) < 0]$, this work has proven that a neuron will evolve to an asymptotically stable state that is appropriate for a given visual environment (in agreement with what has been observed experimentally). Importantly, this occurs without the need to assume any modification of the inhibitory synapses in the network.

There is an interesting consequence of assuming the neuron is under the influence of an inhibitory mean field network. Recall that according to the theory of Bienenstock *et al.* (26), MD leads to convergence of geniculocortical synapses to a state where stimulation of the deprived eye input results in an output that equals zero ($c = 0$). However, with average network inhibition, the evolution of the cell to this state does not require that the efficacy of deprived-eye synapses be driven completely to zero. Instead, these excitatory synapses will evolve to a state where their influence is exactly offset by intracortical inhibition. Thus, the removal of intracortical inhibition in this network would reveal responses from otherwise ineffective inputs. This result is in accordance with the experimental observation of "unmasking" of synapses when the inhibitory effects of γ-aminobutyric acid (GABA) are antagonized with the blocking agent bicuculline (9).

A Possible Physiological Mechanism

One of the consequences of the network theory discussed in the previous section is that the experimental results obtained in visual cortex over the last generation can be explained by modification of excitatory synapses, with minimal changes in intracortical inhibition. The balance of available experimental evidence supports this conclusion. For example, Singer (29) found using intracellular recording that geniculocortical synapses on inhibitory interneurons are more resistant to MD than are synapses on pyramidal cell dendrites. Moreover, recent work suggests that the density of inhibitory synapses that use GABA in kitten striate cortex is also unaffected by MD during the critical period (30). Taken together, these theoretical and experimental results indicate that the search for mechanism should be focused on the excitatory synapses that impinge on excitatory cells in visual cortex. Interestingly, this type of synapse is formed exclusively on dendritic spines, a feature that distinguishes it from other types of cortical synapse (31). This suggests that experience-dependent modifications in striate cortex occur primarily at axospinous synapses.

What mechanisms support the experience-dependent modification of axospinous synapses? According to the theory, when the postsynaptic cell is depolarized beyond θ_M, then active synapses will be strengthened. Depolarization beyond θ_M minimally requires the synchronous activation of converging excitatory afferents. When postsynaptic activity fails to reach θ_M, then the active synapses will be weakened. The identification of the physiological basis of θ_M is therefore central to an understanding of the modification mechanism.

Work on long-term potentiation (LTP) in the hippocampal slice preparation has provided an important insight into the nature of the modification threshold that may be applicable to the visual cortex. LTP is a long-lasting increase in the synaptic strength of excitatory afferents that have been tetanically stimulated (32). The induction of LTP depends on the coactivation of converging excitatory afferents [input cooperativity (33)], the depolarization of the postsynaptic neuron (34), the activation of N-methyl-D-aspartate (NMDA) receptors (35), and the postsynaptic entry of Ca^{2+} ions (36). A current

working hypothesis is that the synchronous tetanic activation of converging afferents depolarizes the target dendrite beyond the threshold for postsynaptic Ca^{2+} entry through gates linked to the NMDA receptor (37). Elevated dendritic calcium then triggers the intracellular changes that lead to enhanced synaptic efficacy (38).

NMDA receptors are a subclass of excitatory amino acid receptor, and glutamic acid or a closely related substance is thought to be the transmitter of excitatory axospinous synapses at many locations in the forebrain (39). These receptors are widely distributed in the cerebral cortex, including the visual areas (40). It appears that NMDA receptors normally coexist postsynaptically with quisqualate or kainate receptors or both (41). The "non-NMDA" receptors are thought to mediate the classical excitatory postsynaptic potential that normally results from electrical stimulation of axospinous synapses (35). NMDA receptors, on the other hand, appear to be linked to a membrane channel that will pass Ca^{2+} ions. Dingledine (42) reported that activation of NMDA receptors leads to calcium flux only when the cell is concurrently depolarized. This voltage dependency is apparently due to a blockage of the NMDA channel by Mg^{2+} ions that is alleviated only when the membrane is depolarized sufficiently (43). Thus, calcium entry through channels linked to the NMDA receptor could specifically signal pre- and postsynaptic coactivation (44).

Recently Kleinschmidt *et al.* (45) have obtained results that suggest that NMDA receptor–mediated Ca^{2+} entry also contributes to the synapse modifications that underlie OD plasticity in striate cortex. Specifically, they have found that intracortical infusion of 2-amino-5-phosphonovaleric acid (APV), a selective antagonist of the NMDA receptor (46), prevents the OD shift that would normally occur after MD. Moreover, this pharmacological treatment also resulted in a striking loss of orientation selectivity and a decrease in visual responsiveness. These data support Singer's hypothesis that dendritic calcium entry is a crucial variable for synaptic modifications in the striate cortex (17).

Our theoretical analyses lead us to suggest several refinements of the Singer hypothesis. Specifically, we propose the following: (i) θ_M is the membrane potential at which NMDA-receptor activation by sensory fiber activity results in dendritic calcium entry; (ii) increased calcium flux across the dendritic spine membrane results specifically in an increase in the synaptic gain; and (iii) activated synapses accompanied by no postsynaptic calcium signals will be weakened over time.

This physiological model is consistent with the theory of Bienenstock *et al.* (26). According to the model, the value and sign of ϕ are determined by the Ca^{2+} movement into dendritic spines. Synaptic efficacy will increase when presynaptic activity evokes a large postsynaptic calcium signal ($\phi > 0$). This will occur only when the membrane potential exceeds the level required to open the NMDA receptor–activated calcium channels ($c > \theta_M$). When the amplitude of the evoked Ca^{2+} signal falls below a certain critical level, corresponding to $\phi = 0$ and $c = \theta_M$, then active synapses will be weakened over time. Application of an NMDA-receptor blocker theoretically would increase the value of θ_M, such that it would take a greater level of depolarization to achieve the critical calcium flux. In accordance with the experimental observations of Kleinschmidt *et al.* (45), the theoretical consequence would be a loss of orientation selectivity, a prevention of OD plasticity, and a decrease in visual responsiveness.

Changes in synaptic efficacy that depend on postsynaptic calcium have been observed directly in *Hermissenda* (47). In this invertebrate, a classically conditioned response will result from the repeated pairing of light (the conditioned stimulus) with rotation of the animal (the unconditioned stimulus). The crucial modification occurs at the level of the "type B" photoreceptor, which is both depolarized by light and synaptically activated by inputs from the vestibular system. The pairing of light with rotation depolarizes the cell beyond the threshold for Ca^{2+} entry. Elevated intracellular Ca^{2+} leads to a long-term change that leaves the cell more excitable to light than before conditioning. In this case, as in hippocampal LTP, postsynaptic calcium entry leads specifically to an increase in the efficacy of the active synapses.

Our physiological model makes no statement about the actual locus where the modification is stored. Calcium ion acts as an intracellular second messenger to activate a host of enzymes including protein kinases (48), phosphatases (49), and proteases (38). In *Hermissenda* the synaptic efficacy appears to be increased by the covalent modification of potassium channels in the postsynaptic membrane (47). The essential modification that underlies LTP in the hippocampus is still controversial. The candidates range from alterations in the morphology of dendritic spines (38) to changes in the amount of transmitter released presynaptically (50). The weakening of synapses whose activity is not coincident with postsynaptic calcium entry could be explained by receptor desensitization (51). Any of these changes could contribute to the experience-dependent modification of neuronal response properties in visual cortex.

However, the model does make some explicit predictions about the regulation of the calcium messenger system that is linked to the NMDA receptors on cortical dendrites. Recall that θ_M depends on the average activity of the cell. If the average activity decreases, as during binocular deprivation, then θ_M decreases and it should take less dendritic depolarization to maintain synaptic efficacy. One way this could occur in our model would be to alter the voltage or transmitter sensitivity of the NMDA receptors with the result that less synaptic activity (depolarization) would be required to evoke the necessary calcium signal. It is well documented that receptor supersensitivity occurs as a consequence of postsynaptic inactivity at many locations in the nervous system (52). Alternatively, a weak calcium signal could be amplified at points further downstream, for example, by increasing the activity of calcium-activated enzymes.

Dendritic spines obviously play an important role in this model. We speculate that these postsynaptic structures are specialized to isolate high levels of intracellular calcium. The morphological organization of spines appears to be ideally suited for this task. Most mature spines are physically separated from the dendrite by a narrow neck, and in many cases contain an organelle called the "spine apparatus," which is thought to be a type of endoplasmic reticulum that can sequester free Ca^{2+} (53). The length of the spine neck may be constantly changing in the living brain (54), but electron microscopic examination has led repeatedly to the conclusion that in the fixed brain the longest spine necks on cortical pyramidal cells are found at the ends of apical dendrites, whereas the shortest spine necks are a consistent feature of the part of the dendrite near the cell body (55).

The unusual morphology of dendritic spines raises some interesting questions with regard to the nature of NMDA receptor–mediated Ca^{2+} influx. Numerous modeling studies have shown that the high electrical resistance of the spine neck should amplify the depolarization evoked by synaptic activity within the spine head (56). Consequently, NMDA-activated Ca^{2+} entry should occur more readily in spines with longer necks (higher resistance). Synapses on a spine with high neck resistance might even be capable of evoking significant Ca^{2+} entry without concurrent dendritic depolarization. This raises the possibility that the modifiability of axospinous synapses might depend on spine shape. In this context, we note that total light deprivation leads to the development in visual cortex of truncated spines without a constricted neck region (57). A period of rearing in the dark is also known to leave kitten striate cortex unusually modifiable by visual experience (7).

Conclusions

We have presented an algorithm for synaptic modification that reproduces classical experimental results in visual cortex. These include the relation of cell tuning and response to various visual environments experienced during the critical period: normal rearing, binocular deprivation, monocular deprivation, and reverse suture. A molecular model for this form of modification has been proposed based on the NMDA receptors. In this model θ_M as proposed by Bienenstock *et al.* (26) is identified with the voltage-dependent unblocking of the NMDA receptor channels. A consequence of this relation is that the membrane potential at which Ca^{2+} enters through NMDA channels should vary depending on the history of prior cell activity.

Stated in this language, many questions become of obvious interest. Among these are: How long does it take θ_M to adjust to a new average firing rate? What is the molecular basis for this adjustment? How do the putative global modulators of cortical plasticity, such as acetylcholine and norepinephrine, interact with the second-messenger systems linked to NMDA receptors on cortical dendrites? Can we provide direct evidence that those cells that modify are or are not those acted on by the modulators? Are the known morphological features of dendritic spines causally related to the modifiability of synaptic strength? Do the same rules apply to reorganization in adults as apply in the developmental period?

There has been much discussion in recent years about possible modification of synapses between neurons as the physiological basis of learning and memory storage. Molecular models for learning at the single-synapse level have been presented (11, 17, 38, 47, 58), various learning algorithms have been proposed that show some indication of appropriate behavior (14, 24–26, 59), and a mathematical structure for networks of neurons is rapidly evolving (60). A concerted effort to unite these approaches has begun, and the close interaction between theory and experiment has greatly enriched both endeavors. Theory has been anchored to experimental observations and experiments have been focused onto those issues most relevant to sorting out the various possible hypotheses. Further, this interaction has enabled us to pose new questions with precision and clarity.

REFERENCES AND NOTES

1. S. M. Sherman and P. D. Spear, *Physiol. Rev.* **62**, 738 (1982); Y. Frégnac and M. Imbert, *ibid.* **64**, 325 (1984).
2. D. H. Hubel and T. N. Wiesel, *J. Physiol. (London)* **160**, 106 (1962).
3. ——, *ibid.* **206**, 419 (1970).
4. S. LeVay, T. N. Wiesel, D. H. Hubel, *J. Comp. Neurol.* **191**, 1 (1980).
5. C. J. Shatz and M. P. Stryker, *J. Physiol. (London)* **281**, 267 (1978).
6. J. Presson and B. Gordon, *Soc. Neurosci. Abstr.* **8**, 5.10 (1982).
7. G. D. Mower *et al.*, *J. Comp. Neurol.* **235**, 448 (1985).
8. L. Mioche and W. Singer, personal communication.
9. F. H. Duffy, S. R. Snodgrass, J. L. Burchfiel, J. L. Conway, *Nature (London)* **260**, 256 (1976).
10. T. N. Wiesel and D. H. Hubel, *J. Neurophysiol.* **28**, 1029 (1965).
11. G. S. Stent, *Proc. Natl. Acad. Sci. U.S.A.* **70**, 997 (1973).
12. W. Singer, J. Rauschecker, R. Werth, *Brain Res.* **134**, 568 (1977).
13. J. P. Rauschecker and W. Singer, *Nature (London)* **280**, 58 (1979).
14. D. O. Hebb, *The Organization of Behavior* (Wiley, New York, 1949).
15. T. Kasamatsu and J. D. Pettigrew, *J. Comp. Neurol.* **185**, 139 (1979); R. D. Freeman and A. B. Bonds, *Science* **206**, 1093 (1979); P. Buisseret and W. Singer, *Exp. Brain Res.* **51**, 443 (1983); M. F. Bear and W. Singer, *Nature (London)* **320**, 172 (1986).
16. W. Singer, in *The Neurosciences Fourth Study Program*, F. O. Schmitt and F. G. Worden, Eds. (MIT Press, Cambridge, MA, 1979), pp. 1093–1109.
17. W. Singer, in *The Neural and Molecular Bases of Learning*, J.-P. Changeux and M. Konishi, Eds. (Dahlem Konferenzen, Wiley, New York, 1987), pp. 301–336.
18. It is possible, however to condition single neurons in anesthetized preparations by pairing visual stimulation with the iontophoretic application of K^+ or a cocktail of putative facilitatory transmitters. See: Y. Frégnac, S. Thorpe, D. Shulz, E. Bienenstock, *Soc. Neurosci. Abstr.* **10**, 314.5 (1984); J. Greuel, H. Luhman, W. Singer, in preparation.
19. W. Singer and J. Rauschecker, *Exp. Brain Res.* **47**, 223 (1982).
20. H. Geiger and W. Singer, *Int. J. Dev. Neurosci. Suppl.* R328 (1984).
21. C. Blakemore, L. J. Garey, Z. B. Henderson, N. V. Swindale, F. Vital-Durand, *J. Physiol. (London)* **307**, 25P (1980).
22. M. D. Jacobson, H. O. Reiter, B. Chapman, M. P. Stryker, *Soc. Neurosci. Abstr.* **11**,

31.8 (1985). However, see also, J. M. Greuel, H. J. Luhmann, W. Singer, *Dev. Brain Res.*, in press.

23. J. Rauschecker and W. Singer, *Dev. Brain Res.* **4**, 495 (1982).

24. L. N Cooper, in *Proceedings of the Nobel Symposium on Collective Properties of Physical Systems*, B. Lindquist and S. Lindquist, Eds. (Academic Press, New York, 1973), vol. 24, pp. 252–264; M. M. Nass and L. N Cooper, *Biol. Cybern.* **19**, 1 (1975).

25. L. N Cooper, F. Liberman, E. Oja, *Biol. Cybern.* **33**, 9 (1979).

26. E. L. Bienenstock, L. N Cooper, P. W. Munro, *J. Neurosci.* **2**, 32 (1982).

27. When presynaptic fiber is spontaneously active, $d = 0$.

28. L. N Cooper and C. Scofield, in preparation.

29. W. Singer, *Brain Res.* **134**, 508 (1977).

30. M. F. Bear, D. M. Schmechel, F. F. Ebner, *J. Neurosci.* **5**, 1262 (1985); G. D. Mower, W. F. White, R. Rustad, *Brain Res.* **380**, 253 (1986).

31. M. Colonniér, in *The Organization of the Cerebral Cortex*, F. O. Schmitt, F. G. Worden, G. Adelman, S. G. Dennis, Eds. (MIT Press, Cambridge, MA, 1981), pp. 125–152.

32. T. Lømo, *Acta Physiol. Scand. Suppl.* **68**, 128 (1966); T. V. P. Bliss and T. Lømo, *J. Physiol. (London)* **232**, 331 (1973).

33. B. L. McNaughton, R. M. Douglas, G. V. Goddard, *Brain Res.* **157**, 277 (1978).

34. S. R. Kelso, A. H. Ganong, T. H. Brown, *Proc. Natl. Acad. Sci. U.S.A.* **83**, 5326 (1986).

35. G. L. Collingridge, S. L. Kehl, H. McLennan, *J. Physiol. (London)* **334**, 33 (1983); E. W. Harris, A. H. Ganong, C. W. Cotman, *Brain Res.* **323**, 132 (1984).

36. G. Lynch, J. Larson, S. Kelso, G. Barrionuevo, F. Schottler, *Nature (London)* **305**, 719 (1983).

37. H. Wigström and B. Gustafsson, *Acta Physiol. Scand.* **123**, 519 (1985); P. O. Andersen, in *The Neural and Molecular Bases of Learning*, J.-P. Changeux and M. Konishi, Eds. (Dahlem Konferenzen, Wiley, New York, 1987), pp. 239–262.

38. G. Lynch and M. Baudry, *Science* **224**, 1057 (1984).

39. C. W. Cotman, A. Foster, T. Lanthorn, in *Glutamate as a Neurotransmitter*, G. DiChiara and G. Gessa, Eds. (Raven, New York, 1981), pp. 1–27; D. G. Nicholls and T. S. Sihra, *Nature (London)* **321**, 772 (1986); T. P. Hicks, D. Lodge, H. McLennan, Eds., *Excitatory Amino Acid Transmission* (Liss, New York, 1987).

40. D. T. Monaghan and C. W. Cotman, *J. Neurosci.* **5**, 2909 (1985).

41. A. C. Foster and G. E. Fagg, *Brain Res. Rev.* **7**, 103 (1985). NMDA receptors are so named because of their high affinity for N-methyl-D-aspartate. Foster and Fagg have proposed a formal nomenclature for the acidic amino acid receptors: A1 for the "NMDA" receptor, A2 for the "quisqualate" receptor, and A3 for the "kainate" receptor. However, because this scheme has not yet been widely adopted, we use the term "NMDA receptor" in this article.

42. R. Dingledine, *J. Physiol. (London)* **343**, 385 (1983).

43. L. Nowak, P. Bregestovski, P. Ascher, A. Herbet, A. Prochiantz, *Nature (London)* **307**, 462 (1984).

44. Recent work indicates that NMDA receptors contribute significantly to the response generated by trains of presynaptic action potentials. This contribution is probably due to the temporal summation of fast excitatory postsynaptic potentials, which leads to a postsynaptic depolarization sufficient to relieve the Mg^{2+} block of the ionophore. This contribution of the NMDA receptor–mediated Ca^{2+} current to "normal" synaptic transmission in no way diminishes the importance of its novel role as a signal of pre- and postsynaptic coactivation. See: T. E. Salt, *Nature (London)* **322**, 263 (1986); C. E. Herron, R. A. J. Lester, E. J. Coan, G. L. Collingridge, *ibid.*, p. 265.

45. A. Kleinschmidt, M. F. Bear, W. Singer, *Neurosci. Lett. Suppl.* **26**, S58 (1986); in preparation.

46. J. C. Watkins and R. H. Evans, *Annu. Rev. Pharmacol. Toxicol.* **21**, 165 (1981).

47. D. L. Alkon, *Science* **226**, 1037 (1984).

48. E. J. Nestler, S. I. Walaas, P. Greengard, *ibid.* **225**, 1357 (1984).

49. A. A. Stewart, T. S. Ingebritsen, A. Manalan, C. B. Klee, P. Cohen, *FEBS Lett.* **137**, 80 (1982).

50. A. C. Dolphin, M. L. Errington, T. V. P. Bliss, *Nature (London)* **297**, 496 (1982); D. V. Agoston and U. Kuhnt, *Exp. Brain Res.* **62**, 663 (1986).

51. T. Heidmann and J.-P. Changeux, *C. R. Acad. Sci. Ser. C.* **285**, 665 (1982).

52. W. B. Cannon and A. Rosenblueth, *The Supersensitivity of Denervated Structures* (Macmillan, New York, 1949); J. R. Sporn, B. B. Wolfe, T. R. Harden, T. Kendall, P. B. Molinof, *Mol. Pharmacol.* **13**, 1170 (1977).

53. F. Joó, A. Mihály, A. Párducz, *Curr. Top. Res. Synapses* **1**, 119 (1984); E. Fifkova, J. A. Markham, R. J. Delay, *Brain Res.* **266**, 163 (1983).

54. F. Crick, *Trends Neurosci.* **5**, 44 (1982).

55. E. G. Jones and T. P. S. Powell, *J. Cell Sci.* **5**, 509 (1969); R. H. Laatsch and W. M. Cowan, *J. Comp. Neurol.* **128**, 359 (1966); F. F. Ebner and M. Colonniér, *ibid.* **160**, 51 (1975).

56. R. G. Coss and D. H. Perkel, *Behav. Neural Biol.* **44**, 151 (1985).

57. M. Friere, *J. Anat.* **126**, 193 (1978).

58. E. R. Kandel and J. H. Schwartz, *Science* **218**, 433 (1982).

59. C. Von der Marlsburg, *Kybernetik* **14**, 85 (1973); H. Frohn, H. Geiger, W. Singer, *Biol. Cybern.* **55**, 333 (1987).

60. T. Kohonen, *Associative Memory: A System Theoretical Approach* (Springer-Verlag, Berlin, 1977); J. J. Hopfield and D. W. Tank, *Science* **233**, 625 (1986).

61. Supported by U.S. Office of Naval Research contract N000-14-86-K-0041. We thank W. Singer, J. Greuel, and N. Intrator for valuable discussions.

129

Molecular Mechanisms for Synaptic Modification in the Visual Cortex: Interaction Between Theory and Experiment

(with M. F. Bear)

Neuroscience and Connectionist Theory, eds. M. Gluck and D. Rumelhart
(Lawrence Erlbaum Associates, Hillsdale, New Jersey, 1990) p. 65

In the following book chapter Mark and I try to summarize our thinking on the consequences of the BCM theory as well as various possible underlying mechanisms. Again, although some of the ideas are speculative and some have been superseded, they have provided valuable directions for further research while some have come to the forefront of current thinking on molecular mechanisms.

Neuroscience and Connectionist Theory, eds. M. Gluck and D. Rumelhart
© Lawrence Erlbaum Associates, New Jersey, 1990

2

Molecular Mechanisms for Synaptic Modification in the Visual Cortex: Interaction Between Theory and Experiment

Mark F. Bear and Leon N. Cooper
Center for Neural Science and Physics Department
Brown University

INTRODUCTION

As this volume attests, in the last several years we have witnessed an explosion of interest in computational neural network models of learning and memory. In each of these models information is stored in the "synaptic" coupling between vast arrays of converging inputs. Such distributed memories can be shown to display many properties of human memory: recognition, association, generalization, and resistance to the partial destruction of elements within the network. An interesting feature of these models is that the performance is constrained by the patterns of connectivity within the network. This reinforces the view, long held by neurobiologists, that an understanding of neural circuitry holds a key to elucidating brain function. Hence, modern neural network models attempt to incorporate the salient architectural features of the brain regions of interest. However, another crucial aspect of network function concerns the way that the synaptic junctions are modified to change their strength of coupling. Most models have assumed a form of modification based on Hebb's (1949) proposal that synaptic coupling increases when the activity of converging elements is coincident. Variations on this venerable "learning rule" have been enormously successful in simulations of various forms of animal learning. However, this work has also shown that just as network behavior depends on connectivity, the capabilities of the network vary profoundly with different modification rules. What forms of synaptic modification are most appropriate? Again, we must look to the brain for the answer.

Concurrent with the recent developments in neural network theories of

65

learning and memory has been the experimental demonstration of experience-dependent synaptic plasticity at the highest level of the mammalian nervous system, the cerebral neocortex. A neurobiological problem of extraordinary interest is to identify the molecular mechanisms that underlie this process of cortical modification. For the complex forms of plasticity evoked in neocortex by changes in the sensory environment, an essential first step in sorting out the various possibilities is to derive a set of rules that can adequately account for the observed modifications. These rules serve as a guide toward identifying candidate mechanisms that can then be tested experimentally. Hence, it can be seen that two lines of inquiry—one concerning neural network theory, the other concerning molecular mechanisms of synapse modification—converge at the level of the modification rule.

We have proposed such a modification rule to explain the rich body of experimental evidence available on the experience-dependent plasticity of the feline visual cortex during early postnatal development. This theoretical form of modification is able to account for the results of a wide variety of visual deprivation experiments, and has led to a number of predictions that appear to have been confirmed by more recent experiments. In this chapter we illustrate how this theory has interacted with experiment to suggest a possible molecular basis for synapse modification in the visual cortex.

ANALYSIS OF VISUAL CORTICAL PLASTICITY

The central visual pathway arises at the two retinae with the axonal projections of retinal ganglion cells into the optic nerves (Fig.2.1). The first central synaptic relay occurs in the lateral geniculate nucleus (LGN) of the dorsal thalamus. The lateral geniculate projects via the optic radiation to the primary visual cortex, otherwise known as striate cortex or area 17.

Neurons in the striate cortex of normal adult cats respond selectively to the visual presentation of oriented bars of light and most are activated by stimulation of either eye (Hubel & Wiesel, 1962). Both of these properties—orientation selectivity and binocularity—can be modified by visual experience during a critical period of early postnatal development that, in the cat, extends from approximately 3 weeks to 3 months of age (Frégnac & Imbert, 1984; Sherman & Spear, 1982). The problem of visual cortical plasticity can be divided into three parts. First, what controls the onset and duration of the critical period? The answer to this question is unknown at present, but some interesting possibilities include specific patterns of gene expression (Neve & Bear, 1988; Sur, Frost, & Hockfield, 1988) which may be under hormonal control (Daw, Sato, & Fox, 1988). Second, within the plastic period, what factors enable synaptic modification to proceed? This ques-

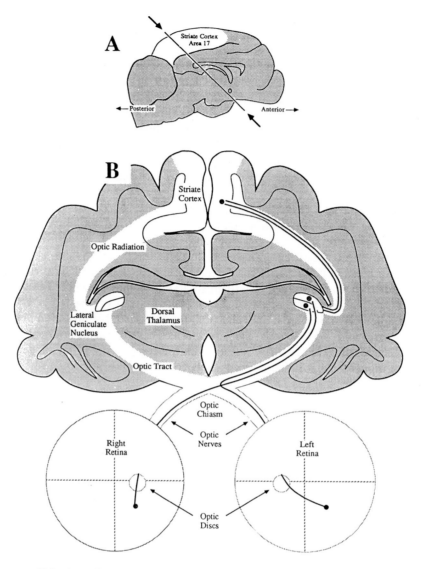

FIG. 2.1. The retino-geniculo-cortical pathway in the cat. (a) Mid-sagittal view of a cat brain. The primary visual cortex, also called striate cortex or area 17, lies on the medial wall of the postlateral gyrus. If the brain were cut in the plane indicated by the two arrows, the resulting cross-section would reveal the major components of the primary visual pathway, as illustrated in (b). Visual information arising at homotypic points in the two retinae (points viewing the same region of visual space) remains segregated in the lateral geniculate nucleus. Cortical neurons are responsive to stimulation of either eye due to the convergence of geniculocortical projections.

67

tion is prompted by the observation that experience-dependent modifications of visual cortex seem to require that animals attend to visual stimuli and use vision to guide behavior (Singer, 1979). The best candidates for "enabling factors" are the neuromodulators acetylcholine and norepinephrine that are released in visual cortex by fibers arising from neurons in the basal forebrain and brain stem (Bear & Singer, 1986). Third, when modifications are allowed to occur during the critical period, what controls their direction and magnitude? This is where the interaction between theory and experiment has been most fruitful, and this is the question we address in this chapter.

Kitten striate cortex, illustrated in Fig.2.2, is a well-differentiated structure with six layers and an intricate intracortical connectivity whose details remain only poorly understood (Martin, 1987). Nonetheless, it is known that the large majority of neurons in layers III, IV and VI receive direct monosynaptic input from the lateral geniculate nucleus (Ferster & Lindstrom, 1983; Martin, 1987; Toyama, Matsunami, Ohmo, & Tokashiki, 1974). The receptive fields of LGN neurons resemble those of retinal ganglion cells: they are monocular (i.e., they respond to stimulation of only one eye) and, for the most part, lack selectivity for stimulus orientation. Hence, cortical binocularity results from the convergence of lateral geniculate inputs onto cortical neurons. This convergence is not equal for every

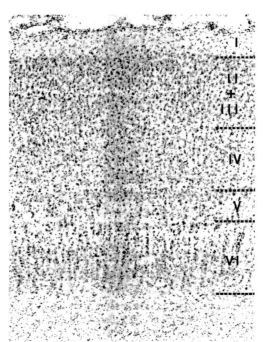

FIG. 2.2. Photomicrograph of a cross section of kitten area 17. This section was stained with Cresyl violet, which reveals the distribution of neuronal cell bodies. The six cortical layers are indicated. Magnification = 50x.

neuron and the term *ocular dominance* is used to describe the relative contribution of the two eyes to the cell's response. Although intracortical inhibition is acknowledged to play an important role in the refinement of orientation selectivity (Ramoa, Paradiso, & Freeman, 1988; Sillito et al., 1980), there is evidence that this property is also generated by the pattern of convergence of LGN inputs onto cortical neurons (Ferster, 1986; Hubel & Wiesel, 1962). Thus, in the first stage of the theoretical analysis, there is some justification for stripping away much of the complexity of the striate cortex, and considering a single cortical neuron receiving converging inputs from the two eyes via the LGN (Fig.2.1).

Cortical binocularity can be disrupted by a number of manipulations of visual experience during the critical period. For example, if the eyes are misaligned by severing one of the extraocular muscles, then cortical neurons lose their binocularity and become responsive only to one eye or the other (Blakemore & van Sluyters, 1974). Likewise, if one eye is deprived of patterned visual input (usually by suturing the eyelid closed), the ocular dominance of cortical neurons shifts such that most cells become responsive exclusively to stimulation of the open eye (Wiesel & Hubel, 1965). These changes in cortical binocularity can occur quite rapidly and are presumed to reflect the modification of the synaptic effectiveness of the converging inputs from the two eyes.

The consequences of binocular deprivation on visual cortex stand in striking contrast to those observed after monocular lid closure. First, binocular deprivation leads to a loss of orientation selectivity, an effect never seen after monocular deprivation. Second, although a week of monocular deprivation during the second postnatal month leaves few neurons in striate cortex responsive to stimulation of the deprived eye, most cells remain responsive to visual stimulation through either eye after a comparable period of binocular deprivation (Wiesel & Hubel, 1965). Thus, it is not merely the absence of patterned activity in the deprived geniculo-cortical projection that causes the decrease in synaptic efficacy after monocular deprivation.

Gunther Stent (1973) pointed out that one difference between monocular and binocular deprivation is that only in the former instance are cortical neurons active. This consideration led him to hypothesize that evoked postsynaptic activity is a necessary condition for synaptic modification in the striate cortex, and the sign of the change (+ or −) depends on the concurrent level of presynaptic input activity. Synaptic disconnection of afferent inputs deprived of patterned activity occurs only after monocular deprivation because only under these conditions are cortical neurons still driven by visual stimulation (through the open eye). Subsequent work suggested, however, that the generation of action potentials in a cortical neuron does not ensure that ocular dominance modifications will occur after monocular deprivation (Bear & Singer, 1986; Kasamatsu & Pettigrew, 1979; Singer, 1982). To reconcile these data with the Stent model, Wolf Singer (1979) introduced the idea that there is a critical level of postsynaptic activation that must be reached before experience-dependent modifications will occur, and that this threshold is higher than the depolarization required for somatic sodium-spikes. A similar type of modification rule has been proposed for the activity-dependent synaptic changes in the dentate gyrus (Levy, Colbert & Desmond, chap. 5 in this volume). According to this hypothesis, the "enabling factors" mentioned could be any inputs that render cortical neurons more excitable and hence more likely to exceed this "plasticity threshold" (Bear & Singer, 1986; Greuel, Luhman, & Singer, 1987).

This hypothesis is challenged by the finding that the effects of monocular deprivation can be rapidly reversed by opening the deprived eye and suturing closed the other eye. Such a "reverse suture" leads to a robust ocular dominance shift back to the newly opened eye, even though visually evoked postsynaptic activity is low or absent at the time of the reversal (because the only source of patterned visual input to cortical neurons is the functionally disconnected afferents from the unsutured eye). The effects of reducing cortical inhibition on ocular dominance modification are also difficult to explain by this hypothesis. Intracortical infusion of bicuculline, an antagonist of the inhibitory neurotransmitter γ-aminobutyric acid (GABA) that decreases orientation selectivity and generally increases cortical responsiveness, retards rather than facilitates the functional disconnection of the deprived eye after monocular deprivation (Ramoa et al., 1988).

Reiter and Stryker (1988) recently performed a direct test of the hypothesis that postsynaptic activation is simply permissive to the process of synaptic modification. They continuously infused muscimol into striate cortex as kittens were monocularly deprived for 7 days. Muscimol is a GABA receptor agonist that prohibits cortical neurons from firing, presumably by clamping the membrane near the chloride equilibrium potential. With the muscimol still present in cortex, they mapped the cortex to determine the extent of activity blockade. They found that all cortical cell responses were eliminated within several millimeters of the infusion cannula, even though LGN fiber activity was readily demonstrated. When the muscimol wore off, they measured the ocular dominance of neurons in the zone of cortex whose activity had been blocked. They observed an unexpected ocular dominance shift toward the deprived eye; that is, most neurons were no longer responsive to stimulation of the retina that had been more active during the period of monocular deprivation.

Although all experiments involving chronic intracortical drug infusion must be interpreted with extreme caution, these muscimol results seem to indicate that ocular dominance modifications can occur in the absence of evoked action potentials. Furthermore, the data suggest that patterned presynaptic activity can lead to either an increase or a decrease in synaptic strength, depending on whether or not the target neurons are allowed to respond.

An alternative theoretical solution to the problem of visual cortical plasticity, proposed by Cooper, Liberman, and Oja (1979), is able to account for these varied results. According to this theory, the synaptic efficacy of active inputs increases when the postsynaptic target is concurrently depolarized beyond a "modification threshold", θ_M. However, when the level of postsynaptic activity falls below θ_M, then the strength of active synapses decreases.

An important additional feature was added to this theory in 1982 by Bienenstock, Cooper, and Munro (BCM). They proposed that the value of the modification threshold is not fixed, but instead varies as a nonlinear function of the average output of the postsynaptic neuron. This feature provides the stability properties of the model, and is necessary to explain, for example, why the low level of postsynaptic activity caused by binocular deprivation does not drive the strengths of all cortical synapses to zero.

This form of synaptic modification can be written:

$$dm_j / dt = \phi (c, \bar{c}) d_j$$

where m_j is the efficacy of the jth LGN synapse onto a cortical neuron, d_j is the level of presynaptic activity of the jth LGN afferent, c is the level of activation of the postsynaptic neuron,[1] and $\bar{c}$ is the time average of postsynaptic neuronal activity (d_j and c are viewed as averages over about half a second; $\bar{c}$ is the average over a period that could be several hours). The crucial function, ϕ, is shown in Fig.2.3.

One significant feature of this model is the change of sign of ϕ at the modification threshold, θ_M. When the jth synapse is active ($d_j > 0$) and the level of postsynaptic activation exceeds the modification threshold ($c > \theta_M$), then the sign of the modification is positive and the strength of the synapse increases. However, when the jth synapse is active and the level of postsynaptic activation slips below the modification threshold ($c < \theta_M$), then the sign of the modification is negative and the strength of the synapse decreases. Thus, effective synapses are strengthened and ineffective synapses are weakened, where synaptic effectiveness is determined by whether or not the presynaptic pattern of activity is accompanied by the simulta-

---

[1]In a linear approximation, $c = \mathbf{m}^l \cdot \mathbf{d}^l + \mathbf{m}^r \cdot \mathbf{d}^r$, where $\mathbf{d}$ and $\mathbf{m}$ are vectors representing the total input activity and synaptic weight of the array of fibers carrying information from the left or right eyes.

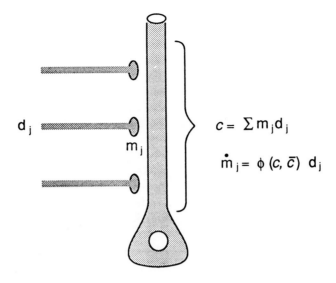

$$c = \sum m_j d_j$$

$$\dot{m}_j = \phi(c, \bar{c})\, d_j$$

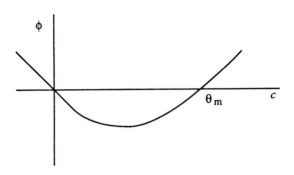

FIG. 2.3. TOP: Cartoon of a cortical neuron receiving LGN input to illustrate the BCM theoretical notation. BOTTOM: The BCM modification function.

neous depolarization of the target dendrite beyond the modification threshold.

According to this model, synaptic weakening requires that the postsynaptic membrane potential falls below the modification threshold. Thus, during monocular deprivation, the deprived-eye synapses will decrease in strength each time the open-eye input activity does not strongly depolarize the cortical neuron. This occurs when the input patterns conveyed by the open-eye afferents fail to match the stimulus selectivity of the neuron.

Therefore, the theory predicts a relationship between the ocular dominance shift after monocular deprivation and the degree of orientation tuning of cortical neurons.

The application of bicuculline to cortex, by reducing the stimulus selectivity of cortical neurons, increases the probability that the unstructured activity from the deprived eye correlates with cortical activation at or beyond the modification threshold. Therefore, in agreement with experimental results (Ramoa, Paradiso, & Freeman, 1988), the theory predicts that no synaptic disconnection of deprived-eye afferents would occur when cortex is disinhibited. On the other hand, muscimol treatment would suppress the postsynaptic response well below the modification threshold regardless of the afferent input. In accordance with experimental observations of Reiter and Stryker (1988) the theory would predict an ocular dominance bias toward the less active eye.

Another significant feature of this theory is that the value of the modification threshold (θ_M) is not fixed, but instead varies as a nonlinear function of the average output of the cell ($\bar{c}$). In a simple situation:

$$\theta_M = (\bar{c})^2$$

This feature allows neuronal responses to evolve to selective and stable "fixed points" (Bienenstock, Cooper, & Munro, 1982). However, more importantly in the context of the present discussion, it is this feature of the theory that accounts for the differences between monocular and binocular deprivation. Deprivation of patterned input leads to synaptic disconnection after monocular deprivation because open-eye input activity continues to drive cortical neurons sufficiently to maintain θ_M at a high value. However, because average cortical activity falls during binocular deprivation, the value of θ_M approaches zero (Fig.2.4). In this case, the unstructured input activity causes synaptic strengths to perform a "random walk" (Bienenstock et al., 1982). Consequently, the theory also predicts the loss of orientation selectivity that has been observed after binocular deprivation.

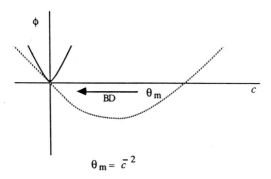

FIG. 2.4. A period of binocular deprivation (BD) decreases the value of the modification threshold, and therefore changes the shape of ϕ.

The sliding modification threshold also permits a theoretical explanation for the effects of reverse suture. The output of a cortical neuron in area 17 approaches zero just after the reversal because its only source of patterned input is through the eye whose synapses had been weakened as a consequence of the prior monocular deprivation. However as $\bar{c}$ diminishes, so does the value of θ_M. Eventually, the modification threshold attains a value below the small output that is evoked by stimulation of the unsutured eye, allowing these active synapses to increase in strength. If θ_M does not adjust to the new average firing rate too rapidly, the cell's response to the previously open eye will diminish before its response to the newly opened eye increases.

Analysis and computer simulations using this theoretical form of synaptic modification are able to reproduce the classical results of manipulating visual experience during the critical period (Bienenstock et al., 1982; Clothiaux, Bear, & Cooper, 1988). The theory can account for the acquisition of orientation selectivity with normal visual experience as well as the effects of monocular deprivation, binocular deprivation, and reverse suture. In addition, as we have seen, this form of modification offers a solution to the seemingly paradoxical effects of pharmacologically manipulating cortical activity during monocular deprivation. It is worthwhile to note that there need not be any modification of inhibitory circuitry to account for the experience-dependent modifications in striate cortex using this theory (Cooper & Scofield, 1988). This is reassuring because experimental efforts to uncover modifications of inhibitory circuits in visual cortex have consistently yielded negative results (Bear, Schmechel, & Ebner, 1985; Mower, White, & Rustad, 1987; Singer, 1977).

The success of this theory encourages us to ask whether this form of synaptic modification has a plausible neurobiological basis. The remainder of this chapter summarizes the progress we have made in answering this question (as of October 1988).

A Molecular Mechanism
for Increasing Synaptic Strength in Visual Cortex

According to the theory, synaptic strength increases when presynaptic inputs are active ($d > 0$) and the target dendrite is depolarized beyond the modification threshold ($c > \theta_M$). The relevant measure of input activity is likely to be the rate of transmitter release at the geniculocortical synapses. Although the exact identity of this transmitter substance is still not known with certainty, available evidence indicates strongly that it acts via excitatory amino acid receptors (Tsumoto, Masui, & Sato, 1986). This leads to the following question: When excitatory amino acid receptors are activated, what distinguishes the response at depolarized membrane potentials ($c > \theta_M$) from the response at the resting potential ($c < \theta_M$)?

As elsewhere, cortical excitatory amino acid receptors fall into two broad categories: NMDA and non-NMDA. Both types of excitatory amino acid receptor are thought to coexist subsynaptically (Fig.2.5). The ionic conductances activated by non-NMDA receptors at any instant depend only on the input activity, and are independent of the postsynaptic membrane potential. However, the ionic channels linked to NMDA receptors are blocked with Mg^{++} at the resting potential, and become effective only upon membrane depolarization (Mayer & Westbrook, 1987; Nowak et al., 1984). Another distinctive feature of the NMDA receptor channel is that it will conduct calcium ions (Dingledine, 1984; MacDermott et al., 1986). Hence, the

POSTSYNAPTIC RESPONSE TO GLUTAMATE

(1) AT RESTING MEMBRANE POTENTIAL

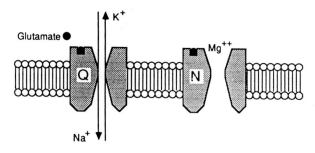

(2) AT DEPOLARIZED MEMBRANE POTENTIALS

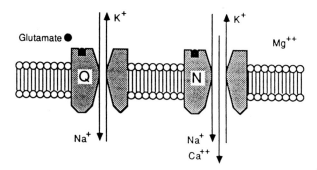

FIG. 2.5. Cartoon to illustrate the two types of excitatory amino acid (EAA) receptor and the ionic conductances they activate. Glutamate acts at all subtypes of EAA receptor and is likely to be the neurotransmitter at geniculocortical synapses. Note that the binding of glutamate to the NMDA (N) receptor activates a calcium conductance, but only when the Mg^{++} block is lifted at depolarized membrane potentials. Q is meant to represent the quisqualate subtype of EAA receptor.

passage of Ca^{++} through the NMDA channel could specifically signal when pre- and postsynaptic elements are concurrently active. These considerations have led us to propose that θ_M relates to the dendritic membrane depolarization at which presynaptic activity leads to a critical postsynaptic Ca^{++} flux, and that the Ca^{++}, acting as a second messenger, leads to an enhancement of synaptic strength (Bear, Cooper, & Ebner, 1987).

Data from *in vitro* brain slice experiments lend strong support to this hypothesis. Long-term potentiation (LTP) of synaptic effectiveness that normally results from tetanic afferent stimulation cannot be induced in either the CA1 subfield of the hippocampus (Collingridge, Kehl, & McLennan, 1983; Harris, Ganong, & Cotman, 1984) or the visual cortex of rats (Artola & Singer, 1987; Kimura, Tsumoto, Nishigori, & Shirokawa, 1988) and kittens (Connors & Bear, 1988) when NMDA receptors are blocked (see also Granger, Ambrose-Ingerson, Staubli, & Lynch, Chap. 3 in this volume). On the other hand, the application of N-methyl-D-aspartate (the selective agonist that gives the NMDA receptor its name) to hippocampal slices can induce a form of synaptic potentiation that can last for 30 minutes (Kauer, Malenka, & Nicoll, 1988) or longer (Thibault et al., 1988). The idea that elevations in postsynaptic Ca^{++} trigger the increase in synaptic strength is supported by the finding that intracellular injection of the Ca^{++} chelator EGTA, which essentially removes ionic calcium, blocks the induction LTP in CA1 pyramidal cells (Lynch et al., 1983). Furthermore, the intracellular release of Ca^{++} from the photolabile calcium chelator nitr-5 produces a long-lasting potentiation of synaptic transmission that resembles LTP (Malenka, Kauer, & Nicoll, 1988). Taken together, these data indicate that the calcium conductance mediated by the NMDA receptor plays a special role in strengthening synaptic relationships in the cortex.

The possible involvement of NMDA receptors in the experience-dependent modification of visual cortex was examined in a recent series of experiments carried out in Wolf Singer's laboratory (Kleinschmidt, Bear, & Singer, 1987). The selective NMDA receptor antagonist 2-amino-5-phosphonovaleric acid (APV) was infused continuously into striate cortex as kittens were monocularly deprived (Fig.2.6). After 7 days the APV treatment was stopped and the cortex 3-6 mm away from the infusion cannula was assayed electrophysiologically for changes in ocular dominance and orientation selectivity. The APV treatment was found to produce a concentration-dependent increase in the percentage of neurons with binocular, unoriented receptive fields (Fig.2.7). Qualitatively, the results resembled those expected in visual cortex after binocular deprivation. Yet, electrophysiological recordings during the week of APV infusion revealed that NMDA receptor blockade did not eliminate visual responsiveness in striate cortex.

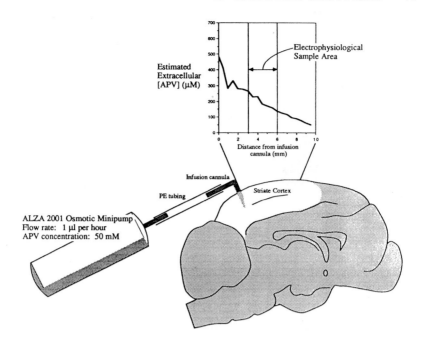

FIG. 2.6.   Method used by Kleinschmidt, Bear, and Singer (1987) to block cortical NMDA receptors *in vivo*. Osmotic minipumps were implanted that could deliver the NMDA receptor antagonist APV to the visual cortex at a rate of 50 nmol per hour for 7 days. As indicated, the extracellular APV concentration in the cortex varies as a function of distance from the infusion cannula. As a rule, sites were studied electrophysiologically between 3 and 6 mm from the infusion cannula where APV concentrations were approximately 200 μM.

According to the "NMDA hypothesis" (Bear et al., 1987) APV infusion should, in effect, raise the value of θ_M. If cortical neurons remain moderately responsive to visual stimulation, but are unable to achieve θ_M, then a theoretical consequence will be a modified "random walk".[2] This will result in a loss of orientation selectivity and a slow loss of synaptic efficacy—a result similar to that of binocular deprivation.

However, if the postsynaptic response is low, the predicted effect of NMDA receptor blockade during monocular deprivation is a loss of

---

[2]This effect requires that **d** have positive and negative components. According to Bienenstock, Cooper, and Munro (1982), when input fibers are spontaneously active, **d** = 0; when they carry noise (an effect of lid suture or dark rearing), **d** averages to zero. In a patterned input environment, **d** has positive and negative components, but the average is likely to be greater than zero.

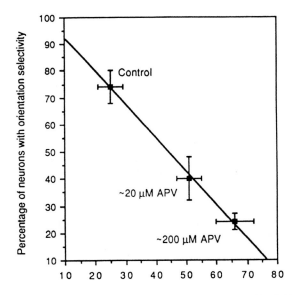

FIG. 2.7. The effects of NMDA receptor blockade on the response of visual cortex to monocular deprivation. Increasing extracellular concentrations of APV (estimated concentrations are indicated) increases the percentage of neurons with binocular, unoriented receptive fields. Data from "Blockade of 'NMDA' Receptors Disrupts Experience-Dependent Modifications of Kitten Striate Cortex" by A. Kleinschmidt, M. F. Bear, and W. Singer, 1987, *Science, 238.* Reprinted with permission.

synaptic strength at a rate proportional to the level of presynaptic activity. This could explain the observations of Reiter and Stryker (1988), assuming that the hyperpolarization produced by muscimol treatment renders cortical NMDA receptors ineffective. In a recent study, APV infusion was also found to have this effect (Bear et al., 1989). In this experiment, the cortex was studied at various distances from the infusion cannula after 2 days of monocular deprivation and APV treatment. The ocular dominance of units studied within 3 mm of the cannula, where APV concentrations are highest (Fig.2.6), was found to be strongly biased toward the deprived eye (Fig.2.8). Most neurons were binocular at sites ≥4 mm from the cannula.

To summarize, available experimental evidence supports the hypothesis that the consolidation or strengthening of at least some geniculocortical synapses depends on the activation of NMDA receptors. This molecular mechanism is consistent with the BCM theory assuming that NMDA receptors become sufficiently active to increase synaptic strength only

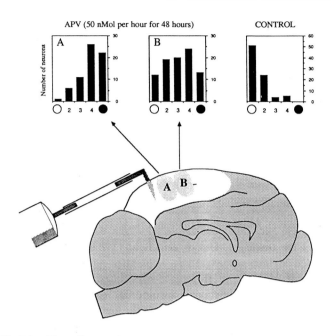

FIG. 2.8. The effects of high (APV) on the cortical response to monocular deprivation. At the top of this figure are ocular dominance histograms; bar height indicates the number of neurons in each ocular dominance category. Filled and open circles indicate the ocular dominance categories containing neurons responsive only to deprived-eye or open-eye stimulation, respectively. Neurons in category 3 were equally responsive to stimulation of either eye; neurons in categories 2 and 4 were binocularly activated, but showed a preference for stimulation of the open or deprived eyes, respectively. Note that within 3 mm of the infusion cannula (A), most neurons responded preferentially to stimulation of the eye that had been deprived of normal patterned vision.

when the postsynaptic target is depolarized beyond the modification threshold, θ_M.

A Molecular Mechanism
for Decreasing Synaptic Strength in Visual Cortex

The theory states that when the postsynaptic depolarization falls below the modification threshold then synaptic strengths decrease at a rate proportional to input activity. What signals input activity when the membrane is hyperpolarized and the NMDA channel is fully blocked with Mg^{++}?

Certainly the activity of non-NMDA receptors reflects the amount of transmitter release regardless of whether or not NMDA receptors are effective. This has inspired us to search for an intracellular second messenger other than Ca^{++} that depends solely on the activation of non-NMDA receptors. One possibility has been suggested by recent investigations of excitatory amino acid mediated phosphoinositide (PIns) turnover in the cerebral cortex (Fig.2.9). This work has shown that during a finite period of postnatal development, stimulation of rat hippocampus (Nicoletti et al., 1986) or neocortex (Dudek, Bowen, & Bear, 1988) with glutamate or ibotenate (but not NMDA) leads to the hydrolysis of phosphatidyl inositol-4,5 biphosphate to produce inositol triphosphate (IP_3) and diacyl glycerol (DG). Both IP_3 and DG function as intracellular second messengers (Berridge, 1984).

Of particular interest is the age-dependence of the excitatory amino acid stimulated PIns turnover. Dudek and Bear (1989) found very recently that in the kitten striate cortex, there is a striking correlation between the developmental changes in ibotenate-stimulated PIns hydrolysis and the susceptibility of visual cortex to monocular deprivation (Fig.2.10). It is difficult to resist the suggestion that this mechanism is likely to play a central role in the modification of cortical synapses during the critical period.

At present there is not a shred of evidence to indicate the nature of this role. However, the theory suggests one interesting possibility. Namely, that the stimulation of PIns hydrolysis by non-NMDA receptor activation leads to a decrease in synaptic strength. According to this hypothesis, changes in synaptic efficacy would result from changes in a balance between NMDA receptor mediated Ca^{++} entry and non-NMDA receptor mediated PIns turnover (Bear, 1988). Synaptic strength would increase when the NMDA

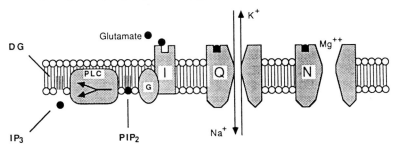

FIG. 2.9. Recently characterized EAA receptor site (I) that is linked to PIns turnover. The receptor is linked via a G protein to the enzyme phospholipase C (PLC) which hydrolyzes phosphatidyl inositol-4,5 biphosphate (PIP$_2$) to produce inositol triphosphate (IP$_3$) and diacyl glycerol (DG). Both IP$_3$ and DG function as intracellular second messengers.

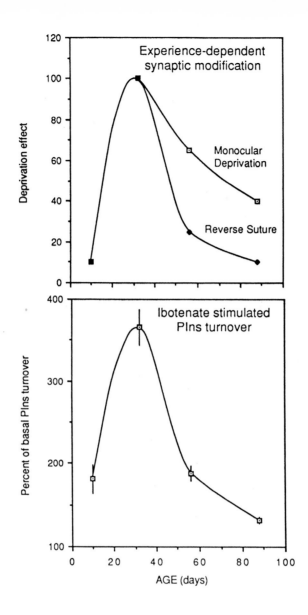

FIG. 2.10. TOP: Visual cortical plasticity as a function of age, estimated by Blakemore and van Sluyters (1974) using reverse suture and Olson and Freeman (1980) using monocular deprivation. BOTTOM: Postnatal changes in ibotenate stimulated PIns turnover in kitten striate cortex (from Dudek & Bear, submitted for publication).

81

signal exceeds the non-NMDA signal. This occurs when the input activity is coincident with strong depolarization ($c > \theta_M$). Synaptic strength would decrease when input activity consistently correlates with insufficient membrane depolarization ($c < \theta_M$) because the non-NMDA signal exceeds the NMDA signal.

Although this hypothesis was formulated purely on theoretical grounds (Bear, 1988), some recent work in Carl Cotman's laboratory supports the idea that the second messenger systems linked to NMDA and non-NMDA receptors might be antagonistic (Palmer et al., 1988). They find in the neonatal hippocampus that NMDA inhibits excitatory amino acid stimulated PIns turnover in a Ca^{++} dependent fashion.

To summarize, the theory states that synaptic strength decreases when input activity fails to coincide with postsynaptic depolarization beyond the modification threshold. In the previous section, we argued that this occurs when NMDA receptor activation falls below a critical level. Recent experimental evidence suggests that input activity in the absence of NMDA receptor activation is a favorable condition for the hydrolysis of membrane inositol phospholipids. These considerations lead us to suggest that phosphoinositide hydrolysis might provide the biochemical trigger for decreases in synaptic efficacy.

A Molecular Mechanism
for the Sliding Modification Threshold

A critical feature of this theory of synapse modification is that θ_M, the level of dendritic depolarization at which the sign of the synaptic modification changes, floats as a nonlinear function of average cell activity ($\bar{c}$). θ_M is "quasi-local," in the sense that it has the same value at all synapses on a given neuron (Bear, et al., 1987; Bienenstock et al., 1982). Thus, we search for a molecular mechanism that would provide a signal that is (a) uniformly available throughout the dendritic tree and (b) is regulated by average neuronal activity.

One mechanism that fits this description is the activity-dependent expression of specific neuronal genes (Black et al., 1987). Indeed, Neve and Bear (1988) recently demonstrated that visual experience can regulate gene expression in the kitten striate cortex. For example, the mRNA transcript for the neuronal growth associated protein GAP43 (Benowitz & Routtenberg, 1987) was found to be increased by rearing kittens in complete darkness. Moreover, this increase in GAP43 gene expression was reversed by only 12 hours of light exposure (Fig.2.11).

According to the molecular model developed so far, adjustments of the modification threshold conceivably could occur by changing the balance between the synaptic reward and punishment signals generated by the

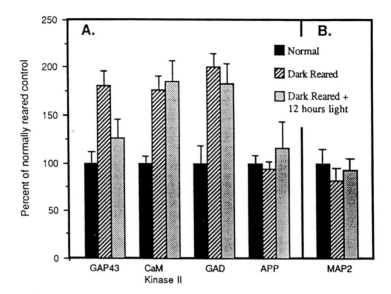

FIG. 2.11. Expression of growth associated protein GAP43, calcium calmodulin dependent protein kinase II (CaM kinase II), glutamic acid decarboxylase (GAD) and Alzheimer amyloid precursor protein (APP) genes in striate cortex of dark-reared and age-matched normal kittens (postnatal day 40–50). In order to reduce the variance, all individual values in (A) are normalized against those for the MAP2 gene, which did not change significantly under the conditions tested (B). Data from "Visual Experience Regulates Gene Expression in the Developing Striate Cortex" by R. L. Neve and M. F. Bear, 1988. *Proc Nat Acad Sci USA.*

NMDA and non-NMDA receptors, respectively. Therefore, we have focused our search on the products of activity-dependent gene expression that could potentially affect this balance. Calcium-calmodulin dependent protein kinase II (CaM kinase II) is one such molecule. CaM kinase II is a major constituent of the postsynaptic density, and is a critical link in the biochemical cascade of events that is triggered by Ca^{++} entry. It is not difficult to imagine how changes in the level of CaM kinase might alter the effectiveness of NMDA receptor mediated Ca^{++} signals. Indeed, in the striate cortex of dark-reared kittens, Neve and Bear (1988) found that the CaM kinase transcript is elevated over control levels. Similarly, Hendry and Kennedy (1988) found in primate visual cortex that immunoreactive CaM kinase II is increased in columns deprived of normal input after monocular deprivation.

Another way to change the balance between NMDA and non-NMDA receptors is to alter the effectiveness of the receptors in generating second messengers. For the NMDA receptor this could be accomplished in several ways, including changes in the properties of the ion channel (MacDonald, Salter, & Mody, 1988) or of the binding sites (Lodge, Aram, & Fletcher, 1988). Due to this abundance of potential regulatory mechanisms, we decided to address this issue using a functional assay of receptor effectiveness: NMDA-stimulated uptake of ^{45}Ca into slices of kitten visual cortex maintained *in vitro* (Sherin, Feldman, & Bear, 1988).

We predicted that under conditions where the modification threshold had a low value, slices of visual cortex should show heightened sensitivity to applied NMDA. One such condition is binocular deprivation. In Fig.2.12, the NMDA stimulated calcium uptake of slices from normal kittens is compared with that measured in slices from animals binocularly deprived for 4 days. There is a significant decrease in the maximum Ca^{++} uptake evoked by saturating concentrations of NMDA in slices from binocularly deprived kittens (Fig.2.12a). One simple explanation for this result is a decrease in the total number of NMDA receptors in binocularly deprived striate cortex, perhaps reflecting a global loss of synaptic strength. However, if this is the case, then uptake in slices from binocularly deprived animals should be lower at all concentrations of NMDA. Yet, at low concentrations (12.5-25 μM) the measured uptake is the same for normal and binocularly deprived cortex. Thus, it is possible that the NMDA receptors that remain in BD cortex might be relatively more effective in generating a calcium signal. This is illustrated in Fig.2.12b, where uptake from control and BD kittens is expressed as a percentage of maximal uptake. It is clearly premature to draw any firm conclusions from these data because calcium uptake depends on a complex interaction between NMDA and voltage-gated calcium entry, as well as on calcium extrusion and sequestration. Nonetheless, this work indicates that this question is worth exploring in more detail, perhaps now with receptor binding techniques.

Although still in its infancy, this work has already been able to show that the changes in average cortical activity produced by visual deprivation lead to alterations in gene expression and NMDA stimulated calcium uptake. Thus, the biological precedent for a sliding threshold mechanism in striate cortex is now established. Future work will be aimed at teasing out which changes are relevant for experience-dependent synaptic modification.

A molecular model that captures the essence of the BCM theory is presented in Fig.2.13. According to this model, the efficacy of an active synapse increases when the postsynaptic signal generated by NMDA receptor activation ("N," probably Ca^{++}) exceeds the signal produced by the activation of non-NMDA receptors ("Q," possibly a product of PIns turnover). This occurs when the summed postsynaptic depolarization ($\Sigma m_j d_j$) is greater than the modification threshold (θ_M). When the level of postsynaptic depolarization falls below the modification threshold, N < Q and the synapse weakens. Considered in this way, the modification

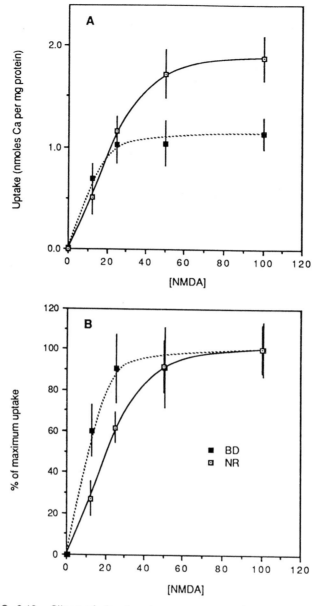

FIG. 2.12. Slices of visual cortex were prepared and maintained *in vitro,* and uptake of ^{45}Ca was monitored as different concentrations of NMDA were applied for 2 minutes. Shown here is NMDA stimulated ^{45}Ca uptake by slices of striate cortex from normally reared 4- to 6-week-old kittens and age-matched kittens that had been binocularly deprived for 4 days prior to sacrifice. In A uptake is expressed in nmoles per mg protein; in B uptake is expressed as the percentage of maximum uptake. Data from "NMDA-evoked Calcium Uptake by Slices of Visual Cortex Maintained In Vitro" by J. E. Sherin, D. Feldman, and M. F. Bear, 1988, *Neuroscience Abstracts,* 14.

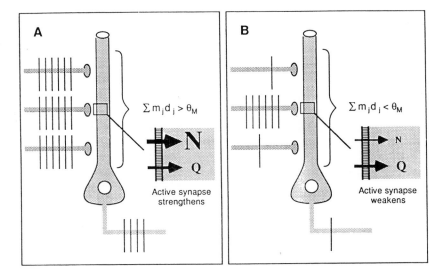

FIG. 2.13.   A molecular model for synapse modification in the striate cortex. According to this model, changes in the efficacy of an active synapse depend on the balance between postsynaptic signals linked to activation of NMDA ("N") and non-NMDA ("Q") receptors. See text for further explanation.

threshold becomes the critical level of postsynaptic depolarization at which the NMDA receptor dependent Ca^{++} flux is sufficient to balance the synaptic "punishment" produced by activation of non-NMDA receptors. We imagine that whether or not a given Ca^{++} flux is sufficient possibly depends on the availability of postsynaptic Ca^{++}-activated enzymes that, in turn, depends on the regulation of gene expression by average neuronal activity (Bear, 1988).

Generalization to a Many-neuron System

The BCM theory of synaptic modification deals with a single cortical neuron receiving input from the lateral geniculate nucleus only. The second stage of the theoretical analysis requires that relevant intracortical connections be incorporated into the model. Consider the simple network illustrated in Fig.2.14a in which cortical neurons (both excitatory and inhibitory) receive input from the LGN and from each other. The integrated output of the ith neuron (in the linear region) may be written

$$c_i = \mathbf{m}_i^l \cdot \mathbf{d}^l + \mathbf{m}_i^r \cdot \mathbf{d}^r + \Sigma L_{ij} c_j \qquad (3)$$

where the term $\Sigma L_{ij} c_j$ is the sum of the output from other cells in the network multiplied by the strength of their synapses on the ith cell. It is assumed that the intracortical synapses do not modify, or modify only slowly, and that the net influence of the intracortical connections is inhibitory.

Analysis of geniculocortical modification in this network leads to a very complex set of coupled nonlinear stochastic equations (Scofield & Cooper, 1985). However a mean-field approximation permits dramatic simplification of these equations (Cooper & Scofield, 1988). In a manner similar to the theory of magnetism, the individual effects of other cortical neurons are replaced by their average effect. The integrated output of the ith cortical neuron now becomes

$$c_i = (\mathbf{m}_i^1 - \alpha^1) \cdot \mathbf{d}^1 + (\mathbf{m}_i^r - \alpha^r) \cdot \mathbf{d}^r \qquad (4)$$

$$= (m_i - \alpha)d \qquad (5)$$

where $-\alpha$ represents the average inhibitory influence of the intracortical connections (Fig.2.14b).

There is an interesting theoretical consequence of assuming that each cortical neuron is under the influence of an inhibitory mean field. According to the BCM theory, monocular deprivation leads to convergence of geniculocortical synapses to a state where stimulation of the deprived eye input results in an output that equals zero ($c = 0$). However, with average network inhibition, the evolution of the cell to this state does not require that the efficacy of deprived-eye synapses be driven completely to zero. Instead, these excitatory synapses will evolve to a state where their influence is exactly offset by intracortical inhibition. Thus, the removal of intracortical inhibition in this network would reveal responses from otherwise ineffective inputs. This result is in accordance with the experimental observation of "unmasking" of synapses when the inhibitory effects of GABA are antagonized with bicuculline (Duffy, Snodgrass, Burchfiel, & Conway, 1976).

No revisions in the molecular model (Fig.2.13) are required to incorporate the mean field theory, although it is clear that the balance between NMDA and non-NMDA receptor activation will vary depending on network inhibition. α depends on the average connection strengths of intracortical synpases (L_o), which are assumed to not be modified, and the spatial average of the geniculocortical synapses "viewing" the same point in visual space, which changes only slowly (in comparison with the modification of m_j). Hence, in simulations of the evolution of the cortical network, α remains relatively constant from iteration to iteration.

However, it is interesting to note that if the value of α were to vary as a

88

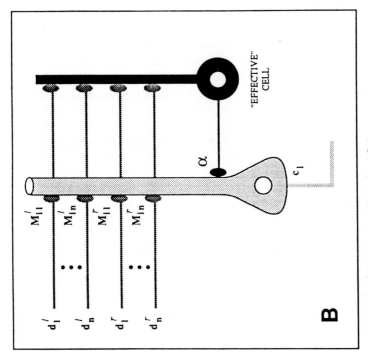

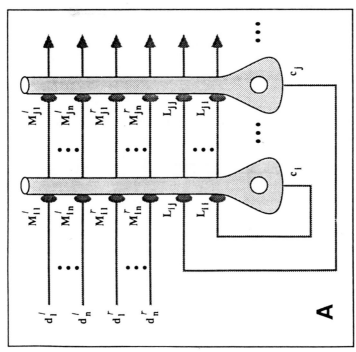

FIG. 2.14. A neural network in which every neuron receives inputs from the LGN and from each other. In B, using a mean field approximation, all other cortical neurons are replaced by an "effective cell," and the individual effects of the intracortical connections are replaced by their average effect (α), assumed to be inhibitory.

function of the timing of coherent inputs, the model could account for the changes in hippocampal synapses induced by patterned electrical stimulation (see Granger, Ambros-Ingerson, Staubli & Lynch, chap. 3 in this volume). Input activity coincident with $\alpha = 0$ (which, according to Larson and Lynch [1986], occurs when hippocampal inputs are stimulated at theta frequency) would be more likely to depolarize the neuron beyond θ_M and consequently would increase synaptic strength (Fig.2.15a). Conversely, input activity patterned in such a way as to coincide with strong inhibition ($\alpha > > 0$) should yield a depression of synaptic strength (Fig.2.15b). Such an effect has been reported very recently by Stanton and Sejnowski (1988) in hippocampus.

CONCLUDING REMARKS

We have presented a theoretical model for synaptic modification that can explain the results of normal rearing and various deprivation experiments in visual cortex. Further, we have shown that crucial concepts of the theory have a plausible molecular basis. Although some of this work is in a preliminary state, it provides an excellent illustration of the benefit of the interaction of theory with experiment.

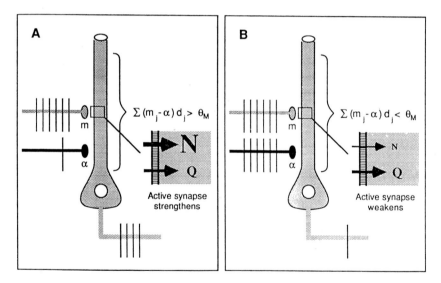

FIG. 2.15. Incorporation of inhibition into the model of Fig. 2.13. When input activity is coincident with strong inhibition (B), the balance of postsynaptic second messenger signals favors the non-NMDA receptors, and the active geniculocortical synapses weaken.

Theory enables us to follow a long chain of arguments and to connect, in a fairly precise way, various hypotheses with their consequences. It forces us to refine our language so that questions can be formulated with clarity and precision. Experiment focuses our attention on what is real; it separates what might be from what is; it tells us what must be explained and what is possible among explanations.

The theoretician who develops his arguments with close attention to the experimental results may thereby create a concrete structure of sufficient clarity so that new questions, of great interest and amenable to experimental verification, become apparent. The sliding threshold provides an excellent example. The concept of the modification threshold was introduced to account for such classical results as the development of neuron selectivity in normal visual environments and the various deprivation experiments. This led to unexpected theoretical consequences such as the correlation between ocular dominance and selectivity (now supported experimentally) and is sufficient to explain the results of the various pharmacological experiments.

Once convinced of the utility of this concept, the question of its physical basis became of great interest. This led us to the efforts concerning NMDA receptors, phosphoinositide turnover and regulation of gene expression in visual cortex (on a grander scale, Gregor Mendel's concept of the gene, introduced to explain the color of the sweet peas in his garden, was sufficiently attractive to provoke the activity that finally resulted in our present understanding of gene structure). And, as is almost always the case for an idea of richness, when the physical basis of the abstract concept is finally delineated, it contains a wealth of detail, subtlety, and possibility for manipulation that would have been not only impossible but ludicrous as part of the original proposal.

ACKNOWLEDGMENTS

The work on which this article is based was supported in part by the Office of Naval Research, the National Eye Institute, and the Alfred P. Sloan Foundation

REFERENCES

Artola, A., & Singer, W. (1987). Long-term potentiation and NMDA receptors in rat visual cortex. *Nature, 330,* 649–652.

Bear, M. F. (1988) Involvement of excitatory amino acid receptors in the experience-dependent development of visual cortex. In E. A. Cavalheiro, J. Lehman, & L. Turski (Eds.), *Recent advances in excitatory amino acid research.* New York: Alan R. Liss.

Bear, M. F., Cooper, L. N., & Ebner, F. F. (1987). A physiological basis for a theory of synapse modification. *Science, 237,* 42–48.

Bear, M. F., Kleinschmidt, A., Gu, Q., & Singer, W. (1989). Disruption of experience-dependent synaptic modifications in striate cortex by infusion of an NMDA receptor antagonist. Manuscript submitted for publication.

Bear, M. F., Schmechel, D. E., & Ebner, F. F. (1985). Glutamic acid decarboxylase in the striate cortex of normal and monocularly deprived kittens. *Journal of Neuroscience, 5,* 1262-1275.

Bear, M. F., & Singer, W. (1986). Modulation of visual cortical plasticity by acetylcholine and noradrenaline. *Nature, 320,* 172-176.

Benowitz, L. I., & Routtenberg, A. (1987). A membrane phosphoprotein associated with neural development, axonal regeneration, phospholipid metabolism, and synaptic plasticity. *Trends in Neuroscience 10,* 527-532.

Berridge, M. J. (1984). Inositol trisphosphate and diacylglycerol as second messengers. *Biochemical Journal, 220,* 345-360.

Bienenstock, E. L., Cooper, L. N., & Munro, P. W., (1982). Theory for the development of neuron selectivity: Orientation specificity and binocular interaction in visual cortex. *Journal of Neuroscience, 2,* 32-48.

Black, I. B., Adler, J. E., Dreyfus, C. F., Friedman, W. F., LaGamma, E. F., & Roach, A. H. (1987) Biochemistry of information storage in the nervous system. *Science, 236,* 1263-1268.

Blakemore, C., & van Sluyters, R. C. (1974). Experimental analysis of amblyopia and strabismus. *British Journal of Ophthalmology, 58;*176-182.

Clothiaux, E., Bear, M. F., & Cooper, L. N (1988). Experience-dependent synaptic modifications in the visual cortex studied using a neural network model. *Neuroscience Abstracts 14,* 81.13.

Collingridge, G. L., Kehl, S. L., & McLennan, H. (1983). Excitatory amino acids in synaptic transmission in the Schaffer collateral-commisural pathway of the rat hippocampus. *Journal of Physiology, 334,*33-46.

Connors, B. W., & Bear, M. F. (1988). Pharmacological modulation of long term potentiation in slices of visual cortex. *Neuroscience Abstracts, 14,* 298.8.

Cooper L. N, Lieberman, F., & Oja, E. (1979). A theory for the acquisition and loss of neuron specificity in visual cortex. *Biological Cybernetics, 33,* 9-28.

Cooper L. N., & Scofield, C. L. (1988). Mean-field theory of a neural network. *Proceedings of the National Academy of Sciences, USA, 85,* 1973-1977.

Daw, N. W., Sato, H., & Fox, K. (1988). Effect of cortisol on plasticity in the cat visual cortex. *Neuroscience Abstracts, 14,* 81.11.

Dingledine, R. (1983). N-methylaspartate activates voltage-dependent calcium conductance in rat hippocampal pyramidal cells. *Journal of Physiology 343,* 385-405.

Dudek, S. M. & Bear, M. F. (1989). Ibotenate-stimulated phosphoinositide turnover: A biochemical correlate of the critical period for synaptic modification in visual cortex. Manuscript submitted for publication.

Dudek, S. M., Bowen, W. D., & Bear, M. F. (1988). Postnatal changes in glutamate stimulated PI turnover in rat neocortical synaptoneurosomes. *Developmental Brain Research, 47,* 123-128.

Duffy, F. H., Snodgrass, S. R., Burchfiel, J. L., & Conway, J. L. (1976). Bicuculline reversal of deprivation amblyopia in the cat. *Nature, 260,* 256-257.

Ferster, D. (1986). Orientation selectivity of synaptic potentials in neurons of cat primary visual cortex. *Journal of Neuroscience, 6,* 1284-1301.

Ferster, D., & Lindström, S. (1983). An intracellular analysis of geniculo-cortical connectivity in area 17 of the cat. *Journal of Physiology, 342,* 181-215.

Frégnac, Y., & Imbert, M. (1984). Development of neuronal selectivity in the primary visual cortex of the cat. *Physiological Review, 64,* 325-434.

Greuel, J. M., Luhman, H. J., & Singer, W. (1987). Evidence for a threshold in experience-dependent long-term changes of kitten visual cortex. *Developmental Brain Research, 34,* 141–149.

Harris, E. W., Ganong, A. H., & Cotman, C. W. (1984). Long-term potentiation involves activation of N-methyl D-aspartate receptors. *Brain Research, 323,* 132–137.

Hebb, D. O. (1949). *The organization of behavior.* New York: Wiley.

Hendry, S. H., & Kennedy, M. B. (1986). Immunoreactivity for a calmodulin-dependent protein kinase is selectively increased in macaque striate cortex after monocular deprivation. *Proceedings of the National Academy of Sciences USA, 83,* 1536–1541.

Hubel, D. H., & Wiesel, T. N. (1962). Receptive fields, binocular interactions and functional architecture in the cat's visual cortex. *Journal of Physiology (Lon.), 160,* 106–154.

Hubel, D. H., & Wiesel, T. N. (1970). The period of susceptibility to the physiological effects of unilateral eye closure in kittens. *Journal of Physiology (Lon.) 206;* 419–436.

Kasamatsu, T., & Pettigrew, J. D. (1979). Preservation of binocularity after monocular deprivation in the striate cortex of kittens treated with 6-hydroxydopamine. *Journal of Comparative Neurology 185,* 139–162.

Kauer, J. A., Malenka, R. C., & Nicoll, R. A. (1988). NMDA application potentiates synaptic transmission in the hippocampus. *Nature, 334,* 250–252.

Kimura, F., Tsumoto, T., Nishigori, A., & Shirokawa, T. (1988). Long-term synaptic potentiation and NMDA receptors in the rat pup visual cortex. *Neuroscience Abstracts 14,* 81.10.

Kleinschmidt, A., Bear, M. F., & Singer, W. (1987). Blockade of "NMDA" receptors disrupts experience-dependent modifications of kitten striate cortex. *Science, 238,* 355–358.

Larson, J., & Lynch, G. (1986). Synaptic potentiation in hippocampus by patterned stimulation involves two events. *Science 232,* 985–988.

Lodge D., Aram J. A., & Fletcher, E. J. (1988). Modulation of N-methylaspartate receptor-channel complexes: An overview. In E. A. Cavalheiro, J. Lehman, & L. Turski (Eds.), *Recent advances in excitatory amino acid research.* New York: Alan R. Liss.

Lynch, G., Larson, J., Kelso, S., Barrionuevo S., & Schottler, F. (1983). Intracellular injections of EGTA block induction of hippocampal long-term potentiation. *Nature, 305,* 719–721.

MacDermott, A. B., Mayer, M. L., Westbrook, G. L., Smith, S. J., & Barker, J. L. (1986). NMDA receptor activation increases cytoplasmic calcium concentration in cultured spinal cord neurones. *Nature, 321,* 519–522.

MacDonald, J. F., Salter , M. W., & Mody, I. (1988). Intracellular regulation of the NMDA receptor. In E. A. Cavalheiro, J. Lehman, & L. Turski (Eds.), *Recent advances in excitatory amino acid research.* New York: Alan R. Liss.

Malenka, R. C., Kauer, J. A., & Nicoll, R. A. (1988). Postsynaptic calcium is sufficient for potentiation of hippocampal synaptic transmission. *Science, 242,* 81–84

Martin, K. A. C. (1987). Neuronal circuits in cat striate cortex. In E. G. Jones & A. Peters (Eds,), *The cerebral cortex.* New York: Plenum Press.

Mayer, M. L., & Westbrook, G. L. (1987). The physiology of excitatory amino acids in the vertebrate central nervous system. *Prog. Neurobiol, 28,* 197–276.

Mower, G. D., White, W. F., & Rustad, R. (1986). [³H] Muscimol binding of GABA receptors in the visual cortex of normal and monocularly deprived cats. *Brain Research, 380,* 253–260.

Neve, R. L., & Bear, M. F. (in press). Visual experience regulates gene expression in the developing striate cortex. *Proceedings of the National Academy of Sciences USA.*

Nicoletti, F., Iadarola, M. J., Wroblewski, J. T., & Costa, E. (1986). Excitatory amino acid recognition sites coupled with inositol phospholipid metabolism: Developmental changes and interaction with α1-adrenoreceptors. *Proceedings of the National Academy of Sciences USA, 83,* 1931–1935.

Nowak, L., Bregostovski, P., Ascher, P., Herbert, A., & Prochiantz, A. (1984). Magnesium gates glutamate-activated channels in mouse central neurones. *Nature, 307,* 462-465.

Olson, C. R., & Freeman, R. D. (1980). Profile of the sensitive period for monocular deprivation in kittens. *Experimental Brain Research, 39,* 17-21.

Palmer, E., Monaghan, D. T., Kahle, J., & Cotman, C. W. (1988). Bidirectional regulation of phosphoinositide metabolism by glutamate receptors: Stimulation by two classes of QA receptors and inhibition by NMDA receptors. *Neuroscience Abstracts, 14,* 37.16.

Ramoa, A. S., Paradiso, M. A., & Freeman, R. D. (1988). Blockade of intracortical inhibition in kitten striate cortex: Effects on receptive field properties and associated loss of ocular dominance plasticity. *Experimental Brain Research, 73,* 285-296.

Reiter, H. O., & Stryker, M. P. (1988). Neural plasticity without action potentials: Less active inputs become dominant when kitten visual cortical cells are pharmacologically inhibited. *Proceedings of the National Academy of Sciences USA, 85,* 3623-3627.

Scofield, C. L., & Cooper, L. N. (1985). Development and properties of neural networks. *Contemporary Physics 26,* 125-145.

Sherin, J. E., Feldman, D., & Bear, M. F. (1988). NMDA-evoked calcium uptake by slices of visual cortex maintained in vitro. *Neuroscience Abstracts, 14,* 298.7.

Sherman, S. M., & Spear, P. D. (1982). Organization in visual pathways in normal and visually deprived cats. *Physiological Review, 62,* 738-855.

Sillito, A. M., Kemp, J. A., Milson, J. A., & Berardi, N. (1980). A re-evaluation of the mechanisms underlying simple cell orientation selectivity. *Brain Research, 194,* 517-520.

Singer, W. (1977). Effects of monocular deprivation on excitatory and inhibitory pathways in cat striate cortex. *Experimental Brain Research, 134,* 508-518.

Singer, W. (1979). Central core control of visual cortex functions. In F. O. Schmitt & F. G. Worden (Eds.), *The Neurosciences Fourth Study Program* (pp. 1093-1109). Cambridge, MA: MIT Press.

Singer, W. (1982). Central core control of developmental plasticity in the kitten visual cortex: I. Diencephalic lesions. *Experimental Brain Research, 47,* 209-222.

Stanton, P. K., & Sejnowski, J. J. (in press). Associative long-term depression in the hippocampus: Induction of synaptic plasticity by Hebbion Covariance. *Nature.*

Stent, G. S. (1973). A physiological mechanism for Hebb's postulate of learning. *Proceedings of the National Academy of Sciences USA, 70,* 997-1001.

Sur, M., Frost, D. O., & Hockfield, S. (1988). Expression of a surface-associated antigen on Y-cells in the cat lateral geniculate nucleus is regulated by visual experience. *Journal of Neuroscience, 8,* 874-882.

Thibault, O., Joly, M., Müller, D., Schottler, F., Dudek, S., & Lynch, G. (1989). Long-lasting physiological effects of bath applied N-methyl-D-aspartate. *Brain Research, 476,* 170-173.

Toyama, K. K., Matsunami, Ohno, T., & Tokashiki, S. (1974). An intracellular study of neuronal organization in the visual cortex. *Experimental Brain Research, 21,* 45-66.

Tsumoto, T., Masui, H., & Sato, H. (1986). Excitatory amino acid neurotransmitters in neuronal circuits of the cat visual cortex. *Journal of Neurophysiology, 55,* 469-483.

Wiesel, T. N., & Hubel, D. H. (1965). Comparison of the effects of unilateral and bilateral eye closure on cortical unit responses in kittens. *Journal of Neurophysiology, 28,* 1029-1040.

Synaptic Plasticity in Visual Cortex: Comparison of Theory with Experiment

(with E. E. Clothiaux and M. F. Bear)

J. Neurophysiol. **66**, 1785 (1991)

People grow older but computers seem to grow younger. New shiny powerful machines replace previous shiny models before they have a chance to grow dull. Our laboratory now had workstations capable of running simulations out of the question several years before. So it became possible to test our theoretical ideas and compare them with experiment in a detailed manner that would have been totally impractical previously.

This was the project we began when Eugene Clothiaux joined us as a graduate student. About Eugene there are many stories. One concerns his habit of working all night and sleeping through most of the morning. Nothing extraordinary except that he often slept on a little couch in our laboratory. The project we concocted to lift the couch while he was asleep and transport it to another room never materialized.

In the following paper we attempt, using the BCM theory with a fixed set of parameters, to reproduce both the kinetics and the equilibrium states of the experience dependent modifications seen in kitten visual cortex. We restricted ourselves to what we call "the classical experiments", those that involved manipulations without the addition of pharmacological agents.

One of the more interesting experiments we try to explain is that of reverse suture. This manipulation seems most clearly to require a moving threshold. In fact it seems to give a direct measure of how rapidly this threshold moves.

In the course of our analysis we came face-to-face with the fact that we needed a more detailed knowledge of how the threshold moves. At that time we had several theoretical possibilities (including the one mentioned in the paper with Nathan Intrator later). Further, there is no real reason that the threshold should increase at the same rate as it decreases (since these very likely are produced by different underlying mechanisms). However to avoid an explosion of possibilities, we tried to keep things as simple as possible.

In the end we accomplished what we were after; but, in retrospect, more complexity might have been the wiser course.

JOURNAL OF NEUROPHYSIOLOGY
Vol. 66, No. 5, November 1991. *Printed in U.S.A.*

Synaptic Plasticity in Visual Cortex: Comparison of Theory With Experiment

EUGENE E. CLOTHIAUX, MARK F. BEAR, AND LEON N. COOPER

Physics Department and Center for Neural Science, Brown University, Providence, Rhode Island 02912

SUMMARY AND CONCLUSIONS

1. The aim of this work was to assess whether a form of synaptic modification based on the theory of Bienenstock, Cooper, and Munro (BCM) can, with a fixed set of parameters, reproduce both the kinetics and equilibrium states of experience-dependent modifications that have been observed experimentally in kitten striate cortex.

2. According to the BCM theory, the connection strength of excitatory geniculocortical synapses varies as the product of a measure of input activity (d) and a function (ϕ) of the summed postsynaptic response. For all postsynaptic responses greater than spontaneous but less than a critical value called the "modification threshold" (θ), ϕ has a negative value. For all postsynaptic responses greater than θ, ϕ has a positive value. A novel feature of the BCM theory is that the value of θ is not fixed, but rather "slides" as a nonlinear function of the average postsynaptic response.

3. This theory permits precise specification of theoretical equivalents of experimental situations, allowing detailed, quantitative comparisons of theory with experiment. Such comparisons were carried out here in a series of computer simulations.

4. Simulations are performed by presenting input to a model cortical neuron, calculating the summed postsynaptic response, and then changing the synaptic weights according to the BCM theory. This process is repeated until the synaptic weights reach an equilibrium state.

5. Two types of geniculocortical input are simulated: "pattern" and "noise." Pattern input is assumed to correspond to the type of input that arises when a visual contour of a particular orientation is presented to the retina. This type of input is said to be "correlated" when the two sets of geniculocortical fibers relaying information from the two eyes convey the same patterns at the same time. Noise input is assumed to correspond to the type of input that arises in the absence of visual contours and, by definition, is uncorrelated.

6. By varying the types of input available to the two sets of geniculocortical synapses, we simulate the following types of visual experience: *1*) normal binocular contour vision, *2*) monocular deprivation, *3*) reverse suture, *4*) strabismus, *5*) binocular deprivation, and *6*) normal contour vision after a period of monocular deprivation.

7. The constraints placed on the set of parameters by each type of simulated visual environment, and the effects that such constraints have on the evolution of the synaptic weights, are investigated in detail.

8. It was discovered that the exact form and dependencies of the sliding modification threshold are critical in obtaining a set of simulations that are consistent with the experimentally observed kinetics of synaptic modification in visual cortex. In particular, to account for observed changes during reverse suture and binocular deprivation, the value of θ could approach zero only when the synaptic strengths were driven to very low values. In the present model, this was achieved by including in the calculation of θ the postsynaptic responses generated by spontaneous input activity.

9. It is concluded that the modification of excitatory geniculocortical synapses according to rules derived from the BCM theory can account for both the outcome and kinetics of experience-dependent synaptic plasticity in kitten striate cortex. The understanding that this theory provides should be useful for the design of neurophysiological experiments aimed at elucidating the molecular mechanisms in play during the modification of visual cortex by experience.

INTRODUCTION

Experiments performed over the last three decades indicate that the response properties of neurons in striate cortex of the cat can be modified by manipulating the visual experience of the animal during a critical period of postnatal development. Although these experiments do not provide detailed information about the molecular mechanisms in play during synapse modification, they can shed light on the dynamics of synaptic change that these mechanisms produce. A theory that can account for the observed dynamics may yield vital insight in the search for the underlying molecular mechanisms. In addition, the understanding such a theory might provide could make possible the use of visual cortex as a preparation for the study of various complex interactions between neurotransmitters and receptors that lead to learning and memory storage.

Such a theory has been developed in our laboratory to account for the wide variety of experience-dependent modifications that have been observed in kitten visual cortex (reviewed by Bear et al. 1987). Originally, Nass and Cooper (1975) explored a theory in which the modification of visual cortical synapses was purely Hebbian; that is, a change to a synapse was based on the multiplication of the pre- and postsynaptic activities, and stabilization of the synaptic weights was produced by limiting modification to cortical responses below a maximum. Cooper et al. (1979) incorporated the idea that the sign of the modification should be based on whether the postsynaptic response is above or below a threshold. Responses above the threshold lead to strengthening of the active synapses; responses below the threshold lead to weakening of the active synapses. To stabilize the synapses without having to impose external constraints on them, θ was allowed to slide as a nonlinear function of the recent history of the cell's postsynaptic response by Bienenstock et al. (1982) (referred to as BCM theory). BCM is a "single-cell" theory, according to which modifications occur at the synapses of fibers from the lateral geniculate nucleus (LGN) onto a single cortical neuron. Scofield and Cooper (1985) extended this to a network of interconnected neurons, such as that in kitten striate cortex. To

1785

incorporate the finding that some synapses may be more resistant to change than others (e.g., Singer 1977), they introduce two types of cells into the network: cells with modifiable synapses and cells with nonmodifiable synapses. The fully connected network was later simplified by Cooper and Scofield (1988) with the introduction of a mean-field theory, which in effect replaces all of the individual intra-cortical connections to a cortical neuron from every other cell in the network by one set of "effective synapses" that conveys to the cell the average activity of all of the other cells in the network. Cooper and Scofield (1988) showed that the evolution of LGN-cortical synapses is similar for either a single neuron receiving only LGN input or a neuron embedded in a mean-field network.

Several other theoretical attempts have been made to model various aspects of synaptic plasticity in visual cortex (e.g., Linsker 1986a–c; Malsburg 1973; Miller et al. 1989; Pérez et al. 1975). We reserve detailed comparison with these different approaches for another publication. However, we note that one crucial distinction between BCM and many other theories is the means of synaptic stabilization. In BCM this is accomplished by a dynamically varying modification threshold, whereas these other treatments (e.g., Malsburg 1973; Miller et al. 1989; Pérez et al. 1975) require some sum over synaptic strengths to remain constant. Without entering into a discussion here of which mechanism is more plausible, we note that it becomes an important experimental issue to distinguish between these two very different mechanisms for synaptic stabilization. In addition, the BCM theory (as will be made clear below) allows for precise specification of theoretical equivalents of experimental situations, allowing detailed and quantitative comparisons of theory with experiment.

The aim of the present effort is to provide such a comparison of the BCM theory with experiment for what we call "classical" rearing conditions. These include normal binocular vision (normal rearing or NR), monocular deprivation (MD), reverse suture (RS), strabismus (ST), and binocular deprivation (BD), as well as the restoration of normal binocular vision after various forms of deprivation (recovery or RE). Comparisons with the various pharmacological manipulations that affect visual cortical plasticity (e.g., Bear et al. 1990; Greuel et al. 1987; Reiter and Stryker 1988) will be considered elsewhere.

Review of relevant experiments

Because the literature on visual cortical plasticity is not without some controversy, it is useful to state our understanding of the consequences of the various visual environments on visual cortical organization. The modifications of interest are those that occur in kitten visual cortex during the second postnatal month after brief (<2 wk) manipulations of visual experience.

ACQUISITION OF ORIENTATION SELECTIVITY AND BINOCULAR RESPONSIVENESS DURING NR.   Important characteristics of the visual responses of most neurons in adult striate cortex are that they *1*) are binocular and *2*) show a strong preference for contours of a particular orientation (Hubel and Wiesel 1962). In the early literature there is some controversy concerning the acquisition of orientation selectivity

in young animals. According to one extreme view, the development of this property occurs entirely postnatally and requires patterned binocular visual experience (Barlow and Pettigrew 1971; Pettigrew 1974). The opposing view is that this property is present in a rudimentary form at birth and elaborates even in the absence of patterned visual input (Hubel and Wiesel 1963). We adopt the view that has emerged in the more recent literature; although some orientation selectivity is present in newborn animals and may improve without visual experience during the first few weeks, maturation to adult levels of specificity and responsiveness requires contour vision during the first 2 mo of life (Albus and Wolf 1984; Barlow 1975; Blakemore and Van Sluyters 1975; Bonds 1979; Braastad and Heggelund 1985; Buisseret and Imbert 1976; Frégnac and Imbert 1978, 1984; Movshon and Van Sluyters 1981). Evidently, the increase in selectivity and responsiveness can occur quite rapidly; after dark-rearing, mature response properties can develop with only 6 h of binocular contour vision (Buisseret et al. 1978, 1982; Imbert and Buisseret 1975).

MD AFTER A PERIOD OF NR.   The synaptic connections that generate binocular, selective responses in visual cortex remain sensitive to manipulations of visual experience during the second and third postnatal months. One manipulation is the deprivation of pattern vision through one eye, usually produced by suturing the eyelid closed. This results in a loss of responsiveness to stimulation of the deprived eye (Blakemore and Van Sluyters 1974b; Hubel and Wiesel 1970; Movshon and Dürsteler 1977; Olson and Freeman 1975; Wiesel and Hubel 1963). This change in cortical ocular dominance can occur very rapidly, with noticeable effects occurring in as little as 4–8 h (Freeman and Olson 1982). Recent chronic recording studies indicate that the loss of deprived-eye responsiveness is not necessarily accompanied by an immediate increase in responsiveness through the open eye (Mioche and Singer 1989). It appears, however, that eventually there is an increase in open-eye responsiveness and that cortical neurons responding to the open eye retain their original selectivity.

The effects of lid suture evidently are caused by the loss of pattern vision rather than a diminution of retinal activity per se (Blakemore 1976). For example, diffusing contact lenses are as effective as lid suture in producing the ocular dominance shift (Blakemore 1976; Wiesel and Hubel 1963). It should also be noted that, in spite of the lack of responsiveness, synaptic connections from the deprived eye are still physically present in cortex (Burchfiel and Duffy 1981; Duffy et al. 1976; Freeman and Ohzawa 1988; Kratz et al. 1976; Sillito et al. 1981; Tsumota and Suda 1978).

RS.   Blakemore and Van Sluyters (1974b) first showed that if, after a period of MD, the deprived eye was opened and the formerly open eye sutured closed, then cortical neurons would become responsive to the newly opened eye and lose responsiveness to the newly closed eye. Acute studies indicated that, as the ocular dominance of cortical responses shifts from one eye to the other during RS, there is never a period when a substantial number of cells can be strongly, and equally, activated by both eyes (Movshon 1976). Indeed, recent chronic recording experiments of Mioche and Singer (1989) demonstrate that neurons lose responsiveness to the newly deprived eye before the newly opened eye shows any recovery, thereby preventing cells from showing strong binocular responses. For the majority of neurons, lost responsiveness occurs within 24 h of the RS and is followed by a recovery of newly opened-eye responsiveness over the next 48–72 h. These experiments also indicate that, as the initially deprived-eye synapses to visual cortex recover, the original orientation selectivity emerges. However, in their acute RS experiments, Blakemore and Van Sluyters (1974b) and Movshon (1976) found binocular neurons with different optimal orientations in tests of the two eyes, leaving open the possibility that during RE the initially deprived eye can acquire an optimal orientation different from its originally preferred input.

ST. Hubel and Wiesel (1965) first showed that misaligning the two eyes by sectioning an extraocular muscle results in a loss of binocular responsiveness. Subsequent experiments have repeatedly verified Hubel and Wiesel's results; either divergent or convergent squint produces cortical neurons that are generally responsive to one eye or the other, but not both (Blakemore and Eggers 1978; Ikeda and Tremain 1977; Van Sluyters and Levitt 1980; Yinon 1976; Yinon et al. 1975). The critical feature in producing the loss of binocularity is the lack of correlated input in the two eyes. Numerous results indicate that a loss of correlated input, resulting from the researchers' either alternating occlusion, rotating the image in one eye relative to the other eye, or producing different patterns of illumination on corresponding regions of the two retinas simultaneously, eventually leads to a loss of binocularity (Blakemore 1976; Blakemore and Van Sluyters 1974a; Blasdel and Pettigrew 1979; Cynader and Mitchell 1977).

BD. There is general agreement that the consequences of depriving both eyes of normal vision are substantially different from the effects of MD. Whereas several days of MD are sufficient to cause a functional disconnection of the deprived eye, an equivalent period of BD leaves most cortical neurons responsive to stimulation of either eye (Wiesel and Hubel, 1965). However, quantitative studies indicate that brief periods (≤1 wk) of BD (achieved by placing the animals in a dark room) produce a 50% drop in peak neuronal responsiveness to the preferred orientation and a slight broadening of orientation selectivity (Freeman et al. 1981). Longer periods of BD lead to a further decrease in responsiveness and selectivity. [The effects of long-term BD seemingly differ depending on whether the deprivation is produced by suturing both eyelids or by placing the animal in complete darkness. After binocular lid suture there appears to be a loss of binocularity (Kratz and Spear 1976; Mower et al. 1981; Watkins et al. 1978), that is not seen during dark-rearing (Freeman et al. 1981; Mower et al. 1981).]

RE. The extent of a neuron's recovery of its normal receptive-field properties during binocular vision after a period of deprivation appears to depend on the duration of the deprivation and its possible effects on eye alignment. Dark-rearing generally produces a population of cortical neurons that are poorly responsive to stimulation of either eye. Brief periods of binocular vision after weeks of dark-rearing lead to the development of substantial levels of selectivity, responsiveness, and binocularity in a very short time (Buis-

seret et al. 1978, 1982), whereas binocular vision after prolonged periods of dark-rearing often produces cortical neurons that are responsive but not binocular (Cynader 1983; Cynader et al. 1976). As Cynader (1983) points out, the longer periods of dark-rearing can lead to eye misalignment and the loss of correlated input when vision is restored, which is sufficient to cause a breakdown in binocularity. Similarly, if a brief period of MD is followed by a period of patterned binocular vision, then cortical neurons can regain their binocular receptive fields (Blasdell and Pettigrew 1978; Freeman and Olson 1982). Binocular vision after longer periods of MD leads to the recovery of some cortical cells' receptive-field properties in the deprived eye without an attendant rise in binocularity (Hubel and Wiesel 1970; Mitchell et al. 1977; Olson and Freeman 1978). The lack of binocular neurons in these kittens, again, may be due to the development of a ST during the longer periods of MD (Mitchell et al. 1977; Olson and Freeman 1978).

In this paper we present a detailed development of the BCM theory that is able to account for these varied experimental results. This work confirms and extends previous mathematical analyses and yields predictions that should be experimentally verifiable.

METHODS

A mean-field approximation to the full complexity of visual cortex has been described previously (Cooper and Scofield 1988; Scofield and Cooper 1985). In this approximation, it is proposed that visual cortex (and possibly other regions of cortex) can be treated by a combination of statistical and single-cell methods, allowing the evolution of some of the synapses (e.g., the LGN-cortical synapses) to be analyzed in detail.

Consider a single cortical neuron receiving an array of LGN fibers from the left and right eyes (Fig. 1). This neuron can be

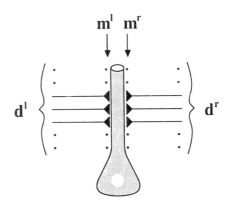

$$c = (\mathbf{m}^l \cdot \mathbf{d}^l) + (\mathbf{m}^r \cdot \mathbf{d}^r)$$

FIG. 1. Cartoon of a cortical neuron receiving LGN input from the left and right eyes to illustrate the theoretical notation. Horizontal lines represent LGN-cortical fibers; solid triangles represent corresponding synaptic contacts. Cell response c is a linear function of left- and right-eye LGN-cortical input activities, represented by vectors $\mathbf{d}^l$ and $\mathbf{d}^r$, and left- and right-eye synaptic weights, represented by vectors $\mathbf{m}^l$ and $\mathbf{m}^r$. Value of c is obtained by multiplying the LGN activity d_j of each fiber by its corresponding synaptic weight m_j and summing over all synapses to the cell.

viewed as the ith cell in a highly interconnected network with both excitatory and inhibitory connections. In the mean-field approximation of this intricate network, the intracortical connections to the ith cell are replaced by a set of synaptic weights that convey to the cell the average activity of the other cells in the network. A further simplification that we employ in this paper is to treat the mean-field connections to the cell in the adiabatic limit; that is, the mean-field connections to the ith cell are assumed to remain constant while the cell's LGN-cortical synapses seek their equilibrium values. According to the analysis of Cooper and Scofield (1988), this does not change the position and/or the stability properties of the equilibrium states.

Postsynaptic cortical cell response

We denote the response of the single cell at time t by the scalar $c(t)$, which is dependent on the LGN–cortical afferent activity $d(t)$ and the weights, or efficacy $m(t)$, of the LGN-cortical synapses. In general, the output $c(t)$ is a nonlinear function of the input $d(t)$ and the weights $m(t)$. However, we assume that there is a region of linear dependence of $c(t)$ on $d(t)$ and $m(t)$. The existence of such a linear region appears to have some experimental justification (Ohzawa and Freeman 1986a,b). Therefore, for the purposes of the present paper, $c(t)$ is written as

$$c(t) = m(t) \cdot d(t) \qquad (1)$$

i.e., the sum over all of the synapses to the postsynaptic target cell of the activity of each LGN–cortical afferent fiber multiplied by the strength of the synaptic connection of that fiber with the postsynaptic target cell.

LGN input

To model the input to visual cortex that arises from the regions of the two retinas that view the same point in visual space, we assume that LGN activity is a precise mapping of retinal ganglion cell activity. (This means that a specific image results in a specific distribution of LGN activity.) Two types of LGN-cortical input are of particular interest: 1) activity elicited when visual contours are presented to the retinas ("pattern" input) and 2) activity that arises in the absence of visual contours ("noise"; Fig. 2). From our point of view, the important distinction between pattern and noise input is the degree of correlation that the two types of input produce in the LGN activity. For a specific input pattern, the activity of one LGN neuron is assumed to have a precise relation to the activity of other LGN neurons (that is, a given pattern in the visual field maps into a given set of LGN neuron activities), whereas for noise the activity of one LGN neuron is independent of the activity of the other LGN neurons. Differences between distinct patterns (for example, between various stimulus orientations) are mapped into different distributions of activity across the LGN. The extent to which this is a reasonable model of a visual environment is currently under investigation (D. Sheinberg, work in progress).

It is possible that the activity of neighboring ganglion cells may be correlated in the absence of contour vision (Maffei and Galli-Resta 1990; Mastronarde 1983a,b, 1989; Meister et al. 1991). These correlations are possibly important for aspects of cortical development that occur before birth (e.g., Miller et al. 1989). However, it is not necessary to incorporate these local correlations to capture the results of the postnatal deprivation experiments using the BCM theory, nor would incorporation of these correlations substantially alter the outcome of the simulations. Our work does incorporate the fact that retinal ganglion cells and LGN neurons fire action potentials spontaneously in the absence of light (Barlow and Levick 1969; Hubel and Wiesel 1961; Kuffler et al. 1957).

As illustrated in Fig. 2, the salient features of the input environment can be modeled by use of a one-dimensional array of LGN fibers from each eye as the input to the cortical neuron. Accordingly, apart from noise fluctuations, a specific input pattern always corresponds to the same distribution of activity across the array of LGN fibers from each eye. The consistent and unique distribution of activity corresponding to a specific input pattern (not the actual shape of the distribution of activity) is the critical feature. For example, in the *third row* of Fig. 2, the pattern of LGN activity corresponding to a horizontal bar of light on the retina is represented as a unimodal distribution of activity with a peak activity on one fiber and slightly lower activities on the immediately adjacent fibers. This pattern of activity is the signature for a horizontal bar of light: each time a horizontal bar of light is imaged on the retina, this same pattern of LGN activity is generated (apart from superimposed fluctuations or noise); no other pattern of retinal stimulation leads to this pattern of activity. The choice of a unimodal distribution of activity is for convenience; it is in no way essential. Any distribution of activity could be chosen to represent a horizontal bar of light, as long as 1) this distribution occurs whenever a horizontal bar of light is on the retina and 2) this distribution differs from that arising when bars of other orientations are presented.

Activity along an array of N_{gl} fibers is represented as a vector in an N_{gl}-dimensional space. For example, the array of LGN–cortical input fiber activity from the left eye at time t is written as a single vector $d^l(t)$. (As a rule, symbols appearing in boldface roman represent vectors, and symbols appearing in light roman represent scalars. Boldface italic symbols represent vectors that have both a left- and right-eye component, designated by the superscripts l and r, respectively.) The array of synaptic weights formed by the LGN-cortical fibers from the left eye to the neuron is denoted by the vector $m^l(t)$. The jth component $d^l_j(t)$ of the vector $d^l(t)$ is a scalar representing the fiber activity of one (the jth) left-eye LGN-cortical afferent; the jth component $m^l_j(t)$ of the vector $m^l(t)$ is a scalar representing one (the jth) left-eye LGN–cortical afferent synaptic weight. For the corresponding right-eye quantities, we have the vector $d^r(t)$ and $m^r(t)$ with components $d^r_j(t)$ and $m^r_j(t)$, respectively. Therefore the total input to the cell with synapses $m(t) = [m^l(t), m^r(t)]$ at time t is $d(t) = [d^l(t), d^r(t)]$.

Input to the simulated neuron at time t, i.e., $d(t) = [d^l(t), d^r(t)]$, is measured with respect to the average spontaneous activity of each of the LGN-cortical fibers. For example, the input from the left eye at time t to the jth synapse of the simulated cortical neuron is

$$d^l_j(t) = d^l_{a,j}(t) - d^l_{s,j} \qquad (2)$$

where $d^l_{a,j}(t)$ is the actual firing frequency and $d^l_{s,j}$ is the average spontaneous activity of the jth left-eye LGN afferent fiber. For simplicity, we assume that $d^l_{s,j}$ is independent of time, the eye, and the fiber; that is, $d^l_{s,j} = d^r_{s,j} = d_s$ for all of the LGN fibers at each time t. Therefore

$$d^l_j(t) = d^l_{a,j}(t) - d_s \qquad (3a)$$

and

$$d^r_j(t) = d^r_{a,j}(t) - d_s \qquad (3b)$$

The corresponding vector equation for all of the fibers is

$$d(t) = d_a(t) - d_s \qquad (4)$$

which implies that

$$d^l(t) = d^l_a(t) - d^l_s \qquad (4a)$$

165

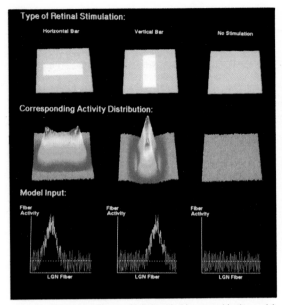

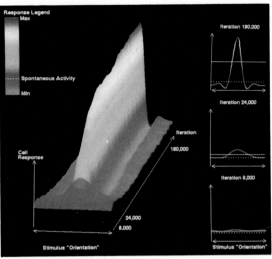

FIG. 2. Illustration of types of LGN-cortical input used in the model. *Top row:* horizontal bar of light on the retina (*left*), a vertical bar of light on the retina (*middle*), and the retina in the absence of any stimulation (*right*). *Middle row:* distribution of activity in a 2-dimensional array of LGN neurons that might arise from these different types of retinal stimulation. The important point of this figure is that different types of retinal stimulation are mapped onto distinct distributions of LGN activity. We call LGN activity elicited when visual contours are presented to the retina "pattern" input to the visual cortex. We call LGN activity that arises in the absence of visual contours "noise" input to the cortex. As illustrated in this figure, the important difference between pattern and noise input is that for the former there are predictable relationships in the activities of the LGN cells, whereas in the latter the activity of one LGN cell is independent of the activity of the other LGN cells. Different patterns that arise from contours of different orientations, for example, are distinguished by different distributions of correlated activity in the LGN cells. Salient features of the LGN-cortical input can be captured in a 1-dimensional array of LGN fibers, as shown in the *bottom row*. Pure patterns are described mathematically by $d_{peak} e^{-\gamma \{1 - \cos [(2\pi/N_{p})(j - j_{\omega})]\}} + d_{base}$, where j denotes the fiber for which activity is being calculated, j_{ω} is the fiber in which the peak activity occurs for pattern ω, d_{base} is the minimum fiber activity in a pattern, d_{peak} is the peak displacement in activity from d_{base}, and γ describes the rate at which the activity falls off with increasing fiber distance from the fiber with peak activity. In the absence of retinal stimulation, the LGN-cortical input activity to the network (*3rd row, right*) is assumed to be randomly distributed about its spontaneous level, represented by the dashed line.

for the left-eye fibers and

$$\mathbf{d}^{r}(t) = \mathbf{d}_{a}^{r}(t) - \mathbf{d}_{s}^{r} \qquad (4b)$$

for the right-eye fibers.

During simulations of BD or MD, we assume that the deprived eyes are receiving noise input. For example, to model noise input to the left eye (the right eye would be exactly the same), assume that $d_{a,j}^{l}(t)$ fluctuates randomly around d_{s}, implying that $d_{j}^{l}(t)$ fluctuates around 0. We call these fluctuations of $d_{j}^{l}(t)$ around 0 noise and label them $n_{j}^{l}(t)$. The input to the cortical neuron in the absence of patterned activity is then simply

$$d_{j}^{l}(t) = n_{j}^{l}(t) \qquad (5)$$

for the *j*th fiber of the left eye (Tables 1 and 2).

FIG. 4. Simulation of NR for a cortical cell receiving 1 set of LGN–cortical afferent fibers. The 3-dimensional figure shows the evolution of the cell's response to different input patterns (stimulus "orientations") as a function of time or iteration number. Cross-sections through this figure at a given iteration can be viewed as the tuning curve of the cortical neuron to stimuli of various orientations. There are N_{p} patterns (stimulus "orientations") in the training and testing set (*Eq. 22*); to produce the continuous curves in the 3-dimensional graphs the cell's responses to the N_{p} patterns at different times are connected in a smooth manner. Graded color scale (*top left*) is the legend for the strength of the response: red corresponds to the maximum response, purple corresponds to the minimum response, and blue is at the level of spontaneous activity. This same legend applies to all of the 3-dimensional graphs in Figs. 6–13; therefore the strength of the cell response in one figure can be compared directly with the cell response in any other figure. To illustrate how responses above θ increase in strength and those below θ decrease in strength, 3 slices through the 3-dimensional graph are illustrated to the *right* of the figure. Dashed white horizontal line represents the cell's level of spontaneous activity and solid white line represents the value of θ. As the tuning curve becomes more selective, θ increases and eventually stabilizes the synaptic weights.

In NR or MD the open eyes of the animal are constantly viewing regularly shaped objects, which are assumed to give rise to correlated patterns of LGN activity. For example, left-eye contour vision of a specific object gives rise to a pattern of LGN–cortical fiber activity, which we label by $\mathbf{d}_{a}^{l}(t) = \mathbf{d}_{a}^{\omega}$, where the ω represents a particular pattern of correlated activity corresponding to a specific visual stimulus at time t. Now the input to the *j*th left-eye synapse of the neuron, $d_{j}^{l}(t)$, measures the departure of $d_{a,j}^{l}(t) = d_{a,j}^{\omega,l}$ from the spontaneous rate d_{s}

$$d_{j}^{l}(t) = d_{a,j}^{l}(t) - d_{s} = d_{a,j}^{\omega,l} - d_{s} \qquad (6)$$

Note that even though $d_{a,j}^{\omega,l}$ is positive, $d_{j}^{l}(t)$ may have positive and negative values. Because in an actual visual environment there generally will be random fluctuations in $d_{j}^{l}(t)$ for multiple presentations of the same visual stimulus labeled by ω, we add a noise term to the input, which we again call $n_{j}^{l}(t)$ (Tables 1 and 2)

$$d_{j}^{l}(t) = d_{a,j}^{\omega,l} - d_{s} + n_{j}^{l}(t) \qquad (7a)$$

Defining $d_{j}^{\omega,l} = d_{a,j}^{\omega,l} - d_{s}$, the input to the *j*th left-eye synapse of the neuron for the pattern labeled by ω presented at the time t becomes

$$d_{j}^{l}(t) = d_{j}^{\omega,l} + n_{j}^{l}(t) \qquad (7b)$$

Spontaneous level of the cell response:

$$c_s(t) = \boldsymbol{m}(t) \cdot \boldsymbol{d}_s = \mathbf{m}^l(t) \cdot \mathbf{d}_s^l + \mathbf{m}^r(t) \cdot \mathbf{d}_s^r$$

where $\mathbf{d}_s^l = \mathbf{d}_s^r$ = Spontaneous activity of the LGN-cortical afferent fibers.

Actual level of the cell response:

$$c_a(t) = \boldsymbol{m}(t) \cdot \boldsymbol{d}_a(t) = \mathbf{m}^l(t) \cdot \mathbf{d}_a^l(t) + \mathbf{m}^r(t) \cdot \mathbf{d}_a^r(t)$$

where $\boldsymbol{d}_a(t)$ = Actual activity in the LGN-cortical afferent fibers.

Cell response measured with respect to the spontaneous level with non-LGN noise superimposed:

$$c(t) = c_a(t) - c_s(t) + c_{noise}(t) = \mathbf{m}^l(t) \cdot \mathbf{d}^l(t) + \mathbf{m}^r(t) \cdot \mathbf{d}^r(t) + c_{noise}(t)$$

where $\mathbf{d}^l(t) = \mathbf{d}_a^l(t) - \mathbf{d}_s^l$, $\mathbf{d}^r(t) = \mathbf{d}_a^r(t) - \mathbf{d}_s^r$ and $c_{noise}(t)$ represents non-LGN noise.

Components of $\mathbf{d}^l(t)$ and $\mathbf{d}^r(t)$ (patterned input):

$$d_j^l(t) = d_j^{\omega,l} + n_j^l(t) = d_{a,j}^{\omega,l} - d_s + n_j^l(t)$$

$$d_j^r(t) = d_j^{\omega,r} + n_j^r(t) = d_{a,j}^{\omega,r} - d_s + n_j^r(t)$$

where ω represents one of the N_p patterns of activity.

Components of $\mathbf{d}^l(t)$ and $\mathbf{d}^r(t)$ (noise input):

$$d_j^l(t) = n_j^l(t)$$

$$d_j^r(t) = n_j^r(t)$$

Rule for synaptic modification:

$$\frac{d\boldsymbol{m}(t)}{dt} = \eta\phi(c(t), \theta(t))\boldsymbol{d}(t)$$

where

$$\theta(t) = \frac{\overline{\overline{[c_a(t)]}}^p}{c_0}$$

$$\overline{\overline{c_a(t)}} = \frac{1}{\tau} \int_{-\infty}^{t} c_a(t') \, e^{-[(t-t')/\tau]} \, dt'$$

These are the patterns presented during the developmental portion of the simulations; for "tests" of responses the noiseless patterns $d_j^{\omega,l}$ are presented to the cell. With the same arguments for the right eye, a normal, binocular viewing experience at some time t leads to a specific correlated pattern of activity $\boldsymbol{d}^\omega = (\mathbf{d}^{\omega,l}, \mathbf{d}^{\omega,r})$ with noisy fluctuations $\boldsymbol{n}(t) = [\mathbf{n}^l(t), \mathbf{n}^r(t)]$, so that

$$\boldsymbol{d}(t) = [\mathbf{d}^l(t), \mathbf{d}^r(t)] = [\mathbf{d}^{\omega,l} + \mathbf{n}^l(t), \mathbf{d}^{\omega,r} + \mathbf{n}^r(t)] \tag{8}$$

Synaptic modification

Synaptic modification is governed by the rules of the BCM theory. According to BCM, modification of the jth synaptic junction at time t $[dm_j(t)/dt]$ is proportional to the product of the input activity of the jth LGN-cortical fiber $[d_j(t)]$ and a function ϕ, which is defined in terms of the postsynaptic response.

The behavior of ϕ at two values of the postsynaptic response is particularly important (Fig. 3A). The first is the average postsynaptic cortical cell activity due to the spontaneous firing of action potentials in the presynaptic LGN–cortical afferent fibers. Recall that the average level of this activity is dependent on d_s for each LGN fiber of each eye. Letting $\boldsymbol{d}_s = (\mathbf{d}_s^l, \mathbf{d}_s^r)$ represent the spontaneous activity of all of the incoming LGN fibers to the cell, the average "spontaneous" activity of the postsynaptic neuron is (Tables 1 and 2)

$$c_s(t) = \boldsymbol{m}(t) \cdot \boldsymbol{d}_s = \mathbf{m}^l(t) \cdot \mathbf{d}_s^l + \mathbf{m}^r(t) \cdot \mathbf{d}_s^r \tag{9}$$

$c_s(t)$ serves as the baseline for measuring the impact of the current LGN–cortical afferent fiber activity on the prevailing state

of the postsynaptic cell. If the LGN–cortical afferent fiber activity goes above its average spontaneous level, the postsynaptic cell is depolarized compared with its average spontaneous state. On the other hand, if the LGN–cortical afferent fiber activity drops below its average spontaneous level, the postsynaptic cell is hyperpolarized compared with its average spontaneous state. This departure of the postsynaptic response from its average spontaneous state is the measure of the postsynaptic cell response in the model. The ϕ function is zero at $c_s(t)$ with a negative slope (ϵ_0) at postsynaptic response levels above $c_s(t)$.

The second important level of postsynaptic cell response occurs at a value designated by $c_s(t) + \theta$. Here ϕ changes sign from negative to positive with a positive slope (ϵ_θ). According to the BCM theory, the value of θ changes as a power (larger than 1) of some measure of the average postsynaptic response. This important feature of θ ensures the boundedness of the synaptic weights without placing artificial constraints on them and guarantees that $\theta \geq 0$. Therefore the threshold defined by $c_s(t) + \theta$ is never at a level of postsynaptic response lower than $c_s(t)$. Furthermore, when all of the synaptic weights approach 0 and $c_s(t) \approx 0$ for all t, θ must also eventually approach 0, thereby allowing for the possible recovery of synaptic strength.

For synaptic modification, the departure of the postsynaptic cell response from $c_s(t)$ is the measure of interest. Therefore we define

$$c(t) = c_a(t) - c_s(t) \tag{10}$$

where

$$c_a(t) = \boldsymbol{m}(t) \cdot \boldsymbol{d}_a(t) = \mathbf{m}^l(t) \cdot \mathbf{d}_a^l(t) + \mathbf{m}^r(t) \cdot \mathbf{d}_a^r(t) \tag{10a}$$

Parameter	Significance	Theoretical Constraint	Value Used
N_{gl}	Number of LGN-cortical input fibers from 1 eye	None	12
N_p	Number of training patterns	None	12
$d_{a,j}^{\omega,l(r)}$	ωth input pattern*	None	$d_{peak} = 1.0$ $d_{base} = d_s$ $\gamma = 4.0$
d_s	Spontaneous activity weighting factor†	None	$d_s = 5.0$
$\overline{\overline{n}}$	Average value of the noise input along the LGN-cortical fibers	None	0
$\overline{\overline{n^2}}$	Average of the square of the noise input along the LGN-cortical fibers	None	0.03
$m_j^{l(r)}(0)$	Starting values of the synaptic weights for normal rearing	None	0.0–0.15
$\overline{\overline{c_{noise}}}$	Average value of the non-LGN postsynaptic noise	None	0
$\overline{\overline{c_{noise}^2}}$	Average of the square of the non-LGN postsynaptic noise	None	33.3
η	Modification step size	Upper bound that decreases with increasing τ	0.005
τ	Time constant in the definition of θ	Must lead to an adequate sampling of inputs	1000
p	Nonlinearity in the definition of θ	$p > 1$	2
c_0	Normalization constant in the definition of θ	None	50

LGN, lateral geniculate nucleus. *See legend of Fig. 2. †Relative magnitude of d_s compared with d_{peak} and d_{base} should be interpreted as reflecting the frequency of occurrence of patterned LGN activity against a continuous background of noise.

A

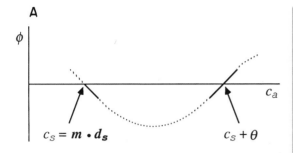

B

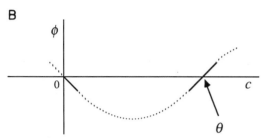

FIG. 3 The ϕ function. A: two important values of the cell response [$c_a(t)$] are the average level of the spontaneous activity (c_s) and θ above the average spontaneous level ($c_s + \theta$). Essential properties of ϕ (i.e., the negative slope, called ϵ_0, for values of c_a just above c_s and the positive slope, called ϵ_θ, through $c_s + \theta$) are illustrated by solid curves. In the simulations we set $\epsilon_0 = -3$ and $\epsilon_\theta = 3$. Dashed lines complete the definition of ϕ in terms of c_a but are not rigidly specified by experiment. B: because the departure of the cell response from its spontaneous level is the quantity of interest, ϕ in Eq. 13 is defined in terms of $c(t)$, which equals $c_a(t) - c_s$.

which is equivalent to

$$c(t) = \mathbf{m}^l(t) \cdot \mathbf{d}^l(t) + \mathbf{m}^r(t) \cdot \mathbf{d}^r(t) \qquad (11)$$

To incorporate non-LGN input to the cell that is independent of the LGN input, we add a noise term to $c(t)$

$$c(t) = \mathbf{m}^l(t) \cdot \mathbf{d}^l(t) + \mathbf{m}^r(t) \cdot \mathbf{d}^r(t) + c_{\text{noise}}(t) \qquad (12)$$

where $c_{\text{noise}}(t)$ is assumed to be uniformly, randomly distributed about 0 (Tables 1 and 2). Thus for ϕ the two important values of postsynaptic response are at $c(t) = 0$ and $c(t) = \theta$. The BCM rule for synaptic modification can now be written (Tables 1 and 2)

$$\frac{d\mathbf{m}(t)}{dt} = \eta\phi[c(t), \theta(t)]\mathbf{d}(t) \qquad (13)$$

where ϕ is a function such that

$$\phi(c = 0, \theta) = \phi(c = \theta, \theta) = 0 \qquad (13a)$$

and the slope of ϕ at c greater than zero (ϵ_0) is negative and the slope of ϕ at $c = \theta$ (ϵ_θ) is positive (Fig. $3B$). The factor η, which we call the modification "step size," is a positive constant that determines the magnitude of the synaptic modification at each iteration.

One of the most striking characteristics of the BCM theory is the moving threshold. It is this moving modification threshold that provides the stability of the system. This contrasts with other synaptic modification proposals, such as those of Malsburg (1973), Pérez et al. (1975), and Miller et al. (1989), in which stability is produced by the requirement that some sum over the strengths of the synapses be constant.

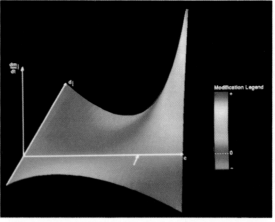

FIG. 5. Sign and magnitude of the change to the jth LGN-cortical fiber (dm_j/dt) for different values of the fiber activity (d_j) and the cell response (c). Notice that relative sizes of the 4 regions change as θ (represented by the arrow) moves.

In their original paper Bienenstock et al. (1982) chose as a candidate for θ the average value of the postsynaptic firing rate

$$\theta(t) = \left(\frac{\overline{\overline{c(t)}}}{c_0}\right)^p \qquad (14)$$

where $\overline{\overline{c(t)}}$ represents a time average of the departure of the postsynaptic firing rate from its spontaneous level, that is

$$\overline{\overline{c(t)}} = \frac{1}{\tau}\int_{-\infty}^{t} c(t')e^{-[(t-t')/\tau]}dt' \qquad (14a)$$

p sets the degree of nonlinearity (p > 1), and c_0 is a normalization constant. This produced a workable system with the desired stability properties. However, the precise value of p and the interpretation of the cell response remained open. For example, should one use the actual firing rate of the cell or the average depolarization of the postsynaptic dendrite? If one used the average depolarization, was the appropriate quantity the total depolarization or its deviation from its spontaneous level? Such questions remain unanswered because no experiments exist to fully distinguish between their consequences.

However, recent experimental innovations, such as the chronic recording technique developed by Mioche and Singer (1988), now place us in a position to investigate much more closely several key features of the moving threshold. Consider two: 1) How rapidly does the moving threshold adjust to changes in the cell response brought on by a manipulation of the visual environment? 2) How should the time averaging (represented by the double bars) and the nonlinearity p relate to each other?

Addressing the second question first, Bienenstock et al. (1982) time-averaged the cell response first and then raised the time average to the power p. Recent work (Intrator 1990; Intrator and Cooper 1991) indicates that a θ based on a time average of the squared cell response, that is

$$\theta(t) = \overline{\overline{c(t)^p}} \qquad (15)$$

with p = 2, has some useful mathematical properties; this form also appears to produce an evolution of the synapses in agreement with the experimental results discussed in the present work.

With regard to the first question, the two elements that control the rate at which θ moves are τ and the feature of the cell response that sets θ. As τ goes to infinity (i.e., as the memory of the time average goes to infinity), the threshold adjusts more slowly to changes in the cell response. Therefore, if θ is high and the cell response suddenly drops, then for large τ, the threshold, θ, drops slowly to a level consistent with the prevailing cell response.

The exact feature of the cell response that should be used to set θ is not known at present. However, inspection of Eq. 10 does provide two distinct possibilities: 1) the total response of the cell (c_a) or 2) the original BCM choice: the deviation of the cell response from its spontaneous level (c). This distinction is important because a θ dependent on c is much more sensitive to changes in the environment than a θ dependent on c_a. When no patterns are present in the input (Eq. 5), $c(t) \approx 0$ on average, whereas $c_a(t) \approx$ $m(t) \cdot d_s$ on average. In this situation, after the first τ iterations, θ based on c is near 0, whereas θ based on c_a is at some positive level. If θ was nonzero at the start of the simulation, then the percentage change for θ dependent on c is much larger than the percentage change for θ dependent on c_a. All of the key features of θ that are currently unspecified provide a rich area for further investigation. In this paper we choose θ to be

$$\theta(t) = \left(\frac{\overline{\overline{c_a(t)}}}{c_0} \right)^p \tag{16}$$

where

$$c_a(t) = m(t) \cdot d_a(t) \tag{16a}$$

is the actual cell response; the double bar, as usual, represents a time average of the underlying quantity, that is

$$\overline{\overline{c_a(t)}} = \frac{1}{\tau} \int_{-\infty}^{t} c_a(t') e^{-[(t-t')/\tau]} dt' \tag{16b}$$

and c_0 is again a normalization constant (Tables 1 and 2). With this choice of θ we are able to model the classical deprivation experiments described above.

Simulations

The visual deprivation experiments we attempt to model are NR, MD, RS, ST, BD, and RE. Each of the experimental paradigms can be simulated with some combination of two types of input to the two eyes: noise, i.e., $n^l(t)$ and $n^r(t)$, and patterns with noise superimposed, i.e., $d^{\omega,l} + n^l(t)$ and $d^{\omega,r} + n^r(t)$ ($1 \leq \omega \leq N_p$).

For example, during NR the input for the two eyes at time t is assumed to be patterned for each eye and correlated between the two eyes, with each pattern of activity having the same probability of occurring

$$d_j^l(t) = d_j^{\omega,l} + n_j^l(t) \tag{17a}$$

and

$$d_j^r(t) = d_j^{\omega,r} + n_j^r(t) \tag{17b}$$

where ω is chosen randomly from the set of N_p patterns at each iteration and $n_j^l(t)$ and $n_j^r(t)$ are independent but statistically equivalent noise terms. MD is similar to NR except that patterned activity is absent from the eye (left) simulated as closed

$$d_j^l(t) = n_j^l(t) \tag{18a}$$

and

$$d_j^r(t) = d_j^{\omega,r} + n_j^r(t) \tag{18b}$$

RS is accomplished by changing the eye that is closed after an initial period of MD. Therefore, if, for the initial period of MD, the left eye is closed, then for RS the input is

$$d_j^l(t) = d_j^{\omega,l} + n_j^l(t) \tag{19a}$$

and

$$d_j^r(t) = n_j^r(t) \tag{19b}$$

which corresponds to the right eye closed. BD is simulated by a total lack of patterned input to the two eyes

$$d_j^l(t) = n_j^l(t) \tag{20a}$$

and

$$d_j^r(t) = n_j^r(t) \tag{20b}$$

Finally, in a simulation of ST, patterns are input to both eyes, but the patterns are uncorrelated

$$d_j^l(t) = d_j^{\omega,l} + n_j^l(t) \tag{21a}$$

and

$$d_j^r(t) = d_j^{\omega',r} + n_j^r(t) \tag{21b}$$

where ω and ω' are independently and randomly chosen from the set of N_p patterns at each time step.

With this choice of the set of patterned and noise inputs and the synaptic modification rule, the sequence of events in a simulation of any experimental paradigm is relatively straightforward (Saul and Clothiaux 1986). First, at each step $t - dt$ to t, a pattern of activity $d(t)$ is generated that represents the activity of the LGN-cortical afferents during the interval dt. The activity assignments to $d(t)$ must be consistent with the particular visual experience paradigm that is being simulated. Once the vector $d(t)$ is constructed, the resulting cell response $c(t)$ is calculated according to Eq. 12 and $dm(t)/dt$ is then determined according to Eq. 13. The whole process is repeated until the synaptic weights $m(t)$ reach equilibrium.

The dynamics of the evolution of the cell-response tuning curve to its final equilibrium value is the information of interest. To follow the change in cell response selectivity and ocular dominance during a simulation, we periodically interrupt the modification process to assess the synaptic weight vector $m(t)$. With a record of $m(t)$ at various intervals during the simulation, a history of the changes in cell response selectivity and ocular dominance can be easily reconstructed. The cell responses generated by stimulation of the left and right eyes separately with all N_p of the noiseless input patterns $d^{\omega,l}$ and $d^{\omega,r}$, respectively, are taken to be the "tuning curves" of the two eyes

Left-Eye Tuning Curve(t)

$$= [c^{l,\omega}(t) = m^l(t) \cdot d^{\omega,l} | \omega = 1, \cdots, N_p] \tag{22a}$$

Right-Eye Tuning Curve(t)

$$= [c^{r,\omega}(t) = m^r(t) \cdot d^{\omega,r} | \omega = 1, \cdots, N_p] \tag{22b}$$

Because patterns labeled by ω and $\omega + 1$ are most similar in their distribution of activity across the LGN-cortical afferents, plotting $c^{l,\omega}(t)$ and $c^{r,\omega}(t)$ versus ω (in ascending order) leads to graphs that are interpreted as analogous to visual cortical neuron response curves obtained experimentally by presenting a light bar across the retina at different orientations.

In this paper a single set of parameters is used to model the kitten visual deprivation experiments. This choice of parameters leads to an evolution of the synapses in agreement with experiment. In the DISCUSSION the effects of changing the parameters on the evolution of the synapses are considered in more detail.

RESULTS

A primary motivation for the original BCM theory was to account for the acquisition of selectivity by cortical neurons in a normal patterned environment. How this comes about is illustrated by the simulation in Fig. 4. At the beginning of the simulation the LGN-cortical synaptic weights start with

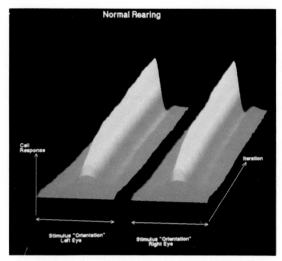

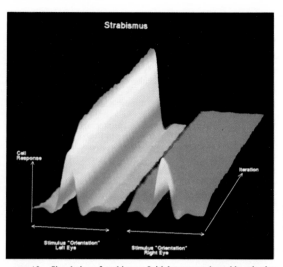

FIG. 6.   Simulation of normal rearing with binocular input. Initial state for this simulation was randomized synaptic weights near 0. It ran for 200,000 iterations. Note that the cell response becomes stable, binocular, and selective. See Fig. 4 for response legend.

FIG. 10.   Simulation of strabismus. Initial state was the stable, selective equilibrium state after normal rearing (Fig. 6). This simulation ran for 200,000 iterations. Note that the cell response becomes monocular. See Fig. 4 for response legend.

random values between 0 and 0.15. Because there is no bias in the synaptic weights, all of the input patterns initially produce a small cell response that may be either slightly above or below θ. During training, when the various patterns are presented at random to the cell, the synaptic weights associated with positive input activity ($d_j > 0$) strengthen when the response goes above θ (implying $\phi > 0$) because, according to *Eq. 13, $d m_j/dt = \eta \phi d_j$ and $\eta \phi d_j > 0$* (Fig. 5). On the other hand, the synaptic weights asso-

ciated with high input activity ($d_j > 0$) weaken when the postsynaptic response fails to reach θ (implying $\phi < 0$). As pointed out by Bienenstock et al. (1982), this results in competition between patterns of activity, with the consequence that the synapses that consistently participate in the activation of the cell above θ increase in strength, whereas the synapses for which activity fails to correlate with target activation beyond θ decrease in strength.

Growth in the cell response to some of the input patterns leads to an increase in θ, decreasing the likelihood of further

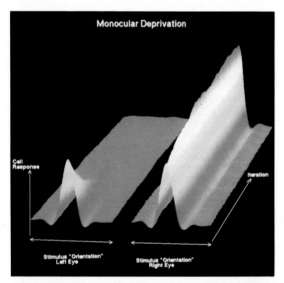

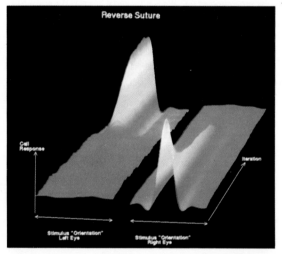

FIG. 7.   Simulation of monocular deprivation. Initial state for this simulation was the stable, binocular, selective equilibrium state achieved during normal rearing (Fig. 6). This simulation ran for 200,000 iterations. Note the loss of responsiveness of the closed eye. See Fig. 4 for response legend.

FIG. 9.   Simulation of reverse suture. Beginning state for this simulation was the stable equilibrium state after monocular deprivation (Fig. 7). This simulation ran for 200,000 iterations. Note the loss of reponsiveness of the newly closed eye before the recovery of responsiveness of the newly opened eye. See Fig. 4 for response legend.

increases in synaptic strength. Eventually, θ becomes large enough to stabilize the synaptic weights; that is, for θ sufficiently large, the average change to the synaptic weights over all of the input patterns is 0, and the cell response to any particular pattern no longer changes significantly. Notice that this stability is dynamic: for each iteration of the simulation there are small changes in the synaptic weights; over many iterations, however, the changes to the weights average to 0.

The format of Fig. 4 is used to illustrate the results of the various simulations in Figs. 6 through 13. The x-axis, labeled by "stimulus 'orientation'," represents the different patterns in the testing set; and the y-axis represents the response of the cell to these patterns. The z-axis of the graph represents time or, more precisely, the number of iterations. The evolution of the cell's tuning curve is illustrated by the plot of "cell response" versus stimulus orientation at each iteration. As Fig. 4 shows, presentation of patterned input to a cell quickly leads to robust responses to a select number of the input patterns. Because θ adjusts to changes in the cell response, the peak cell response eventually stabilizes at some nonzero value. (This value depends on various parameters, such as p, c_0, and τ.)

NR

The NR simulation starts from poorly developed left- and right-eye synapses (i.e., random values between 0 and 0.15). The input to the two eyes is taken to be correlated patterns with noise superimposed (*Eq. 17*). As expected, the cell acquires responsiveness and selectivity (Fig. 6). With a growing cell response, θ increases and eventually stabilizes the synaptic weights. Notice that the cell becomes selective to the same input patterns through the LGN–cortical fibers from each eye; this is because apart from noise the left- and right-eye patterns are identical at each iteration. Although we assume no initial orientation preference, if some orientation preference is present initially, the cell will almost certainly become selective to this orientation.

As discussed before, the value of τ determines the rate at which θ moves. The rate at which θ moves, in turn, affects

the choice of a modification step size η. The modification step size η (*Eq. 13*) must stay below an upper bound that is a function of the memory τ (*Eq. 16b*). As τ increases, implying a longer memory of $\overline{c_a(t)}$ in the calculation of θ, the upper bound on η decreases. If η is too large for a given value of τ, then the modification threshold θ cannot adjust rapidly enough to stop initial large increments in the cell response. When θ finally does increase, it overcompensates and drives the cell response back to zero. The synaptic weights and threshold oscillate in an unrealistic manner. Thus we are restricted to a domain of τ and η that leads to an acceptable evolution of the system.

Once τ and η are fixed at values that lead to a realistic evolution of the synaptic weights, the next important parameter that affects the outcome of the NR simulation is the normalization constant c_0, which is used in the calculation of the value of θ (*Eq. 16*): c_0 sets the level of θ with respect to the initial cell response. For example, as c_0 increases, θ decreases and the cell's initial responses have a higher probability of going above θ; hence, the synaptic weights have a higher probability of initially increasing in strength. Notice also that increasing c_0 causes the system to stabilize at higher synaptic weights.

Finally, the magnitude of the nonlinearity p must be set. The only stringent criterion that p must satisfy is p > 1. In most of our work we set p = 2. This produces a sizable nonlinearity that leads to an acceptable evolution of the system. Generally, as p increases, θ moves more rapidly as the average cell response changes; therefore, increasing the value of p leads to smaller stabilized synaptic weights during NR.

MD

After the period of NR has driven $\boldsymbol{m}(t)$ to a stable selective state, the simulation of MD begins by "closing the left eye." That is, the input to the left eye loses its patterned component while the right eye does not (*Eq. 18*). As Fig. 7 illustrates, the newly closed-eye LGN–cortical synapses immediately begin to drop in strength, with a consequent loss

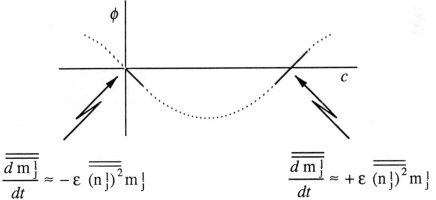

FIG. 8. Average change to the jth left eye synaptic weight (i.e., m_j^l) when the left-eye set of LGN-cortical fibers carries noise and θ is sufficiently far from 0. For cell responses scattered around 0, the average change to m_j^l is given by *Eq. 23a*, whereas for cell responses scattered around θ, the average change to m_j^l is given by *Eq. 23b*. Note that $\overline{[n_j^l]^2} = \overline{n^2}$ for all j.

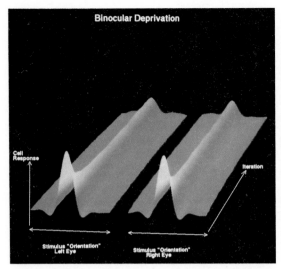

FIG. 11. Simulation of binocular deprivation. Initial state was the stable, selective equilibrium state after normal rearing (Fig. 6). This simulation ran for 200,000 iterations. Note that, on the same time scale as monocular deprivation (Fig. 7), the loss of responsiveness is not as severe during binocular deprivation. See Fig. 4 for response legend.

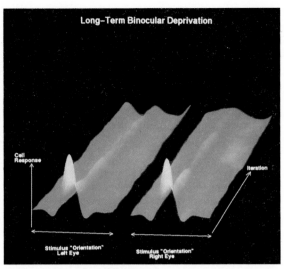

FIG. 12. Same binocular deprivation simulation as in Fig. 11, but graphed over 2,000,000 iterations. For such longer simulations, the cell loses its selectivity but retains some binocular responsiveness. See Fig. 4 for response legend.

of closed-eye response. After the closed-eye input is almost completely ineffectual in evoking a response, the open eye actually increases in efficacy and eventually stabilizes at an enhanced level; the amount of this open-eye enhancement is sensitive to the choice of parameters.

The immediate drop in the cell response after the onset of MD causes a gradual drop in θ. However, θ remains far from 0 after its initial drop because the open-eye input re-

mains effective in driving the cell. When θ is high and the total postsynaptic response is near 0, the closed-eye LGN-cortical synapses receiving noise input decrease in strength. During MD, the postsynaptic response is near 0 whenever the open eye is presented a nonpreferred pattern. On the other hand, when the total postsynaptic response is near θ,

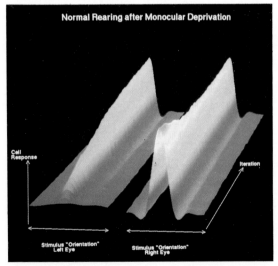

FIG. 13. A simulation of recovery after a period of monocular deprivation. This simulation ran for 200,000 iterations. Note that there is a recovery because the patterned input to the 2 eyes is correlated. See Fig. 4 for response legend.

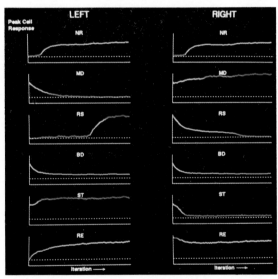

FIG. 14. Peak cell response vs. iteration for 200,000 iterations of the normal rearing, monocular deprivation, reverse suture, binocular deprivation, strabismus, and recovery simulations. These graphs represent the evolution of the left- and right-eye peak cell responses taken from Figs. 6, 7, 9, 11, 10, and 13, respectively. Dashed line is at the cell's average spontaneous level ($c = 0$). See Fig. 4 for response legend.

as would occur when a preferred pattern is shown to the open eye, the closed-eye LGN-cortical synapses receiving noise input actually increase in synaptic strength. Therefore, during MD, the ratio of preferred to nonpreferred open-eye patterns determines whether the closed-eye synaptic weights increase or decrease in strength. (This effect was analyzed by BCM; a detailed treatment is given in the APPENDIX).

The relationship between open-eye selectivity and the evolution of the closed-eye synaptic weights is an integral feature of Bienenstock et al. (1982). To illustrate this aspect of the theory in more detail, consider N consecutive iterations of the simulation, where N is much greater than the number of patterns in the training set but much less than the number of iterations necessary for a significant change in the synaptic weights. For those iterations for which nonpreferred patterns are presented to the open eye so that the cell response is close to 0 (N_0 in number), the average change to the jth closed-eye synapse is

$$\frac{\overline{\overline{dm_j^1(t)}}}{dt} \approx \eta(\epsilon_0)\overline{\overline{n^2}}m_j^1(t) \qquad (23a)$$

where ϵ_0 is the slope of ϕ at the origin and $\overline{\overline{n^2}}$ represents the average of the square of the noise input to the jth left-eye fiber. On the other hand, for all of those iterations for which preferred patterns are presented to the open eye so that the cell response is close to θ (N_θ in number), the average change to the jth closed-eye synapse is

$$\frac{\overline{\overline{dm_j^1(t)}}}{dt} \approx \eta(\epsilon_\theta)\overline{\overline{n^2}}m_j^1(t) \qquad (23b)$$

where ϵ_0 is the slope of ϕ at θ and $\overline{\overline{n^2}}$ again represents the average of the square of the noise input to the jth left-eye fiber. Combining Eqs. 23a and 23b and setting $\epsilon_\theta = -\epsilon_0 = \epsilon$ for simplicity, the average change to the jth closed-eye synaptic weight during each of the N iterations is

$$\frac{\overline{\overline{dm_j^1(t)}}}{dt} \approx \eta\left(\frac{N_\theta - N_0}{N}\right)\epsilon\overline{\overline{n^2}}m_j^1(t) \qquad (23c)$$

which has the solution

$$m_j^1(t) \approx m_j^1(0)\exp\left[\eta\left(\frac{N_\theta - N_0}{N}\right)\epsilon\overline{\overline{n^2}}t\right] \qquad (24)$$

where $m_j^1(0)$ is the synaptic strength at the beginning of the MD (see Fig. 8 and the APPENDIX). Therefore the relative number of optimal to nonoptimal cell responses through stimulation of the open eye determines whether the closed-eye synaptic weights increase or decrease. For a selective set of LGN-cortical synapses $N_\theta \ll N_0$; therefore the closed-eye synaptic weights go to 0 during MD if the cell is selective. This is the source of the correlation between ocular dominance and selectivity during MD.

As Eq. 24 indicates, the rate of disconnection of the closed eye during MD also depends on η, ϵ, and the level of the noise to the closed eye ($\overline{\overline{n^2}}$). Because η and ϵ are set to provide appropriate behavior for NR, the ratio of optimal to nonoptimal open-eye patterns and the level of the closed-eye noise are the only remaining quantities that can be varied during MD. One has the nonintuitive, experimentally testable prediction that higher levels of noise lead to a faster disconnection of the closed-eye synapses, whereas broadening of cortical orientation selectivity slows the disconnection of the closed-eye synapses.

The open-eye synaptic weights eventually increase during MD because the initial drop in θ is large enough to bias the cell response to open-eye patterned input toward the positive region of ϕ above θ. Therefore the open-eye efficacy goes up until it drives θ sufficiently high to stop further increases in the synaptic weights. The amount of this increase is sensitive to the choice of parameters.

RS

After a period of MD, RS consists of closing the open eye and opening the closed eye. For the simulation, if the MD input is given by Eq. 18, then for RS the input is switched to Eq. 19. With such a change of the input to the simulated cortical cell, the newly closed-eye LGN-cortical synaptic weights immediately begin to weaken (Fig. 9). After the newly closed eye is almost completely disconnected, the newly opened eye finally begins to show a recovery. As the newly opened eye continues to recover, the newly closed-eye synapses are eventually driven to 0.

The evolution of the synaptic weights at the beginning of a RS simulation depends critically on the behavior of θ. In this simulation θ remains high at the beginning of the RS for a relatively long time. During this time the newly closed-eye synapses are receiving noise, and the newly opened-eye synapses are weak and ineffective in driving the cell. Therefore the newly closed-eye synapses initially weaken at a rate governed by Eq. 24 with $N_\theta = 0$. Furthermore, the newly opened-eye synapses cannot recover initially because they are unable to produce responses above θ for any of the input patterns. As the simulation proceeds, θ eventually drops low enough for the poor open-eye responses to exceed it, and the open eye begins to recover. At this point the closed eye has almost no influence on the cortical cell.

If θ drops slowly at the beginning of a RS simulation, then the preceding argument indicates that the closed-eye synapses always weaken before the open eye recovers. However, if θ were to drop quickly, the closed eye would not weaken until the open eye had recovered and θ had increased. Mioche and Singer (1989) found that the previously closed eye always weakens before the newly opened eye recovers. These results indicate that θ should not drop too quickly at the beginning of a RS.

ST

Suppose that, after a period of NR, the two eyes become strabismic. The moment the input to the two eyes becomes uncorrelated (Eq. 21) in a simulation, the eye that has the dominant input to a cortical cell holds the cell in a winner-take-all situation (Fig. 10). The weaker eye becomes completely disconnected on a time scale comparable with MD. Note that the results of a ST simulation do not depend critically on any of the parameters. As long as the input is given by Eq. 21, there are no binocular, stable, selective equilibrium states; the only stable, selective equilibrium states are monocular. This result does not depend on the initial state.

BD

For BD after NR the starting point is the stable, selective equilibrium state of $m(t)$ that developed during the prior period of NR. The important aspect of the input for BD is that it does not contain patterns of correlated activity (*Eq. 20*). Once the correlated activity is removed from the input, the cell's responsiveness immediately drops. The loss in responsiveness during BD leads to a weakening of the synaptic weights; however, on the same time scale as MD, the drop in the weights during BD is not as precipitous (Fig. 11). On a longer time scale, however, the average responsiveness to stimulation of both eyes is reduced to low levels, the exact value of which is dependent on the parameters (Fig. 12). Furthermore, the original orientation tuning of the cell is completely lost. Thus, unless there were a built-in preference, due perhaps to innate nonmodifiable synapses or network effects, the return of patterned input could lead to selectivity to a different orientation.

When BD begins, the situation for both eyes' LGN-cortical synapses is similar to the newly closed eye's synapses on the cortical cell during the initial stages of a RS. In both cases there are LGN-cortical fibers carrying noise at a time when the cell response is always low and θ is high. Therefore, during the initial stages of BD after NR, *Eq. 24* with $N_\theta = 0$ is applicable to the synapses of both eyes, and both sets of closed eye synaptic weights start dropping. Because there is no patterned input to keep θ high, however, θ also starts to drop, causing N_θ to increase from 0 and thereby slowing down the rate of decrease of the synaptic weights.

The final state of the synapses depends critically on the average value of the noise along each fiber ($\bar{n}$), the average level of the noise fluctuations along each fiber ($\overline{n^2}$), and the level of the postsynaptic noise $c_{noise}(t)$. When $\bar{n} = 0$ and θ is nonzero, the only equilibrium state for the system is at $m(t) = 0$. However, when θ eventually approaches 0 in response to $m(t)$ going to 0, the synapses start undergoing a random walk that takes $m(t)$, and hence θ, away from their respective origins. The degree to which $m(t)$ departs from the origin depends on the magnitude of $\overline{n^2}$ and the amplitude of the postsynaptic noise $c_{noise}(t)$. As both $\overline{n^2}$ and the amplitude of $c_{noise}(t)$ increase, thereby increasing the scatter of the cell response about its spontaneous level, the random walk around the origin becomes more noticeable in the sense that $m(t)$ can spend more time further from the origin.

For $\bar{n} \neq 0$ the synaptic weight vector $m(t)$ is not constrained to remain at the origin on average. As $\bar{n}$ departs from 0, simulations and preliminary analyses suggest that a number of equilibrium states appear in which distance from the origin increases as the magnitude of $\bar{n}$ increases. These points are usually, but not always, unselective with respect to the original set of training patterns used during NR. Furthermore, for reasonable values of $\bar{n}$, the cell's response remains poor compared with its NR levels. These results were obtained using *Eq. 16* for θ; further investigation is required to determine the evolution of $m(t)$ for $\bar{n} \neq 0$ and different forms of θ.

In the current simulation $\bar{n} = 0$. Therefore the average value for $m(t)$ is the origin. Because the amplitude of $c_{noise}(t)$ is high, $m(t)$ fluctuates noticeably around the origin with a loss of its original selectivity (Fig. 12).

RE

RE after a period of MD is simulated by changing the input to the cell to that of NR. In this case the eye that was closed during the MD regains the response properties that it acquired during the initial period of NR (Fig. 13). However, if the two eyes are assumed to have become misaligned during the period of MD, then during the recovery period the input must be given by *Eq. 21*. Uncorrelated input to a cell that is already completely shifted to one eye does not produce any significant changes, and the initially deprived eye never recovers. Because ST is sometimes a result of MD, this may explain why recovery of normal binocularity is sometimes, but not always, observed.

As already mentioned, the loss of binocularity during uncorrelated patterned input to the two eyes is not strongly dependent on any of the parameter settings and is solely due to a lack of correlation in the input from the two eyes. However, if the eyes are assumed to be aligned during a recovery period after MD, the properties of the noise $n_j(t)$ become important. As the variance of the noise increases (i.e., as $\overline{n^2}$ goes up), the rate of recovery of the initially deprived-eye synapses increases. Because $\overline{n^2}$ is always assumed to be nonzero, the use of *Eq. 17* for the input to the cell always leads to a stable, selective equilibrium state that is binocular.

DISCUSSION

Comparison with experiment

Our simulations show that the BCM theory leads to an evolution of the cell response to equilibrium states in agreement with "classical" visual deprivation experiments in kitten striate cortex. The simulations were produced with a single set of parameter settings, and each simulation was started using the equilibrium state of a preceding simulation (with the exception of NR, which started from randomized synaptic weights). Starting a simulation from the equilibrium state of a preceding simulation is not necessary; however, this facilitates comparison of simulations with experimental results.

Of particular interest is the relative timing for various effects during the simulated forms of deprivation (Fig. 14). To facilitate comparisons of the timing of synaptic modification among the various simulations, as well as comparisons of simulations with actual experimental observations, we can convert simulation time (i.e., iteration) to real time. It is understood, of course, that this conversion is primarily for comparative purposes, and would be affected by such things as the age of the animal, periods of sleep, etc. These caveats notwithstanding, we can use MD to calculate a conversion of iteration to real time. At the height of the critical period, MD leads to a loss of responsiveness to the closed eye in 24 hours (Mioche and Singer 1989). In the simulation, with our parameter choice, the closed eye disconnects in 67,000 iterations. Therefore, with these parameters, one iteration corresponds to 1.3 s of real time.

Thus, in the simulation of RS, the deprived eye disconnects significantly in 24 h and almost completely by the 48th h. During the BD simulation, there is a significant loss in responsiveness during the first 24 h; however, thereafter some responsiveness is retained. ST simulations lead to a disconnection of one eye in 14 h, which is more severe than MD. During NR simulations, once the cell response begins to increase after 10 h of patterned experience, it climbs rapidly during the next 14 h to near its final state. The recovery of the previously closed eye during RS simulations also occurs on the order of 1 day. Notice in the RS simulation, however, that the previously closed eye does not begin to recover until θ is sufficiently low to allow the poor-cell response to exceed it; this takes ~24 h. By this time the currently closed eye is almost completely disconnected. When normal, binocular input is restored after a period of MD, the recovery of the closed eye in the simulation takes place on a much longer time scale, showing gradual gains over 3 days.

The kinetics of our simulations are in good agreement with experimental results. The loss of closed-eye responsiveness during simulations of MD and RS occurs over approximately the same time scales of 24 and 24–48 h, respectively, agreeing with the experimental findings of Mioche and Singer (1989). Mioche and Singer (1989) hypothesize that the same mechanism is at play in the disconnection of the closed eye during these two experiments; this idea can be accounted for in the BCM theory by the theoretical result that noise input to a set of synapses at a time when the modification threshold is high leads to a decrement in their respective synaptic weights. The rapid loss of one eye's synaptic efficacy during the ST simulation is actually more severe than the deterioration of the closed-eye synaptic weights during MD. (This is one illustration of a general consequence of the theory that the rate of disconnection increases with increased fluctuations of input to the deteriorating eye.) Although we know of no experimental evidence that specifies the absolute rate at which one eye's responsiveness is lost during ST, the final monocular state of $m(t)$ obtained during the ST simulation agrees with all of the experimental findings. The loss of responsiveness during the BD simulation also has many features in agreement with experiment: it is less severe than MD in the short term and leads to a generally poor, unselective response in the long term. Finally, the relatively rapid acquisition of selectivity and responsiveness is consistent with results observed in kittens reared in complete darkness and then exposed to binocular patterns during the second postnatal month.

The evolution of open-eye synaptic weights during the simulation of MD and the effective-eye synaptic weights during the simulation of ST leads to an enhancement of the cell response to that eye. In the case of the MD simulation, the enhancement in the open-eye response does not occur until after the closed eye is almost completely disconnected, which appears to be in keeping with the experimental results of Mioche and Singer (1989) in those instances in which an enhancement of the open-eye response was observed. Experimental confirmation of the increase in effective-eye response during the ST simulation is more difficult

to determine because there are apparently no studies that have directly investigated this point. However, we emphasize that, unlike the disconnection of the closed eye during MD and RS, this increased responsiveness depends on parameters, the form of θ, and the properties of the equilibrium state obtained during the prior period of NR.

The largest potential discrepancy between the kinetics of the simulations and experimental findings is the quick recovery of the currently open eye during RS. Experimentally, the recovery of this eye is a slow and incomplete process, taking from 48 to 72 h to exhibit just partial recovery (Mioche and Singer 1989). In the simulation, on the other hand, the recovery is rapid and complete within 24 h; once the cell response to patterned input begins to exceed θ, the synapses receiving the patterned input recover in a robust fashion. To slow down the recovery of the open eye during RS, other features would likely have to be incorporated; these might include changes in modifiability with increasing age (i.e., the critical period) and the potential irreversible loss of some connections that are deprived for too long. It is also possible that Mioche and Singer (1989) underestimate the speed of recovery of a healthy cell because they are recording from neurons that may have deteriorated after several days of chronic recording.

Sensitivity of results to parameter values

We now discuss the dependence of our results on the choice of parameters (see Table 3). Consider first RS. As mentioned, Mioche and Singer (1989) find that the newly closed eye's ability to drive the postsynaptic cell significantly lessens before the currently open eye input to the cell strengthens. For us to obtain this result, the modification threshold θ must drop sufficiently slowly with the advent of the RS while $\overline{\overline{n^2}} \neq 0$ along the closed-eye fibers (Eq. 23a). Keeping θ high guarantees both the disconnection of the closed eye from the cell, as long as $\overline{\overline{n^2}} \neq 0$, and the lack of a recovery in the open eye until the closed eye disconnects.

TABLE 3. *Sensitivity of results to parameter values*

Parameter	NR	MD	RS	BD	ST	RE
N_{gl}	0	0	0	0	0	0
N_p	0	0	0	0	0	0
$d_{a,j}^{o,l(r)}$	0	0	0	0	0	0
d_s	0	0	++	++	0	0
$\overline{\overline{n}}$	0	+	+	++	0	0
$\overline{\overline{n^2}}$	0	++	++	++	0	++
$m_j^{l(r)}(0)$	0	0	0	0	0	0
$\overline{\overline{c}}_{noise}$						
$\overline{\overline{c^2}}_{noise}$	+	+	+	++	+	0
η	++	0	0	0	0	0
τ	++	0	++	++	0	0
p	++					
c_0	+					

Value of the parameter either has a critical effect on the results of the simulation (++), affects the outcome of the simulation but is not critical (+), or does not affect the outcome of the simulation (0). Blank spaces indicate that the effect of the parameter on the simulation has not been studied in detail. NR, normal rearing; MD and BD, monocular and binocular deprivation, respectively; RS, reverse suture; ST, strabismus; RE, recovery.

Comparisons with experiment indicate that θ should drop from its pre-RS level to the low level of the postsynaptic response during the initial stages of a RS in $\sim$24–28 h.

The rate of change of θ is determined by the memory of the cell response time average (τ) and the definition of θ. In these simulations τ corresponds to 22 min of real time. As we have discussed before, τ cannot be arbitrarily increased (for a given value of η) without introducing unwanted oscillations. Setting τ approximately equal to 24 h would produce the desired evolution of the synapses during RS; however, τ would now be much too large for the current value of η to lead to acceptable behavior during NR (RE-SULTS, *NR*). Therefore, for θ to stay above the prevailing low postsynaptic response (c) for >22 min during the initial stages of a RS, it must have a dependence on something other than just the response c itself. In this paper we have defined θ in terms of $c_a(t)$; as the results indicate, this form of θ produces the desired evolution of the synapses during RS.

The critical difference between $c(t)$ and $c_a(t)$ in the time average for θ is the average level of the spontaneous activity (d_s). As d_s increases, $c_a(t)$ depends less on the cell response to patterned input per se and more on the overall average fiber activity, including the spontaneous activity, as well as on the synaptic weights themselves. Making θ dependent on total depolarization rather than on deviations of the depolarization from its average level reduces its sensitivity to sudden changes in the visual environment. When removal of patterned input to a cell leads to a radical change in the cell response, θ adjusts more slowly to the change with larger values of d_s. In the current simulations $d_s = 5.0$, which keeps θ sufficiently high to prevent the open eye from recovering during RS until the closed eye is almost completely disconnected.

With our parameter choice, the maximum displacement from spontaneous LGN fiber activity for pattern input is 1.0 ($d_{peak} + d_{base} - d_s = 1.0$), which is much smaller than d_s in Table 2. It must be emphasized that this does not mean that the level of spontaneous activity in a given iteration is 80% of the total LGN response. The magnitude of d_{peak} relative to d_s should be interpreted as reflecting the frequency of occurrence of patterned LGN input activity against a continuous background of noise. From this point of view, increasing d_s amounts to decreasing the frequency of occurrence of patterned LGN input compared with noisy LGN input. Simulations with $d_{peak} = 100$, $d_s = 10$, and patterned input occurring every sixth iteration between five consecutive iterations of noise input lead to an evolution of the synaptic weights that is equivalent to Figs. 6–13. Finally, we note that for d_s even as small as 0.2, there is still a significant ($\approx$60%) drop in the closed-eye response before the open eye recovers during RS.

For $\overline{\overline{n^2}} \neq 0$ the closed eye is guaranteed to disconnect during MD because the open-eye ratio of optimal to nonoptimal patterns remains small (*Eq. 24*). Therefore, for the simulations to be consistent with the MD experimental findings, $\overline{\overline{n^2}} \neq 0$. Notice that the mechanism of the closed-eye disconnection is the same during MD and RS, as long as θ drops slowly during RS and the cell response to open-eye input remains selective during MD.

Two other experiments that help set the magnitude of $\overline{\overline{n^2}}$

are ST and RE. In the ST simulation, one of the two eyes completely disconnects from the cell in $\sim$14 h, whereas it takes $\sim$24 h for a disconnection of the closed-eye input during MD. If the rates of change of the two sets of synapses actually occur on the same time scale during ST and MD, then the time for disconnection in MD could be decreased to 14 h by increasing the magnitude of $\overline{\overline{n^2}}$. However, changing $\overline{\overline{n^2}}$ also changes the rate at which the closed eye recovers during RE (RESULTS, *RE*). Therefore detailed quantitative knowledge of the time rates of cell response changes during experiments of MD, ST, and RE would lead to more stringent constraints on $\overline{\overline{n^2}}$.

With η, τ, and $\overline{\overline{n^2}}$ set by NR, MD, RS, ST and RE, and with a given form of θ, the evolution of the synapses during the initial stages of BD is fixed. With the onset of BD, the cell response drops rapidly. In the current simulations both eyes' LGN-cortical synaptic weights significantly weaken as θ drops to the prevailing postsynaptic cell response over the first 24–48 h. The situation is reminiscent of RS, in which the cell response is low and θ is initially high and drops slowly due to the large value of d_s; however, in the BD simulation there is no patterned input to an eye that eventually increases cell response and therefore θ. It follows that, once θ reaches the low response level during BD, the rate of weakening of the synaptic weights gradually diminishes.

As detailed in RESULTS, the long-term evolution of the synapses during BD depends on parameters such as $\overline{n}$ and $\overline{c_{noise}^2}$, in addition to $\overline{\overline{n^2}}$. The parameters $\overline{n}$ and $\overline{c_{noise}^2}$ have subtle effects on the evolution of the synapses during simulations of NR, MD, RS, and ST, as well. For example, with the onset of patterned binocular input to "immature" synapses (i.e., NR), the high level of $\overline{c_{noise}^2}$ prevents both eyes from immediately acquiring selectivity and responsiveness (Fig. 14); with the current parameter settings it takes $\sim$10 h for the cell to sort out the LGN-evoked response from the high levels of postsynaptic noise. The eventual acquisition of selectivity in the presence of such large levels of noise ($\overline{c_{noise}^2} = 33.3$, Table 2) demonstrates that the system is robust to large fluctuations in the postsynaptic response. In the ST simulation, the slight retention of response to stimulation of the disconnected eye is also due to the large level of $\overline{c_{noise}^2}$; if $\overline{c_{noise}^2}$ is set to 0, then the weak eye immediately loses its ability to evoke any response from the cell. Finally, large levels of $\overline{c_{noise}^2}$ generally decrease the rate at which the closed eye synapses disconnect during MD and RS because the assumptions leading to *Eq. 23a* are weakened.

Apart from BD, other effects of changing $\overline{n}$ appear during MD and RS. As $\overline{n}$ departs from 0, the approximations leading to *Eq. 23a* become less valid (see APPENDIX). Therefore the rate of disconnection of the closed eye synapses begins to decrease for $\overline{n}$ sufficiently large in both the MD and RS cases.

The parameters p and c_0 are not rigidly set by simulations of the "classical" experiments. In the current simulations we choose p = 2; this produces a convenient nonlinearity in the calculation of θ consistent with the theory. c_0 is chosen sufficiently large to ensure that θ is below the majority of the cell's responses at the start of NR, leading to an initial enhancement of the cell's responsiveness. Other than these considerations, there are as yet no binding experimental constraints on p and c_0.

The slopes of ϕ at the origin (ϵ_0) and at θ (ϵ_θ) are constrained, so that ϵ_0 must be <0 and ϵ_θ must be >0. As *Eq. 23* indicates, their magnitudes provide another degree of freedom in controlling the disconnection of the closed eye, or eyes, during MD, RS, and BD. However, there are not yet any experimental results that directly determine these magnitudes. The uncertainty in the magnitudes ϵ_0 and ϵ_θ suggests a certain freedom in specifying the exact functional form of ϕ. Although ϕ must be <0 for $0 < c < \theta$ and ϕ must be > 0 for $c > \theta$, as many simulations show, the exact form of ϕ in these two regions does not appear to be critical. Perhaps of more interest is the dependence of ϕ on c for $c < 0$. The only theoretical constraint on ϕ is that it must not be too negative for $c < 0$ (see APPENDIX).

In summary, NR places constraints on p, τ, η, and c_0. RS, with the closed eye disconnecting before the open eye recovers, places constraints on τ and the form of θ. Once τ and the form of θ are set by RS, the evolution of the synaptic weights during BD is fixed. The closed-eye disconnection during MD, RS, and BD indicates that $\overline{n^2} > 0$. The rate of disconnection of the closed eye during MD and RS, compared with the loss of cell response to one eye's input during ST together with the recovery of the previously closed eye during RE, places constraints on the magnitude of $\overline{n^2}$. The parameter settings $\overline{\overline{n}}$ and $\overline{c_{noise}^2}$ have secondary consequences that affect the equilibrium states and the rapidity with which they are reached during simulations of several of the paradigms, especially BD. In this paper we have set the non-LGN postsynaptic fluctuations about the cell's spontaneous level, so that $\overline{c_{noise}}$ is 0. With the constraints on the parameters just outlined, a major implication of this investigation is that the modification threshold θ changes from its value acquired during monocular deprivation to near 0 on the order of 24–48 h of RS.

Alternate forms for θ

We emphasize that the definition of θ in *Eq. 16* is just one of several possibilities that has the properties outlined in METHODS (*Synaptic modification*). Another possibility, for example, is θ defined by *Eq. 15*, where the nonlinearity is applied to the cell response before the averaging takes place. Although the different dependences of θ have relatively small effects on the evolution of synapses during the different paradigms described in the present work, they have very different mathematical properties and strikingly different physiological implications. Preliminary investigation indicates that they might also have different consequences on the simulated outcomes of various pharmacological experiments, such as the blockade of cortical *N*-methyl-D-aspartate (NMDA) receptors (Bear et al. 1990). Regardless of its exact form, the RS finding of Mioche and Singer (1989) implies that the threshold slides from its NR value to near 0 in ~24–48 h.

Extension to networks

The single-cell results presented here can be interpreted within the framework of a cortical network of cells. As mentioned in the introductory paragraphs of METHODS, placing the single cell in a mean-field network does not change the position and/or stability properties of the equilibrium states; however, there may be changes in the dynamics of the evolution of the synapses during simulations of the paradigms considered here that may refine some of the constraints on the parameters. In this context the validity of the mean-field approximation becomes an interesting experimental question, for, if the intracortical connections can be described as an average effect on a cell, then the single-cell simulations are an accurate description of what is happening in a network.

Placing the single cell in a network of cells has other important consequences. For example, in the single-cell simulations the spontaneous activity increases and decreases with the cell's synaptic weights (*Eq. 9*). However, in a network of cells the spontaneous activity of any one cell is not necessarily tied so closely to its LGN input. In fact, all of the influence to a cell from other excitatory and inhibitory cells in a network could possibly make the spontaneous activity of a cell to some extent independent of its LGN input. This is a complicating factor not explicitly considered in the present work.

Possible mechanisms

We are now led to the question of the biological basis of the theoretical form of modification employed successfully in this paper. Our work suggests that the modifications of interest may occur mainly at excitatory geniculocortical synapses; the postsynaptic responses at these synapses are mediated by excitatory amino acid receptors. Thus mechanisms linked to excitatory amino acid receptors seem to be an appropriate place to look for the molecular basis of visual cortical plasticity.

Theory requires that input activity that coincides with postsynaptic activation beyond θ leads to an enhancement of synaptic efficacy. Work on the hippocampus and visual cortex in vitro has shown that the pairing of input activity with postsynaptic depolarization can lead to a long-term potentiation (LTP) of the active synapses (Artola et al. 1990; Frégnac et al. 1990; Gustafsson et al. 1987; Kelso et al. 1986). In these locations LTP induction appears to depend on a voltage-dependent Ca^{2+} conductance that is mediated by NMDA receptors. Thus Bear et al. (1987) have proposed that θ might be related to the membrane potential at which the NMDA-receptor–dependent Ca^{2+} flux reaches the threshold for inducing synaptic long-term potentiation.

One consequence of the association of θ with NMDA-receptor mechanisms is that input activity that consistently fails to correlate with postsynaptic activation sufficient to recruit an NMDA-receptor–mediated Ca^{2+} flux should lead to a long-term depression (LTD) of synaptic efficacy. Such a form of modification has been observed recently in both hippocampus (Chattarji et al. 1989; Stanton and Sejnowski 1989; Staubli and Lynch 1990) and visual cortex (Artola and Singer 1990; Bear et al. 1990; Frégnac et al. 1990; Kimura et al. 1990). The mechanism of this form of plasticity is unknown; however, recent experimental evidence suggests that activation of a class of non-NMDA receptors (the "metabotropic" quisqualate receptors) stimulates the hydrolysis of membrane inositol phospholipids in kitten striate cortex (Dudek and Bear 1989). These considerations have led to the suggestion that phosphoinositide hydrolysis might provide the biochemical trigger for use-dependent decreases in synaptic efficacy (Bear 1988; Bear and Cooper 1990; Dudek and Bear 1989). In this context it is interesting to note that the developmental time course of excitatory amino acid–stimulated phosphoinositide turnover correlates precisely with the critical period for synaptic modification in kitten visual cortex (Dudek and Bear 1989). As yet, however, the precise mechanism of this form of plasticity is unknown.

Another consequence of the association of θ with NMDA-receptor mechanisms is that the effectiveness of NMDA-receptor activation in triggering synaptic modification should be a function of the recent history of cortical cell activity (Bear et al. 1987; Bear and Cooper 1990). In principle, this could occur in a number of ways, including the activity-dependent regulation of NMDA-receptor sensitivity, postsynaptic Ca^{2+} buffers or pumps, Ca^{2+}-activated enzymes, opposing mechanisms of synaptic weakening, etc.

Our work suggests that the value of θ is low after ~48 h of BD in kitten striate cortex. Recent data suggest that, although the density of NMDA receptors is unaffected by 4–6 days of BD (I. J. Reynolds and M. F. Bear, unpublished data), NMDA-stimulated $^{45}Ca^{2+}$ accumulation is significantly decreased in visual cortical slices prepared from BD (but not MD) animals (Feldman et al. 1990). One explanation (among many) for this result is that cortical inactivity causes a decrease in intracellular Ca^{2+}-binding proteins. This hypothesis is particularly attractive in light of work by Holmes and Levy (1990) and Zador et al. (1990), who suggested that induction of LTP by NMDA-receptor activation might be particularly sensitive to changes in Ca^{2+} buffers in dendritic spines. In fact, Lowenstein et al. (1991) have reported very recently that the genetic expression of calbindin, a high-affinity Ca^{2+} buffer is increased in dentate gyrus by focal activation of the perforant path. Regardless of the mechanism, however, these $^{45}Ca^{2+}$ uptake experiments indicate that cortical Ca^{2+} homeostasis can vary significantly as a function of activity.

Detailed discussions of these hypothetical molecular mechanisms have been published elsewhere (e.g., Bear and Cooper 1990; Bear and Dudek 1991). Regardless of whether these specific hypotheses ultimately prove to be correct, this work demonstrates that the theory developed here can serve as a bridge between molecular mechanisms and a description of visual cortical plasticity. We are able to follow a long chain of arguments and to connect, in a fairly precise way, various hypotheses with their consequences. The understanding this provides, we believe, makes visual cortex an ideal preparation with which to study the mechanisms of experience-dependent synaptic modification and its relationship to behavior.

APPENDIX

To show why the closed eye disconnects during MD, we analyze in more detail the original arguments given by Bienenstock et al. (1982). According to Eq. 13, the modification of the synaptic weights occurs as the product of the input activity d_j and ϕ. When nonoptimal patterns are presented to the open (right) eye and

noise to the deprived (left) eye, the cell response falls near zero. At this low level of cell response, ϕ can be approximated by a line with a negative slope $-\epsilon$; that is, $\phi \approx -\epsilon c$ (Fig. 8). Therefore the jth left-eye synaptic weight modifies as the product of $-\epsilon c$ and its input activity $n_j^l(t)$

$$\frac{dm_j^l(t)}{dt} \approx \eta(-\epsilon c)n_j^l(t) \qquad (A1)$$

Substituting the definition of c (Eq. 12) into Eq. A1 gives

$$\frac{dm_j^l(t)}{dt} \approx \eta(-\epsilon)[\boldsymbol{m}^l(t) \cdot \boldsymbol{n}^l(t)n_j^l(t) + \boldsymbol{m}^r(t) \cdot \boldsymbol{d}^r(t)n_j^l(t)] \quad (A2a)$$

or

$$\frac{dm_j^l(t)}{dt} \approx \eta(-\epsilon)\{[\sum_{k=1}^{N_{sl}} m_k^l(t)n_k^l(t)]n_j^l(t) + [\sum_{k=1}^{N_{sl}} m_k^r(t)d_k^r(t)]n_j^l(t)\} \qquad (A2b)$$

Averaging Eq. A2b over N_0 nonoptimal open eye patterns, where N_0 is small compared with the number of iterations necessary for a significant change in the synaptic weights, yields

$$\frac{\overline{\overline{dm_j^l(t)}}}{dt} \approx \eta(-\epsilon)\{\sum_{k=1}^{N_{sl}} m_k^l(t)\left[\frac{1}{N_0}\sum_{s=1}^{N_0} n_k^l(t_s)n_j^l(t_s)\right] + \sum_{k=1}^{N_{sl}} m_k^r(t)\left[\frac{1}{N_0}\sum_{s=1}^{N_0} d_k^r(t_s)n_j^l(t_s)\right]\} \qquad (A3)$$

where the double bar represents the average change to the jth left-eye synaptic weight over the N_0 nonoptimal patterns. Note that t_s, $s = 1, \ldots, N_0$ denotes N_0 iterations of the simulation centered at time t; the synaptic weights are assumed to change negligibly during the N_0 iterations. Because the noise inputs to the different left-eye fibers are independent of each other

$$\frac{1}{N_0}\sum_{s=1}^{N_0} n_k^l(t_s)n_j^l(t_s) = \left[\frac{1}{N_0}\sum_{s=1}^{N_0} n_k^l(t_s)\right]\left[\frac{1}{N_0}\sum_{s=1}^{N_0} n_j^l(t_s)\right] = \left(\overline{\overline{n}}\right)^2 \quad (A4a)$$

for $k \neq j$ and

$$\frac{1}{N_0}\sum_{s=1}^{N_0} n_k^l(t_s)n_j^l(t_s) = \frac{1}{N_0}\sum_{s=1}^{N_0} [n_j^l(t_s)]^2 = \overline{\overline{n^2}} \qquad (A4b)$$

for $k = j$. Using Eq. 8 for $d_k^r(t_s)$

$$\frac{1}{N_0}\sum_{s=1}^{N_0} d_k^r(t_s)n_j^l(t_s) = \frac{1}{N_0}\sum_{s=1}^{N_0} d_k^{w,r}n_j^l(t_s) + \frac{1}{N_0}\sum_{s=1}^{N_0} n_k^r(t_s)n_j^l(t_s) \qquad (A5a)$$

Because the noise on the right-eye fibers is independent of the noise and patterns on the left-eye fibers, Eq. A5a is equivalent to

$$\left\{\left[\frac{1}{N_0}\sum_{s=1}^{N_0} d_k^{w,r}\right]\left[\frac{1}{N_0}\sum_{s=1}^{N_0} n_j^l(t_s)\right]\right\} + \left\{\left[\frac{1}{N_0}\sum_{s=1}^{N_0} n_k^r(t_s)\right]\left[\frac{1}{N_0}\sum_{s=1}^{N_0} n_j^l(t_s)\right]\right\} = \overline{d^w}\overline{\overline{n}} + \overline{\overline{n}}^2 \quad (A5b)$$

Assuming that $\overline{\overline{n}} \approx 0$, Eq. A3 becomes

$$\frac{\overline{\overline{dm_j^l(t)}}}{dt} \approx \eta(-\epsilon)\overline{\overline{n^2}}m_j^l(t) \qquad (A6)$$

for the average change to the jth left-eye synaptic weight when nonoptimal patterns are presented to the open right eye.

Now when optimal patterns are presented to the responsive and selective open (right) eye and noise to the deprived (left) eye, the cell response falls near θ. At this high level of cell response, ϕ can be approximated by a line with a positive slope ϵ; that is, $\phi \approx \epsilon(c - \theta)$ (Fig. 8). Therefore the jth left-eye synaptic weight modifies as the product of $\epsilon(c - \theta)$ and its input activity $n_j^l(t)$

$$\frac{dm_j^l(t)}{dt} \approx \eta\epsilon(c - \theta)n_j^l(t) \tag{A7}$$

With using the same averaging procedures and assumptions as for the nonoptimal patterns, the average change to the jth left-eye synaptic weight over the optimal open-eye patterns becomes

$$\frac{\overline{dm_j^l(t)}}{dt} \approx \eta(\epsilon)\overline{\overline{n^2}}m_j^l(t) \tag{A8}$$

The closed-eye synaptic weights weaken when the open eye receives nonpreferred input ($Eq.\ A6$) and strengthen when the open eye receives preferred input ($Eq.\ A8$). To produce the weakening signal (i.e., $Eq.\ A6$), we assumed that $\phi \approx -\epsilon c$ near 0. However, we are not confined to this choice. In fact

$$\phi = \begin{cases} -\epsilon c & c \geq 0 \\ 0 & c < 0 \end{cases} \tag{A9}$$

is also a viable form for ϕ near the origin. In this case $Eqs.\ A1–A6$ still hold for positive responses near the origin. Because $\phi = 0$ for $c < 0$, any responses <0 lead to no change in any of the synaptic weights. Let N_{0+} represent the number of iterations in which $c \geq 0$ and N_{0-} represent the number of iterations in which $c < 0$ ($N_0 = N_{0+} + N_{0-}$). The average change to the jth left-eye synaptic weight is simply $Eq.\ A6$ weighted by the ratio N_{0+}/N_0

$$\frac{\overline{dm_j^l(t)}}{dt} \approx \eta\left(\frac{N_{0+}}{N_0}\right)(-\epsilon)\overline{\overline{n^2}}m_j^l(t) \tag{A10}$$

Therefore, for cell responses near the origin, the closed-eye synaptic weights still decrease, but at a rate reduced by N_{0+}/N_0. Because $N_{0+} \approx N_{0-}$ for random, uniformly distributed noise, the closed-eye synaptic weights weaken at one-half the rate compared with $Eq.\ A6$.

There are limitations, however, on the form of ϕ near the origin. For example

$$\phi' = \begin{cases} -\epsilon c & c \geq 0 \\ \epsilon c & c < 0 \end{cases} \tag{A11}$$

is an example of a function that does not produce a disconnection of the closed-eye synaptic weights for responses near the origin. As expected, for the N_{0+} iterations in which $c \geq 0$, $Eqs.\ A1–A6$ remain valid. The analysis for the N_{0-} iterations in which $c < 0$ is identical to $Eqs.\ A1–A6$ but with $-\epsilon$ replaced by ϵ. Therefore the average change to the jth left-eye synaptic weight after N_0 iterations is

$$\frac{\overline{dm_j^l(t)}}{dt} \approx \eta\left(\frac{N_{0-} - N_{0+}}{N_0}\right)(-\epsilon)\overline{\overline{n^2}}m_j^l(t) \tag{A12}$$

For random, uniformly distributed noise in this case (i.e., $N_{0+} \approx N_{0-}$), the closed-eye synaptic weights would not weaken when the open eye receives nonpreferred input. On average, they would increase in strength because of the open-eye preferred input. Therefore, to guarantee that the closed-eye synaptic weights decrease during MD, we take $\phi \geq 0$ for $c < 0$.

We thank J. I. Gold and D. G. Aliaga for invaluable assistance in constructing the color figures used in this paper.

This work was supported in part by the Office of Naval Research (Contract N00014-86-K-0041), the National Science Foundation (Contracts EET-8719102 and DIR-8720084) and the Army Research Office (Contract DAAL03-88-K-0116).

Address for reprint requests: L. N. Cooper, Dept. of Physics and Center for Neural Science, PO Box 1843, Brown University, Providence, RI 02912.

Received 27 March 1991; accepted in final form 10 July 1991.

REFERENCES

ALBUS, K. AND WOLF, W. Early post-natal development of neuronal function in the kitten's visual cortex: a laminar analysis. *J. Physiol. Lond.* 348: 153–185, 1984.

ARTOLA, A., BRÖCHER, S., AND SINGER, W. Different voltage-dependent thresholds for inducing long-term depression and long-term potentiation in slices of rat visual cortex. *Nature Lond.* 347: 69–72, 1990.

ARTOLA, A. AND SINGER, W. The involvement of N-methyl-D-aspartate receptors in induction and maintenance of long-term potentiation in rat visual cortex. *Eur. J. Neurosci.* 2: 254–269, 1990.

BARLOW, H. B. Visual experience and cortical development. *Nature Lond.* 258: 199–204, 1975.

BARLOW, H. B. AND LEVICK, W. R. Changes in the maintained discharge with adaptation level in the cat retina. *J. Physiol. Lond.* 202: 699–718, 1969.

BARLOW, H. B. AND PETTIGREW, J. D. Lack of specificity of neurones in the visual cortex of young kittens (Abstract). *J. Physiol. Lond.* 218: 98P–100P, 1971.

BEAR, M. F. Involvement of excitatory amino acid receptors in the experience-dependent development of visual cortex. In: *Frontiers in Excitatory Amino Acid Research*, edited by E. Cavalheiro, J. Lehmann, and L. Turski. New York: Liss, 1988, vol. 46, p. 393–401.

BEAR, M. F. AND COOPER, L. N. Molecular mechanisms for synaptic modification in the visual cortex: interaction between theory and experiment. In: *Neuroscience and Connectionist Theory*, edited by M. A. Gluck and D. E. Rumelhart. Hillsdale NJ: Erlbaum, 1990, p. 65–93.

BEAR, M. F., COOPER, L. N., AND EBNER, F. F. A physiological basis for a theory of synapse modification. *Science Wash. DC* 237: 42–48, 1987.

BEAR, M. F. AND DUDEK, S. M. Excitatory amino acid stimulated phosphoinositide turnover: pharmacology, development and role in visual cortical plasticity. *Ann. NY Acad. Sci.* 627: 42–56, 1991.

BEAR, M. F., KLEINSCHMIDT, A., GU, Q., AND SINGER, W. Disruption of experience-dependent synaptic modifications in striate cortex by infusion of an NMDA receptor antagonist. *J. Neurosci.* 10: 909–925, 1990.

BIENENSTOCK, E. L., COOPER, L. N., AND MUNRO, P. W. Theory for the development of neuron selectivity: orientation specificity and binocular interaction in visual cortex. *J. Neurosci.* 2: 32–48, 1982.

BLAKEMORE, C. The conditions required for the maintenance of binocularity in the kitten's visual cortex. *J. Physiol. Lond.* 261: 423–444, 1976.

BLAKEMORE, C. AND EGGERS, H. M. Effects of artificial anisometropia and strabismus on the kitten's visual cortex. *Arch. Ital. Biol.* 116: 385–389, 1978.

BLAKEMORE, C. AND VAN SLUYTERS, R. C. Experimental analysis of amblyopia and strabismus. *Br. J. Ophthalmol.* 58: 176–182, 1974a.

BLAKEMORE, C. AND VAN SLUYTERS, R. C. Reversal of the physiological effects of monocular deprivation in kittens: further evidence for a sensitive period. *J. Physiol. Lond.* 237: 195–216, 1974b.

BLAKEMORE, C. AND VAN SLUYTERS, R. C. Innate and environmental factors in the development of the kitten's visual cortex. *J. Physiol. Lond.* 248: 663–716, 1975.

BLASDEL, G. G. AND PETTIGREW, J. D. Effect of prior visual experience on cortical recovery from the effects of unilateral eyelid suture in kittens. *J. Physiol. Lond.* 274: 601–619, 1978.

BLASDEL, G. G. AND PETTIGREW, J. D. Degree of interocular synchrony required for maintenance of binocularity in kitten's visual cortex. *J. Neurophysiol.* 42: 1692–1710, 1979.

BONDS, A. B. Development of orientation tuning in the visual cortex of kittens. In: *Developmental Neurobiology of Vision*, edited by R. D. Freeman. New York: Plenum, 1979, p. 31–41.

BRAASTAD, B. O. AND HEGGELUND, P. Development of spatial receptive-field organization and orientation selectivity in kitten striate cortex. *J. Neurophysiol.* 53: 1158–1178, 1985.

BUISSERET, P. AND IMBERT, M. Visual cortical cells: their developmental properties in normal and dark reared kittens. *J. Physiol. Lond.* 255: 511–525, 1976.

BUISSERET, P., GARY-BOBO, E., AND IMBERT, M. Ocular motility and recovery of orientational properties of visual cortical neurones in dark-reared kittens. *Nature Lond.* 272: 816–817, 1978.

BUISSERET, P., GARY-BOBO, E., AND IMBERT, M. Plasticity in the kitten's visual cortex: effects of the suppression of visual experience upon the orientational properties of visual cortical cells. *Dev. Brain Res.* 4: 417–426, 1982.

BURCHFIEL, J. L. AND DUFFY, F. H. Role of intracortical inhibition in deprivation amblyopia: reversal by microiontophoretic bicuculline. *Brain Res.* 206: 479–484, 1981.

CHATTARJI, S., STANTON, P. K., AND SEJNOWSKI, T. J. Commissural synapses, but not mossy fiber synapses, in hippocampal field CA3 exhibit associative long-term potentiation and depression. *Brain Res.* 495: 145–150, 1989.

COOPER, L. N., LIBERMAN, F., AND OJA, E. A theory for the acquisition and loss of neuron specificity in visual cortex. *Biol. Cybern.* 33: 9–28, 1979.

COOPER, L. N. AND SCOFIELD, C. L. Mean-field theory of a neural network. *Proc. Natl. Acad. Sci. USA* 85: 1973–1977, 1988.

CYNADER, M. Prolonged sensitivity to monocular deprivation in dark-reared cats: effects of age and visual exposure. *Dev. Brain Res.* 8: 155–164, 1983.

CYNADER, M., BERMAN, N., AND HEIN, A. Recovery of function in cat visual cortex following prolonged deprivation. *Exp. Brain Res.* 25: 139–156, 1976.

CYNADER, M. AND MITCHELL, D. E. Monocular astigmatism effects on kitten visual cortex development. *Nature Lond.* 270: 177–178, 1977.

DUDEK, S. M. AND BEAR, M. F. A biochemical correlate of the critical period for synaptic modification in the visual cortex. *Science Wash. DC* 246: 673–675, 1989.

DUFFY, F. H., SNODGRASS, S. R., BURCHFIEL, J. L., AND CONWAY, J. L. Bicuculline reversal of deprivation amblyopia in the cat. *Nature Lond.* 260: 256–257, 1976.

FELDMAN, D., SHERIN, J. E., PRESS, W. A., AND BEAR, M. F. N-methyl-D-aspartate–evoked calcium uptake by kitten visual cortex maintained *in vitro. Exp. Brain Res.* 80: 252–259, 1990.

FREEMAN, R. D., MALLACH, R., AND HARTLEY, S. Responsivity of normal kitten striate cortex deteriorates after brief binocular deprivation. *J. Neurophysiol.* 45: 1074–1084, 1981.

FREEMAN, R. D. AND OHZAWA, I. Monocularly deprived cats: binocular tests of cortical cells reveal functional connections from the deprived eye. *J. Neurosci.* 8: 2491–2506, 1988.

FREEMAN, R. D. AND OLSON, C. Brief periods of monocular deprivation in kittens: effects of delay prior to physiological study. *J. Neurophysiol.* 47: 139–150, 1982.

FRÉGNAC, Y. AND IMBERT, M. Early development of visual cortical cells in normal and dark-reared kittens: relationship between orientation selectivity and ocular dominance. *J. Physiol. Lond.* 278: 27–44, 1978.

FRÉGNAC, Y. AND IMBERT, M. Development of neuronal selectivity in primary visual cortex of cat. *Physiol. Rev.* 64: 325–434, 1984.

FRÉGNAC, Y., SMITH, D., AND FRIEDLANDER, M. J. Postsynaptic membrane potential regulates synaptic potentiation and depression in visual cortical neurons. *Soc. Neurosci. Abstr.* 16: 331, 1990.

GREUEL, J. M., LUHMANN, H. J., AND SINGER, W. Evidence for a threshold in experience-dependent long-term changes of kitten visual cortex. *Dev. Brain Res.* 34: 141–149, 1987.

GUSTAFSSON, B., WIGSTRÖM, H., ABRAHAM, W. C., AND HUANG, Y. Y. Long-term potentiation in the hippocampus using depolarizing current pulses as the conditioning stimulus to single volley synaptic potentials. *J. Neurosci.* 7: 774–780, 1987.

HOLMES, W. R. AND LEVY, W. B. Insights into associative long-term potentiation from computational models of NMDA receptor-mediated calcium influx and intracellular calcium concentration changes. *J. Neurophysiol.* 63: 1148–1168, 1990.

HUBEL, D. H. AND WIESEL, T. N. Integrative action in the cat's lateral geniculate body. *J. Physiol. Lond.* 155: 385–398, 1961.

HUBEL, D. H. AND WIESEL, T. N. Receptive fields, binocular interaction and functional architecture in the cat's visual cortex. *J. Physiol. Lond.* 160: 106–154, 1962.

HUBEL, D. H. AND WIESEL, T. N. Receptive fields of cells in striate cortex of very young, visually inexperienced kittens. *J. Neurophysiol.* 26: 994–1002, 1963.

HUBEL, D. H. AND WIESEL, T. N. Binocular interaction in striate cortex of kittens reared with artificial squint. *J. Neurophysiol.* 28: 1041–1059, 1965.

HUBEL, D. H. AND WIESEL, T. N. The period of susceptibility to the physiological effects of unilateral eye closure in kittens. *J. Physiol. Lond.* 206: 419–436, 1970.

IKEDA, H. AND TREMAIN, K. E. Different causes for amblyopia and loss of binocularity in squinting kittens. (Abstract). *J. Physiol. Lond.* 269: 26P–27P, 1977.

IMBERT, M. AND BUISSERET, P. Receptive field characteristics and plastic properties of visual cortical cells in kittens reared with or without visual experience. *Exp. Brain Res.* 22: 25–36, 1975.

INTRATOR, N. A neural network for feature extraction: In: *Advances in Neural Information Processing Systems,* edited by D. S. Touretzky. San Mateo, CA: Kaufmann, 1990, vol. 2, p. 719–726.

INTRATOR, N. AND COOPER, L. N. Objective function formulation of the BCM theory of cortical plasticity: statistical connections, stability conditions. *Neural Networks.* In press.

KELSO, S. R., GANONG, A. H., AND BROWN, T. H. Hebbian synapses in hippocampus. *Proc. Natl. Acad. Sci. USA* 83: 5326–5330, 1986.

KIMURA, F., TSUMOTO, T., NISHIGORI, A., AND YOSHIMURA, Y. Long-term depression but not potentiation is induced in Ca^{2+}-chelated visual cortex neurons. *Neuro report.* In press.

KRATZ, K. E. AND SPEAR, P. D. Effects of visual deprivation and alterations in binocular competition on responses of striate cortex neurons in the cat. *J. Comp. Neurol.* 170: 141–151, 1976.

KRATZ, K. E., SPEAR, P. D., AND SMITH, D. C. Postcritical-period reversal of effects of monocular deprivation on striate cortex cells in the cat. *J. Neurophysiol.* 39: 501–511, 1976.

KUFFLER, S. W., FITZHUGH, R., AND BARLOW, H. B. Maintained activity in the cat's retina in light and darkness. *J. Gen. Physiol.* 40: 683–702, 1957.

LINSKER, R. From basic network principles to neural architecture. I. Emergence of orientation-selective cells. *Proc. Natl. Acad. Sci. USA* 83: 7508–7512, 1986a.

LINSKER, R. From basic network principles to neural architecture. II. Emergence of orientation columns cells. *Proc. Natl. Acad. Sci. USA* 83: 8390–8394, 1986b.

LINSKER, R. From basic network principles to neural architecture. III. Emergence of spatial-opponent cells. *Proc. Natl. Acad. Sci. USA* 83: 8779–8783, 1986c.

LOWENSTEIN, D. H., MILES, M. F., HATAM, F., AND MCCABE, T. Up regulation of calbindin-D28K mRNA in the rat hippocampus following focal stimulation of the perforant path. *Neuron* 6: 627–633, 1991.

MAFFEI, L. AND GALLI-RESTA, L. Correlation in the discharges of neighboring rat retinal ganglion cells during prenatal life. *Proc. Natl. Acad. Sci. USA* 87: 2861–2864, 1990.

MALSBURG, CH. VON DER. Self-organization of orientation sensitive cells in the striate cortex. *Kybernetik* 14: 85–100, 1973.

MASTRONARDE, D. N. Correlated firing of cat retinal ganglion cells. I. Spontaneously active inputs to X- and Y-cells. *J. Neurophysiol.* 49: 303–324, 1983a.

MASTRONARDE, D. N. Correlated firing of cat retinal ganglion cells. II. Responses of X- and Y-cells to single quantal events. *J. Neurophysiol.* 49: 325–349, 1983b.

MASTRONARDE, D. N. Correlated firing of cat retinal ganglion cells. *Trends Neurosci.* 12: 75–80, 1989.

MEISTER, M., WONG, R. O. L., BAYLOR, D. A., AND SHATZ, C. J. Synchronous bursts of action potentials in ganglion cells of the developing mammalian retina. *Science Wash. DC* 252: 939–943, 1991.

MILLER, K. D., KELLER, J. B., AND STRYKER, M. P. Ocular dominance column development: analysis and simulation. *Science Wash. DC* 245: 605–615, 1989.

MIOCHE, L. AND SINGER, W. Long-term recordings and receptive field measurements from single units of the visual cortex of awake unrestrained kittens. *J. Neurosci. Methods* 26: 83–94, 1988.

MIOCHE, L. AND SINGER, W. Chronic recordings from single sites of kitten striate cortex during experience-dependent modifications of receptive-field properties. *J. Neurophysiol.* 62: 185–197, 1989.

MITCHELL, D. E., CYNADER, M., AND MOVSHON, J. A. Recovery from the effects of monocular deprivation in kittens. *J. Comp. Neurol.* 176: 53–64, 1977.

MOVSHON, J. A. Reversal of the physiological effects of monocular deprivation in the kitten's visual cortex. *J. Physiol. Lond.* 261: 125–174, 1976.

MOVSHON, J. A. AND DÜRSTELER, M. R. Effects of brief periods of unilateral eye closure on the kitten's visual system. *J. Neurophysiol.* 40: 1255–1265, 1977.

MOVSHON, J. A. AND VAN SLUYTERS, R. C. Visual neural development. *Annu. Rev. Psychol.* 32: 477–522, 1981.

MOWER, G. D., BERRY, D., BURCHFIEL, J. L., AND DUFFY, F. H. Comparison of the effects of dark rearing and binocular suture on development and plasticity of cat visual cortex. *Brain Res.* 220: 255–267, 1981.

NASS, M. M. AND COOPER, L. N. A theory for the development of feature detecting cells in visual cortex. *Biol. Cybern.* 19: 1–18, 1975.

OHZAWA, I. AND FREEMAN, R. D. The binocular organization of simple cells in the cat's visual cortex. *J. Neurophysiol.* 56: 221–242, 1986a.

OHZAWA, I. AND FREEMAN, R. D. The binocular organization of complex cells in the cat's visual cortex. *J. Neurophysiol.* 56: 243–259, 1986b.

OLSON, C. R. AND FREEMAN, R. D. Progressive changes in kitten striate cortex during monocular vision. *J. Neurophysiol.* 38: 26–32, 1975.

OLSON, C. R. AND FREEMAN, R. D. Monocular deprivation and recovery during sensitive period in kittens. *J. Neurophysiol.* 41: 65–74, 1978.

PÉREZ, R., GLASS, L., AND SHLAER, R. Development of specificity in the cat visual cortex. *J. Math. Biol.* 1: 275–288, 1975.

PETTIGREW, J. D. The effect of visual experience on the development of stimulus specificity by kitten cortical neurones. *J. Physiol. Lond.* 237: 49–74, 1974.

REITER, H. O. AND STRYKER, M. P. Neural plasticity without action potentials: less active inputs become dominant when kitten visual cortical cells are pharmacologically inhibited. *Proc. Natl. Acad. Sci. USA* 85: 3623–3627, 1988.

SAUL, A. B. AND CLOTHIAUX, E. E. Modeling and simulation. III. Simulation of a model for development of visual cortical specificity. *J. Electrophysiol. Tech.* 13: 279–306, 1986.

SCOFIELD, C. L. AND COOPER, L. N. Development and properties of neural networks. *Contemp. Physiol.* 26: 125–145, 1985.

SILLITO, A. M., KEMP, J. A., AND BLAKEMORE, C. The role of GABAergic inhibition in the cortical effects of monocular deprivation. *Nature Lond.* 291: 318–320, 1981.

SINGER, W. Effects of monocular deprivation on excitatory and inhibitory pathways in cat striate cortex. *Exp. Brain Res.* 30: 25–41, 1977.

STANTON, P. K. AND SEJNOWSKI, T. J. Associative long-term depression in the hippocampus induced by hebbian covariance. *Nature Lond.* 339: 215–218, 1989.

STAUBLI, U. AND LYNCH, G. Stable depression of potentiated synaptic responses in the hippocampus with 1–5 Hz stimulation. *Brain Res.* 513: 113–118, 1990.

TSUMOTO, T. AND SUDA, K. Evidence for excitatory connections from the deprived eye to the visual cortex in monocularly deprived kittens. *Brain Res.* 153: 150–156, 1978.

VAN SLUYTERS, R. C. AND LEVITT, F. B. Experimental strabismus in the kitten. *J. Neurophysiol.* 43: 686–699, 1980.

WATKINS, D. W., WILSON, J. R., AND SHERMAN, S. M. Receptive-field properties of neurons in binocular and monocular segments of striate cortex in cats raised with binocular lid suture. *J. Neurophysiol.* 41: 322–337, 1978.

WIESEL, T. N. AND HUBEL, D. H. Single-cell responses in striate cortex of kittens deprived of vision in one eye. *J. Neurophysiol.* 26: 1003–1017, 1963.

WIESEL, T. N. AND HUBEL, D. H. Comparison of the effects of unilateral and bilateral eye closure on cortical unit responses in kittens. *J. Neurophysiol.* 28: 1029–1040, 1965.

YINON, U. Age dependence of the effect of squint on cells in kittens' visual cortex. *Exp. Brain Res.* 26: 151–157, 1976.

YINON, U., AUERBACH, E., BLANK, M., AND FRIESENHAUSEN, J. The ocular dominance of cortical neurons in cats developed with divergent and convergent squint. *Vision Res.* 15: 1251–1256, 1975.

ZADOR, A., KOCH, C., AND BROWN, T. H. Biophysical model of a Hebbian synapse. *Proc. Natl. Acad. Sci. USA* 87: 6718–6722, 1990.

Objective Function Formulation of the BCM Theory of Visual Cortical Plasticity: Statistical Connections, Stability Conditions

(with N. Intrator)

Neural Networks **5**, 3 (1992)

Nathan Intrator joined our laboratory as a graduate student in 1986. Since he had already served in the Israeli Army, he was a bit older; he has a wry sense of humor but overall is rather serious; so much so that although we trade many jokes, I can't think of any funny story associated with him. It reminds me of the difference between the opening of talks to Japanese and American audiences. For Japanese one begins with an apology; for Americans, one begins with a joke. In my talks to Japanese audiences I often begin with an apology for not having a joke. So Nathan, I apologize.

Given Nathan's statistical background and his considerable mathematical abilities, we decided to explore the statistical connections of the BCM theory; in particular to attempt to formulate it in terms of an objective function. This, we felt, would reveal its statistical properties as well as make the derivation of the various stability properties easier.

In addition, a trend in neural network theory (discussed in the neural network section of this collection) had moved toward exploring the statistical connections of neural networks. It therefore seemed an appropriate time to undertake this effort.

The results of this paper show that it is in fact possible to formulate the BCM modification algorithm in terms of an objective function. One of the more interesting results is the connection between the unsupervised BCM algorithm and the method known in statistics as projection pursuit, which is useful in reducing the dimensionality of large input spaces by generating interesting features. In the neural network section of this collection some applications are presented. One is, of course, tempted to conjecture that the evolutionary development of this unsupervised learning rule had something to do with its efficacy in aiding visual information processing.

In addition, as a natural consequence of this analysis, another definition of the modification function appears that results in improved stability. This, with the results of the previous paper (Clothiaux *et al.*), underlined that other than a few qualitative features, the precise form of both the modification function and the variation of the modification threshold had not yet been determined by experiment. It remains an open question what detailed form experiment really requires.

Neural Networks, Vol. 5, pp. 3–17, 1992
Printed in the USA. All rights reserved.

INVITED ARTICLE

Objective Function Formulation of the BCM Theory of Visual Cortical Plasticity: Statistical Connections, Stability Conditions*

Nathan Intrator and Leon N Cooper

Brown University

(Received and accepted 26 July in press)

Abstract—*In this paper, we present an objective function formulation of the Bienenstock, Cooper, and Munro (BCM) theory of visual cortical plasticity that permits us to demonstrate the connection between the unsupervised BCM learning procedure and various statistical methods, in particular, that of Projection Pursuit. This formulation provides a general method for stability analysis of the fixed points of the theory and enables us to analyze the behavior and the evolution of the network under various visual rearing conditions. It also allows comparison with many existing unsupervised methods. This model has been shown successful in various applications such as phoneme and 3D object recognition. We thus have the striking and possibly highly significant result that a biological neuron is performing a sophisticated statistical procedure.*

Keywords—Unsupervised learning, Feature extraction, Dimensionality reduction.

1. INTRODUCTION

In the past decade, much work has been done on a theory of synaptic plasticity in visual cortex (Bienenstock, Cooper, & Munro (BCM), 1982). This theory accounts in a precise and quantitative fashion for the modification of response properties of neurons in striate cortex obtained by manipulating the visual experience of the animal during a critical period of postnatal development. It allows a precise specification of theoretical equivalents of experimental situations and makes possible detailed and quantitative comparison of theory with experiment in what are called classical rearing conditions. These include normal rearing, monocular deprivation, reverse suture, strabismus, binocular deprivation, as well as the restoration of normal binocular vision after various forms of deprivation. In detailed simulations, Clothiaux, Cooper, and Bear (in press) find

quantitative agreement of theory and experiment both for equilibrium states and the kinetics by which they are reached.

In this paper, we present an objective function formulation of the BCM theory of visual cortical plasticity. This permits us to demonstrate the connection between the unsupervised BCM learning procedure and various statistical methods, in particular, that of Projection Pursuit (PP). This analysis has led us to slightly modify our learning rule resulting in improved stability and statistical properties. It also provides a general method for stability analysis of the fixed points of the theory and enables us to analyze the behavior and the evolution of the network under various visual rearing conditions. This new model has some advantages over the original exploratory projection pursuit model (Friedman, 1987). Due to its computational efficiency, it can extract several features in parallel, taking into account the interaction between the different extracted features via a lateral inhibition network. Feature extraction based on this model have been applied to various real-world problems such as phoneme recognition of a small-speaker database (Intrator, in press), multispeaker phoneme recognition from the TIMIT database (Intrator & Tajchman, in press) using the Lyon's cochlear model (Slaney, 1988) and 3D object recognition (Intrator & Gold, in press; Intrator, Gold, Bülthoff, & Edelman, in press). We thus

---

* This work was supported in part by the National Science Foundation, the Office of Naval Research, and the Army Research Office.

We wish to thank Geoff Hinton for improving the clarity of the statistical part and Charles Bachmann, Eugene Clothiaux, and Mike Perrone who provided many helpful comments.

Requests for reprints should be sent to Nathan Intrator, Physics Department and Center for Neural Science, Brown University, Providence, RI 02912.

3

have the striking and possibly highly significant result that a biological neuron is performing a sophisticated statistical procedure.

Section 2 reviews the evolution of the BCM theory and its relevance to modeling of the primary visual cortex, area 17. Section 3 describes the statistical motivation behind unsupervised learning. This is used to motivate the objective function formulation of the modified BCM model given in Section 4. Based on statistical considerations, this formulation is further extended to a nonlinear neuron in a lateral inhibition network. In Section 5 we analyze the limiting behavior of the synaptic modification equations, using the formulation described in Section 4 and a connection established between the solution of the averaged deterministic differential equations and the solution of the random version of the equations (Appendix A). Analysis of this model in several situations related to visual experiments is given in Section 6.

2. REVIEW OF BCM THEORY

In this section, we briefly review relevant experimental observations and the BCM theory. This will serve to introduce relevant notation and biological terms.

2.1. Visual Cortical Plasticity: Experimental Results

Neurons in the primary visual cortex, area 17, of normal adult cats are sharply tuned to the orientation of an elongated slit of light and most are activated by stimulation of either eye (Hubel & Wiesel, 1959). Both of these properties—orientation selectivity and binocularity—depend on the type of visual environment experienced during a critical period of early postnatal development.

Monocular deprivation (MD) has profound and reproducible effects on the functional connectivity of striate cortex during the critical period, extending from approximately 3 weeks to 3 months of age in the cat (Frègnac & Imbert, 1984; Sherman & Spear, 1982). Brief periods of MD will result in a dramatic shift in the ocular dominance (OD) of cortical neurons so that most will be responsive exclusively to the open eye. The OD shift after MD is the best known and most intensively studied type of visual cortex plasticity.

When MD is initiated late in the critical period (Presson & Gordon, 1982) or after a period of rearing in the dark (Mower, Caplan, Christen, & Duffy, 1985), it will induce clear changes in cortical OD without a corresponding anatomic change in the geniculocortical projections. Long-term recordings from awake animals also indicate that OD changes can be detected within a few hours of monocular experience; this seems too rapid to be explained by the formation or elimination of axon terminals. Moreover, deprived-eye responses in visual cortex may be restored within minutes to hours under some conditions (Duffy, Sonodgrass, Burchfiel, & Conway, 1976), which suggests that synapses deemed functionally *disconnected* are nonetheless physically present. Therefore, it is reasonable to assume that changes in the functional binocularity may be explained by changes in the efficacy of individual cortical synapses.

The consequences of binocular deprivation (BD) on visual cortex stand in striking contrast to those observed after MD. First, MD leads to a loss of orientation selectivity in the deprived eye much faster than in BD. Second, although 7 days of MD during the second postnatal month leave few neurons in the striate cortex responsive to stimulation of the deprived eye, most cells remain responsive to stimulation through either eye after a comparable period of BD (Wiesel & Hubel, 1965). Thus it is not merely the absence of patterned activity in the deprived geniculate projection that causes the decrease in synaptic efficacy after MD.

The result of a reversed suture (RS) experiment is even more striking. In this experiment, the kitten is first exposed to normal visual environment, then one eye is sutured closed for a few days until the sutured eye becomes functionally disconnected. At that time the sutured eye is opened and exposed to normal visual environment again and the previously opened eye is closed. The result from this experiment is that the newly opened eye does not recover before the previously opened eye becomes disconnected.

2.2. Single Neuron Theory

A theoretical solution to the problem of visual cortical plasticity, was presented by Cooper, Liberman, and Oja (1979). According to this theory, the synaptic efficacy of active inputs increases when the postsynaptic target is concurrently depolarized beyond a *modification threshold*, Θ_m. However, when the level of postsynaptic activity falls below Θ_m, then the strength of active synapses decreases.

An important feature was added to this theory in 1982 by Bienenstock, Cooper, and Munro. They proposed that the value of the modification threshold is not fixed, but instead varies as a nonlinear function of the average output of the postsynaptic neuron. This provided stability properties and explained, for example, why the low level of postsynaptic activity during binocular deprivation does not drive the strengths of all cortical synapses to zero. Their form of synaptic modification can be written as:

$$\dot{m}_j = \phi(c, \Theta_m)d_j \qquad (2.1)$$

where m_j is the efficacy of the j-th Lateral Geniculate Nucleus (LGN) synapse onto a cortical neuron, d_j is the level of presynaptic activity of the j-th LGN afferent, c is the level of activation of the postsynaptic activity of the postsynaptic neuron, which is given (in the linear region), by $m \cdot d$, and Θ_m is a nonlinear function of some time averaged measure of cell activity that in the original BCM formulation was proposed as

$$\Theta_m = (\bar{c})^2. \qquad (2.2)$$

(In BCM, this time average is replaced, for simplicity, by a spatial average over the environmental inputs ($\bar{c} \to m \cdot \bar{d}$). The shape of the function ϕ is given in Figure 1 for two different values of the threshold Θ_m.

Further discussion of the biological relevance of the theory can be found in (Saul & Daniels, 1986; Bear, Cooper, & Ebner, 1987; Bear & Cooper, 1988; Clothiaux et al., in press).

2.3. Lateral Inhibition Network: Mean Field Theory

An extension of the single cell BCM neuron to a lateral inhibition network was presented by (Scofield & Cooper, 1985) and a mean field approximation of this network by (Cooper & Scofield, 1988).

The activity of neuron j in such a network is affected by its input vector d and by the adjacent neurons in the network and can be written

$$c_i = m_i \cdot d + \sum_j L_{ij} c_j. \qquad (2.3)$$

In the context of visual cortex, the first term is due to the input from LGN and the second due to input from other cortical cells. Define $\bar{c}$ as the spatially averaged activity of all the cortical cells in the network: $\bar{c} = 1/N \sum_i c_i$. The mean field approximation is obtained by replacing the inhibitory contribution of cell j, c_j by its average value so that c_i becomes:

$$c_i = m_i \cdot d + \bar{c} \sum_j L_{ij}. \qquad (2.4)$$

From a consistency condition it follows that $\bar{c} = \bar{m} \cdot d + \bar{c}L_0 = (1 - L_0)^{-1}\bar{m} \cdot d$, where $\bar{m} = 1/N \sum_i m_i$, and $L_0 = 1/N \sum_{ij} L_{ij}$, so that $c_i = (m_i + (1 - L_0)^{-1}\bar{m} \sum_j L_{ij}) d$.

If we assume that the lateral connection strengths are function only of the relative distance $i - j$, then L_{ij} becomes a circular matrix so that $\sum_i L_{ij} = \sum_j L_{ij} = L_0$, and

$$c_i = (m_i + L_0(1 - L_0)^{-1}\bar{m})d. \qquad (2.5)$$

In the mean field approximation, one can therefore write $c_i(\alpha) = (m_i - \alpha)d$, with $\alpha = |L_0|(1 + |L_0|)^{-1}\bar{m}$.

When analyzing the position and stability of the fixed points using this approximation, it follows under some mild assumption on the evolution of the average synaptic weights, that there is a mapping

$$m_i' \leftrightarrow m_i(\alpha) - \alpha$$

such that for every neuron in such a network with synaptic weight vector m_i there is a corresponding neuron with weight vector m_i' that undergoes the same evolution (around the fixed points) subject to a translation α.

3. EXTRACTION OF OPTIMAL UNSUPERVISED FEATURES

When a classification of high dimensional vectors is sought, the *curse of dimensionality* (Bellman, 1961) becomes the main factor affecting the classification performance. The curse of dimensionality is due to the inherent sparsity of high dimensional spaces; thus the amount of training data needed to get reasonably low variance estimators becomes ridiculously high. This has led many researchers in recent years to construct methods that specifically avoid this problem. In those cases in which important structure in the data actually lies in a much smaller dimensional space, it becomes reasonable to try to reduce the dimensionality before attempting the classification. This approach can be successful if the dimensionality reduction/feature extraction method loses as little information as possible in the transformation from the high dimensional space to the low dimensional one.

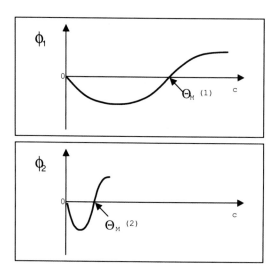

FIGURE 1. The ϕ function for two different Θ_m's.

At a first glance, it seems that a supervised feature extraction method, such as multiple discriminant analysis (see review in Bryan, 1951; Sebestyen, 1962) will always be superior to an unsupervised one, because if one has more information about the problem, it is natural to expect that finding the solution is easier. However, due to the global constraint imposed by the supervision, when the number of parameters (i.e., the dimensionality and number of nodes) is large, the network often will get stuck in a local minimum which is far from an optimal solution. Unsupervised methods, however, use local objective functions which may lead to less sensitivity to the number of parameters in the estimation, and therefore have the potential to avoid the curse of dimensionality (Barron & Barron, 1988).

For the purpose of pattern classification, it is important to devote our attention to those dimensionality reduction methods that allow discrimination between classes and not faithful representations of the data. This leaves out the class of methods such as factor analysis (see review in Harman, 1967) which tend to combine features that seem to have high correlation.

A general class of unsupervised dimensionality reduction methods, called exploratory projection pursuit, is based on seeking *interesting* projections of high dimensional data points (Kruskal, 1969; Switzer, 1970; Kruskal, 1972; Friedman & Tukey, 1974; Friedman, 1987; Huber, 1985, for review). The notion of interesting projections is motivated by an observation made by Diaconis and Freedman (1984) that for most high-dimensional clouds, most low-dimensional projections are approximately normal. This finding suggests that the important information in the data is conveyed in those directions whose single dimensional projected distribution is far from Gaussian. Various projection indices differ on the assumptions about the nature of deviation from normality, and in their computational efficiency. Friedman (1987) argues that the most computationally efficient measures are based on polynomial moments. However, although many synaptic plasticity models are based on second order statistics and lead to extraction of the principal components (Sejnowski, 1977; von der Malsburg, 1973; Oja, 1982; Miller, 1988; Linsker, 1988), second order polynomials are not sufficient to characterize the important features of a distribution (see examples in Duda and Hart (1973), p. 212, and the example in Figure 2). This suggests that in order to use polynomials for measuring deviation from normality, higher order polynomials are required and care should be taken in order to avoid their oversensitivity to outliers. From our earlier discussion, it follows that these polynomial moments should be of higher order than two. In some special cases where the data

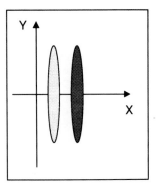

FIGURE 2. Two data clusters which can be separated by projecting to the x axis, cannot be separated by projecting to the y axis, although the variance in the y axis is larger.

is known in advance to be bimodal, it is relatively straightforward to define a good projection index (Hinton & Nowlan, 1990), however, when the structure is not known in advance, it is still valid to seek multimodality in the projected data.

Despite the computational attractiveness, projection indices based on polynomial moments are not directly applicable since they heavily emphasize departure from normality in the tails of the distribution (Huber, 1985). Friedman (1987) addresses this issue by introducing a nonlinear transformation that compresses the projected data from R to $[-1, 1]$ using a normal distribution function. We address the problem by applying a sigmoidal function to the projections, and then applying an objective function based on polynomial moments.

4. FORMULATION OF THE BCM THEORY USING AN OBJECTIVE FUNCTION

With the intuitive idea discussed above, we now present an objective function formulation of the synaptic modification theory of Bienenstock, Cooper and Munro. This yields a statistically plausible objective function whose minimization finds those projections having a single dimensional projected distribution that is far from Gaussian.

This formulation allows us to interpret the biological neuron's behavior from a statistical point of view. In addition, it provides a more powerful means of investigating the kinetics of synaptic development as well as the location and stability of the fixed points under various environmental conditions.

4.1. Single Neuron

We first informally describe the statistical formulation that leads to this objective function. Using a metaphor motivated by statistical decision theory, a neuron is considered as capable of deciding whether to fire or not for a given input and vector of synaptic weights. A loss function is attached to each decision. The neuron's task is then to choose the decision that minimizes the loss. Since the loss function depends on the synaptic weight vector, in addition to the input vector, it is natural to seek a synaptic weight vector that will minimize the sum of the losses associated with every input, or more precisely, the average loss (also called the risk). The search for such a vector, which yields an optimal synaptic weight vector under this formulation, can be viewed as learning or parameter estimation. In those cases where the risk is a smooth function, its minimization can be accomplished by gradient descent.

The ideas presented so far make no specific assumptions regarding the loss function, and it is clear that different loss functions will yield different learning procedures. For example, if the loss function is related to the inverse of the projection variance (including some normalization) then minimizing the risk will yield directions that maximize the variance of the projections, i.e., will find the principal components.

Before presenting a loss function, let us more precisely define the neuronal input, and two useful functions: We consider a neuron with input vector $x = (x_1, \ldots, x_n)$, synaptic weight vector $m = (m_1, \ldots, m_n)$, both in R^n, and activity (in the linear region) $c = x \cdot m$. The input x is assumed to be a bounded, and piecewise constant stochastic process. We allow some time dependency in the presentation of the training patterns, by requiring that x is of Type II mixing.[1] These assumptions are plausible since they represent the closest continuous approximation to the usual training algorithms in which training patterns are presented at random. They are needed for the approximation of the resulting deterministic gradient descent by a stochastic one (Intrator, 1990). For this reason we use a *learning rate* μ that has to decay in time so that this approximation is valid. Define the threshold $\Theta_m = E[(x \cdot m)^2]$, and the functions $\hat{\phi}(c, \Theta_m) = c^2 - \frac{2}{3}c\Theta_m$, $\phi(c, \Theta_m) = c^2 - c\Theta_m$.

Our projection index is aimed at finding directions for which the projected distribution is far from Gaussian; more specifically, since high dimensional clusters have a multimodal projected distribution, our aim is to find a projection index (loss function) that emphasizes multimodality. For computational efficiency, we would like to base the projection index on polynomial moments of low degree. Using second degree polynomials, one can get measures of the

mean and variance of the distribution; these, however, do not give information on multimodality; therefore, higher order polynomials are necessary. Further, the projection index should exhibit the fact that bimodal distribution is already interesting, and any additional mode should make the distribution even more interesting.

With this in mind, consider the following family of loss functions that depend on the synaptic weight vector and on the input x

$$L_m(x) = -\mu \int_0^{(x \cdot m)} \hat{\phi}(s, \Theta_m) ds$$
$$= -\mu \left\{ \frac{1}{3}(x \cdot m)^3 - \frac{1}{4}E[(x \cdot m)^2](x \cdot m)^2 \right\}$$

(4.1)

The motivation for this loss function can be seen in Figure 3, which represents the ϕ function and the associated loss function $L_m(x)$. For simplicity, the loss for a fixed threshold Θ_m and synaptic vector m can be written as $L_m(c) = -\mu c^2(c/3 - \Theta_m/4)$, where c represents the linear projection of x onto m.

The graph of the loss function shows that for any fixed m and Θ_m, the loss is small for a given input x, when either $c = x \cdot m$ is close to zero, or when $x \cdot m$ is larger than Θ_m. Moreover, the loss function remains negative for $(x \cdot m) > \Theta_m$, therefore any kind of distribution at the right hand side of Θ_m is possible, and the preferred ones are those which are concentrated further from Θ_m.

It remains to show why it is not possible that a minimizer of the average loss will be such that all the mass of the distribution will be concentrated on one side of Θ_m. This cannot happen because the threshold Θ_m is dynamic and depends on the projections in a nonlinear way, namely, $\Theta_m = E(x \cdot m)^2$. This implies that Θ_m will always move itself to a position such that the distribution will never be concentrated at only one of its sides.

The risk (expected value of the loss) is given by:

$$R_m = -\mu E \left\{ \frac{1}{3}(x \cdot m)^3 - \frac{1}{4}E[(x \cdot m)^2](x \cdot m)^2 \right\}$$
$$= -\mu \left\{ \frac{1}{3}E[(x \cdot m)^3] - \frac{1}{4}E^2[(x \cdot m)^2] \right\}.$$

(4.2)

Since the risk is continuously differentiable, its minimization can be achieved via a gradient descent method with respect to m, namely:

$$\frac{dm_i}{dt} = -\frac{\partial}{\partial m_i} R_m = \mu\{E[(x \cdot m)^2 x_i]$$
$$- E[(x \cdot m)^2]E[(x \cdot m)x_i]\}$$
$$= \mu E[\phi(x \cdot m, \Theta_m)x_i]. \quad (4.3)$$

---

[1] The mixing property specifies the dependency of the future of the process on its past.

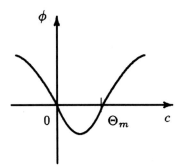
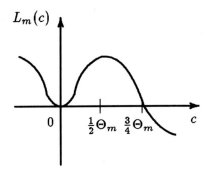

FIGURE 3. The function φ and the loss functions for a fixed m and Θ_m.

The resulting differential equations give a somewhat different version of the law governing synaptic weight modification of the BCM theory. The difference lies in the way the threshold Θ_m is determined. In the original form, this threshold was $\Theta_m = E^p(c)$ for $p > 1$, while in the current form $\Theta_m = E(c^p)$ for $p > 1$. The latter takes into account the variance of the activity (for $p = 2$) and therefore is always positive, which ensures stability even when the average of the inputs is zero. It should be noted here, that the original theory (Bienenstock et al., 1982) assumed that the inputs were positive, whereas the present threshold relaxes this assumption and yields stability for a larger class of bounded inputs.

Either form seems consistent with currently available experimental results (Clothiaux et al., in press) but imply quite different underlying physiological mechanisms. The original BCM form requires that a history of activity (likely cell depolarization) be stored and then via a nonlinear process produce the modification threshold. The present form of Θ_m requires that the nonlinear process occur first. When the existence of the moving threshold is established by observation,[2] the next question of great interest will be its precise dependence on cell parameters.

4.2. Extension to a Nonlinear Neuron

The fact that the distribution has part of its mass on both sides of Θ_m makes it a plausible projection index that seeks multimodalities. However, this projection index will be more general if, in addition, the loss is insensitive to outliers and if we allow any projected distribution to be shifted so that the part of the distribution that satisfies $c < \Theta_m$ will have its mode at zero. The oversensitivity to outliers is addressed by considering a nonlinear neuron in which the neuron's activity is defined to be $c = \sigma(x \cdot m)$, where σ usually represents a smooth sigmoidal function. The ability

[2] Some indications of a moving threshold has been already found (Yang & Faber, 1991).

to shift the projected distribution so that one of its modes is at zero is achieved by introducing a threshold β so that the projection is defined to be $c = \sigma(x \cdot m + \beta)$. From a biological viewpoint, β can be considered as spontaneous activity. The modification equations for finding the optimal threshold β are easily obtained by observing that this threshold effectively adds one dimension to the input vector and vector of synaptic weights so that $x = (x_1 \ldots, x_n, 1)$, $m = (m_1, \ldots, m_n, \beta)$, and therefore, β can be found by using the same synaptic modification equations. For the rest of the paper we shall assume that this treshold is added to the projection, without specifically writing it.

For the nonlinear neuron, Θ_m is defined to be $\Theta_m = E[\sigma^2(x \cdot m)]$. The loss function is given by:

$$
\begin{aligned}
L_m(x) &= -\mu \int_0^{\sigma(x \cdot m)} \hat{\phi}(s, \Theta_m)ds \\
&= -\mu \left\{ \frac{1}{3} \sigma^3(x \cdot m) - \frac{1}{4} E[\sigma^2(x \cdot m)]\sigma^2(x \cdot m) \right\}
\end{aligned}
$$

(4.4)

The gradient of the risk becomes:

$$
\begin{aligned}
-\nabla_m R_m &= \mu\{E[\sigma^2(x \cdot m)\sigma'x] \\
&\quad - E[\sigma^2(x \cdot m)]E[\sigma(x \cdot m)\sigma'x]\} \\
&= \mu E[\phi(\sigma(x \cdot m), \Theta_m)\sigma'x],
\end{aligned}
$$

(4.5)

where σ' represents the derivative of σ at the point $(x \cdot m)$. Note that the multiplication by σ' reduces sensitivity to outliers of the differential equation since for outliers σ' is close to zero. The gradient decent procedure is valid, provided that the risk is bounded from below (see Section 5).

4.3. Extension to a Network With Feedforward Inhibition

We now define a network with feedforward inhibition. The activity of neuron k in the network is $c_k = x \cdot m_k$, where m_k is the synaptic weight vector of neuron k. The *inhibited* activity and threshold of the k-th neuron is given by

$$\bar{c}_k = c_k - \eta \sum_{j \neq k} c_j, \qquad \tilde{\Theta}_m^k = E[\bar{c}_k^2]. \qquad (4.6)$$

This feedforward network should be contrasted with a lateral inhibition network (used, for example, by Cooper & Scofield, 1988) in which the inhibited activity is given by $c_k = c_k(0) + \Sigma\ L_{ij}c_j$. The relation between these two networks will be discussed in the next section.

For the feedforward network, the loss function is similar to the one defined in a single feature extraction with the exception that the activity $c = x \cdot m$ is replaced by $\bar{c}$. Therefore the risk for node k is given by:

$$R_k = -\mu \left\{ \frac{1}{3} E[\bar{c}_k^3] - \frac{1}{4} E^2[\bar{c}_k^2] \right\}, \qquad (4.7)$$

and the total risk is given by

$$R = \sum_{k=1}^{N} R_k. \qquad (4.8)$$

To find the gradient of R we write:

$$\frac{\partial \bar{c}_k}{\partial m_j} = -\eta x, \quad \frac{\partial \bar{c}_k}{\partial m_k} = x,$$

$$\frac{\partial R_k}{\partial m_k} = \frac{\partial R_k}{\partial \bar{c}_k}\frac{\partial \bar{c}_k}{\partial m_k} = -\mu\{E[\bar{c}_k^2 x] - E[\bar{c}_k^2]E[\bar{c}_k x]\},$$

$$\frac{\partial R_j}{\partial m_k} = \frac{\partial R_j}{\partial \bar{c}_j}\frac{\partial \bar{c}_j}{\partial m_k} = -\eta\frac{\partial R_j}{\partial m_j},$$

$$\Rightarrow \frac{\partial R}{\partial m_k} = \frac{\partial R_k}{\partial m_k} - \eta \sum_{j \neq k} \frac{\partial R_j}{\partial m_j}$$

$$= \mu[E[\phi(\bar{c}_k, \tilde{\Theta}_m^k)x] - \eta \sum_{j \neq k} E[\phi(\bar{c}_j, \tilde{\Theta}_m^j)x]].$$

$$(4.9)$$

The equation performs a constraint minimization in which the derivative with regard to one neuron can become orthogonal (when $\eta \to 1$) to the sum over the derivatives of all other synaptic weights. Nevertheles, the coupling is very simple to calculate and does not require any matrix inversion. The equation, therefore, demonstrates the ability of the network to perform exploratory projection pursuit in parallel, since the minimization of the risk involves minimization of nodes $1, \ldots, N$, which are loosely coupled.

When the nonlinearity of the neuron is included, the inhibited activity is defined (as in the single neuron case) as $\bar{c}_k = \sigma(c_k - \eta \Sigma_{j \neq k} c_j)$, and R_k are defined as before. However, in this case

$$\frac{\partial \bar{c}_k}{\partial m_j} = -\eta\sigma'(\bar{c}_k)x, \quad \frac{\partial \bar{c}_k}{\partial m_k} = \sigma'(\bar{c}_k)x. \qquad (4.10)$$

Therefore, the total gradient becomes:

$$\dot{m}_k = \frac{\partial R}{\partial m_k} = \mu\{E[\phi(\bar{c}_k, \tilde{\Theta}_m^k)\sigma'(\bar{c}_k)x]$$

$$- \eta \sum_{j \neq k} E[\phi(\bar{c}_j, \tilde{\Theta}_m^j)\sigma'(\bar{c}_j)x]\}. \qquad (4.11)$$

The lateral inhibition network performs a search of k-dimensional projections together; thus may find a richer structure that a stepwise approach might miss (e.g., see example 14.1, Huber, 1985).

4.4. Some Related Statistical and Computational Issues

The proposed method uses low order polynomial moments which are computationally efficient, yet it does not suffer from the main draw back of polynomial moments—sensitivity to outliers. It naturally extends to multidimensional projection pursuit using the feedforward inhibition network. The number of calculations of the gradient grows linearly with the dimensionality and *linearly* with the number of projections sought. The projection index contains a single dimensional scaling (see the contribution of Hastie and Tibshirani to the discussion in Jones & Sibson, 1987), therefore, removing the need for a sphering transformation to the data, however, a sphering transformation will result in a Type III projection index (see Huber, 1985). The projection index has a natural stochastic gradient descent version which further accelerates the calculation by eliminating the need to calculate the empirical expected value of the gradient. All the above lead to a fully parallel algorithm that may be implemented on a multiprocessor machine, and produce a practical feature extractor for very high dimensional problems.

Although the projection index is motivated by the desire to search for clusters in the high dimensional data, the resulting feature extraction method is quite different from other pattern recognition methods that search for clusters. Since the class labels are not used in the search, the projection pursuit is not biased to the class labels. This is in contrast to classical methods such as discriminant analysis (Fisher, 1936; Sebestyen, 1962, and numerous recent publications). The issue of using an unsupervised method vs. supervised for revealing structure in the data has been discussed extensively elsewhere. We would only like to add that it is striking that in various low-dimensional examples (Friedman & Tukey, 1974; Jones, 1983; Friedman, 1987), the exploratory capabilities of PP were not worse than those of supervised methods such as discriminant analysis and factor analysis in discovering structure, thus suggesting that in high dimensions where supervised methods may fail, PP can still find useful structure.

The resulting method concentrates on projections that allow discrimination between clusters and not faithful representation of the data, which is in contrast to principal components analysis or factor analysis which tend to combine features that have high correlation (see review in Harman, 1967).

The method differs from cluster analysis by the fact that it searches for clusters in the low dimensional projection space, thus avoiding the inherent sparsity of the high dimensional space. The search for multimodality is further constrained by the desire to seek those projections that are orthogonal to all but one of the clusters (or have a mode at zero). This constraint simplifies the search since it implies that a set of K linearly independent clusters may have at most K optimal projections as opposed to at most $\binom{K}{2}$ separating hyperplanes.

4.5. Comparison of Linear Feedforward With Lateral Inhibition Network: Mean Field Approximation

For the linear case, using the notation of Cooper and Scofield (1988), neuron activity in the lateral inhibition network is given by

$$c = Md + Lc. \tag{4.12}$$

In the mean field approximation this becomes

$$c = Md + L\bar{c}, \tag{4.13}$$

where M is the synaptic matrix for N neurons, $c = (c_1, \ldots, c_N)^T$, L is the inhibitory connection matrix with norm less than 1 and $\bar{c}$ is the averaged activity over all neurons in the network. In the context of visual cortex, the first term is due to the input from LGN and the second due to input from other cortical cells. If we define $c(0) = Md$, then the averaged inhibited activity can be written as

$$\bar{c} = \bar{c}(0) + L\bar{c}, \tag{4.14}$$

or

$$\bar{c} = (I - L)^{-1}\bar{c}(0)$$
$$= (I + L + L^2 + L^3 + \ldots) \bar{c}(0). \tag{4.15}$$

Using this notation, the activity of a neuron in the feedforward network as defined in Section 4.3 can be written

$$\bar{c} = c(1) = c(0) + Lc(0), \tag{4.16}$$

which leads to an averaged activity of the form

$$\bar{c}(1) = (I + L)\bar{c}(0), \tag{4.17}$$

which is a first order approximation of eqn (4.14); it is useful primarily because it removes the need to invert a matrix, which becomes impossible in the nonlinear neuronal case. In addition, successive approximations

$$\bar{c}(1) = \bar{c}(0) + L\bar{c}(0) = (I + L)\bar{c}(0),$$
$$\bar{c}(2) = \bar{c}(0) + L\bar{c}(1) = (I + L + L^2)\bar{c}(0), \tag{4.18}$$
$$\vdots$$

can be thought of as including monosynaptic, bisynaptic, trisynaptic, etc., events and thus follow the time course of the postsynaptic potentiation. It follows that $\bar{c}(k) \to \bar{c}$, as $k \to \infty$, thus recapturing the lateral inhibition network. Within a scaling factor, the first order feedforward network, as will be shown below, generates the same synaptic modification equations as the lateral inhibition network in the Cooper and Scofield mean field approximation.

For a feedforward network with neuron activity given by $\bar{c}_i = m_i \cdot d + \Sigma_j L_{ij}c_j$, as in Section 4,

$$\dot{m} = -\frac{\partial R}{\partial m_k} = -\left[\frac{\partial R_k}{\partial m_k} + \sum_j L_{kj}\frac{\partial R_j}{\partial m_j}\right]$$

$$= \mu\left[E[\phi(\bar{c}_k, \tilde{\Theta}_m^k)x] + \sum_j L_{kj}E[\phi(\bar{c}_j, \tilde{\Theta}_m^j)x]\right]. \tag{4.19}$$

Let $\bar{m} = 1/N \Sigma_j m_j$ $(m_j \in R^n)$. We assume that the inhibitory contributions are a function only of $i - j$ (not dependent on the absolute position of a cell in the network) so that $\Sigma_i L_{ij} = L_0 = \Sigma_j L_{ij}$, and that $\Sigma_i L_{ij}E[\phi(\bar{c}_i, \tilde{\Theta}_m^i)x] = \Sigma_j L_{ij}E[\phi(\bar{c}_j, \tilde{\Theta}_m^j)x]$. Then we get:

$$N\dot{\bar{m}} = \sum_j \dot{m}$$

$$= \mu\left[\sum_k E[\phi(\bar{c}_k, \tilde{\Theta}_m^k)x] + \sum_j L_0 E[\phi(\bar{c}_j, \tilde{\Theta}_m^j)x]\right]$$

$$= \frac{(1 + L_0)}{L_0}\mu \sum_k \sum_j L_{kj}E[\phi(\bar{c}_k, \tilde{\Theta}_m^k)x].$$

$$\tag{4.20}$$

This implies that

$$\frac{L_0}{1 + L_0}\dot{\bar{m}} = \mu \sum_k L_{jk}E[\phi(\bar{c}_k, \tilde{\Theta}_m^k)x], \tag{4.21}$$

and hence,

$$\dot{m}_k = \mu\left[E[\phi(\bar{c}_k, \tilde{\Theta}_m^k)x] + \frac{L_0}{1 + L_0}\dot{\bar{m}}\right]. \tag{4.22}$$

Compare this with Cooper and Scofield (1988) (eqn A3):

$$\dot{m}_k = \mu[E[\phi(c_k, \Theta_m^k)x] + L_0\dot{\bar{m}}], \tag{4.23}$$

eqns 4.22 and 4.23 differ only in the constant of inhibition. Thus the mean field approximation of the feedforward network yields the lateral inhibition mean field result merely by scaling the average inhibition.

One result of the mean field approximation (Cooper & Scofield, 1988) is that there is a transformation such that

$$m(\alpha) = m' - \alpha \qquad (4.24)$$

and so the gradient with respect to the weights yields two terms $\dot{m}$ and $\dot{\alpha}$. In the adiabatic case, we assume that α varies slowly with respect to each individual m, so that $\dot{\alpha} = 0$ is a reasonable approximation. In this situation the analysis of Section 5 applies for $m(\alpha)$ (the mean field network) as well as for $m(0)$. In addition, the argument given in the appendix of Cooper and Scofield (1988) regarding the nonadiabatic case holds here as well.

From the system 4.9 we can get:

$$\bar{\dot{m}} = \frac{\mu}{N} E \left[\sum_k [\phi(\bar{c}_k, \tilde{\Theta}_m^k)x] \right.$$

$$\left. - \eta \left(\frac{N}{N-1} \right) \sum_j E[\phi(\bar{c}_j, \tilde{\Theta}_m^j)x] \right], \quad (4.25)$$

which implies

$$\dot{\alpha} = \bar{\dot{m}} = \frac{\mu}{N} \left[1 - \eta \left(\frac{N}{N-1} \right) \right]$$

$$\times E \left[\sum_k [\phi(\bar{c}_k, \tilde{\Theta}_m^k)x] \right]. \qquad (4.26)$$

Therefore, at a fixed point, when all of the cells of the network have reached their respective fixed points, $\dot{m}_i' = 0$ implies that $\bar{\dot{m}} = 0$, meaning $\dot{\alpha} = 0$. Thus the position and stability of the fixed points (as given in Section 5) apply for the mean field network with no additional approximations.

5. ANALYSIS OF THE FIXED POINTS IN HIGH DIMENSIONAL SPACE

In Appendix A we show using a general result on random differential equations (Intrator, 1990) that the solution of the random differential equations remains as close as we like, in the L^2 sense, to the solution of the deterministic equations. We have shown in Section 4 that the deterministic equation converges to a local minimum of the risk. This implies that the solution of the random differential equation converges to a local minimum of the risk in L^2. Based on the statistical formulation, we can say that the local minima of the risk are *interesting features* extracted from the data, which correspond to directions in which the single dimensional distribution of the projections is far from a Gaussian distribution by means of penalized skewness measure.

In the following, we attempt to analyze the shape of the high dimensional risk function under specific inputs, namely, we look for the location of the critical points of the risk and locate those which have a local minima given a specific training set. This completely characterizes the solution of the synaptic modification equations and sheds more light on the power of the risk functional in finding interesting directions in the data. In doing so we gain some detailed information on the behavior of the solution of the random differential equations as a model for learning in visual cortex under various rearing conditions.

We consider linear neurons under the mean field assumptions. Furthermore, since the introduction of the threshold β does not pose any mathematical difficulty as was described before, we omit it in the analysis.

First, we analyze the limiting behavior of the solution in the case where we have n linearly independent inputs (not necessarily orthogonal). The introduction of noise into the system will be done in the next sections.

5.1. *n* Linearly Independent Inputs

The random differential equation is given by

$$\dot{m}_\varepsilon = \varepsilon \mu(t) \phi(x \cdot m, \Theta_m)x, \qquad m_\varepsilon(0) = m_0, \quad (5.1)$$

the averaged (batch) version of the gradient descent is given by:

$$\bar{\dot{m}}_\varepsilon = \varepsilon \mu(t) E[\phi(x \cdot \bar{m}, \Theta_{\bar{m}})x] \qquad \bar{m}_\varepsilon(0) = m_0. \quad (5.2)$$

The main tool in establishing the following results is the connection between the solution to the deterministic differential eqn 5.2 and the solution of the random differential eqn 5.1. A general result which yields this connection is given in Intrator (1990) and will be discussed in the appendix. When applied to this specific differential equation, the result says that

$$\sup_{t>T} E|m_\varepsilon - \bar{m}_\varepsilon|^2 \xrightarrow[\varepsilon \to 0]{} 0. \qquad (5.3)$$

PROPOSITION 5.1. *Let* $x_{(1)}, \ldots, x^{(n)}$ *be n linearly independent bounded vectors in* R^n. *Let D be the random process so that* $P[D = x^{(i)}] = p_i$, $p_i > 0$, $i = 1, \ldots, n$, $\Sigma p_i = 1$.

Then the critical points of eqn 5.1 are the 2^n *weight vectors* $m^{(i)} \in R^n$, *each a solution to one of the equations:* $Am^{(i)} = v^{(i)}$, $i = 0, \ldots, 2^n - 1$, *where A is the matrix whose i-th row is the input vector* $x^{(i)}$, *and* $\{v^{(i)}, i = 0, \ldots, 2^n - 1\}$, *is the n dimensional set of vectors of the form:*

$$v^{(0)} = (0, \ldots, 0),$$

$$v^{(1)} = \left(\frac{1}{p_1}, 0, \ldots, 0 \right),$$

$$v^{(2)} = \left(0, \frac{1}{p_2}, 0, \ldots, 0 \right),$$

$$v^{(3)} = \left(\frac{1}{p_1 + p_2}, \frac{1}{p_1 + p_2}, 0, \ldots, 0 \right),$$

$$v^{(4)} = \left(0, 0, \frac{1}{p_3}, 0, \ldots, 0 \right),$$

$$, \ldots,$$

$$v^{(2^n - 1)} = (1, \ldots, 1).$$

Proof. Rewrite eqn 5.1 in the form

$$\dot{m}_\varepsilon = \varepsilon \mu(t)(x \cdot m_\varepsilon)[x \cdot m_\varepsilon - \Theta_m]x, \quad (5.4)$$

where $\Theta_m = E[(x \cdot m_\varepsilon)^2]$. Since $\varepsilon \mu(t) > 0$, then m_ε is a critical point if either $x^{(i)} \cdot m_\varepsilon = 0$, or $x^{(i)} \cdot m_\varepsilon = \Theta_m \neq 0$, $i = 1, \ldots, n$.

There are exactly 2^n possibilities for the set of n numbers $x^{(i)} \cdot m_\varepsilon$ to be either 0, or nonzero. Therefore there are only 2^n possible solutions.

Let $m_\varepsilon^{(1)}$ be such that $m_\varepsilon^{(1)} \cdot x^{(1)} \neq 0$, and $m_\varepsilon^{(1)} \cdot x^{(i)} = 0$, $i > 1$. Then for $m_\varepsilon^{(1)}$,

$$\Theta_m = E[(x \cdot m_\varepsilon)^2] =$$

$$\sum_{i=1}^{N} p_i(x^{(i)} \cdot m_\varepsilon^{(1)})^2 = p_1(x^{(1)} \cdot m_\varepsilon^{(1)})^2. \quad (5.5)$$

When we combine the condition $\Theta_m = x^{(1)} \cdot m_\varepsilon^{(1)}$, we get $x^{(1)} \cdot m_\varepsilon^{(1)} = 1/p_1$.

Now suppose $m = m_\varepsilon^{(3)}$, is such that $x^{(1)} \cdot m_\varepsilon^{(3)} = x^2 \cdot m_\varepsilon^{(3)} = \Theta_m$, and $x^{(i)} \cdot m_\varepsilon^{(3)} = 0$, $i > 2$. In this case, $\Theta_m = p_1(x^{(1)} \cdot m_\varepsilon^{(3)})^2 + p_2(x^2 \cdot m_\varepsilon^{(3)})^2$, which yields, $x^j \cdot m_\varepsilon^{(3)} = 1/p_1 + 1/p_2$, $j = 1, 2$.

The other cases are treated similarly. ◆

5.1.1. Stability of the Solution

Let $m(t)$ be a solution of a random differential equations then m_0 is said to be a stable point if for any $\delta > 0$ there is a $\tau(\delta)$ such that for any $K > 0$, $t \geq \tau$

$$P\{|m(t) - m_0|^2 > K\} \leq \frac{\delta}{K}. \quad (5.6)$$

This roughly says that if m_0 is a stable point, then the probability of finding the solution *far* from this point is *small*.

LEMMA 5.1. *Let m be a critical point for the random differential equation. Then m is a stable (unstable) critical point, if it is a stable (unstable) critical point for the averaged deterministic version. The stability of the stochastic equation is in the L^2 sense.*

Proof. From eqn 5.3 follows that if m_ε is a critical point of the random version, then it is a critical point of the averaged deterministic equation. If this point is a stable (unstable) point of the deterministic equation, then perturbing both equations (i.e., starting

from an initial condition that is close to m_ε), will yield that the deterministic equation will converge back to (diverge from) the original critical point. This is independent of ε and with probability one since it is a deterministic equation. Consequently, the random solution must stay close to the deterministic solution which in this case, implies the stability (instability) of the random solution. ◆

THEOREM 5.1. *Under the conditions of proposition 5.1, the critical points $m_\varepsilon^{(i)}$ that are stable are only those in which the corresponding vector v^i, has one and only one nonzero element in it.*

Proof. From the lemmas it follows that it is enough to check the stability on the deterministic version of the equations at the critical points of the random version.

The gradient is then given by:

$$-\nabla_m R_m = E[(x \cdot m)^2 x] - E[(x \cdot m)^2]E[(x \cdot m)x], \quad (5.7)$$

and the second order derivative is given by:

$$-\nabla_m^2 R_m = 2E[(x \cdot m)x \times x] - E[(x \cdot m)^2]E[x \times x] - 2E[(x \cdot m)x] \times E[(x \cdot m)x]. \quad (5.8)$$

The critical point $m = 0$ is clearly unstable since the second derivative matrix is zero, and changes sign around $m = 0$. For selective solution, we can choose without loss of generality $m^{(1)}$ which is the solution for $v^{(1)}$. Putting $m^{(1)}$ into the gradient equation gives:

$$-\nabla_m R_{|m = m^{(1)}} = p_1(x^{(1)} \cdot m^{(1)})^2 x^{(1)}$$
$$- p_1(x^{(1)} \cdot m^{(1)})^2 p_1(x^{(1)} \cdot m^{(1)})x^{(1)}$$
$$= p_1(x^{(1)} \cdot m^{(1)})^2[1 - p_1](x^{(1)} \cdot m^{(1)})x^{(1)}]. \quad (5.9)$$

Since $m^{(1)}$ is a critical point and $x^{(1)}$ is the preferred input, we get from the fact that the gradient is equal to zero at $m^{(1)}$: $(x^{(1)} \cdot m^{(1)}) = 1/p_1$, $E(x \cdot m)^2 = 1/p_1$.

Define the matrix B to be

$$B = E[x \times x] = \sum_{i=1}^{N} p_i x^{(i)} \times x^{(i)}, \quad (5.10)$$

since the inputs are independent and span the whole space, it follows that B is positive definite. Putting $(x^{(1)} \cdot m^{(1)})$ into eqn 5.8 gives:

$$-\nabla_m^2 R_{m = m^{(1)}} = p_1(x^{(1)} \cdot m^{(1)})$$
$$\times \left(2x^{(1)} \times x^{(1)} - \frac{1}{p_1}B - 2x^{(1)} \times x^{(1)} \right), \quad (5.11)$$

which is negative definite, thus leading to a stable critical point.

Now, assume, without loss of generality, that $m = m^{(3)}$, then

$$-\nabla_m R|_{m=m^{(3)}} = [p_1(x^{(1)} \cdot m^{(3)})^2 x^{(1)} + p_2(x^{(2)} \cdot m^{(3)})^2 x^{(2)}]$$
$$- [p_1(x^{(1)} \cdot m^{(3)})^2 + p_2(x^{(2)} \cdot m^{(3)})^2]$$
$$\times [p_1(x^{(1)} \cdot m^{(3)})x^{(1)} + p_2(x^{(2)} \cdot m^{(3)})x^{(2)}].$$
(5.12)

Since $m^{(3)}$ is a critical point, we have from Proposition 5.1 that $(x^{(1)} \cdot m^{(3)}) = (x^{(2)} \cdot m^{(3)}) = 1/p_1 + 1/p_2$, and $E(x \cdot m)^2 = 1/p_1 + 1/p_2$.

Putting this into eqn 5.8 gives:

$$-\nabla_m^{(3)} R|_{m=m^{(3)}} = 2p_1(x^{(1)} \cdot m^{(3)})x^{(1)} \times x^{(1)}$$
$$+ 2p_2(x^{(2)} \cdot m^{(3)})x^{(2)} \times x^{(2)}$$
$$- \frac{1}{p_1 + p_2} B$$
$$- 2([p_1(x^{(1)} \cdot m^{(3)})x^{(1)} + p_2(x^{(2)} \cdot m^{(3)})x^{(2)}]$$
$$\times [p_1(x^{(1)} \cdot m^{(3)})x^{(1)} + p_2(x^{(2)} \cdot m^{(3)})x^{(2)}])$$
$$= \left(\frac{2p_1}{p_1 + p_2} - \frac{2p_1^2}{(p_1 + p_2)^2}\right) x^{(1)} \times x^{(1)}$$
$$+ \left(\frac{2p_2}{p_1 + p_2} - \frac{2p_2^2}{(p_1 + p_2)^2}\right) x^{(2)} \times x^{(2)}$$
$$- \frac{1}{p_1 + p_2} B$$
$$- \frac{2p_1 p_2}{(p_1 + p_2)^2} (x^{(1)} \times x^{(2)} + x^{(2)} \times x^{(1)}).$$
(5.13)

Denote the above gradient matrix by G. Without loss of generality we may assume that $p_1 \geq p_2$. Then consider a vector y which is orthogonal to all but $x^{(2)}$. Then

$$y^T G y = y^T x^{(2)} \times x^{(2)} y \frac{p_2}{p_1 + p_2}$$
$$\times \left(6 - \frac{6p_2}{p_1 + p_2} - 3\right) \geq 0, \quad (5.14)$$

since $p_2/p_1 + p_2 \leq \frac{1}{2}$. It is easy to see that by replacing $m^{(3)}$ with $\lambda m^{(3)}$, the second derivative along $m^{(3)}$ changes sign at $\lambda = 1$, which implies instability.

The proof for the other critical points follows in exactly the same way. ◆

5.2. Noise With No Patterned Input

This is a special case, which is related to the binocular deprivation environment discussed in Section 2.1, and hence is analyzed separately. In general, we consider input as being composed of pattern and noise. The patterned input represents a highly correlated set of patterns that appear at random and are supposed to mimic key features in visual environment such as edges with different orientation, etc. The noise in an uncorrelated type of input is assumed

to exist in large network of neurons receiving inputs from several parts of cortex. Patterned input is associated with open eyes and pure noise with closed eyes.

When the input contains only noise, the averaged deterministic solution has a stable critical point and the random solution stays close to the deterministic one (see Appendix A). When the input is composed of noise with zero mean only, we find that the averaged version has a stable zero solution (as opposed to the case with patterned input). This implies that the solution of the random version wanders about the origin but stays close to zero in L^2 norm.

5.2.1. Noise With Zero Mean

The crucial property of white noise x is the fact that it is symmetric around zero. This implies that $E(x \cdot m)^3 = 0$, and the risk,

$$R_m = -\{E[(x \cdot m)^3] - E^2[(x \cdot m)^2]\}$$
$$= E^2[(x \cdot m)^2] \geq 0. \quad (5.15)$$

It is easy to see that only for $m = 0$, $R_m = 0$, and this is the only critical point in this case. Since this result is related to binocular deprivation experiments, it should be emphasized again that the solution to the stochastic version of the differential equations will wander around zero in a random manner but with a small magnitude that is controlled by the learning rate μ.

In view of the properties of the risk, we can say that when the distribution of x has zero skewness in every direction, the only stable minima of the risk is $m = 0$. This is not true when the noise has a positive or a negative average as analyzed in the next section.

5.2.2. Noise With Positive Mean

We assume that x is now bounded random noise, with $\bar{x} > 0$, and x has the same single dimensional distribution in all directions, which implies that $\bar{x}_i = \bar{x}_1 > 0$, $i \geq 0$. Let $x = \bar{x} + y$, where y is random noise with zero average. Denote $\text{Var}(y_1) = \lambda$. The following identities can easily be verified:

$$E(x \cdot m)^2 = (\bar{x} \cdot m)^2 + \text{Var}(y \cdot m),$$
$$E(y \cdot m)y = \lambda m,$$
$$E(y \cdot m)^2 y = 0. \quad (5.16)$$

Putting these identities in the first and second gradient (eqn. 5.7, 5.8) we get:

$$-\nabla_m R_m = [(\bar{x} \cdot m)^2 + \text{Var}(y \cdot m)]\bar{x} - [(\bar{x} \cdot m)^2$$
$$+ \text{Var}(y \cdot m)][(\bar{x} \cdot m)\bar{x} + \lambda m]. \quad (5.17)$$

We are looking for critical points of the gradient,

$$\nabla_m R_m = 0 \Rightarrow m_i = \left[\frac{1}{\lambda} - \frac{(\bar{x} \cdot m)}{\lambda}\right]\bar{x}_i. \quad (5.18)$$

Equation 5.18 suggests a consistency condition that has to be filled, namely, if we multiply both sides of this equation by $\bar{x}_i$ and sum over all i's we get:

$$(\bar{x} \cdot m) = \left[\frac{1}{\lambda} - \frac{(\bar{x} \cdot m)}{\lambda}\right]\|\bar{x}\|^2, \quad (5.19)$$

therefore,

$$(\bar{x} \cdot m) = \frac{\|\bar{x}\|^2}{\lambda + \|\bar{x}\|^2}. \quad (5.20)$$

When substituting eqn 5.20 into eqn 5.18 we get the explicit ratio between m_i and $\bar{x}_i$, namely,

$$m_i = \left[\frac{1}{\lambda + \|\bar{x}\|^2}\right]\bar{x}_i. \quad (5.21)$$

The second derivative is given by:

$$-\nabla_m^2 R_m = [2(\bar{x} \cdot m)^2(\bar{x} \cdot m)^2 - \text{Var}(y \cdot m)](\bar{x} \times \bar{x})$$
$$+ E[\{2(y \cdot m) - 2(y \cdot m)(\bar{x} \cdot m)\}(y \times \bar{x})]$$
$$- 2\lambda(\bar{x} \cdot m)(\bar{x} \times m)$$
$$- 2\lambda(y \cdot m)(y \times m)$$
$$- \lambda[(\bar{x} \cdot m)^2 + \text{Var}(y \cdot m)]I. \quad (5.22)$$

Using relations 5.16 and 5.20 of the critical points, we get to a gradient in terms of $\bar{x} \times \bar{x}$, and λ — the variance of the noise. Let $\tau = 1/\lambda + \|\bar{x}\|^2$, then

$$-\nabla_m^2 R_m = \tau^2(\bar{x} \times \bar{x})[(2 - 4)$$
$$- \lambda\|\bar{x}\|^2 + 2\lambda^2\tau$$
$$- 2\lambda^2\|\bar{x}\|^2 - 2\lambda^2]. \quad (5.23)$$

It follows that the gradient is positive definite for any noise with variance $\lambda > 0$. This implies stability of the averaged version and stability in the L^2 sense of the random version.

5.3. Patterned Input With Noise

We now explore the change in the position of critical points under small noise. The result relies on the smoothness of the projection index and on the fact that noise can be presented as a small perturbation.

Let the input $x = d + h$, where d is the patterned input and h is a small random noise with zero mean. If the mean of the noise is nonzero it can always be absorbed in the patterned input and the resulting

noise will have a zero mean. Let $\lambda = \text{Var}(h \cdot m)$ which is small as well. Consider the projection index

$$R_m = -\mu\left\{\frac{1}{3}E[(x \cdot m)^3] - \frac{1}{4}E^2[(x \cdot m)^2]\right\}$$
$$= -\mu\left\{\frac{1}{3}E[(d \cdot m)^3] - \frac{1}{4}E^2[(d \cdot m)^2]\right.$$
$$\left. + \lambda\left\{E(d \cdot m) - \frac{\lambda}{4}[1 + E(d \cdot m)^2]\right\}\right\}. \quad (5.24)$$

Thus $R_m(d + h) = R_m(d) + O(\lambda)$, yielding robustness to small noise.

6. APPLICATION TO VARIOUS REARING CONDITIONS

In the following section, we relate the analysis described above to some visual cortical plasticity experiments. Extensive simulation using the complete set of known experimental results on visual cortical plasticity have shown that the modified version of Θ_m is consistent with the current experimental results.

6.1. Normal Rearing (NR)

This case has been covered by Theorem 5.1 from which it follows that a neuron will become selective to one of the inputs. Note that it also follows that the synaptic weights of both eyes become selective to the same orientation.

6.2. Monocular Deprivation (MD)

From Theorem 5.1 we can get an explicit expression to Θ_m in the case of n linearly independent inputs. Recall that the only stable points in such a case are those in which the synaptic weight m is orthogonal to all but one of the inputs. Assuming that all the K inputs have the same probablity $1/K$, we get: $\Theta_m = E(x \cdot m)^2 = 1/K\sum_{i=1}^{K}(x_i \cdot m)^2 = 1/K(x_{i_0} \cdot m)^2$ where x_{i_0} is the input which is not orthogonal to m. Putting that into the deterministic version of the gradient descent it follows immediately that $x_{i_0} \cdot m = K$, which implies that $\Theta_m = 1/K(x_{i_0} \cdot m)^2 = K$ and $E(x \cdot m) = 1/K(x_{i_0} \cdot m) = 1$. This result will be used in the following MD analysis.

The assumptions in the monocular deprivation case are that the input to the left (right) eye is composed of noise only, namely d^r represents patterned input plus noise, and $d^l = n$. We also assume that the noise has zero average and has a symmetric distribution uniform in all directions. We relax the assumption that d^r has zero mean and instead assume that $E(d^r \cdot m^r) < 1/2E(d^r \cdot m^r)^2$. This is easily achieved when the dimensionality is larger than 2 (following from the calculation at the beginning of this section). We have:

$$R = - \left\{ \frac{1}{3} E(x \cdot m)^3 - \frac{1}{4} E^2(x \cdot m)^2 \right\}$$

$$= - \left\{ \frac{1}{3} E[(d^r \cdot m^r + n \cdot m^l)^3] \right.$$

$$\left. - \frac{1}{4} E^2[(d^r \cdot m^r + n \cdot m^l)^2] \right\}$$

$$= - \left\{ \frac{1}{3} E[(d^r \cdot m^r)^3] + \frac{1}{3} E[(n \cdot m^l)^3] \right.$$

$$+ E[(d^r \cdot m^r)^2(n \cdot m^l)] + E[(d^r \cdot m^r)(n \cdot m^l)^2]$$

$$- \frac{1}{4} [E^2[(d^r \cdot m^r)^2] + E^2[(n \cdot m^l)^2]$$

$$+ 2 E[(d^r \cdot m^r)^2] E[(n \cdot m^l)^2]$$

$$+ 4E[d^r \cdot m^r] E[n \cdot m^l] (E[(d^r \cdot m^r)^2]$$

$$\left. + E[(n \cdot m^l)^2] + E[d^r \cdot m^r] E[n \cdot m^l])] \right\}$$

$$= - \left\{ \frac{1}{3} E[(d^r \cdot m^r)^3] - \frac{1}{4} E^2[(d^r \cdot m^r)^2] \right\}$$

$$+ \frac{1}{4} \text{Var}(n \cdot m^l)[\text{Var}(n \cdot m^l)$$

$$+ 2E[(d^r \cdot m^r)^2] - 4E[d^r \cdot m^r]]. \tag{6.1}$$

The first term of the risk is due to the open eye and is therefore minimized when the neuron becomes selective as in the regular normal rearing case. The second term is non-negative due to the previous assumption, and therefore can be minimized only if $m^l = 0$. Note that in a mean field, this means that $m^l \to \alpha$. It can also be seen that when the right eye becomes selective (implying that the term $2E[(d^r \cdot m^r)^2] - 4E[d^r \cdot m^r]$ becomes larger), then the driving force for $\text{Var}(n \cdot m^l)$ to go to zero becomes larger. This is consistent with the experimental observation which suggests that the synapses of the closed eye do not go down until the open eye becomes selective.

6.3. Binocular Deprivation (BD)

This case has been analysed in Section 5.2. During BD we assume that the input is noise; the conclusion was that either synaptic weights perform a random walk around zero, or in case of positive average noise, a random walk about a positive weight that is a function of the average of the noise and its variance.

6.4. Reversed Suture (RS)

The limiting behavior of RS is similar to that of MD, described above. Computer simulations show that it is possible to achieve a disconnection of the newly closed eye before the newly open eye becomes selective (Clothiaux et al., in press).

6.5. Strabismus

From Theorem 5.1 we infer that a stable fixed point is such that its projection to one of the inputs is positive and it is orthogonal to all the other inputs. Under strabismus we assume that the input to both eyes is uncorrelated, therefore this situation is possible only if the vector of synaptic weights of one eye is orthogonal to all but one of the inputs; thus the vector of synaptic weights of the other eye is orthogonal to all the inputs. Since the inputs span the whole space, this vector must be zero.

7. DISCUSSION

We have presented an objective function formulation of the BCM theory of visual cortical plasticity. This permits us to demonstrate the connection between the unsupervised BCM learning procedure and the statistical method of projection pursuit and provides a general method for stability analysis of the fixed points. Relating this unsupervised learning to statistical theory enables comparison with various other statistical and unsupervised methods for feature extraction.

Analysis of the behavior and the evolution of the network under various visual rearing conditions is in agreement with experimental results. We thus have the result that a biological neuron may be performing a sophisticated statistical procedure. An experimental question of great interest is posed: How does the modification threshold depend on the average activity of the cell $\Theta_m \simeq \bar{c}^2$ as in the original BCM versus $\Theta_m \simeq c^2$ as presented here.

REFERENCES

Barron, A. R., & Barron, R. L. (1988). Statistical learning networks: A unifying view. In Wegman, E., (Ed.), *Computing Science and Statistics*: Proceedings of the 20th Symposium Interface (pp. 192–203). Washington, DC: American Statistical Association.

Bear, M. F., & Cooper, L. N (1988). Molecular mechanisms for synaptic modification in the visual cortex: Interaction between theory and experiment. In M. Gluck and D. Rumelhart (Eds.), *Neuroscience and Connectionist Theory* (pp. 65–94). Hillsdale, NJ: Lawrence Erlbaum.

Bear, M. F., Cooper, L. N, & Ebner, F. F. (1987). A physiological basis for a theory of synapse modification. *Science*, **237**, 42–48.

Bellman, R. E. (1961). *Adaptive Control Processes*. Princeton, NJ: Princeton University Press.

Bienenstock, E. L., Cooper, L. N, & Munro, P. W. (1982). Theory for the development of neuron selectivity: Orientation specificity and binocular interaction in visual cortex. *Journal Neuroscience*, **2**, 32–48.

Bryan, J. G. (1951). The generalized discriminant function: Mathematical foundations and computational routines. *Harvard Education Review*, **21**, 90–95.

Clothiaux, E. E., Cooper, L. N, & Bear, M. F. (in press). Synaptic plasticity in visual cortex: Comparison of theory with experiment. *Journal of Neurophysiology*.

Cooper, L. N, Liberman, F., & Oja, E. (1979). A theory for the acquisition and loss of neurons specificity in visual cortex. *Biological Cybernetics, 33,* 9–28.

Cooper, L. N & Scofield, C. L. (1988). Mean-field theory of a neural network. *Proceedings of the National Academy of Sciences, 85,* 1973–1977.

Diaconis, P. & Freedman, D. (1984). Asymptotics of graphical projection pursuit. *Annals of Statistics, 12,* 793–815.

Duda, R. O. & Hart, P. E. (1973). *Pattern Classification and Scene Analysis.* New York: John Wiley & Sons.

Duffy, F H., Sonodgrass, S. R., Burchfiel, J. L., & Conway, J. L. (1976). Bicuculline reversal of deprivation amblyopia in the cat. *Nature, 260,* 256–257.

Fisher, R. A. (1936). The use of multiple measurements in taxonomic problems. *Annals of Eugenics, 7,* 179–188.

Frègnac, Y., & Imbert, M. (1984). Development of neuronal selectivity in primary visual cortex of cat. *Physiology Review, 64,* 325–434.

Friedman, J. H. (1987). Exploratory projection pursuit. *Journal of the American Statistical Association, 82,* 249–266.

Friedman, J. H., & Tukey, J. W. (1974). A projection pursuit algorithm for exploratory data analysis. *IEEE Transactions on Computers,* C(23), 881–889.

Geman, S. (1977) Averaging for random differential equations. In A. T. Bharucha-Reid (Ed.), *Approximate Solution of Random Equations* (pp. 49–85). New York: North Holland.

Harman, H. H. (1967). *Modern factor analysis* (2nd ed.). Chicago and London: University of Chicago Press.

Hinton, G. E, & Nowlan, S. J. (1990). The bootstrap widrow-hoff rule as a cluster-formation algorithm. *Neural Computation,* 2(3), 355–362.

Hubel, D. H., & Wiesel, T. N. (1959). Integrative action in the cat's lateral geniculate body. *Journal of Physiology 148,* 574–591.

Huber, P. J. (1985). Projection pursuit (with discussion). *Annals of Statistics, 13,* 435–475.

Intrator, N. (1990). *An averaging result for random differential equations* (Tech. Rep. No. 54). Providence, RI: Brown University, Center For Neural Science.

Intrator, N. (in press). Feature extraction using an unsupervised neural network. *Neural Computation* (in press).

Intrator, N., & Gold, J. I. (in press). Three-dimensional object recognition of gray level images: The usefulness of distinguishing features. Submitted.

Intrator, N., Gold, J. I., Bülthoff, H. H., & Edelman, S. (in press). Three-dimensional object recognition using an unsupervised BCM network. In *Proceedings of the 8th Israeli Conference on AIVC.*

Intrator, N., & Tajchman, G. (in press). Unsupervised feature extraction from a cochlear model for speech recognition. In B. H. Juang, S. Y. Kung, & C. A. Kamm ((Eds.), *Neural Networks for Signal Processing—Proceedings of the 1991 IEEE Workshop* (pp. 460–469).

Jones, M. C. (1983). *The projection pursuit algorithm for exploratory data analysis.* Unpublished doctoral dissertation, University of Bath, School of Mathematics.

Jones, M. C., Sibson, R. (1987). What is projection pursuit? (with discussion). *Journal Royal Statistical Society,* Ser. A(150), 1–36.

Kruskal, J. B. (1969). Toward a practical method which helps uncover the structure of the set of multivariate observations by finding the linear transformation which optimized a new 'index of condensation'. In R. C. Milton and J. A. Nelder (Eds.) *Statistical Computation.* New York: Academic Press.

Kruskal, J. B. (1972). Linear transformation of multivariate data to reveal clustering. *Multidimensional Scaling: Theory and Application in the Behavioral Sciences, I, Theory.* New York and London: Seminar Press.

Linsker, R. (1988). Self-organization in a perceptual network. *IEEE Computer, 88,* 105–117.

Miller, K. D. (1988). Correlation-based models of neural development. In M. Gluck and D. Rumelhart (Eds.) *Neuroscience and Connectionist Theory* (pp. 267–353). Hillsdale, NJ: Lawrence Erlbaum.

Mover, G., Caplan, C., Christen, W., & Duffy, F. (1985). Dark rearing prolongs physiological but not anatomical plasticity of the cat visual cortex. *Journal Computational Neurobiology, 235,* 448–466.

Oja, E. (1982). A simplified neuron model as a principal component analyzer. *Journal of Mathematical Biology, 15,* 267–273.

Presson, J., & Gordon, B. (1982). The effects of monocular deprivation on the physiology and anatomy of the kitten's visual system. *Society of Neuroscience Abstracts, 8,* 5–10.

Saul, A., & Daniels, J. D. (1986). Modeling and simulation I: Introduction and guidelines. *Journal of Electrophysiological Techniques, 13,* 95–109.

Scofield, C. L., & Cooper, L. N (1985). Development and properties of neural networks. *Contemporary Physics, 26,* 125–145.

Sebestyen, G. (1962). *Decision Making Processes in Pattern Recognition.* New York: Macmillan.

Sejnowski, T. J. (1977). Storing covariance with nonlinearly interacting neurons. *Journal of Mathematical Biology, 4,* 303–321.

Sherman, S., & Spear, P. (1982). Organization of visual pathways in normal and visually deprived cats. *Physiology Review, 62,* 738–855.

Slaney, M. (1988). Lyon's cochlear model (Technical Report) Cupertino, CA: Apple Corporate Library.

Switzer, P. (1970). Numerical classification. In V. Barnett (Ed.), *Geostatistics.* New York: Plenum Press.

von der Malsburg, C. (1973). Self-organization of orientation sensitivity cells in the striate cortex. *Kybernetik, 14,* 85–100.

Wiesel, T., & Hubel, D. (1965). Comparison of the effects of unilateral and bilateral eye closure on cortial unit responses in kittens. *Journal of Neurophysiology, 28,* 1029–1040.

Yang, X., & Faber, D. S. (in press). Initial synaptic efficacy influences induction and expression of long-term changes in transmission. *Proceedings of the National Academy of Science,* 88(10), 4299–4303.

APPENDIX A: CONVERGENCE OF THE SOLUTION OF THE RANDOM DIFFERENTIAL EQUATIONS

To show the explicit dependency on the learning rate, we rewrite the random modification equations in the form:

$$\dot{m}_\varepsilon = \varepsilon \mu(t) \phi(x \cdot m_\varepsilon, \Theta_m) x, \qquad m_\varepsilon(0) = m_0, \qquad \text{(A.1)}$$

and the deterministic differential equations,

$$\overline{\dot{m}}_\varepsilon = \varepsilon \mu(t) E[\phi(x \cdot \overline{m}_\varepsilon, \Theta_m) x], \qquad \overline{m}_\varepsilon(0) = m_0, \qquad \text{(A.2)}$$

The convergence of the solution will be shown in two steps. First we show that the solution of the averaged deterministic equation converges, and then we use Theorem A.1 to show the convergence of the solution of the random differential equation to the solution of its averaged deterministic equation.

A.1. Convergence of the Deterministic Equation

The deterministic differential equations represent a negative gradient of the risk. Therefore, in order to show convergence of the solution we only need to show that the risk is bounded from below. This will assure that the solution converges to a local minimum of the risk.

We can assume that m the synaptic weight vector lies in the space spanned by the random variable x. When we replace the random variable x with a training set $x^1, \ldots, x^n$, this assumption says that $m \in \text{Span}\{x^1, \ldots, x^n\}$. This implies that there is a $\lambda > 0$, so that $\forall m \ \text{Var}(x \cdot m) \geq \lambda \|m\|^2 > 0$.

To show that the vector $\overline{m}_\varepsilon$ is bounded we assume that none of its components is zero (since zero is definitely bounded), and multiply both sides of the above equation by $\overline{m}_\varepsilon$. This implies:

$$\frac{1}{2} \frac{d}{dt} \|\overline{m}_\varepsilon\|_2^2 = E[(x \cdot \overline{m}_\varepsilon)^3] - E^2[(x \cdot \overline{m}_\varepsilon)^2]$$

$$\leq \|\overline{m}_\varepsilon\|^3 - \text{Var}^2(x \cdot \overline{m}_\varepsilon)$$

$$\leq \|\overline{m}_\varepsilon\|^3 - \lambda^2 \|\overline{m}_\varepsilon\|^4$$

$$= \|\overline{m}_\varepsilon\|^3 \{1 - \lambda^2 \|\overline{m}_\varepsilon\|\}, \tag{A.3}$$

which implies that $\|\overline{m}_\varepsilon\| \leq 1/\lambda^2$.

Using this fact we can now show the convergence of $\overline{m}_\varepsilon$. We observe that $\overline{m}_\varepsilon = -\nabla R$, where $R(\overline{m}_\varepsilon) = -\mu\{\frac{1}{3}E[(x \cdot \overline{m}_\varepsilon)^3] - \frac{1}{4}E^2[(x \cdot \overline{m}_\varepsilon)^2]\}$ is the risk. R is bounded from below since $\|\overline{m}_\varepsilon\|$ is bounded, therefore $\overline{m}_\varepsilon$ converges to a local minimum of R as a solution to the gradient descent.

A.2. Convergence of the Random Equation

Using the averaged deterministic version convergence we shall now show the convergence of the random version. For this we need a general result on random differential equations (Intrator, 1990) which is cited below. This result is an extension of a result by Geman (1977) which says that under some smoothness conditions on the second order derivatives of the differential equations, the solution of the random differential equation remains close (in the L^2 sense) to the deterministic solution *for all times*.

We start with some preliminary notation. Let $H(x, \omega, t)$ be a continuous and mixing R^m valued random process for any fixed x and t, where ω is a sample point in a probability space. Define $G(x, t) = E[H(x, \omega, t)]$, the expected value with respect to ω. Let $\mu(t)$ be a continuous monotone function decreasing to zero, and let $\varepsilon > 0$ be arbitrary. Consider the following random differential equation together with its associated averaged version,

$$\dot{x}_\varepsilon(t, \omega) = \varepsilon\mu(t)H(x_\varepsilon(t, \omega), \omega, t), \quad x_\varepsilon(0, \omega) = x_0 \in R^n.$$

$$\dot{y}_\varepsilon(t) = \varepsilon\mu(t)G(y_\varepsilon(t), t), \quad y_\varepsilon(0) = x_0 \in R^n. \tag{A.4}$$

ε generates a family of solutions x_ε, and y_ε.

THEOREM A.1. *Given the above system of random differential equations, assume:*

1. *$H \in R^n$ is jointly measurable with respect to its three arguments, and is of Type II φ mixing.*

2. *$G(x, t) = E[H(x(s, \omega), t)]$, and for all i and j*

$$\frac{\partial}{\partial x_j} G_i(x, t) \text{ exists, and is continuous in } (x, t).$$

3. (a) *There exists a unique solution, $x(t, \omega)$, on $[0, \infty)$ for almost all ω; and*
   (b) *A solution to*

$$\frac{\partial}{\partial t} g(t, s, x) = G(g(t, s, x), t), \quad g(s, s, x) = x,$$

   exists on $[0, \infty) \times [0, \infty) \times R^n$.

4. *There exist continuous functions $B_1(r)$, $B_2(r)$, and $B_3(r)$, such that for all $i, j, k, \tau \geq 0$, and ω:*
   (a) *$|H_i(x, \omega, t)| \leq B_1(|x|)$;*
   (b) *$|(\partial/\partial x_j)H_i(x, \omega, t)| \leq B_2(|x|)$;*
   (c) *$|(\partial^2/\partial x_j \partial x_k)H_i(x, \omega, t)| \leq B_3(|x|)$.*

5. *$\sup_{\varepsilon > 0, t} |y_\varepsilon(t)| \leq B_4$ for some $B_4 \forall t$*

6. *$\exists \gamma > 0, c > 0$, such that $\varphi(\delta) \leq \delta^{-\gamma}$, and $\tilde{\nu} \leq t^{-(1/\gamma + 1 + c)}$, for a monotone decreasing ν. Then under conditons 1–6:*

$$\lim_{\varepsilon \to 0} \sup_{t \geq 0} E|x_\varepsilon - y_\varepsilon|^2 = 0, \tag{A.5}$$

To use this result, we need only to show that the deterministic and the random solutions are bounded, which will ensure conditions 2–5. Then under the mixing conditions 1 and 6 on the input x, we get the desired result.

Verifying that the random solution is bounded for every ω can be done by multiplying both sides of the random differential equations by m_ε, assuming its components are not zero and applying the assumptions made above on $\text{Var}(x \cdot m_\varepsilon)$, we get

$$\frac{1}{2} \frac{d}{dt} \|m_\varepsilon\|^2 = (x \cdot m_\varepsilon)^3 - (x \cdot m_\varepsilon)^2 E[(x \cdot m_\varepsilon)^2]$$

$$= (x \cdot m_\varepsilon)^2 \{(x \cdot m_\varepsilon) - E[(x \cdot m_\varepsilon)^2]\}$$

$$\leq (x \cdot m_\varepsilon)^2 \{(x \cdot m_\varepsilon) - \text{Var}(x \cdot m_\varepsilon)\}$$

$$\leq (x \cdot m_\varepsilon)^2 \{(x \cdot m_\varepsilon) - \lambda\|m_\varepsilon\|^2\}$$

$$\leq (x \cdot m_\varepsilon)^2 \{\|m_\varepsilon\| - \lambda\|m_\varepsilon\|^2\}, \tag{A.6}$$

which implies that the derivative of the norm will become negative whenever $\|m_\varepsilon\| > \lambda$, therefore $\|m_\varepsilon\| \leq 1/\lambda$.

Finally, since the random solution remains close to a converging deterministic solution, it remains close (in the L^2 sense) to its limit for large enough t. δ is arbitrary, which implies that

$$E|m_\varepsilon(t) - \dot{m}|^2 \xrightarrow[\mu_0 \to 0]{} 0 \tag{A.7}$$

◆

Homosynaptic Long-Term Depression in Area CA1 of Hippocampus and Effects of *N*-Methyl-D-Aspartate Receptor Blockade

(with S. M. Dudek and M. F. Bear)

Proc. Natl. Acad. Sci. USA **89**, 4363 (1992)

For years work in hippocampus on the phenomenon of LTP and work in visual cortex on various developmental phenomena had proceeded along separate tracks. When we ran a joint seminar at Brown in the spring of 1992 with the participation of Richard Granger, our earlier conviction of the close connection of these two phenomena was reinforced. For example, the suggestion that the theta rhythm might be the key rhythm by which information was put into the nervous system, if also true in visual cortex, would provide us with a natural means of synchronizing the input data, a means not necessarily required but one that would make the theoretical analysis much simpler.

About that time, Mark, working with Serena Dudek who had come from Gary Lynch's laboratory, began a systematic search for evidence of what has come to be called long-term depression (LTD) in hippocampus. There had been previous suggestions that such a phenomenon might occur but, for various technical reasons, it seemed difficult to nail down and no one had come up with a reproducible protocol.

Mark's understanding of the theoretical need for such an effect as well as what it should depend on led him to design the protocol that finally tied it down (as he once told me, *"This was one case in which one had to believe in order to see"*). The following paper gives their results.

Proc. Natl. Acad. Sci. USA
Vol. 89, pp. 4363–4367, May 1992
Neurobiology

Homosynaptic long-term depression in area CA1 of hippocampus and effects of N-methyl-D-aspartate receptor blockade

(long-term potentiation/hippocampal slice/synaptic plasticity/learning/memory)

Serena M. Dudek and Mark F. Bear*

The Center for Neural Science, Brown University, Providence, RI 02912

Communicated by Leon N Cooper, January 28, 1992 (received for review December 30, 1991)

ABSTRACT We tested a theoretical prediction that patterns of excitatory input activity that consistently fail to activate target neurons sufficiently to induce synaptic potentiation will instead cause a specific synaptic depression. To realize this situation experimentally, the Schaffer collateral projection to area CA1 in rat hippocampal slices was stimulated electrically at frequencies ranging from 0.5 to 50 Hz. Nine hundred pulses at 1–3 Hz consistently yielded a depression of the CA1 population excitatory postsynaptic potential that persisted without signs of recovery for >1 hr after cessation of the conditioning stimulation. This long-term depression was specific to the conditioned input, ruling out generalized changes in postsynaptic responsiveness or excitability. Three lines of evidence suggest that this effect is accounted for by a modification of synaptic effectiveness rather than damage to or fatigue of the stimulated inputs. First, the effect was dependent on the stimulation frequency; 900 pulses at 10 Hz caused no lasting change, and at 50 Hz a synaptic potentiation was usually observed. Second, the depressed synapses continued to support long-term potentiation in response to a high-frequency tetanus. Third, the effects of conditioning stimulation could be prevented by application of NMDA receptor antagonists. Thus, our data suggest that synaptic depression can be triggered by prolonged NMDA receptor activation that is below the threshold for inducing synaptic potentiation. We propose that this mechanism is important for the modifications of hippocampal response properties that underlie some forms of learning and memory.

A cardinal feature of neuronal responses in the central nervous system is stimulus selectivity. Work on a number of preparations, including the developing visual cortex (1), adult somatosensory cortex (2), and hippocampus (3), has shown convincingly that neuronal selectivity can be modified by experience. It is a widely held belief that such modifications are a likely basis for learning and memory in mammals.

Experience-dependent shifts in selectivity require that neurons acquire responsiveness to new stimuli and lose responsiveness to previously effective stimuli. A critical question concerns the mechanisms whereby this is accomplished. According to a theory developed for visual cortex (4–6), stimulus selectivity will evolve if excitatory synaptic inputs are potentiated when they consistently yield postsynaptic responses greater than a critical value (called the modification threshold, θ_m) and are depressed when they consistently yield responses greater than zero (the average "spontaneous" value) but less than θ_m. Because the value of the modification threshold is a function of the average postsynaptic cell activity (5, 6), the selectivity of a neuron can be modified when the probability of different patterns of input

activity shifts, as it would when the animal is experiencing a novel environment.

On the basis of work done on long-term potentiation (LTP) in area CA1 of the hippocampus (7), the proposal was made that this theoretical modification threshold θ_m related to the postsynaptic membrane potential at which N-methyl-D-aspartate (NMDA) receptors became sufficiently active to trigger increases in synaptic strength (8, 9). It follows from this hypothesis that patterns of input activity that consistently fail to activate NMDA receptors in a manner that is sufficient to trigger LTP should produce a long-term depression (LTD) of synaptic effectiveness. We have tested this theoretical prediction in the Schaffer collateral–CA1 pathway in rat hippocampal slices. These data were first presented at the 1991 Society for Neuroscience meeting in New Orleans (10).

MATERIALS AND METHODS

The experiments described in this paper were performed on transverse slices prepared from the hippocampi of adult male albino rats (body weight ≈ 200 g). Each animal was given an overdose of sodium pentobarbital (≈75 mg/kg, i.p.) and was decapitated soon after the disappearance of any corneal reflexes. The brain was rapidly removed and immersed in ice-cold artificial cerebrospinal fluid (ACSF) containing, in mM, NaCl, 124; KCl, 5; NaH_2PO_4, 1.25; $MgCl_2$, 1.5; $CaCl_2$, 2.5; $NaHCO_3$, 26; and dextrose, 10. The hippocampus was dissected free and sectioned into 0.4-mm-thick slices by using a vibrating microtome. These slices were collected in ice-cold ACSF and gently transferred to an interface slice chamber. Here, the slices were maintained in an atmosphere of humidified 95% O_2/5% CO_2 and superfused at a rate of 1 ml/min with 35°C ACSF saturated with 95% O_2/5% CO_2.

After at least 1 hr of equilibration in the slice chamber, a concentric bipolar stimulating electrode was placed in the trajectory of the Schaffer collateral fibers projecting to the stratum radiatum of area CA1. In some experiments a second stimulating electrode was placed on the opposite (subicular) side of the recording location to activate a second converging input. The recording pipette was filled with 1 M NaCl and was placed in the apical dendritic layer of CA1. Population excitatory postsynaptic potentials (EPSPs) were evoked by using 10- to 30-μA stimuli of 0.2-msec duration. These responses were digitized at 20 kHz and stored on a computer. The initial slope of the population EPSP was extracted as a measure of the magnitude of the response.

At the start of each experiment, a full input–output curve was constructed. A stimulation intensity was selected for baseline measurements that yielded between ½ and ⅔ of the maximal response. In general, slices were studied only if the

Abbreviations: LTP, long-term potentiation; LTD, long-term depression; NMDA, N-methyl-D-aspartate; EPSP, excitatory postsynaptic potential; LFS, low-frequency stimulation; AP5, DL-2-amino-5-phosphonovaleric acid.
*To whom reprint requests should be addressed.

4363

maximal response amplitude was ≥ 2 mV and, as a rule, slices were studied for no longer than 6 hr *in vitro*. Baseline measurements were collected by using single shocks every 30 sec. The conditioning stimulation consisted of 900 pulses delivered at frequencies ranging from 1 to 50 Hz. In every experiment an *F* test was used to confirm that any change in the response after conditioning could not be explained by a linearly drifting baseline (11). The summary graphs in Figs. 1–3 were generated as follows: (*i*) the EPSP slope data for each experiment were expressed as percentages of the preconditioning baseline average; (*ii*) the time scale in each experiment was converted to time from the onset of conditioning; and (*iii*) the time-matched normalized data were averaged across experiments and expressed as the means $\pm$ SEM.

RESULTS

In an effort to provide a high level of presynaptic input activity without producing a postsynaptic response so large that it yielded LTP, we tried extended periods of low-frequency stimulation (LFS) in the range of 0.5–5 Hz. The lasting consequences of 900 pulses delivered at 1 Hz are illustrated in Fig. 1. The response was often facilitated immediately after the onset of conditioning stimulation, but

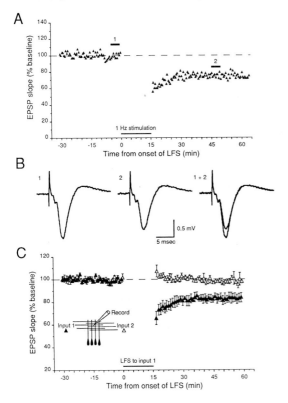

FIG. 1. (*A*) Record of a representative experiment in which LTD was induced by LFS. Each point represents a single measure of the initial slope of the population EPSP evoked by stimulation of the Schaffer collaterals at 0.03 Hz. The horizontal bar represents the period of 1-Hz conditioning stimulation. (*B*) Means of 10 consecutive sweeps before and after LFS conditioning at the times indicated by the numbers in *A*. (*C*) Normalized means ($\pm$SEM) of five experiments in which the response to two independent inputs was monitored (*Inset* shows the stimulating/recording configuration). LFS (1-Hz) conditioning stimulation given to input 1 produced a depression only of the response to input 1 ($\blacktriangle$).

this was always followed by a progressive decline in the response magnitude during conditioning to values below the initial baseline (data not shown). When the baseline measurements were resumed (at 0.03 Hz), the first response was always depressed, even below the value attained during the LFS (Fig. 1*A*). Although there was usually some recovery in response magnitude over the next several minutes, the population EPSP slope always reached a plateau at a value that was significantly depressed as compared with the pre-LFS control period (Fig. 1*B*).

There are several possible explanations for these results that do not involve plasticity of the stimulated synapses. For example, it is possible that the prolonged LFS produces excitotoxic damage in the target neurons or some generalized loss of postsynaptic excitability. If this were the case, then LFS of one input to a postsynaptic neuron would be expected to produce LTD of a second input that did not undergo conditioning stimulation. However, we found that 1-Hz LFS applied to the Schaffer collaterals had no lasting effect on the response to a second, independent, input (Fig. 1*C*; *n* = 5). As it is likely that at least some fraction of the fibers in the two paths converged onto the same postsynaptic cells, these data can be taken as evidence that the LTD is "input specific" and thus confined to the stimulated synapses.

Input specificity of LTD does not rule out the possibility of permanent damage or neurotransmitter depletion of the conditioned pathway. We attempted to address this concern in several ways. First, we confirmed that the depressed input could still undergo LTP after a patterned high-frequency tetanus (data not shown). This observation does not rule out presynaptic damage or depletion, but it does mean that the depressed synapses are sufficiently viable to support the mechanisms that give rise to LTP. A second approach was to investigate the effect of varying the stimulation frequency. We reasoned that if the LTD was simply explained by the damage caused by prolonged stimulation, then the effect should be relatively independent of stimulation frequency. However, to the contrary, the lasting effects of conditioning stimulation showed a marked frequency dependence. As illustrated in Fig. 2, 900 pulses at 3 Hz resulted in LTD (81% $\pm$ 2% of control at 30 min after conditioning; *n* = 5); however, the same number of pulses at 10 Hz produced (on average) no lasting change (101% $\pm$ 4% of control; *n* = 6), and the pulses at 50 Hz produced a variable potentiation (121% $\pm$ 7% of control; *n* = 5).

The possibility remains that the frequency dependence of the effect is explained by the superimposition of LTP onto the presynaptic damage caused by the conditioning stimulation. However, this explanation is ruled out by a third line of evidence. In six experiments, 200 μM DL-2-amino-5-phosphonovaleric acid (AP5) was added to the bath and then 900 pulses were delivered at 1 Hz. This drug treatment markedly inhibited the induction of LTD (Fig. 3). However, in every case LTD was induced by the same stimulation immediately following wash-out of the AP5. AP5 is known to block postsynaptic NMDA receptors (7) and is not thought to have any significant presynaptic actions. Preliminary work indicates that MK801, an NMDA channel blocker, will also interfere with induction of LTD. Therefore, these data argue strongly that the LTD we report here is a real manifestation of synaptic plasticity in the conditioned pathway, and they suggest a possible involvement of postsynaptic NMDA receptors in the induction mechanism.

DISCUSSION

The major finding of this study is that repetitive stimulation of the Schaffer collateral input to CA1 at frequencies below the threshold for inducing LTP can lead to a reliable and long-lasting synaptic depression. This LTD is specific for the

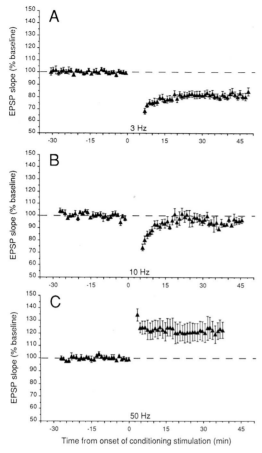

FIG. 2. Normalized averages of experiments in which 900 pulses were delivered at different frequencies. (A) 3 Hz, $n = 5$. (B) 10 Hz, $n = 6$. (C) 50 Hz, $n = 5$. Conventions are as for Fig. 1.

stimulated input and may be triggered, at least in part, by activation of NMDA receptors.

Most of the previous accounts of activity-dependent decreases in synaptic effectiveness in hippocampus have con-

cerned a phenomenon usually called "heterosynaptic" depression (see ref. 12 for review). This refers to a generalized decrease in the response to synaptic inputs that are silent during episodes of strong postsynaptic activation (13–15). Even the postsynaptic activity that accompanies low-frequency (1–15 Hz) antidromic stimulation of CA1 neurons can produce this effect (16). Therefore, it is of significance that the LTD observed in the present study was input specific (Fig. 1C). This makes it unlikely that the LTD reported here is simply the manifestation of an orthodromically induced, but nonetheless heterosynaptic, change in synaptic effectiveness.

Depression of synaptic transmission after prolonged periods of repetitive stimulation is not a new observation. For example, at the neuromuscular junction, del Castillo and Katz (17) showed that 2-Hz stimulation causes a progressive decline in the quantal release of acetylcholine. This "post-tetanic depression" is often ascribed to depletion of neurotransmitter, and it has been observed and analyzed at central synapses (cf. ref. 18) recently, including those in CA1 (19). Although this effect is of considerably shorter duration than the LTD we report here, the question nonetheless arises of whether our results can be explained by a lasting depletion of transmitter at the stimulated synapses. In addressing this issue it is instructive to consider that, at the neuromuscular junction, post-tetanic depression is unaffected by postsynaptic acetylcholine receptor antagonists (17). In marked contrast, however, we have found that NMDA receptor antagonists can prevent the induction of LTD in CA1. Because the release of neurotransmitter at the Schaffer collateral–CA1 synapse is not directly affected by these drugs (7), this observation suggests that depletion of neurotransmitter is not a likely explanation for LTD. The blockade of LTD by NMDA receptor antagonists also distinguishes the present work from a previous report of homosynaptic depression in CA1 (20).

As NMDA receptors are thought to be exclusively postsynaptic in CA1, our data suggest that LTD is triggered by a postsynaptic event. Although the nature of this event remains to be determined, it seems likely that postsynaptic Ca^{2+} entry is involved (28). In this light it is interesting to note that Lisman (21) has presented a model that shows the feasibility of using quantitative differences in the activity-dependent changes in postsynaptic Ca^{2+} to determine whether synaptic weights will increase or decrease. Another possibility is that metabotropic glutamate receptor activation in conjunction with an elevation in postsynaptic Ca^{2+} serves as a trigger for LTD (22, 29).

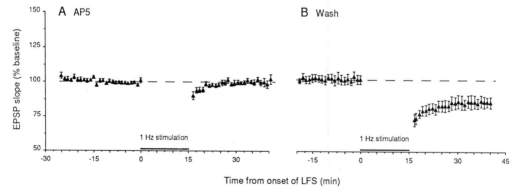

FIG. 3. Effects of an NMDA receptor antagonist on the lasting effect of 900 pulses at 1 Hz. (A) AP5 was used in the concentration of 200 μM to overcome any competition for the NMDA receptors that might result from the surge of endogenous transmitter released during stimulation. The EPSP slope was unaffected when the AP5 was introduced, suggesting that the drug at this concentration was still selective for NMDA receptors. (B) Removal of antagonist: 30–60 min elapsed between the last point in A and the first point in B to allow time for wash-out of the drug. All data are expressed as percentage of the baseline prior to the first LFS in AP5, $n = 6$. Conventions are as for Fig. 1.

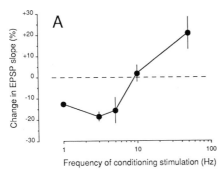

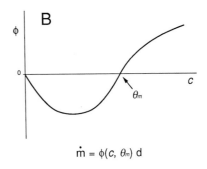

$$\dot{m} = \phi(c, \theta_m) \, d$$

FIG. 4. Comparison of theory and experiment. (*A*) Mean (±SEM) effect of 900 pulses of conditioning stimulation delivered at various frequencies on the response measured 30 min after conditioning; $n \geq 5$ for each point. (*B*) Modification function ϕ of Bienenstock *et al.* (5). Please refer to the text for definition of terms.

Although the LTD documented here is of relatively small magnitude and requires prolonged repetitive stimulation at low frequencies, we believe that it is of physiological significance. The magnitude of the depression can be increased by increasing the number of pulses; for example, we observe LTD of >30% with 1800 pulses at 1 Hz. And, although this may appear to be an extreme type of stimulation, we note that hippocampal electroencephalogram (EEG) recordings in awake rats are characterized by 1- to 10-Hz oscillations that reflect the synchronous firing of hippocampal neurons. Particularly interesting in light of the frequency dependence of the present results (Fig. 2) is the reported shift in the peak of the EEG power spectrum from 1 to 7 Hz as rats go from a state of stillness to active exploration of their environment (23).

Finally, the results of this study may be compared with the theoretical form of modification that inspired it (Fig. 4). According to the theory of Bienenstock *et al.* (5), the modification of excitatory synaptic weight (m) is assumed to proceed as the product of the input activity (d) and a function ϕ of the postsynaptic response (c). For situations where d is a positive constant, then the curve in Fig. 4*B* can be interpreted as the sign and magnitude of the synaptic weight change at different levels of postsynaptic response. If the effects of varying stimulation frequency in the present study are explained by different values of postsynaptic response (perhaps the integrated postsynaptic depolarization or Ca^{2+} level) during the conditioning stimulation, then it can be seen from Fig. 4*A* that our data are in striking agreement with assumptions of this theory.

Another central feature of the theory is the sliding modification threshold (5, 6). That is, the value of θ_m in Fig. 4*B* is assumed to vary as a function of the average postsynaptic response. In this context it is interesting to compare the present results with previous investigations of the effects of LFS in hippocampus. Although earlier reports (24–26) failed to show (or overlooked) LTD in the "naive" hippocampus, they did demonstrate that 1-Hz stimulation applied within several hours of inducing LTP would cause a substantial reduction in the potentiated response. Because this effect is also AP5 sensitive (26), it is possible that "depotentiation" and LTD share a common mechanism, and differ only by degree. This then suggests that the effectiveness of LFS in producing LTD may be dependent on the recent history of activation, or strength, of the synapse. Indeed, although we observe a clear LTD with 1-Hz stimulation in the naive hippocampal slice, we also find that the magnitude of the effect is greatly enhanced after induction of LTP (10). Thus, these data appear to suggest that the other critical assumption of the theory—that of a sliding modification threshold—also may have a plausible physiological basis in the hippocampus (see also ref. 27).

We thank Joel Gold for his assistance in the early phases of this study and Dr. Leon Cooper for many helpful discussions. We also thank the reviewers, whose suggestions improved this paper. This work was supported by an Office of Naval Research Young Investigator Award to M.F.B.

1. Frégnac, Y. & Imbert, M. (1984) *Physiol. Rev.* **64,** 325–434.
2. Merzenich, M. M. (1987) in *The Neural and Molecular Bases of Learning*, eds. Changeux, J.-P. & Konishi, M. (Wiley, Chichester, U.K.), pp. 337–358.
3. McNaughton, B. L. & Nadel, L. (1990) in *Neuroscience and Connectionist Theory*, eds. Gluck, M. A. & Rumelhart, D. E. (Lawrence Erlbaum, Hillsdale, NJ), pp. 1–63.
4. Cooper, L. N, Lieberman, F. & Oja, E. (1979) *Biol. Cybern.* **33,** 9–28.
5. Bienenstock, E. L., Cooper, L. N & Munro, P. W. (1982) *J. Neurosci.* **2,** 32–48.
6. Clothiaux, E. E., Bear, M. F. & Cooper, L. N (1991) *J. Neurophys.* **66,** 1785–1804.
7. Collingridge, G. L. & Davies, S. N. (1989) in *The NMDA Receptor*, eds. Watkins, J. C. & Collingridge, G. L. (IRL, Oxford), pp. 123–136.
8. Bear, M. F., Cooper, L. N & Ebner, F. F. (1987) *Science* **237,** 42–48.
9. Bear, M. F. & Cooper, L. N (1990) in *Neuroscience and Connectionist Theory*, eds. Gluck, M. A. & Rumelhart, D. E. (Lawrence Erlbaum, Hillsdale, NJ), pp. 64–93.
10. Bear, M. F., Dudek, S. M. & Gold, J. T. (1991) *Soc. Neurosci. Abstr.* **17,** 1329.
11. Bear, M. F., Press, W. A. & Connors, B. W. (1992) *J. Neurophys.*, in press.
12. Bindman, L., Christofi, G., Murphy, K. & Nowicky, A. (1991) in *Aspects of Synaptic Transmission*, ed. Stone, T. W. (Taylor & Francis, London), pp. 3–25.
13. Lynch, G. S., Dunwiddie, T. & Gribkoff, V. (1977) *Nature (London)* **266,** 737–739.
14. Levy, W. B. & Steward, O. (1979) *Brain Res.* **175,** 233–245.
15. Pockett, S., Brookes, N. H. & Bindman, L. J. (1990) *Exp. Brain Res.* **80,** 196–200.
16. Dunwiddie, T. & Lynch, G. (1978) *J. Physiol. (London)* **276,** 353–367.
17. del Castillo, J. & Katz, B. (1954) *J. Physiol. (London)* **124,** 574–585.
18. Korn, H., Faber, D. S., Burnod, Y. & Triller, A. (1984) *J. Neurosci.* **4,** 125–130.
19. Larkman, A., Stratford, K. & Jack, J. (1991) *Nature (London)* **350,** 344–347.
20. Stanton, P. K. & Sejnowski, T. J. (1989) *Nature (London)* **339,** 215–218.
21. Lisman, J. (1989) *Proc. Natl. Acad. Sci. USA* **86,** 9574–9578.
22. Bear, M. F. & Dudek, S. M. (1991) *Ann. N.Y. Acad. Sci.* **627,** 42–56.
23. Green, E. J., McNaughton, B. L. & Barnes, C. A. (1990) *J. Neurosci.* **10,** 1455–1471.
24. Barrionuevo, G., Schottler, F. & Lynch, G. (1980) *Life Sci.* **27,** 2385–2391.
25. Staubli, U. & Lynch, G. (1990) *Brain Res.* **513,** 113–118.

Neurobiology: Dudek and Bear

Proc. Natl. Acad. Sci. USA 89 (1992)   4367

26. Fujii, S., Saito, K., Miyakawa, H., Ito, K. & Kato, H. (1991) *Brain Res.* **555,** 112–122.
27. Huang, Y.-Y., Colino, A., Selig, D. K. & Malenka, R. C. (1992) *Science* **255,** 730–733.
28. Bröcher, S., Artola, A. & Singer, W. (1992) *Proc. Natl. Acad. Sci. USA* **89,** 123–127.
29. Linden, D. J., Dickinson, M. H., Smeyne, M. & Connor, J. A. (1991) *Neuron* **7,** 81–89.

Common Forms of Synaptic Plasticity in the Hippocampus and Neocortex in Vitro

(by A. Kirkwood, S. M. Dudek, J. T. Gold, C. D. Aizenman, and M. F. Bear)

Science **260**, 1518 (1993)

Having established the existence of LTD in hippocampus, the Brown experimental group turned its attention to the investigation of whether similar effects occur in other regions of cortex. The following paper suggests that common forms of plasticity do exist in various regions of cortex. In addition, we see here an indication that specific cortical circuitry may play an important role in determining the critical period and in shaping the details of cortical plasticity.

Reprint Series
4 June 1993, Volume 260, pp. 1518–1521

Common Forms of Synaptic Plasticity in the Hippocampus and Neocortex in Vitro

Alfredo Kirkwood, Serena M. Dudek, Joel T. Gold,
Carlos D. Aizenman, and Mark F. Bear

Common Forms of Synaptic Plasticity in the Hippocampus and Neocortex in Vitro

Alfredo Kirkwood, Serena M. Dudek, Joel T. Gold,
Carlos D. Aizenman, Mark F. Bear

Activity-dependent synaptic plasticity in the superficial layers of juvenile cat and adult rat visual neocortex was compared with that in adult rat hippocampal field CA1. Stimulation of neocortical layer IV reliably induced synaptic long-term potentiation (LTP) and long-term depression (LTD) in layer III with precisely the same types of stimulation protocols that were effective in CA1. Neocortical LTP and LTD were specific to the conditioned pathway and, as in the hippocampus, were dependent on activation of N-methyl-D-aspartate receptors. These results provide strong support for the view that common principles may govern experience-dependent synaptic plasticity in CA1 and throughout the superficial layers of the mammalian neocortex.

Activity-dependent synaptic plasticity in the mammalian brain is best understood in the CA1 region of the adult hippocampus, where conditioning stimulation of the Schaffer collateral pathway in vitro can induce N-methyl-D-aspartate (NMDA) receptor-dependent LTP (1, 2) and LTD (3, 4). An important question is whether what has been learned about the hippocampus can be applied generally to synaptic plasticity in the cerebral cortex. Although LTP has been demonstrated in the visual cortex (5), in the mature neocortex in vitro it has been reported to occur with low probability (6–8) and to require for induction pharmacological treatments to reduce inhibition (9–11) and stimulation patterns that vary substantially from those that are effective in the hippocampus (6, 7, 12–14). Similarly, the conditions for evoking LTD in the hippocampus may be very different from

Department of Neuroscience, Brown University, Providence, RI 02912.

those in the neocortex. For example, in the hippocampus low-frequency stimulation evokes NMDA-dependent LTD (3, 4), whereas the same type of stimulation in the neocortex can yield LTP (6). Furthermore, strong stimulation in the presence of NMDA receptor antagonists causes LTD in the neocortex (15, 16) but no change in the CA1 region (17). In order to reconcile these disparate results, we directly compared the plasticity of synaptic responses evoked in adult rat hippocampal field CA1 with those evoked in adult rat and immature cat visual cortical layer III. To more closely approximate the stimulation-recording arrangement in the hippocampus (stimulating the Schaffer collaterals and recording in CA1), in neocortical preparations we stimulated the direct input to layer III from layer IV rather than using the traditional approach of stimulating the white matter (18).

In CA1, brief high-frequency (100 Hz) bursts of stimulation delivered to the Schaffer collaterals at the theta rhythm (5 to 7

Hz) consistently lead to induction of LTP that is specific to the conditioned pathway and is sensitive to the pharmacological blockade of NMDA receptors (19) (Fig. 1A). Consistent with previous observations (20), we observed that the potentiated response often decayed over the 20 to 40 min immediately after the tetanus until a stable value was reached, which in our experiments was 143 ± 7% of the base line control (n = 4). LTD was produced by 900 pulses delivered at 1 to 3 Hz, either from a potentiated (Fig. 1A, row 2) or a naïve (Fig. 1A, row 5) state. Like LTP, this synaptic depression persisted for many hours in vitro, was input-specific, and depended on NMDA receptor activation for its induction (3, 4). From the naïve state, 900 pulses at 1 Hz reduced the response to 80 ± 4% of the base line control, measured 45 min after the cessation of conditioning stimulation (n = 5).

Despite the fact that the superficial lay-ers of the sensory neocortex are rich in NMDA receptors (21), attempts to evoke plasticity in layer III with protocols identical to those in hippocampal experiments have generally met with limited success (14). The standard approach has been to stimulate ascending corticipetal fibers at the white matter–layer VI border and to record either intracellularly or extracellularly in layer III (5–14). However, this arrangement may not be optimal because much of the activity evoked in layer III by white matter stimulation is relayed by layer IV neurons (22). Therefore, to better approximate the situation in CA1, we attempted to evoke synaptic plasticity in visual neocortex by stimulating the more direct path from layer IV to layer III.

Stimulation of layer IV in slices of adult rat visual cortex evoked a large negative field potential in layer III with a peak latency of ~4 ms (Fig. 1B, row 3) (23). As with the hippocampus, theta-burst stimula-tion (TBS) potentiated the layer III re-sponse to test stimulation, and low-frequen-cy stimulation (LFS) caused LTD from ei-ther a potentiated or a naïve state (Fig. 1B). In contrast to the hippocampus, there was no evidence for a decaying form of potentiation in rat visual cortex immediate-ly after TBS (Fig. 1B, row 4); rather, the response increased in the first 15 min and reached a plateau value of 132 ± 4% of the base line control (n = 43). However, just like hippocampus (3), LFS depressed the naïve synaptic response in a frequency-dependent manner; 900 pulses at 1 Hz reduced the response to 71 ± 7% of the base line control (n = 5; Fig. 1B, row 5) but at 4 Hz had no lasting effect (98 ± 10%, n = 4). These forms of plasticity are not peculiar to the rat cortex; layer III responses in visual cortex prepared from 5- to 7-week-old kittens also exhibited LTP and LTD after TBS and LFS, respectively, of layer IV (Fig. 1C). To confirm that changes in the field potential truly reflect modifications of synaptic potentials, we simultaneously re-corded intracellular excitatory postsynaptic potentials (EPSPs) and extracellular field potentials in rat visual cortical layer III (n = 5). In all cases, a change in the field potential amplitude correlated with a change in the EPSP (Fig. 2, A and B).

In the hippocampus, both LTP and LTD are input-specific; only the conditioned in-puts demonstrate the plasticity (1, 3). We addressed this issue in our neocortical prep-aration by monitoring the response to a second input originating in layer III just lateral to the recording site (Fig. 2C). If the plasticity induced by layer IV stimulation was not input-specific, the LTP and LTD would have been reflected in the response to layer III stimulation. However, conditioning stimulation of layer IV caused a change only in the base line response evoked from layer IV; responses evoked from adjacent layer III were unaffected (Fig. 2D).

Previous work in CA1 has shown that induction of LTD by low-frequency stimu-lation can be blocked by antagonists of NMDA receptors (3). In contrast, a form of LTD has been reported for the neocortex, whose induction by patterned high-fre-quency stimulation (400 to 500 pulses de-livered in 50- to 100-Hz bursts of 1 to 2 s in duration) is actually promoted by NMDA receptor blockade (15, 16). This difference could be accounted for by the different stimulation protocols or, alternatively, by differences in the two types of cortex. To address this question, we applied the NMDA receptor antagonist D,L-2-amino-5-phosphonovaleric acid (AP5; 100 μM) to slices of rat visual cortex and attempted to induce LTD with 1-Hz stimulation. Block-ade of NMDA receptors in the visual cortex inhibited the induction of LTD by LFS, as

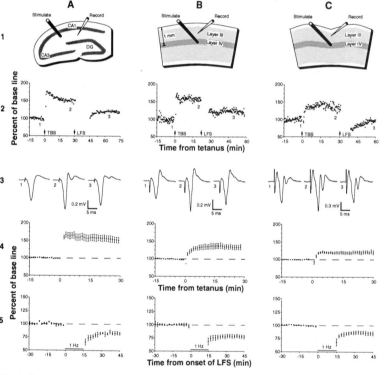

Fig. 1. Similar forms of synaptic plasticity in slices of adult rat hippocampus (**A**), adult rat visual cortex (**B**), and immature cat visual cortex (**C**). Row 1 shows the stimulation-recording configura-tions (DG, dentate gyrus). Row 2 shows changes in the extracellular field potential induced by TBS and by LFS [900 pulses were delivered at 1 Hz in (A) and (C) and at 3 Hz in (B)]. Response magnitude was measured as the change in the initial slope of the negative field potential in (A) and as the peak negativity in (B) and (C). Row 3 shows averages of four consecutive field potentials taken in each preparation before conditioning stimulation, after TBS, and after LFS for the experiments in row 2. Row 4 shows the average change in response magnitude after TBS [n = 4 for (A); n = 19 for (B); and n = 9 for (C)]. Row 5 shows the average change in response after LFS (900 pulses at 1 Hz), starting from an unpotentiated state [n = 5 for (A); n = 5 for (B); and n = 6 for (C)].

in the hippocampus (Table 1). Also, induction of LTP by TBS in both rat and kitten visual cortex was blocked by AP5, as has been reported for CA1 (24) (Table 1).

The plasticity we report here for the neocortex occurs with high probability and does not require the use of drugs such as bicuculline. Our success may be attributable to the fact that we stimulated layer IV rather than using the traditional approach of stimulating the fibers at the layer VI–white matter border. Indeed, a direct comparison of the effects of TBS delivered to layers VI and IV (without the use of bicuculline)

confirms that these two configurations have different effects on the layer III evoked response (Fig. 3A). The average change in response magnitude 20 min after conditioning stimulation of layer VI was only 108 ± 4% of the base line control, which was less than that obtained after layer IV stimulation ($P < .001$; t test). Although the half-maximal responses collected for the base line are greater when layer IV is stimulated (683 ± 62 μV) than when layer VI is stimulated (318 ± 59 μV), this does not appear to be sufficient to account for the difference in LTP (Fig. 3B). We believe that the key

difference lies in the distinct patterns of cortical activation that result from stimulation of the two sites (25).

Our work demonstrates that when similar preparations of neocortex and hippocampus are studied in vitro, they can yield similar forms of synaptic plasticity. Although some differences remain in the details (26), our data suggest the existence of a common substrate, across both phylogeny and postnatal age, for activity-dependent synaptic plasticity in CA1 and in the superficial layers of the neocortex. The substrate we describe here involves the modification of excitatory synaptic effectiveness according to the pattern or amount of NMDA receptor activation. One implication of this finding is that pharmacological manipulations of NMDA receptors in vivo are likely to interfere equally with mechanisms of synaptic enhancement and depression. Recently, NMDA receptor–independent forms of synaptic plasticity have been reported for both visual cortex (15, 16, 27, 28) and CA1 (29). If these similarly prove to be governed according to common principles, it may be possible to construct a

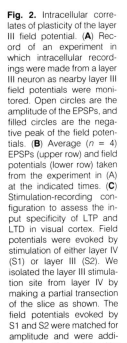

Fig. 2. Intracellular correlates of plasticity of the layer III field potential. (**A**) Record of an experiment in which intracellular recordings were made from a layer III neuron as nearby layer III field potentials were monitored. Open circles are the amplitude of the EPSPs, and filled circles are the negative peak of the field potentials. (**B**) Average ($n = 4$) EPSPs (upper row) and field potentials (lower row) taken from the experiment in (A) at the indicated times. (**C**) Stimulation-recording configuration to assess the input specificity of LTP and LTD in visual cortex. Field potentials were evoked by stimulation of either layer IV (S1) or layer III (S2). We isolated the layer III stimulation site from layer IV by making a partial transection of the slice as shown. The field potentials evoked by S1 and S2 were matched for amplitude and were additive at all stimulus intensities, which suggests that the responses were generated by independent inputs converging on the same target population of postsynaptic neurons. (**D**) Input-specific LTP and LTD with the stimulation-recording arrangement in (C). TBS stimulation of S1 produced an enhancement of the response to base line stimulation of S1 only (filled circles); LFS of S1 produced a depression of the response to base line stimulation of S1 only. Each data point is the average from four experiments; error bars were omitted for clarity. Inset shows average of four consecutive field potentials evoked from layer IV (filled circle) or from layer III (open circle) before and after TBS in one of these cases at the indicated times. Scale bars of inset: 5 ms, 0.25 mV.

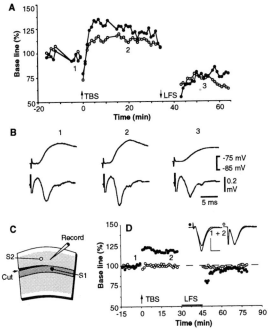

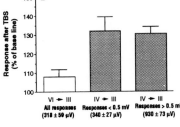

Fig. 3. Differential effects of TBS on the layer III field potential evoked by stimulation of layers IV and VI. (**A**) The fraction of cases in which TBS induced a change in the field amplitude (percent of base line) when layer VI was stimulated (open bars, $n = 18$) and when layer IV was stimulated (hatched bars, $n = 43$). These distributions were significantly different ($P < 0.001$, t test). (**B**) Differences in LTP are not accounted for by the differences in field amplitude. The layer IV cases were divided according to whether the half-maximal field amplitude was greater or less than 0.5 mV. The low-amplitude layer IV group, with a mean response size comparable to the layer VI group, still yielded significantly greater LTP ($P < 0.01$, t test).

Table 1. Effects of conditioning stimulation on synaptic responses (mean ± SEM) in the visual cortex in the presence or absence of 100 μM AP5. The AP5 condition was different than the wash group in all cases ($P \leq 0.01$, t test).

Group	Change in synaptic responses					
	Theta burst in rat		Theta burst in cat		1 Hz in rat	
	% of base line control	n	% of base line control	n	% of base line control	n
Controls	132 ± 4	43	122 ± 4	14	71 ± 7	5
AP5	100 ± 5	4	96 ± 3	4	99 ± 4	4
Wash	128 ± 13	4	128 ± 13	4	87 ± 8	4

general theory for experience-dependent synaptic modification in the mammalian cerebral cortex.

REFERENCES AND NOTES

1. T. V. P. Bliss and T. Lømo, *J. Physiol.* (*London*) **232**, 331 (1973).
2. B. E. Alger and T. J. Teyler, *Brain Res.* **110**, 463 (1976).
3. S. M. Dudek and M. F. Bear, *Proc. Natl. Acad. Sci. U.S.A.* **89**, 4363 (1992).
4. R. M. Mulkey and R. C. Malenka, *Neuron* **9**, 967 (1992).
5. A. Artola and W. Singer, *Nature* **330**, 649 (1987).
6. Y. Komatsu, K. Fujii, J. Maeda, H. Sakaguchi, K. Toyama, *J. Neurophysiol.* **59**, 124 (1988).
7. A. T. Perkins and T. J. Teyler, *Brain Res.* **439**, 222 (1988).
8. F. Kimura, A. Nishigori, T. Shirokawa, T. Tsumoto, *J. Physiol.* (*London*) **414**, 125 (1989).
9. A. Artola and W. Singer, *Eur. J. Neurosci.* **2**, 254 (1990).
10. N. Kato, A. Artola, W. Singer, *Dev. Brain Res.* **60**, 43 (1991).
11. M. F. Bear, W. A. Press, B. W. Connors, *J. Neurophysiol.* **67**, 1 (1992).
12. R. L. Berry, T. J. Teyler, H. Taizhen, *Brain Res.* **481**, 221 (1989).
13. T. Tsumoto, *Prog. Neurobiol.* **39**, 209 (1992).
14. M. F. Bear and A. Kirkwood, *Curr. Opinion Neurobiol.*, in press.
15. A. Artola, S. Bröcher, W. Singer, *Nature* **347**, 69 (1990).
16. J. C. Hirsch and F. Crepel, *Exp. Brain Res.* **85**, 621 (1991).
17. R. S. Goldman, L. E. Chavez-Noriega, C. F. Stevens, *Proc. Natl. Acad. Sci. U.S.A.* **87**, 7165 (1990).
18. Brain slices (400 μm) were prepared and maintained in an interface chamber (Medical Systems, Greenvale, NY) as described (*3, 11*). Microelectrodes were filled with 1 M NaCl (1 to 2 megohms) for extracellular recording or 3 M potassium acetate (80 to 120 megohms) for intracellular recording. Synaptic responses were evoked with 0.02-ms pulses (amplitude, 10 to 200 μA) delivered with a bipolar concentric stimulating electrode (outside diameter, 200 μm). In every experiment, a full input-output curve was generated, and base line responses were obtained at 0.07 to 0.03 Hz with a stimulation intensity that yielded a half-maximal response. To induce LTP, we delivered two to five episodes of TBS at 0.1 Hz. TBS consisted of 10 to 13 stimulus trains delivered at 5 to 7 Hz; each train consisted of four pulses at 100 Hz. To induce LTD, we delivered 900 pulses at 1 to 3 Hz. All rats used were adults (>40 days old).
19. J. Larson, D. Wong, G. Lynch, *Brain Res.* **368**, 347 (1986).
20. R. C. Malenka, *Neuron* **6**, 53 (1991).
21. D. T. Monaghan and C. W. Cotman, *J. Neurosci.* **5**, 2909 (1985).
22. Anatomical experiments indicate that fibers ascending through layer VI terminate densely in layer IV but relatively sparsely in layer III (*30*), and current-source density analysis shows that the prominent current sink in layer III that is evoked by stimulation of the white matter is mostly di- and polysynaptic and always follows activation of layer IV synapses (*6, 31*). Accordingly, the EPSPs evoked by white matter stimulation fail to exhibit a monosynaptic component in half the layer III neurons recorded (*6, 27*). Electrical stimulation of layer IV, besides activating the projection from layer IV neurons, also recruits any direct inputs to layer III that ascend from the white matter.
23. This negative peak was maximal at a depth of ~300 μm from the pia, and current-source density analysis indicates that it reflects a prominent layer III synaptic current sink; therefore, the amplitude of this field potential was used routinely as a measure of synaptic effectiveness.
24. J. Larson and G. Lynch, *Brain Res.* **441**, 111 (1988).
25. We caution, however, that these consequences of layer IV stimulation need not be accounted for solely by activation of the projection of layer IV neurons onto the dendrites in layer III. Clearly, the effects of intracortical stimulation can be quite complex and, in principle, could include the antidromic activation of some layer III neurons and orthodromic activation of intracortical and corticocortical fibers passing through layer IV [cells in the infragranular layers are not part of the essential circuit as these can be further cut away and LTP is left intact (131 ± 6%; *n* = 7)]. However, these complexities are not unique to experiments where layer IV is stimulated. Indeed, stimulation of the cortical white matter may be even more complex considering that in addition to activation of corticipetal fibers and synaptic activation of layer IV neurons, the cells in layers VI, V, and III could all be activated antidromically. The effects of white matter stimulation are further complicated by the requisite use of bicuculline, which, at the concentrations usually used, often promotes widespread epileptiform activation of cortical slices (*32*). Thus, despite some potential complications, layer IV stimulation appears to be an advantageous preparation for the investigation of synaptic plasticity in the superficial layers of the neocortex.
26. One of these residual differences is an early time-course of the responses immediately after the TBS. In the hippocampus, a potentiated response typically is observed immediately after the tetanus and declines to a stable value in about 30 min. The decaying form of potentiation in the hippocampus has been termed short-term potentiation (STP) and may be mechanistically distinct from LTP (*33*). STP typically is not observed in the neocortex.
27. Y. Komatsu, S. Nakajima, K. Toyama, *J. Neurophysiol.* **65**, 20 (1991).
28. V. A. Aroniadou and T. J. Teyler, *Brain Res.* **584**, 169 (1992).
29. L. M. Grover and T. J. Teyler, *Nature* **347**, 477 (1990).
30. A. Burkhalter, *J. Comp. Neurol.* **279**, 171 (1989).
31. K. M. Bode-Greuel, W. Singer, J. B. Aldenhoff, *Exp. Brain Res.* **69**, 213 (1987).
32. Y. Chagnac-Amitai and B. W. Connors, *J. Neurophysiol.* **61**, 747 (1989).
33. C. F. Stevens, *Neuron* **10** (suppl.), 55 (1993).
34. Supported by grants from the National Eye Institute and the U.S. Office of Naval Research. A.K. was supported in part by a grant from the Human Frontiers Science Program. We thank J. Lisman, L. Cooper, and B. Connors for critically reading an earlier version of the manuscript and J. P. Donoghue for helpful discussions.

4 December 1992; accepted 25 February 1993

210

Formation of Receptive Fields in Realistic Visual Environments According to the Bienenstock, Cooper, and Munro (BCM) Theory

(with C. C. Law)

Proc. Natl. Acad. Sci. USA **91**, 7797 (1994)

As discussed previously, our theoretical work in visual cortex is founded on three major assumptions: the first concerns the actual modification algorithm for a single cell. The second, the cortical architecture. And the third, the representation of the visual environment. We have dealt with the first two in some of the preceding papers. The following paper deals with the third. Various aspects of this project have and are being handled by some of my current and recent graduate students and/or post-docs: Charlie Law, Harel Shouval, and Brian Blais.

Our previous representations of the visual environment attempt particularly to distinguish between patterned and nonpatterned environments which produce such dramatic differences in selectivity. (This goes back to the work of Imbert and Buisseret, one of the original motivations for the construction of the theory.) An underlying notion is that an animal in a normal visual environment sees many scenes. Since the receptive field of individual cortical cells (i.e., the portion of the visual environment that is input to a particular cortical cell through the retina and LGN) is relatively small, we conjectured that, in effect, what would be repeated in visual scenes were edges. (Complex curves come in all varieties, but if one views some small portion of these, one is likely to pick out an edge — an effect increased by the edge enhancement properties of retina.) Thus our assumption was that we could represent the normal visual scene by edges distorted by noise. On the other hand, we assumed that a nonpatterned vision environment, for example that produced when the eyes are occluded by a translucent lens or the discharges that occur during dark rearing, could be represented by pure noise. Thus in our prior analysis and simulations the overall distinction between a normal environment and a pattern-deprived environment was that of a fixed number of repeated noise distorted patterns as opposed to pure noise. (One could also fine tune by using various types of correlations in the noise.)

In the next paper with Charlie Law, we replace previous abstractions of the visual environment with real visual images (some of which might have been provided by the Rhode Island Tourism Office – but were actually the result of Harel's photographic skills).

Charlie is very gifted and quick – in particular in getting programs to run. So much so one has to be particularly careful to check assumptions since he will do almost anything to get a simulation to work. (A recurring problem since the motivation of graduate students is to get whatever project they're working on to succeed.) In the following paper, I hope we got it right.

Proc. Natl. Acad. Sci. USA
Vol. 91, pp. 7797–7801, August 1994
Neurobiology

Formation of receptive fields in realistic visual environments according to the Bienenstock, Cooper, and Munro (BCM) theory

(synaptic modification/visual cortex/natural images)

C. Charles Law and Leon N Cooper*

Department of Physics and Neuroscience, Institute for Brain and Neural Systems, Brown University, Box 1843, Providence, RI 02912

Contributed by Leon N Cooper, April 4, 1994

ABSTRACT The Bienenstock, Cooper, and Munro (BCM) theory of synaptic plasticity has successfully reproduced the development of orientation selectivity and ocular dominance in kitten visual cortex in normal, as well as deprived, visual environments. To better compare the consequences of this theory with experiment, previous abstractions of the visual environment are replaced in this work by real visual images with retinal processing. The visual environment is represented by 24 gray-scale natural images that are shifted across retinal fields. In this environment, the BCM neuron develops receptive fields similar to the fields of simple cells found in kitten striate cortex. These fields display adjacent excitatory and inhibitory bands when tested with spot stimuli, orientation selectivity when tested with bar stimuli, and spatial-frequency selectivity when tested with sinusoidal gratings. In addition, their development in various deprived visual environments agrees with experimental results.

In 1982 Bienenstock, Cooper, and Munro (BCM) proposed a concrete synaptic-modification hypothesis in which two regions of modification (Hebbian and anti-Hebbian) are stabilized by the addition of a sliding modification threshold (1). The theory was created originally to explain the development of orientation selectivity of visual cortical neurons in various visual environments. This theory has since proven capable of also explaining ocular dominance and selective response properties in the most diverse visual environments, one of the most thoroughly studied areas in neuroscience. In this paper we examine the consequences of the replacement of previous abstractions of the visual environment by real visual images with retinal processing.

During a critical period of postnatal development, the response properties of neurons in striate cortex of the kitten can be modified by manipulating the visual experience of the animal (2–4). Clothiaux, Bear, and Cooper (CBC) (5) showed that simulations based on the BCM theory, with a fixed set of parameters, reproduce both the kinetics and equilibrium states of experience-dependent modifications that are observed experimentally in these neurons. The rearing conditions that they simulated include normal rearing, monocular deprivation, reverse suture, strabismus, and binocular deprivation.

An important simplification used in all previous BCM simulations is the assumed activity of neurons in the lateral geniculate nucleus (LGN) resulting from visual experience. In CBC, for example, an abstract set of 12 patterns represents the activity on geniculo–cortical afferents supposedly resulting from contoured stimuli with 12 different orientations. LGN activity in a normal visual environment is modeled as this patterned input distorted by noise. In the absence of visual contours, LGN–cortical input activity is assumed to be

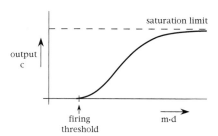

FIG. 1. Sigmoidal activation function. m, Strength of the synapses that connect the input d to the BCM neuron and the sigmoidal function σ.

uncorrelated noise distributed around spontaneous activity. The justification for this simplification is that the visual fields of neurons in primary visual cortex are small, so that reproducible visual contours when viewed through this aperture are most likely to be noise-like or edges distorted by noise, allowing visual input to be represented by a pattern set with a single variable representing the orientation of the stimulus.

In this paper we replace this simplification by a more realistic visual environment. By so doing we can investigate the behavior of BCM neurons subjected to input from LGN more like that actually experienced in normal and deprived situations. In addition, we can examine the validity of previous assumptions.

METHODS

Modification of synaptic weights in the BCM theory is given by

$$\dot{m}_i = \eta\phi(c, \Theta)d_i, \qquad [1]$$

where cell activity measured relative to spontaneous activity is

$$c = \sigma(\mathbf{m}\cdot\mathbf{d}). \qquad [2]$$

The vector **m** represents the strength of the synapses that connect the input **d** to the BCM neuron and the sigmoidal function σ (Fig. 1). The constant η determines the learning speed, and the function φ is

$$\phi(c, \Theta) = (\Theta^{-1})c(c - \Theta). \qquad [3]$$

This equation differs from that used in ref. 6 by the factor Θ^{-1}, which does not affect the position or the stability of fixed points but does affect the rate at which these fixed points are reached. Because Θ changes very slowly, the term acts as a

Abbreviations: BCM, Bienenstock, Cooper, and Munro; CBC, Clothiaux, Bear, and Cooper; LGN, lateral geniculate nucleus.
*To whom reprint requests should be addressed.

7797

Table 1. Parameters of the model

Parameter	Significance	Value
R_r, units	Radius of the retinal patches	5.0
STD_c, units	SD of a ganglion excitatory-center Gaussian distribution	1.0
STD_s, units	SD of a ganglion inhibitory-surround Gaussian distribution	3.0
d_s, Hz	Level of ganglion spontaneous activity	20.0
c_s, Hz	Level of cortical-neuron spontaneous activity	1.0
$\overline{n^2}$, Hz²	Average square of the ganglion noise from a sutured eye	3.0
$m_j(0)$	Starting value of the synaptic weights for normal rearing	0.0–0.001
η	Modification step size	0.000001
τ, iterations	Time constant in the definition of Θ	1000

variable learning rate, speeding up the simulation when the neuron is far from a fixed point (Θ small) and slowing the simulation when Θ and the weights grow large. Thus, the simulations are more stable and less dependent on initial conditions.[†] Following ref. 6 the sliding threshold Θ is defined as the second moment of the cell activity.

$$\Theta = (\tau^{-1} \int_{-\infty}^{t} dt' c^2 e^{\tau^{-1}(t' - t)}),\qquad [4]$$

where the time constant τ determines how rapidly the threshold moves. The parameter set used for the simulations described here are given in Table 1.

[†]We note the obvious: Only some qualitative features of ϕ (such as the negative and positive regions, the zero crossings, and the moving threshold) are required to explain present experimental data. The detailed form of this function will be determined by further sophisticated experiment, analysis, and simulation.

REALISTIC VISUAL ENVIRONMENT

An added feature of the simulations described below is the method by which the visual environment is represented. Real two-dimensional natural images processed using simple retinal inputs are used as visual inputs. Circular regions from the left and the right retinas are used to generate activity in the thalamocortical projections. The LGN is assumed to simply relay the signal generated by the retina to the visual cortex.

To model the visual environment, we use 24 gray-scale images with dimensions of 256×256 pixels. The retinas are composed of square arrays of receptors and ganglion cells spaced one unit apart. Each ganglion cell has an antagonistic center–surround receptive field that approximates a difference of two Gaussian distributions. The SD of the surround Gaussian distribution is three times larger than the SD of the center Gaussian distribution. The resulting receptive field center has a radius of 2.22 units. The ganglion-cell receptive

FIG. 2. Some of the 24 natural images used to represent the visual environment. The receptive field circle determines the current input and moves randomly over the images.

213

Neurobiology: Law and Cooper

Proc. Natl. Acad. Sci. USA 91 (1994)   7799

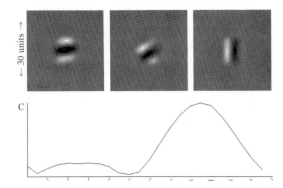

FIG. 3. (*Upper*) Receptive fields developed by a BCM neuron in a realistic visual environment. The neuron had input from a patch of the retina with a diameter of 10.0 units. The light areas of the receptive field map represent excitatory regions, and the dark areas represent inhibitory regions. (*Lower*) Orientation tuning curve generated with bars of light as described in the text. C, output.

fields are balanced so that uniform illumination of any intensity results in spontaneous activity. Cell activity is measured relative to spontaneous activity, and the nonlinearity in σ restricts the *absolute* ganglion-cell activity to be positive.

For each cycle of the simulation, the activity of the receptors in the retina is determined by randomly picking one of the 24 images and randomly shifting the receptive field circle (Fig. 2). The shift is restricted so that none of the ganglion-cell receptive field centers fall within 10 units of the image border. The activity of each receptor in the model is determined by the intensity of a pixel in the image. Obviously, this generates a very large training set and, in contrast to what was done previously, we need make no assumptions concerning relative frequency of patterns and noise.

METHOD OF TESTING THE RECEPTIVE FIELD OF THE BCM NEURON

At selected times during the simulations, the properties of the BCM neurons are tested separately through the left and right eyes. Two-dimensional maps of the receptive field are generated by "shining" small (radius = 0.8 unit) spots of light on the retina at many locations and recording the cortical neuronal activity generated for each spot. This technique is similar to the process used by Jones and Palmer (7) to generate two-dimensional receptive-field profiles of simple cells in cat striate cortex. For simplicity, low-

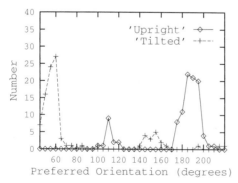

FIG. 4. Orientation preferences developed by 100 individual simulations of normal rearing trained with natural images and 100 simulations trained with natural images rotated by 45°.

contrast spots are used to avoid both the ganglion-cell and the BCM-neuron nonlinearities. Orientation tuning of the neuron is determined by presenting bars of light. For each orientation several bar stimuli (width = 2 units) are generated at different spatial locations. The best response to the set of stimuli determines the amplitude of the tuning curve for that orientation. The maxima of the left- and right-eye tuning curves are then used to determine the binocularity of the BCM neuron. Spatial-frequency selectivity is tested by using sinusoidal gratings at the best orientation of the neurons. For each frequency, many stimuli with various spatial phases are used to determine the maximal response of the neurons.

RESULTS

Most important is the fact that replacement of previously employed simplifications by a realistic visual environment does not significantly alter simulation results. As in the simulations of CBC, normal rearing produces a selective binocular neuron that is equally responsive to stimulation through the left and right eyes; in binocular deprivation, the BCM neuron becomes less responsive and less selective but remains binocular; in monocular deprivation, the sutured eye disconnects from the BCM neuron, whereas in reverse suture the newly closed eye disconnects from the BCM neuron before the newly opened eye reconnects. Fig. 8 shows the results of the reverse suture simulation with natural images. These results, which will be described in more detail elsewhere, suggest that the 12 abstract patterns of CBC were adequate representations of visual experience for these deprivation simulations.

In simulations of normal rearing both eyes receive the same pattern input from the natural images. At the beginning of the simulation synaptic weights are initialized to small random values. Even though at this stage the cortical neuron is relatively unresponsive to visual stimuli, the initialization of the weights adds a small bias that can influence the final selectivity of the neuron. Fig. 3 shows the results of a single simulation that is initialized with random weights. At the end of normal rearing, the BCM neuron is binocular and has properties similar to simple cells found in striate cortex. Its receptive field has adjacent excitatory and inhibitory bands that produce selectivity to the orientation of bar stimuli. In our simulations, the BCM neuron most often becomes selective to horizontal and vertical orientations. The preference for horizontal and vertical stimuli is a property of the natural images and is not a

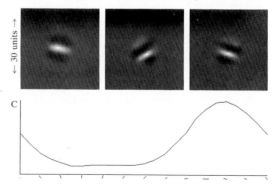

FIG. 5. Receptive field (*Upper*) and orientation selectivity (*Lower*) developed in the natural visual environment without retinal preprocessing. Addition of retinal preprocessing makes little difference to the selectivity developed by the BCM neuron. C, output.

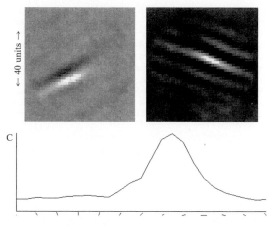

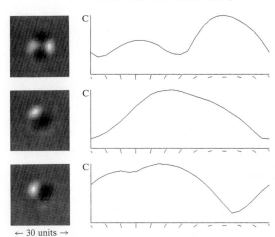

FIG. 6. Receptive fields (*Upper*) and typical selectivity (*Lower*) generated by the cortical neuron. The size of the retinal patches that connect to the BCM neuron is four times larger than previous simulations. Except for an elongated receptive field, the properties of the BCM neuron are similar to simulations with input that has a smaller spatial extent. C, output.

← 30 units →

FIG. 7. Receptive fields (*Left*) and orientation selectivity (*Right*) for three single-cell simulations in an environment of correlated noise. The only difference between the simulations is the initialization of synaptic weights. C, output.

spurious bias introduced by the retinal or cortical models. When the environment is rotated by rotating the natural images, the orientation preference of the model also shifts. Fig. 4 shows orientation-preference histograms generated in two ways: the first uses the normal visual environment, whereas the second uses a visual environment rotated by 45°. Each histogram represents data from 100 simulations.

The results of these simulations are very robust. Changing the parameters of the retina does not significantly alter the simulations. Even completely bypassing the retina has little

effect on the selectivity developed. However, it does change the rate at which the fixed points are reached. Fig. 5 shows the receptive field developed when the BCM neuron is trained directly from the natural images.

The receptive fields developed by the BCM neuron also are not significantly dependent on the size of the retinal patches that project to the cortical neuron. When simulations are run with much larger retinas, the BCM neuron simply disconnects from most of the ganglion cells. Fig. 6 shows the receptive field when the retinal diameter is four times larger than the diameter of the retina in the simulation displayed in Fig. 3.

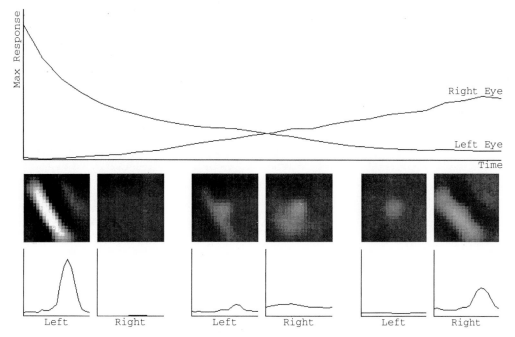

FIG. 8. Reverse suture simulation results. (*Top*) The graph displays the maximum (Max) responsiveness of the BCM neuron to stimulation through the left and right eyes. (*Middle*) The receptive field of the cortical neuron at three different times during the simulation, tested through the left and right eyes separately. (*Bottom*) Orientation tuning as tested through the left and right eyes separately.

Neurobiology: Law and Cooper

Proc. Natl. Acad. Sci. USA 91 (1994)   7801

Mastronarde (8) observed that the activities of neighboring ganglion cells are correlated in the absence of contour vision, and it has been suggested (9) that such activity could play an important role in the formation of receptive fields before any contoured visual experience.

Although the BCM modification algorithm usually is applied to account for cortical neuron response in normal and deprived visual environments, we have simulated the development of BCM neurons in an environment of correlated noise to model spontaneous activity that might occur before visual experience. Small-amplitude, uncorrelated noise is presented as visual input, and the antagonistic center–surround receptive fields in the retina produce positive correlations at small distances (<2.2 units) and negative correlations at medium distances (>2.2 units).

In this environment the BCM neuron develops receptive fields with distinct excitatory and inhibitory regions (Fig. 7). However, the neurons trained with such correlated noise do not become as selective to features such as orientation and spatial frequency as neurons trained with natural images. These results are consistent with the idea that although some selectivity may develop in primary visual cortex before visual experience, maturation to adult levels of selectivity requires contoured visual experience.

DISCUSSION

The development of BCM neurons in realistic visual environments is very similar to that presented previously with various simplifications. Normal rearing produces selective binocular neurons equally responsive to stimulation through left and right eyes. For the various deprived environments we obtain results qualitatively similar to these reported by CBC. What differences occur may actually bring us in closer agreement with observations. For example, in simulations of reverse suture the realistic visual environment eliminates a possible discrepancy between experimental findings and the CBC simulations. It was noted by Mioche and Singer (10) that the recovery of the newly opened eye is slow and incomplete

after reverse suture; in the CBC simulation the neuron has a fast and robust recovery. As suggested in CBC (5), this has several possible explanations. However, the present simulation of reverse suture displayed in Fig. 8, without any additional assumptions, agrees more closely with the experimental findings: the BCM neuron does recover responsiveness to the newly opened eye, but recovery is slow.

We have also shown that the BCM theory is consistent with the hypothesis that correlated retinal spontaneous activity could be responsible for the development of some selectivity before visual experience. Crude orientation selectivity forms in the absence of contoured visual experience when the visual environment after retinal preprocessing is correlated noise. These results are consistent with the view that although some selectivity may develop in the absence of contoured vision, normal visual experience is necessary to develop full selectivity.

We thank the members of the Institute for Brain and Neural Systems for many conversations and, in particular, Ms. Tomoko Ozeki for her assistance in running some of the simulations. This work was supported, in part, by the Office of Naval Research, the Army Research Office, and the National Science Foundation.

1. Bienenstock, E. L., Cooper, L. N & Munro, P. W. (1982) *J. Neurosci.* **2,** 32–48.
2. Buissert, P. & Imbert, M. (1976) *J. Physiol. (London)* **255,** 511–525.
3. Hubel, D. H. & Wiesel, T. N. (1970) *J. Physiol. (London)* **206,** 419–436.
4. Wiesel, T. N. & Hubel, D. H. (1963) *J. Neurophysiol.* **26,** 1003–1017.
5. Clothiaux, E. E., Bear, M. F. & Cooper, L. N (1991) *J. Neurophysiol.* **66,** 1785–1805.
6. Intrator, N. & Cooper, L. N (1992) *Neural Networks* **5,** 3–17.
7. Jones, J. P. & Palmer, L. A. (1987) *J. Neurophysiol.* **58,** 1187–1258.
8. Mastronarde, D. N. (1989) *Trends Neurosci.* **12,** 75–80.
9. Miller, K. D., Keller, J. B. & Stryker, M. P. (1989) *Science* **245,** 605–615.
10. Mioche, L. & Singer, W. (1989) *J. Neurophysiol.* **62,** 185–197.

Effect of Eye Misalignment on Ocular Dominance According to BCM and PCA Synaptic Modification

(with H. Shouval, N. Intrator, and C. C. Law)

The next paper in this sequence is a preprint that gives results attained with the input of realistic images to both eyes with varying degrees of overlap (since it is expected that the left/right inputs to a cortical cell do not exactly match). The results obtained are very satisfying. In addition to the observed selectivity, we can, with very reasonable assumptions, reproduce the observed ocular dominance histogram. We also find a clear distinction between the consequences of two learning rules both of which have stable fixed points. I include it because it illustrates some very current directions of our research. After the usual hassles with reviewers, we expect that it will be published somewhere or other, possibly in a modified form.

Effect of Eye Misalignment on Ocular Dominance according to BCM and PCA Synaptic Modification

Harel Shouval, Nathan Intrator*, C. Charles Law, and Leon N Cooper.
Departments of Physics and Neuroscience and
The Institute for Brain and Neural Systems
Box 1843, Brown University
Providence, R. I., 02912

April 21, 1995

Abstract

In this paper we realistically model a two-eye visual environment and study its effect on single cell synaptic modification. In particular, we study the effect of image misalignment on receptive field formation after eye opening. We show that binocular misalignment effects PCA and BCM learning in different ways. For the BCM learning rule this misalignment is sufficient to produce varying degrees of ocular dominance, whereas for PCA learning binocular neurons emerge in every case. Such differences should help us distinguish between these learning rules.

1 Introduction

It is now generally accepted that receptive fields in the visual cortex of cats are dramatically influenced by the visual environment (For a comprehensive review see, Frégnac and Imbert, 1984) . In normally reared animals, the population of sharply tuned neurons increases monotonically, whereas for dark reared animals it initially increases, but then almost disappears (See, for example, Imbert and Buisseret, 1975) . Ocular dominance is dramatically influenced by such manipulations as monocular deprivation (Wiesel and Hubel, 1963) or reverse suture (Blakemore and Van-Sluyters, 1974; Mioche and Singer, 1989). It has even been shown that preferred orientations can be directly altered by pairing the preferred orientation with a negative current, and the non-preferred orientation with a positive current (Frégnac et al., 1992).

Different models, that attempt to explain how cortical receptive fields evolve, have been proposed over the years (von der Malsburg, 1973; Nass and Cooper, 1975; Perez et al., 1975; Bienenstock et al., 1982; Linsker, 1986; Miller, 1994, e.g.). Such models are composed of several components: the exact nature of the learning rule, the representation of the visual environment, and the architecture of the network.

Most of these models assume a simplified representation of the visual environment (von der Malsburg, 1973, for example), or a second order correlation function of the visual environment (Miller, 1994). Realistic representations of the visual environment have only very recently been considered (Hancock et al., 1992; Law and Cooper, 1994; Liu and Shouval, 1994; Shouval and Liu,

---

*also at Faculty of Exact Sciences Tel-Aviv University

Figure 1: Eight of the natural images used to represent the visual environment.

1995). Furthermore, only in recent years have the statistics of natural images been studied, and compared with neurophysiological findings (Field, 1987; Field, 1989; Baddeley and Hancock, 1991; Atick and Redlich, 1992; Ruderman and Bialek, 1993; Liu and Shouval, 1994; Shouval and Liu, 1995).

Once actual visual scenes are used, it is possible to realistically represent two-eyes input, and account for the fact that the two eyes are not looking exactly at the same visual scene (Li and Atick, 1994). In the past, visual input coming from both eyes has also been modeled using a second order correlation function with various levels of inter and intra-eye correlations. Normal rearing is usually modeled using high intra-eye correlation, while binocular deprivation is modeled using zero inter-eye and intra-eye correlation. Strabismus has been modeled using either zero correlation or negative intra-eye correlation (Miller et al., 1989), with a high inter-eye correlation. In these modeling paradigms there was no account of the various degrees of spatial retinal shift of the images due to depth and to receptive field misalignment. Such spatial shift creates a unique nonsymmetrical inter-eye correlation which varies with the degree of shift between the two eyes.

In this paper, we realistically model a two-eye visual environment and study its effect on single cell synaptic modification. In particular, we study the effect of image misalignment between the two eyes on receptive field formation [1].

We compare the effect of such image misalignment on two different learning rules PCA (Oja, 1982) and BCM (Bienenstock et al., 1982; Intrator and Cooper, 1992). We have chosen to examine these two because they are well defined and and have stable fixed points. Many other proposed learning rules (Sejnowski, 1977; Linsker, 1986; Miller et al., 1989, for example), are closely related to the PCA rule. Their outcome depends only on first and second order statistics. The BCM rule in contrast depends also on third order statistics.

We show that binocular misalignment has very different effects on these two learning rules. For the BCM learning rule misalignment is sufficient to produce varying degrees of ocular dominance, whereas for the PCA learning rule binocular neurons will emerge independent of the misalignment.

---

[1] Postnatal plasticity may be needed in animals with binocular vision, as was already suggested by Blakemore and van Sluyters (1975), in order to overcome an imprecise developmental alignment.

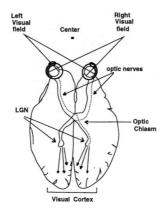

Figure 2: A schematic drawing of the Visual pathway

2 A Binocular Visual Environment Composed of Natural Images

We have used a set of 24 natural scenes, some of which are displayed in Figure 1. These pictures were taken at Lincoln Woods State Park, scanned into a 256 X 256 pixel image. No corrections have been used for the optical distortions of the instruments. We have avoided man-made objects, because they have many sharp edges, and straight lines, which make it easier to achieve oriented receptive fields.

In this paper, we limit ourselves to study receptive field formation near the fovea, and thus do not model the change in resolution which corresponds to the complex log retinotopic mapping.

Figure 2 shows schematically the anatomy of the visual pathway; Light which falls on the retina is encoded by the receptors. The signal is then processed by the retinal circuitry and is projected by the ganglion cells onto the optic nerve. Signals from both eyes cross at the Optic Chiasm and continue to the Lateral Geniculate Nucleus (LGN). In the LGN inputs from both eyes are segregated in different layers. From the LGN, signals project up to the visual cortex.

The receptive fields of both ganglion cells and LGN projections, have a center surround shape (Figure 3). Some are "on center", which means that they are excited by spots of light falling on their centers, and inhibited by light on the surround, others are "off center" and are excited by light falling on their surround, and inhibited by light in their center. In the cat, the most abundant type near the fovea are the X cells which are linear, i.e., their response to an image that is composed of several components is roughly the sum of the response to the independent components, (Orban., 1984).

We have chosen to model the effect of the retinal preprocessing by convolving the images with a difference of Gaussians (DOG) filter, with a center radius of one pixel ($\sigma_1 = 1.0$) and a surround radius of three ($\sigma_2 = 3$)[2]. The effect of this preprocessing is shown in figure 3.

As illustrated in Figure 4, the input vectors from both eyes are chosen as small, partially overlapping, circular regions of the preprocessed natural images; these converge on the same cortical cell.

The input from the right and left eye respectively are denoted by $\mathbf{d}^l$ and $\mathbf{d}^r$, and the output of

---

[2]This ratio between the center and surround in biologically plausible, and enables the PCA rule to produce oriented receptive fields.

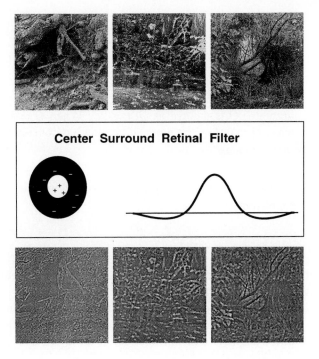

Center Surround Retinal Filter

Figure 3: Three of the natural images used (top) processed by a Difference of Gaussians filter and presented at the bottom.

the cortical neuron then becomes

$$c = \sigma(\mathbf{d}^l \cdot \mathbf{m}^l + \mathbf{d}^r \cdot \mathbf{m}^r), \tag{2.1}$$

where σ is the non linear activation function of each neuron. We have used a nonsymmetric activation function to account for the fact that neuronal activity as measured from spontaneous activity has a longer way to go up than to go down to zero activity.

In order to examine the effect of varying the overlap between the receptive fields we define an overlap parameter $O = s/2a$, where a is the receptive field radius in pixels, and s is the linear overlap in pixels, as shown in Figure 5. When the left and right receptive fields are completely overlapping $O = 1$, when they are completely separate $O \leq 0$.

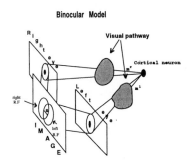

Figure 4: Schematic diagram of the two eye model, including the visual input preprocessing.

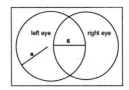

Figure 5: The overlap parameter O is defined as $O = s/2a$. When $O = 1$ the receptive fields are completely overlapping, when $O = 0$ they are non overlapping but touching $O < 0$ means that they are non overlapping and not touching.

In order to asses the degree of cell binocularity, we introduced an ocular dominance measure B based on left and right eye response:

$$B = \frac{L - R}{L + R}. \tag{2.2}$$

B is calculated by first finding the orientation at which the cell has the greatest binocular response to a sinusoidal grading, and then measuring L and R at that orientation.

This measure has been motivated by that used by Albus (1975) in defining the bin boundaries, for a seven bin ocular dominance histogram. Since there is always some activity from both eyes, we have extended bin 1 and 7 slightly. Our bin boundaries are given by:

$$-.085, -0.5, -0.15, 0, 15, 0.5, 0.85$$

3 Cortical plasticity learning rules

We have employed these realistic visual inputs to test two of the leading visual cortical plasticity rules that have been used to model various normal rearing and visual deprivation experiments: Principal components analysis (PCA) and the Bienenstock Cooper and Munro (BCM) model. The two algorithms differ by their information extraction properties as discussed in Intrator and Cooper (1994); PCA extracts second order statistics from the visual environment, while BCM extracts information contained in third order statistics as well. While both models have several versions and modifications, we examine only a single representative of each; different versions of these models may produce somewhat different results.

3.1 Principal Components Analysis

Principal components analysis (PCA) is one of the most widely used feature extraction method for pattern recognition tasks. PCA features are those orthogonal directions which maximize the variance of the projected distribution of the data. They also minimize the mean squared error between the data and a linearly reconstructed version of it based on these projections. Principal components are optimal when the goal is to accurately reconstruct the inputs. They are not necessarily optimal when the goal is classification and the data is not normally distributed (see for example, p. 212, Duda and Hart, 1973).

A simple interpretation of the Hebbian learning rule, is that with appropriate stabilizing constraints it leads to the extraction or approximation of principal components. This has often been modeled (See for example; von der Malsburg,1973; Sejnowsky, 1977; Oja, 1982; Linsker, 1986;

Miller et. al., 1989) . The learning rule that we use has been proposed by Oja (1982), and has the form:

$$\Delta m_i = \eta[d_i c - c^2 m_i] \tag{3.1}$$

where d_i is the presynaptic activity at synapse i, c is the postsynaptic activity, and m_i is the strength of the synaptic efficacy of junction i. η, is a small learning rate. This learning rule has been shown to converge to the principal component of the data.

It was shown by Shouval and Liu (1995) that when trained with preprocessed natural images, this learning rule produces horizontal oriented receptive fields. This result depends on the preprocessing, and on the non radially symmetric portion of the correlation function.

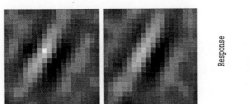 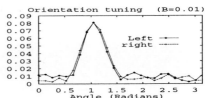

Figure 6: Receptive fields for a BCM neuron with completely overlapping inputs and receptive fields. Tuning curves are selective and similar in both eyes.

3.2   BCM learning rule

The BCM theory (Bienenstock et al., 1982) has been introduced to account for the striking dependence of the sharpness of orientation selectivity on the visual environment. We shall be using a variation due to Intrator and Cooper (1992) for a nonlinear neuron with a nonsymmetric sigmoidal transfer function. Using the above notation, the synaptic modification is governed by

$$\dot{m}_j = \eta\phi(c, \Theta_M)d_j, \tag{3.2}$$

where the neuronal activity is given by $c = \sigma(m \cdot d)$, $\phi(c, \Theta_M) = c(c - \Theta_M)$, and Θ_M is a nonlinear function of some time averaged measure of cell activity, which in its simplest form is given by

$$\Theta_M = E[c^2], \tag{3.3}$$

where E denotes the expectation over the visual environment. The transfer function σ is non symmetric around 0 to account for the fact that cortical neurons show a low spontaneous activity, and can thus fire at a much higher rate relative to the the spontaneous rate, but can go only slightly below it.

It has been shown (Intrator and Cooper, 1992) that this version of the BCM learning rule leads to a neuron which seeks multi-modality in the projected distribution (rather than simply maximizing the variance) and is thus suitable for finding clusters in high dimensional space. Simulations using this learning rule were found to be in agreement with the many experimental results on visual cortical plasticity (Clothiaux et al., 1991; Law and Cooper, 1994). A network implementation of this neuron which can find several projections in parallel while retaining its computational

223

efficiency, was found applicable for extracting features from very high dimensional vector spaces (Intrator et al., 1991; Intrator, 1992).

The modification rule used by Law and Cooper (1994) with a variable learning rate ,

$$\dot{m}_j = \eta \frac{1}{\Theta_M} \phi(c, \Theta_M) d_j, \tag{3.4}$$

produced faster convergence with qualitatively similar results.

4   Results

In all the results reported here we used a fixed circular receptive field with diameter of 20 pixels. We tested the robustness of the results to receptive fields of sizes 10 to 30 pixels and got no qualitative difference in the results.

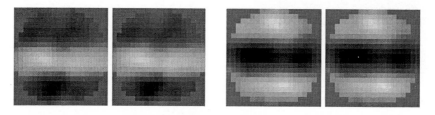

Figure 7: Receptive fields for completely overlapping inputs using the PCA rule. Comparison of linear neuron (right) and a non linear neuron (left). These examples are almost identical up to a sign, due to the inversion symmetry of the linear PCA rule (both **m** and −**m** are eigen-vectors).

4.1   Completely overlapping receptive fields

With completely overlapping receptive fields, BCM neurons develop various orientation preferences all highly selective. A typical example of such receptive fields and orientation selectivity is presented in Figure 6. A less typical result would be a slight ocular preference with high orientation selectivity. It should be noted, that high orientation selectivity is obtained for a single neuron with no need to introduce lateral inhibition between neurons; cells produce receptive fields, of all orientations, similar to simple cell receptive fields observed in visual cortex (Jones and Palmer, 1987; Kandel and Schwartz, 1991, for review). These results are in sharp contrast to those of a PCA neuron; PCA neurons developed a receptive field with horizontal orientation selectivity only (Figure 7). Orientation selectivity in PCA neurons depends on the retinal preprocessing. When PCA neurons are trained on raw natural images, the dominant solution is radially symmetric(Liu and Shouval, 1994). However when retinal preprocessing is included oriented solutions can be attained(Shouval and Liu, 1995). The strong preference to the horizontal direction is due to a slight bias in the correlation function of natural images to this direction (Shouval and Liu, 1995), if the images are rotated by an angle θ so are the preferred orientations of the PCA neurons. Apparently, the effect of this bias is stronger than the effect of any receptive fields misalignment as will be demonstrated below.

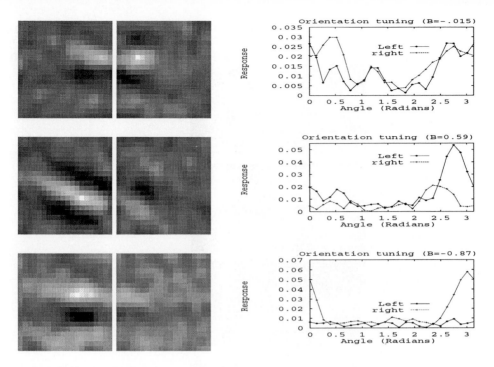

Figure 8: Receptive fields with a small overlap ($O = .2$) using the BCM rule. Results vary from fully binocular cells with a moderate orientation tuning (top) to less binocular cells with well defined receptive fields as well as high orientation selectivity in one eye(middle), and finally monocular highly selective cells (bottom).

4.2   Partially and nonoverlapping receptive fields

BCM neurons acquire selectivity to various orientations in the partial and the nonoverlapping case as well. When receptive fields are misaligned, various ocular dominance preferences may occur (Figure 8) even for the same overlap. This result stands in sharp contrast to the one obtained by PCA neurons; only binocular neurons emerge under for the PCA rule.

The BCM receptive field formation results are summarized in Figures 9 and 10. Receptive field misalignment does not affect orientation selectivity of the dominant eye, but does produce varying degrees of ocular dominance; this depends on the degree of overlap between the receptive fields. The main result is that ocular dominance depends strongly (even for single cell simulations) on the degree of overlap between visual input to the two eyes. Figure 10 presents a summary of 700 runs showing dependence of ocular dominance on visual input overlap. It is evident that binocularity increases when the degree of overlap increases.

The PCA results of partially overlapping receptive fields are presented in Figure 11. As mentioned above, it can be seen that the degree of overlap between receptive fields does not alter the optimal orientation, so that whenever a cell is selective its orientation is in the horizontal direction. The degree of overlap does affect the shape of the receptive fields, and the degree of orientation selectivity that emerges under PCA: orientation selectivity decreases as the amount of overlap decreases. However, when there is no overlap at all, one again gets heigher selectivity. For PCA,

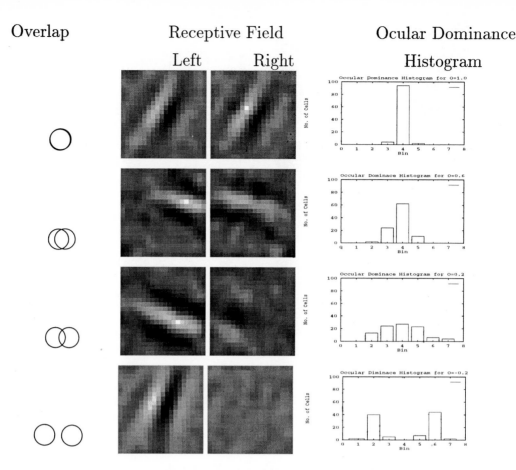

Figure 9: BCM neurons with different overlap values; $O = 1, 0.6, 0.2, -0.2$ from top to bottom. The occular dominance histograms summerise the occular dominance of 100 cells at each overlap value. The dependency of ocular dominance on visual overlap is evident.

there is also a symmetry between the receptive fields of both eyes. This arises from invariance to a parity transformation defined in the appendix, that imposes binocularity.

We also studied the possibility that under the PCA rule, different orientation selective cells would emerge if the misalignment between the two eyes was in the vertical direction. This tests the effect of a shift orthogonal to the preferred orientation. The results (Figure 11, bottom right) show that there is no change in the orientation preference; even in this case only horizontal receptive fields emerge.

To test whether the preference to horizontal orientation was an artifact of the program or the preprocessing, we trained on a visual environment that was rotated by 45 degrees. The results rule out such possibility (Figure 12), since the preferred orientation that emerges under these conditions is rotated by 45 degrees as well.

The PCA results described above were quite robust to introduction of nonlinearity in cell's activity; there was no qualitative difference in the results when a non symmetric sigmoidal transfer

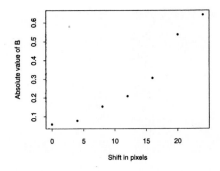

Figure 10: Dependence of ocularity on visual input overlap for BCM learning : Binocularity increases when overlap increases. The value of $|B|$ was obtained by averaging over 100 runs at each overlap value.

function was used.

5 Conclusions

In this paper we contrast the consequences of a real visual environment on two learning rules, PCA and BCM, for a single cell. We show that the BCM neuron develops orientation selective cells to all orientations as well as varying ocular dominance. This is consistent with observation. In contrast the PCA neuron is unable to develop cells selective to all orientations and the cells are always binocular. All PCA neurons develop receptive fields that are symmetric under a parity transformation.

6 Acknowledgments

The authors thank the members of Institute for Brain and Neural Systems for many conversations. This research was supported by the Charles A. Dana Foundation, the Office of Naval Research and the National Science Foundation.

A Appendix: Symmetry properties of the eigenstates of the two eye problem

The evolution of neurons in a binocular environment under the PCA learning rule, according to equation 3.1 reaches a fixed point when

$$\mathbf{Qm} = \lambda \mathbf{m}. \tag{A.1}$$

where $\mathbf{m}^T = (\mathbf{m}^l, \mathbf{m}^r)$, the left and right eye synaptic strengths; the two-eye correlation function $\mathbf{Q}$ has the form:

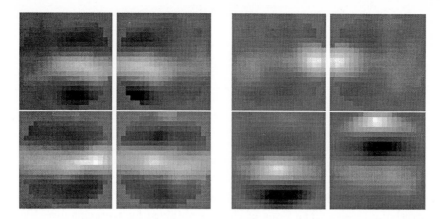

Figure 11: Receptive fields for partially overlapping inputs using the PCA rule, Receptive field for an overlap value of $O = .6$ (top left). Receptive field for a small overlap, $O = .2$ (top right). Receptive field for no overlap , $O = -.2$ (bottom left). Receptive field for shift in the vertical direction between the visual inputs when $O = .5$ (bottom right). In all cases the cell is binocular and horizontal. The symmetry property evident in these receptive fields is analyzed in the appendix.

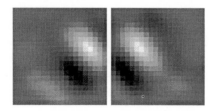

Figure 12: Receptive field formation from rotated images using the PCA rule, for an overlap value $O = 0.5$.

$$\mathbf{Q} = \begin{pmatrix} Q_{ll} & Q_{lr} \\ Q_{rl} & Q_{rr} \end{pmatrix} \tag{A.2}$$

where Q_{ll} and Q_{rr} are the correlation functions within the left and right eyes, Q_{lr} and Q_{rl} are the correlation functions between the left-right and right-left eyes. We denote by upper case $R's$ the coordinates in each receptive field with respect to a common origin, and by lower case $\mathbf{r}'s$ the coordinates from the centers of each of the receptive fields. Thus R_{0l} and R_{0r} are the coordinates of the centers of the left and right eyes, R_l and R_r are the coordinates of points in both receptive fields, and $\mathbf{r}_l$ and $\mathbf{r}_r$ are the coordinates of the same points with respect to the centers of the left and right receptive field centers. For a misalignment $\mathbf{s}$ between receptive field centers, $R_{0l} + \mathbf{s} = R_{0r}$, therefore $R_r - R_l = R_{0r} - R_{0l} + \mathbf{r}_r - \mathbf{r}_l = \mathbf{s} + \mathbf{r}_r - \mathbf{r}_l$ (See figure 13).

Using translational invariance, it is easy to see that

$$\begin{aligned} Q_{ll} &= E\left(d(\mathbf{r}_l)d(\mathbf{r}'_l)\right) = Q(\mathbf{r} - \mathbf{r}') \\ Q_{rr} &= E\left(d(\mathbf{r}_r)d(\mathbf{r}'_r)\right) = E\left(d(\mathbf{r}_l + \mathbf{s})d(\mathbf{r}'_l + \mathbf{s})\right) = Q(\mathbf{r} - \mathbf{r}') \end{aligned}$$

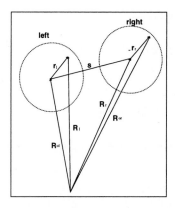

Figure 13: Coordinates for the two eyes. For a shift $\mathbf{s}$ between the two eyes, $R_{0l} + \mathbf{s} = R_{0r}$. Therefore $R_l - R_r = R_{0l} - R_{0r} + \mathbf{r}_r - \mathbf{r}_l = \mathbf{s} + \mathbf{r}_r - \mathbf{r}_l$

$$
\begin{aligned}
Q_{lr} &= E\left(d(\mathbf{r}_l)d(\mathbf{r}'_r)\right) = E\left(d(\mathbf{r}_l)d(\mathbf{r}'_l + \mathbf{s})\right) = Q(\mathbf{r} - \mathbf{r}' + \mathbf{s}) \\
Q_{rl} &= E\left(d(\mathbf{r}_r)d(\mathbf{r}'_l)\right) = E\left(d(\mathbf{r}_l + \mathbf{s})d(\mathbf{r}'_l)\right) = Q(\mathbf{r} - \mathbf{r}' - \mathbf{s})
\end{aligned}
$$

where E denotes an average with respect to the environment and where, occasionally, for simplicity, we replace $\mathbf{r}_l$ by $\mathbf{r}$. Since $Q(\mathbf{r} - \mathbf{r}') = E\left(d(\mathbf{r}_l)d(\mathbf{r}'_l)\right)$ then $Q(\mathbf{r} - \mathbf{r}') = Q(\mathbf{r}' - \mathbf{r})$.

We now introduce a two-eye parity operator $\mathbf{P}$, which inverts the coordinates, as well as the two eyes:

$$
\mathbf{P} : \begin{cases} \mathbf{r}_l \Rightarrow (-\mathbf{r}_l) \\ \mathbf{r}_r \Rightarrow (-\mathbf{r}_r) \\ \mathbf{s} \Rightarrow (-\mathbf{s}) \end{cases} \tag{A.3}
$$

It follows that under $\mathbf{P}$ $\quad R_l - R_r = \mathbf{r}_r - \mathbf{r}_l + \mathbf{s} \Rightarrow -\mathbf{r}_r + \mathbf{r}_l - \mathbf{s}$

The two-eye parity operator can also be written in matrix form in terms of the one eye parity operator P, thus

$$
\mathbf{P} = \begin{pmatrix} 0 & P \\ P & 0 \end{pmatrix}. \tag{A.4}
$$

The effect $\mathbf{P}$ on the two-eye receptive fields $\mathbf{m}$ is

$$
\mathbf{P}\begin{pmatrix} m^l(\mathbf{r}_l) \\ m^r(\mathbf{r}_r) \end{pmatrix} = \begin{pmatrix} m^r(-\mathbf{r}_l) \\ m^l(-\mathbf{r}_r) \end{pmatrix}
$$

Any correlation function that is invariant to a two-eye parity transformation $\mathbf{P}$, has eigen-functions $\mathbf{m}^T(\mathbf{r}) = \left(m^l(\mathbf{r}_r), m^r(\mathbf{r}_l)\right)$, that are also eigen-functions of $\mathbf{P}$. This imposes symmetry constraints on the resulting receptive fields that force them to be binocular.

229

Any correlation function [3] of the form

$$\mathbf{Q} = \begin{pmatrix} Q(\mathbf{r} - \mathbf{r}') & Q'(\mathbf{r} - \mathbf{r}' + \mathbf{s}) \\ Q'(\mathbf{r} - \mathbf{r}' - \mathbf{s}) & Q(\mathbf{r} - \mathbf{r}') \end{pmatrix} \tag{A.5}$$

is invariant to the two-eye parity transform $\mathbf{P}$ (that is $\mathbf{PQP} = \mathbf{Q}$), as long as $Q(\mathbf{x}) = Q(-\mathbf{x})$ and $Q'(\mathbf{x}) = Q'(-\mathbf{x})$.

Thus the eigen-functions of $\mathbf{Q}$, are also eigen-function of $\mathbf{P}$. The eigen-value is ± 1, Since $P^2 = 1$. Therefore we deduce that

$$\begin{pmatrix} \mathbf{m}^l(\mathbf{r}_l) \\ \mathbf{m}^r(\mathbf{r}_r) \end{pmatrix} = \pm \begin{pmatrix} \mathbf{m}^r(-\mathbf{r}_l) \\ \mathbf{m}^l(-\mathbf{r}_r) \end{pmatrix}. \tag{A.6}$$

Thus

$$\mathbf{m}(\mathbf{r}) = \begin{pmatrix} \mathbf{m}^l(\mathbf{r}_l) \\ \pm \mathbf{m}^l(-\mathbf{r}_r) \end{pmatrix}. \tag{A.7}$$

This means that the receptive fields for the two eyes are inverted versions of each other up to a sign. Therefore for this learning rule the receptive fields are always perfectly binocular.

References

Albus, K. (1975). Predominance of monocularly driven cells in the projection area of the central visual field in cat's striate cortex. *Brain Research*, 89:341–347.

Atick, J. J. and Redlich, N. (1992). What does the retina know about natural scenes. *Neural Computation*, 4:196–210.

Baddeley, R. and Hancock, P. (1991). A statistical analysis of natural images matchs psychophysically derived orientation tuning curves. *Proc Roy Soc B*, 246(17):219–223.

Bienenstock, E. L., Cooper, L. N., and Munro, P. W. (1982). Theory for the development of neuron selectivity: orientation specificity and binocular interaction in visual cortex. *Journal Neuroscience*, 2:32–48.

Blakemore, C. and Van-Sluyters, R. R. (1974). Reversal of the physiological effects of monocular deprevation in kittens: further evidence for sensitive period. *J. Physiol. Lond.*, 248:663–716.

Blakemore, C. and Van-Sluyters, R. R. (1975). Innate and environmental factors in the development of the kittens visual cortex. *J. Physiol.*, 248:663–716.

Clothiaux, E. E., Cooper, L. N., and Bear, M. F. (1991). Synaptic plasticity in visual cortex: Comparison of theory with experiment. *Journal of Neurophysiology*, 66:1785–1804.

Duda, R. O. and Hart, P. E. (1973). *Pattern Classification and Scene Analysis*. John Wiley, New York.

Field, D. J. (1987). Relations between the statistics of natural images and the response properties of cortical cells. *Journal of the Optical Society of America*, 4:2379–2394.

Field, D. J. (1989). What the statistics of natrual images tell us about visual coding. *SPIE*, 1077:269–276. Human Vision, Visual Processing, and Digital Display.

---

[3]This class includes the type of correlation function described in equation A.2, as well as the type postulated by Miller (Miller et al., 1989), in which $s = 0$ and $Q' = \eta Q$. There monocularity is attained by choosing $\eta < 0$ and restricting weights to be positive. This works only for non oriented receptive fields.

Frégnac, Y. and Imbert, M. (1984). Development of neuronal selectivity in the primary visual cortex of the cat. *Physiol. Rev.*, 64:325–434.

Frégnac, Y., Thorpe, S., and Bienenstock, E. (1992). Cellular analogs of visual cortical epigenesis. i. plasticity of orientation selectivity. *The Journal of Neuroscience*, 12(4):1280–1300.

Hancock, P. J., Baddeley, R. J., and Smith, L. S. (1992). The principal components of natural images. *Network*, 3:61–70.

Imbert, M. and Buisseret, P. (1975). Receptive field charectaristics and plastic properties of visual cortical cells in kittens reared with or without visual experiance. *Exp. Brain Res.*, 22:25–36.

Intrator, N. (1992). Feature extraction using an unsupervised neural network. *Neural Computation*, 4:98–107.

Intrator, N. and Cooper, L. N. (1992). Objective function formulation of the BCM theory of visual cortical plasticity: Statistical connections, stability conditions. *Neural Networks*, 5:3–17.

Intrator, N. and Cooper, L. N. (1994). Information theory and visual plasticity. In Arbib, M., editor, *The Handbook of Brain Theory and Neural Networks*. MIT Press. [To Appear].

Intrator, N., Gold, J. I., Bülthoff, H. H., and Edelman, S. (1991). Three-dimensional object recognition using an unsupervised neural network: Understanding the distinguishing features. In Feldman, Y. and Bruckstein, A., editors, *Proceedings of the 8th Israeli Conference on AICV*, pages 113–123. Elsevier.

Jones, J. P. and Palmer, L. A. (1987). The two-dimensional spatial structure of simple receptive fields in cat striate cortex. *jnp*, 58(6):1187–1258.

Kandel, E. R. and Schwartz, J. H., editors (1991). Elsevier, New York, third edition.

Law, C. C. and Cooper, L. N. (1994). Formation of receptive fields in realistic visual environments according the the BCM theory. *Proceedings of the National Academy of Science*, 91:7797–7801.

Li, Z. and Atick, J. J. (1994). Efficient stero coding in the multiscale representation. *Network*, 5:157–174.

Linsker, R. (1986). From basic network principles to neural architecture (series). *Proceedings of the National Academy of Science*, 83:7508–7512, 8390–8394, 8779–8783.

Liu, Y. and Shouval, H. (1994). Localized principal components of natural images - an analytic solution. Network., 5.2:317–325.

Miller, K. D. (1994). A model for the development of simple cell receptive fields and the ordered arrangement of orientation columns through activity-dependent competition between on- and off-center inputs. *J. of Neurosci.*, 14.

Miller, K. D., Keller, J. B., and Striker, M. P. (1989). Ocular dominance column development: Analysis and simulation. *Science*, 245:605–615.

Mioche, L. and Singer, W. (1989). Chronic recording from single sites of kitten striate cortex during experiance-dependent modification of synaptic receptive-field properties. *J. Neurophysiol.*, 62:185–197.

Nass, M. N. and Cooper, L. N. (1975). A theory for the development of feature detecting cells in visual cortex. *Biol. Cyb.*, 19:1–18.

Oja, E. (1982). A simplified neuron model as a principal component analyzer. *Math. Biology*, 15:267–273.

Orban., G. A. (1984). *Neuronal Operations in the Visual Cortex.* Springer Verlag.

Perez, R., Glass, L., and shalaer, R. J. (1975). Development of specificity in the cat visual cortex. *J. Math. Biol.*, 1:275.

Ruderman, D. L. and Bialek, W. (1993). Statistics of natural images: Scaling in the woods. In Cowan, J. D., Teaasaauro, G., and Alsppector, J., editors, *Advances in Neural Information Processing Systems*, volume 6. Morgan Kaufmann.

Sejnowski, T. J. (1977). Storing covariance with nonlinearly interacting neurons. *Journal of Mathematical Biology*, 4:303–321.

Shouval, H. and Liu, Y. (1995). Principal component neurons in a realistic visual environment. Submitted.

von der Malsburg, C. (1973). Self-organization of orientation sensitive cells in striata cortex. *Kybernetik*, 14:85–100.

Wiesel, T. N. and Hubel, D. H. (1963). Single-cell responses in striate cortex of kittens deprived of vision in one eye. *J. Neurophysiol.*, 26:1003–1017.

Theory of Synaptic Plasticity in Visual Cortex

(with N. Intrator, M. F. Bear, and M. A. Paradiso)
In: *Synaptic Plasticity: Molecular, Cellular and Functional Aspects*
eds. R. Thompson, M. Baudry, and J. E. Davis, MIT Press (1994)

The previous papers lead to some of our current work in this area, the main thrust of which is to check further consequences of the BCM theory as well as to pursue the underlying physiological mechanisms. A major effort is underway to observe the moving threshold.* When, hopefully, this is done, we will proceed to the complex task of tracking down the precise biological equivalents of such theoretical variables as c [e.g., Is c the summed postsynaptic depolarization — is it the same under all circumstances in determining m as well as θ?], the specific biochemical sequences by which synapses are altered – where the alterations (memory) are stored. This will no doubt illuminate such questions as the sites of long- and short-term memory and the mechanisms that govern the transfer from one to the other.

From the preceding papers we conclude that the BCM form of synaptic modification can account for experimental results in visual cortex and very likely has been directly observed. Related forms of modification were originally introduced to construct distributed and associative memories in cortex. If these arguments turn out to be correct we will have succeeded in one of our original aims: to make experimental contact between the abstract form of modification required to construct distributed and associative memories, and experimental results. In addition, we will have constructed a theoretical structure sufficiently precise so that it can be compared with experimental results. This gives us the opportunity to test specific assumptions and to refine both theory and experiment in detail that would have been out of the question originally. We will in fact have a path from the molecular basis of modifications that occur in the nervous system to some aspects of higher level mental activity.

The following book chapter, with which we conclude this section, provides an almost current overall view of our thinking.

---

* The correct placement of this threshold is very likely critical in achieving sharpness of learning and memory storage.

Synaptic Plasticity: Molecular, Cellular and Functional Aspects, eds. R. Thompson, M. Brudry, and J. E. Davis. © MIT Press, 1994

Theory of Synaptic Plasticity in Visual Cortex

Nathan Intrator, Mark F. Bear, Leon N. Cooper, and Michael A. Paradiso

1 INTRODUCTION

Because of its great complexity, visual cortex would not seem to be an auspicious region of the brain in which to carry out an investigation of synaptic plasticity or of the mechanisms and sites of memory storage. It is, in addition, almost certain that much of the architecture of visual cortex is preprogrammed genetically, leaving a relatively minor percentage to be shaped or modified by experience. However, the fact that visual cortex is accessible to single-cell electrophysiology, so that the output of individual cells can be measured, whereas the inputs can be controlled by varying the visual experience of the animal has made this a preferred area for experimentation and analysis. Thus over the past 30 years, a great deal of experimental and theoretical work has been done to investigate the responses of visual cortical cells, as well as the alterations in these responses under various visual rearing conditions.

It is widely believed that much of the learning and resulting organization of visual cortex as well as other parts of the central nervous system occurs due to modification of the efficacy or strength of at least some of the synaptic junctions between neurons, thus altering the relation between presynaptic and postsynaptic potentials. The vast amount of experimental work done in visual cortex—particularly area 17 of čat and monkey—strongly indicates that one is observing a process of synaptic modification dependent on the information locally and globally available to the cortical cells. Furthermore, it is known that small but coherent modifications of large numbers of synaptic junctions can result in distributed memories. Whether and how such synaptic modification occurs, what precise forms it takes, and what the physiological and/or anatomical bases of this modification are, are among the most interesting questions in this area. There is no need to assume that such mechanisms operate in exactly the same manner in all portions of the nervous system or in all animals. However, one would hope that certain fundamental similarities exist, so that a detailed analysis of the properties of these mechanisms in one preparation would lead to some conclusions that are generally applicable.

It is our hope that such a general form of modifiability manifests itself for at least some cells of visual cortex that are accessible to experiment. If so, one then may be able to distinguish between different cortical plasticity theories with theoretical tools and the aid of sophisticated experimental paradigms. Among the difficulties faced by theoreticians are (1) adequate representation of the visual environment; (2) knowledge of what the actual inputs to cortical cells are; (3) the appropriate rule for synaptic modification; and (4) an adequate representation of the complex architecture of visual cortex.

In this article, we give a short account of the BCM (Bienenstock, Cooper, and Munro) theory of visual cortical plasticity that has been developed over the past 10 years, address the difficulties mentioned above and compare the consequences of the theory with experiment. We discuss recent physiological experiments that seem to provide verification of some of the underlying assumptions of the theory, and finally, we initiate a comparison of the BCM theory with other theories that have been proposed. We assume that the reader has some familiarity with experiments demonstrating plasticity in visual cortex. A brief review may be found in Clothiaux et al. (1991).

2 BCM THEORY

In what follows we give a brief overview of the BCM theory of synaptic plasticity. For a more detailed account the reader is referred to the various references cited below.

2.1 Single Cell

A typical neuron in striate cortex receives thousands of afferents from other cells. Most of these afferents derive from the lateral geniculate nucleus (LGN) and from other cortical neurons. We have approached the analysis of this complex network in several stages. In the first stage we consider a single neuron with inputs from both eyes (i.e., LGN) but without intracortical interactions.

The output of this neuron (in the linear region) can be written

$$c = m^l \cdot d^l + m^r d^r,$$

where d^l (d^r) are the LGN inputs coming from the left (right) eye to the vector of synaptic junctions m^l (m^r). The neuron firing rate (in the linear region) is therefore the sum of the inputs from the left eye multiplied by the appropriate left-eye synaptic weights plus the inputs from the right eye multiplied by the appropriate right-eye synaptic weights. Thus the neuron integrates signals from the left and right eyes. (For simplicity, whenever possible we shall omit the left and right superscripts.) According to the theory presented by Bienenstock, Cooper, and Munro (BCM; 1982), the synaptic weight changes over time as a function of local and global variables: its change in time, $\dot{m}_j$, is given below:

$$\dot{m}_j = F(d_j, \ldots, m_j; d_k, \ldots, c; \bar{\bar{c}}; X, Y, Z).$$

Here variables such as $d_j, \ldots, m_j$ are designated local. These represent information (such as the incoming signal, d_j, and the strength of the synaptic junction,

m_j) available locally at the synaptic junction, m_j. Variables such as $d_k, \ldots, c$ are designated quasi-local. These represent information (such as c, the firing rate or depolarization of the postsynaptic cell, or d_k, the incoming signal to another synaptic junction) that may not be locally available to the junction m_j but is physically connected to the junction by the cell body itself, thus necessitating some form of internal communication between various parts of the cell and its synaptic junctions. Variables such as $\bar{c}$ (the time-averaged output of the cell) are averaged local or quasi-local variables. Global variables are designated $X, Y, Z, \ldots$. These latter represent information (e.g., presence or absence of neurotransmitters such as norepinephrine or the average activity of large numbers of cortical cells) that is present in a similar fashion for all or a large number of cortical neurons (distinguished from local or quasi-local variables presumably carrying detailed information that varies from synapse to synapse or cell to cell). Neglecting global variables, one arrives at the following form of synaptic modification equation:

$$\dot{m}_j = \phi(c, \Theta_m)d_j \qquad (2.1)$$

so that the jth synaptic junction, m_j, changes its value in time as the product of the input activity (the local variable d_j) and a function ϕ of quasi-local and time-averaged quasi-local variables, c and Θ_m. Θ_m is a nonlinear function of some time averaged measure of cell activity that in the original BCM formulation was proposed as

$$\Theta_m = (\bar{c})^2. \qquad (2.2)$$

In the BCM theory, this time average is replaced, for simplicity, by a spatial average over the environmental inputs ($\bar{c} \to m \cdot \bar{d}$). The shape of the function ϕ is given in figure 8.1 for two different values of the threshold Θ_m. The occurrence of negative and positive regions for ϕ results in the cell becoming selectively responsive to subsets of stimuli in the visual environment. This happens because the response of the cell is diminished to those patterns for which the output, c, is below threshold (ϕ negative), while the response is enhanced to those patterns for which the output, c, is above threshold (ϕ positive). The nonlinear variation of the threshold Θ_m with the average output of the cell contributes to the development of selectivity and the stability of the system (Bienenstock et al., 1982; Intrator and Cooper, 1992).

2.2 Cortical Network: Mean Field Theory

The actual cortical network is very complex. It includes different cell types, intracortical interactions, and recurrent collaterals. In what follows we present a method of analyzing this complex system. The first step is to divide the inputs to any cell into those from the LGN and those from all other sources. The activity of neuron i is affected by its input vector d from the LGN, and by the adjacent cortical neurons;

$$c_i = m_i \cdot d + \sum_j L_{ij} c_j, \qquad (2.3)$$

Theory of Synaptic Plasticity in Visual Cortex

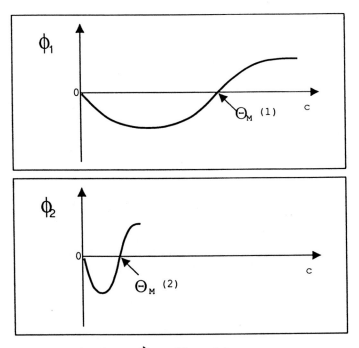

Figure 8.1 The ϕ function for two different Θ_m's.

where L_{ij} are the cortico-cortical synapses. Scofield and Cooper (1985; Cooper and Scofield, 1988) have analyzed a network extension of the single cell theory and a mean field approximation to the full network. Defining $\bar{c} = 1/N \sum_i c_i$, where N is the number of neurons in the network, the mean field approximation is obtained by replacing the inhibitory contribution of cell j, c_j by its average value (i.e., and average on all cells in the network) so that c_i becomes:

$$c_i = m_i \cdot d + \bar{c} \sum_j L_{ij}. \tag{2.4}$$

From a consistency condition it follows that $\bar{c} = \bar{m} \cdot d + \bar{c}L_0 = (1 - L_0)^{-1}\bar{m} \cdot d$, where $\bar{m} = 1/N \sum_i m_i$, and $L_0 = 1/N \sum_{ij} L_{ij}$, so that $c_i = (m_i + (1 - L_0)^{-1}\bar{m} \sum_j L_0)d$.

If we assume that the lateral connection strengths are a function only of the relative distance $i-j$, then L_{ij} becomes a circular matrix so that $\sum_i L_{ij} = \sum_j L_{ij} = L_0$, and

$$c_i = (m_i + L_0(1 - L_0)^{-1}\bar{m})d. \tag{2.5}$$

In the mean field approximation, one can therefore write $c_i(\alpha) = (m_i - \alpha)d$, with $\alpha = |L_0|(1 + |L_0|)^{-1}\bar{m}$.

When analyzing the position and stability of the fixed points using this approximation, it follows under some mild assumption on the evolution of the average synaptic weights, that there is a mapping

$$m_i' \leftrightarrow m_i(\alpha) - \alpha$$

Intrator et al.

such that for every neuron in such a network with synaptic weight vector m_i there is a corresponding neuron with weight vector m_i' that undergoes the same evolution (around the fixed points) subject to a translation α.

Although the averaged inhibition assumption used in the mean field theory is an approximation, the mean field network described above provides a powerful tool to analyze a certain type of network architecture in great detail, and to gain an intuitive understanding of a complex network in terms of the behavior of a single neuron.

2.3 Synapses with Varying Modifiability

In the equations above, all synapses are taken to be modifiable in the same way. However, the behavior of visual cortical cells in various rearing conditions suggests that some cells respond more rapidly to environmental changes than others. In monocular deprivation, for example, some cells remain responsive to the closed eye in spite of the very large shift of most cells to the open eye. Hubel and Wiesel (1959) and Singer (1977), found, using intracellular recording, that geniculo-cortical synapses on inhibitory interneurons are more resistant to monocular deprivation than are synapses on pyramidal cell dendrites. These results suggest that some LGN-cortical synapses modify rapidly, while others modify relatively slowly, with slow modification of some cortico-cortical synapses. Excitatory LGN-cortical synapses onto excitatory cells may be those that modify primarily. Since these synapses are formed exclusively on dendritic spines, this raises the possibility that the mechanisms underlying synaptic modification exist primarily in axo-spinous synapses. To embody these facts we introduce two types of LGN-cortical synapses: those (m_i) that modify according to the modification rule discussed in BCM and those (z_k) that remain relatively constant. In a cortical network with modifiable and nonmodifiable LGN-cortical synapses, and nonmodifiable cortico-cortical synapses L_{ij}, the synaptic evolution equations become

$$\dot{m}_i = \phi(c_i, \Theta_m^i)d,$$

$$\dot{z}_k = 0,$$

$$\dot{L}_{ij} = 0. \tag{2.6}$$

As will be discussed below, such a network is capable of explaining the variety of experiments considered.

3 THE BCM THEORY AND THE NEUROBIOLOGY OF SYNAPTIC MODIFICATION

The BCM theory and its recent extensions originated as an attempt to account for the varied consequences of different visual environments on the developing visual cortex. In cats, the circuitry of the visual cortex can be modified by simple manipulations of visual experience during a "critical period" in the first few months of postnatal development. For example, one such manipulation, monocular deprivation, leads to a disconnection of the inputs from the deprived eye that renders the animal behaviorally blind through that eye. The goal of the BCM theory is to develop a model of synaptic modification that accounts for those striking changes in visual cortex that result from alterations in the patterns and amount of activity arising at the two retinae. While the theory aims to provide a physiologically plausible account of synaptic plasticity, it does not address the mechanism by which plasticity diminishes at the end of the critical period. A number of possible mechanisms have been proposed to account for the short duration of the plastic period, but at present it is not clear that the length of the critical period is determined by the same mechanism as that underlying synaptic change.

The validity of the BCM theory, as with any theory, can be tested in two ways. The first is to derive predictions or consequences of the theory in various situations that can be compared with experimental results. There is a considerable experimental literature on visual cortical plasticity reaching back 30 years which facilitates such comparisons with the BCM theory. The second approach is to attempt to verify the underlying assumptions of the theory, particularly those assumptions that distinguish it from others. In the case of the BCM theory, the most important and unique assumptions concern the form of the synaptic modification function ϕ and the movement of the modification threshold. Over the last 5 years we have made significant progress using both of these approaches, and this work is summarized briefly below.

3.1 Comparison of Theory and Experiment

In work recently published by Eugene Clothiaux and colleagues (Clothiaux et al., 1991) the consequences of the BCM theory were compared in detail with the results of experiments on what were called "classical" rearing conditions. These conditions include normal binocular vision, monocular deprivation, reverse suture, strabismus, binocular deprivation, as well as the restoration of normal binocular vision after various forms of deprivation. Comparisons with the pharmacological manipulations that affect visual cortical plasticity (e.g., Bear et al., 1990; Greuel et al., 1987; Reiter and Stryker, 1988) were not considered and remain an area that is ripe for further work. The modifications considered by Clothiaux et al. were those that occur in kitten visual cortex during the second postnatal month after brief (approx. 2 weeks) changes in visual experience. Particular attention was given to the manner in which the theory predicts that changes in visual experience should affect the binocularity of cortical neurons and the selectivity of these neurons for the stimulus pattern (e.g., its orientation). These properties of binocularity and selectivity distinguish cortical neurons from those in the retina and thalamus. A review of the experimental literature as it relates to the modification of these properties may be found in Clothiaux et al. (1991).

All theories of visual cortical plasticity have to make some assumptions as to how the initial visual scenes are converted into LGN firing rates and how this information reaches visual cortex. We wish to model the input to visual cortex that arises from the regions of the two retinae that view the same point in visual space. For simplicity, Clothiaux et al. assumed that LGN activity is a direct reflection of retinal ganglion cell activity. Two types of LGN-cortical input were modeled: (1) activity elicited when visual contours are presented to the retinae, which we call "pattern" input; and (2) activity that arises in the absence of visual contours, which we call "noise." From our point of view the important distinction between pattern and noise input is the degree of correlation that the two types of input produce in the LGN. For a specific input pattern the activity of one LGN neuron is assumed to have a predictable relationship (i.e., correlation) to the activity of other LGN neurons whereas for noise the activity of one LGN neuron is independent of the activity of the other LGN neurons. (Addition of local correlations such as those suggested by the work of Mastronarde (1989) does not alter the results.) Differences between distinct patterns (for example, between various stimulus orientations) are reflected by the differences in their distribution of activity across the LGN. Using this type of pattern input distorted by noise, and noise alone, Clothiaux et al. were able to reproduce both the outcome and kinetics of synaptic change in visual cortex resulting from normal visual experience and a wide variety of visual deprivation conditions.

As one example of the quantitative nature of these results, consider the simulation of the effects of monocular deprivation (figure 7 in Clothiaux et al., 1991). Beginning from a state in which the simulated neuron is binocularly responsive and selective, substituting pattern input through one eye with noise leads to a rapid synaptic disconnection of the "deprived" eye. Mathematical analysis provides a complete account of the factors on which this result depends if it is governed by the principles of the BCM theory. For example, for this result to be obtained using the BCM theory it is necessary that the neuron be selective (i.e., that it respond vigorously only to a fraction of the patterns that are presented to the "open eye") before the ocular dominance changes, and that the deprived eye inputs carry noise (i.e., that they be active). The prediction that the ocular dominance shift depends on neuronal selectivity was tested by Paradiso and colleagues (Ramoa et al., 1988). They found that cortical infusion of the GABA receptor antagonist bicuculline, which greatly reduces orientation selectivity in visual cortex, eliminates the ocular dominance shift that normally results from monocular deprivation. The second prediction—that the disconnection of the deprived eye depends on noise—has never been tested explicitly, but there are some indications that it is also correct. For example, clinical observations in humans led Jampolsky (1978) to conclude that the effects of monocular diffusion (resulting from lid suture) are more severe than the effects of monocular occlusion (resulting from an opaque eye-patch or contact lens).

To determine the time equivalence of each iteration for the parameters used, the behavior of the model under monocular deprivation can be compared to the results of the corresponding experiment. Equivalence was established between the number of computer iterations and the duration of deprivation required for complete disconnection of the deprived eye

(Clothiaux et al. 1991). Thus, using a fixed set of parameters, one has a direct correspondence between the temporal dynamics of synaptic change in the theory and experiments. This can be used to analyze and compare kinetics and outcome of theory and experiment for other manipulations. For example, in "reverse suture," the deprived eye is opened and the open eye is closed after a period of initial monocular deprivation. Experimentally, it is observed that the newly closed eye shows a greatly reduced response in about 24 hr, but that the recovery of the response to the newly open eye generally does not begin for another 1–2 days (Mioche and Singer, 1989). The same difference in the time required to obtain the initial effect and the reversal is seen with the model. The correspondence of theory and experiment is thus very close. The theoretical explanation for this result is that recovery requires that the modification threshold slide nearly to zero and, using the same parameters that were fixed for monocular deprivation, this requires approximately 24 hr. Similar comparisons for the other experimental manipulations are discussed in detail in Clothiaux et al. (1991). We conclude that when the predictions of the theory have been tested, they are in good agreement with what is seen experimentally.

3.2 Neurobiological Foundations for the Assumptions of the BCM Theory

Recent advances in our understanding of excitatory amino acid (EAA) receptors have suggested a possible physiological basis of the BCM form of synaptic modification. In 1987, Bear and colleagues proposed that the modification threshold Θ_m of the BCM theory related to the membrane potential at which the N-methyl-D-aspartate (NMDA) receptor–dependent Ca^{2+} flux reached the threshold for inducing synaptic long-term potentiation (LTP). In support of the hypothesis that NMDA receptor mechanisms play a role in synaptic plasticity, Bear and co-workers have found that the pharmacological blockade of NMDA receptors with the competitive antagonist AP5 disrupts the physiological (Bear et al., 1990; Kleinschmidt et al., 1987) and anatomical (Bear and Colman, 1990) consequences of monocular deprivation in striate cortex. Although the interpretation of these experiments is compromised by the finding that AP5 reduces visually evoked responses (Fox et al., 1989), the data indicate that activity evoked in visual cortex in the absence of NMDA receptor activation is not sufficient to produce loss of closed-eye responsiveness in MD.

In the past several years our work has focused on the synaptic plasticity that can be evoked in brain slices to better investigate the assumptions of the BCM theory and to address possible underlying mechanisms (Bear et al., 1992; Connors and Bear, 1988; Dudek and Bear, 1992; Kirkwood et al., 1993; Press and Bear, 1990). Hippocampus, particularly CA1 and dentate gyrus, is an advantageous preparation because robust and long-lasting experience-dependent synaptic modifications can be evoked in this structure. Serena Dudek in Bear's lab (1992) recently tested a theoretical prediction that patterns of excitatory input activity that consistently fail to activate target

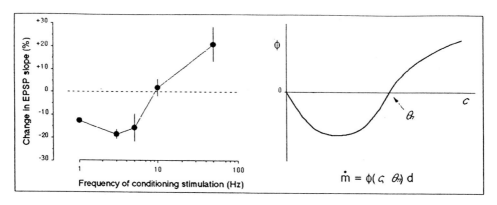

Figure 8.2 Comparison of experimental observations with BCM ϕ function for synaptic modification. (Data replotted from Dudek and Bear 1992.)

neurons sufficiently to induce synaptic potentiation will instead cause a specific synaptic depression. To realize this situation experimentally, the Schaffer collateral projection to CA1 in rat hippocampal slices was stimulated electrically at frequencies ranging from 0.5 to 50 Hz. Nine hundred pulses at 1–3 Hz consistently yielded a depression of the CA1 population EPSP that persisted without signs of recovery for more than 1 hr following cessation of the conditioning stimulation. This long-term depression was specific to the conditioned input and could be prevented by application of NMDA receptor antagonists. This result was surprising in that NMDA receptors are known to participate in the induction of long-term potentiation, an increase in synaptic effectiveness. Indeed, at higher stimulation frequencies the depression was replaced by a potentiation. If the effects of varying stimulation frequency in the experiments of Dudek and Bear are explained by different values of postsynaptic response (perhaps the integrated postsynaptic depolarization or Ca^{2+} level) during the conditioning stimulation, then it can be seen from figure 8.2 that their data are in striking agreement with assumptions of the BCM theory.

Of course, as striking as this similarity is, Dudek's work was performed in hippocampus and the BCM theory was developed for visual cortex. And, although these two forms of synaptic plasticity (depression and potentiation) have been reported in the sensory neocortex (cf. Artola et al., 1990), evidence to date has indicated that they occur with far lower probability, usually require pharmacological treatments for their induction, and are elicited by stimulation patterns that differ dramatically from those that are effective in hippocampus (see discussion in Bear et al., 1992). Together, these data have been taken as support for the view that hippocampus and sensory neocortex may be quite distinct with respect to their capability for synaptic change. However, a direct comparison of plasticity of synaptic responses evoked in adult rat hippocampal field CA1 with those evoked in adult rat and immature cat visual cortical layer III has now been carried out by Alfredo Kirkwood and colleagues in Bear's lab (1993). In the neocortical preparations they have stimulated the direct input to layer III from layer IV rather than using the

Theory of Synaptic Plasticity in Visual Cortex

traditional approach of stimulating the white matter, and find, contrary to the prevailing view, that very similar forms of plasticity, LTP and LTD, are evoked with precisely the same types of stimulation in the three types of cortex without the use of pharmacological treatments. Further, in all three preparations, both LTP and LTD depend on activation of NMDA receptors. These data suggest, first, that hippocampus should not be considered as a privileged site for plasticity in the adult brain and, second, that a common principle may govern experience-dependent synaptic plasticity, both in CA1 and throughout the superficial layers of the neocortex. We believe that this work represents an important advance towards a general theory of experience-dependent synaptic plasticity in the mammalian brain.

It is our opinion that in its entirety this work gives strong justification for a form of modification similar to that assumed by the BCM theory. However, the question of the sliding modification threshold is still open. Although more work remains to be done on this question, we note that two recent studies have shown that the sign and magnitude of a synaptic modification in both hippocampus (Huang et al., 1992) and the Mauthner cell of goldfish (Yang and Faber, 1991) have been shown to depend on the recent history of synaptic activation.

4 REFORMULATION AND EXTENSION OF THE BCM THEORY

In order to compare the BCM theory with other theories of synaptic plasticity as well as to exhibit its information processing and statistical properties, the following formulation proves convenient.

4.1 Objective Function Formulation

In a recent statistical formulation of the BCM theory (Intrator and Cooper, 1992), the threshold Θ_m was defined (In the notation of Intrator and Cooper, d is replaced by x) as

$$\Theta_m = E[(d \cdot m)^2],$$

and an energy function that corresponds to a risk function in statistical decision theory was presented:

$$R_m = -\mu \{ \tfrac{1}{3} E[(d \cdot m)^3] - \tfrac{1}{4} E^2[(d \cdot m)^2] \}. \tag{4.1}$$

It was shown that the differential equations describing synaptic weight modification are a stochastic approximation of the negative gradient of the risk, hence tending to minimize this risk (see Intrator and Cooper, 1992, for review). This formulation permits us to demonstrate the connection between the unsupervised BCM learning procedure and various statistical methods, in particular, that of exploratory projection pursuit (Friedman, 1987). It also provides a general method for stability analysis of the fixed points of the theory and enables us to analyze the behavior and the evolution of the network under various visual rearing conditions. In the next few sections we shall use this formulation to extend the theory to nonlinear neurons, and consequently to a network of feedforward inhibitory neurons.

4.2 Nonlinear Neurons

From statistical considerations that are motivated by the projection pursuit ideas, it is more effective to consider a nonlinear neuron that is less sensitive to possible outliers in the data. This is done by defining the neuron's activity as $c = \sigma(d \cdot m)$, where σ usually represents a smooth sigmoidal function. It is also desirable to have the ability to shift the projected distribution (of the input data) so that one of its peaks is at zero, by introducing a threshold β so that the projection is defined to be $c = \sigma(d \cdot m + \beta)$. From the biological viewpoint, β can be considered as spontaneous activity. The modification equations for finding the optimal threshold β are easily obtained by observing that this threshold effectively adds one dimension to the input vector and the vector of synaptic weights so that $d = (d_1, \ldots, d_n, 1)$, $m = (m_1, \ldots, m_n, \beta)$, and therefore, β can be found by using the same synaptic modification equations. For the rest of the paper we shall assume that this threshold is added to the projection, without specifically writing it.

For the nonlinear neuron, Θ_m is defined to be $\Theta_m = E[\sigma^2(d \cdot m)]$. The gradient of the risk becomes:

$$-\nabla_m R_m = \mu E[\phi(\sigma(d \cdot m), \Theta_m)\sigma' d], \tag{4.2}$$

where σ' represents the derivative of σ at the point $(d \cdot m)$. Note that the multiplication by σ' reduces sensitivity to outliers of the differential equation since for outliers σ' is close to zero. The gradient decent procedure is valid, provided that the risk is bounded from below (cf. Intrator and Cooper, 1992).

4.3 Networks with Feedforward Inhibition: Application to Classification

Intrator and Cooper (1992) have extended the single cell theory to a feedforward inhibition network which does not require the mean field approximation; nor does it require that the cortico-cortical synapses L_{ij} be constant. Thus it is possible to study networks with varying amounts of excitation and inhibition.

The activity of neuron k in the network is $c_k = d \cdot m_k$, where m_k is the synaptic weight vector of neuron k. The *inhibited* activity and threshold of the k'th neuron are given by

$$\tilde{c}_k = c_k - \eta \sum_{j \neq k} c_j, \qquad \tilde{\Theta}_m^k = E[\tilde{c}_k^2]. \tag{4.3}$$

The relation between the feed forward inhibition network and the mean field network is discussed in Intrator and Cooper (1992).

For the feedforward network the risk for node k is given by:

$$R_k = -\mu\{\tfrac{1}{3}E[\tilde{c}_k^3] - \tfrac{1}{4}E^2[\tilde{c}_k^2]\}, \tag{4.4}$$

and the total risk is given by

$$R = \sum_{k=1}^{N} R_k .$$
(4.5)

It follows that the gradient of R becomes:

$$\frac{\partial R}{\partial m_k} = \frac{\partial R_k}{\partial m_k} - \eta \sum_{j \neq k} \frac{\partial R_j}{\partial m_j}$$

$$= \mu \left[E[\phi(\tilde{c}_k, \tilde{\Theta}_m^k) d] - \eta \sum_{j \neq k} E[\phi(\tilde{c}_j, \tilde{\Theta}_{j}) d] \right].$$
(4.6)

The equation performs a constraint minimization in which the derivative with regard to one neuron can become orthogonal (when $\eta \to 1$) to the sum over the derivatives of all other synaptic weights. Nevertheless, the coupling between the neurons is very simple to calculate and does not require any matrix inversion. Equation (4.6) therefore, allows a simple computational algorithm that performs exploratory projection pursuit of several projections in parallel.

When the nonlinearity of the neuron is included, the inhibited activity is defined (as in the single neuron case) as $\tilde{c}_k = \sigma(c_k - \eta \sum_{l \neq k} c_l)$. $\tilde{\Theta}_m^k$, and R_k are defined as before. However, in this case

$$\frac{\partial \tilde{c}_k}{\partial m_j} = -\eta \sigma'(\tilde{c}_k) d, \qquad \frac{\partial \tilde{c}_k}{\partial m_k} = \sigma'(\tilde{c}_k) d.$$
(4.7)

Therefore the total gradient becomes:

$$m_k = \frac{\partial R}{\partial m_k} = \mu \left\{ E[\phi(\tilde{c}_k, \tilde{\Theta}_m^k) \sigma'(\tilde{c}_k) d] - \eta \sum_{j \neq k} E[\phi(\tilde{c}_j, \tilde{\Theta}_m^j) \sigma'(\tilde{c}_j) d] \right\}.$$
(4.8)

This biologically motivated system of equations has many desirable statistical properties and has been applied to various nontrivial feature extraction tasks such as phoneme recognition (Intrator, 1992) and three-dimensional object recognition (Intrator and Gold, 1993).

5 COMPARISON OF BCM WITH OTHER VISUAL CORTICAL PLASTICITY THEORIES

In order to compare ideas concerning visual cortical plasticity, it is important to analyze separately the different components that make up a theory, and to compare theories feature by feature. We consider a theory as being composed of the following three components:

- Synaptic modification equations
- Model of the input environment
- Network architecture

In some cases, there are interactions between these components that are not explicitly defined. For example, several theories are said to have the property of being able to develop orientation selectivity prenatally, that is, using random noise as an input environment (Linsker, 1986; Miller et al., 1989). However, under closer examination, it turns out that they have architectual constraints that actually yield very different input environments. Inputs to the network become strongly locally correlated after the first layer due only to network architecture (the arborization function). The arborization function determines the density of synapses as a function of planar distance from their target cell. This correlated input can then drive the higher level of cells to develop orientation-selective cells. When the arborization function is uniform, all the weights of all layers will become positively saturated; thus no selectivity will develop.

5.1 Comparison Based on Synaptic Modification Equations

To examine the effect of the synaptic modification equations in isolation, we shall fix the network inputs to be the same, and fix the architecture as well. The simplest architecture that would already yield a significant difference between several models is of a single cortical neuron receiving input from a single source (single eye).

In the correlation of activity models (Kammen and Yuille, 1988; Linsker, 1986; Miller et al., 1989; Sejnowski, 1977; Yuille et al., 1989) the input is defined in terms of the correlation of activity in the presynaptic afferents, whereas in the BCM model the input is defined in terms of the presynaptic activity. For reasons that will become clear, it is difficult to transform the correlation of activity models to a presynaptic activity models; however, we can rewrite the BCM model as a correlation of activity model.

To simplify notation and without loss of generality we shall assume that the input activity in each ganglion cell has zero mean. First we show how the transformation from input activity to correlation activity is done by expanding on footnote 15 of Miller et al. (1989). This will be done in the simple case of a single cortical neuron with no interaction between LGN inputs coming from the two eyes. Miller's rule has the following form:

$$\frac{dS(\alpha, t)}{dt} = \lambda A(-\alpha)[c(t) - c_1]a_\sigma(\alpha, t) - \gamma S(\alpha, t) - \varepsilon' A(-\alpha), \tag{5.1}$$

where α is the afferent location, A is the arbor function representing the number (or in the limit, the density of) afferents coming from location α, $a_\sigma(\alpha, t)$ is the afferent activity at location α, the subscript σ represents an addition of threshold and saturation effects, and c_1 is a constant. $c(t)$ is the neuronal activity, which in this simple case of no lateral interactions is given by $c(t) = \sum_\beta S(\beta, t)a_\sigma(\beta, t) + c_2$ where c_2 is some constant. $\gamma S(\alpha, t) + \varepsilon' A(-\alpha)$ are decay functions of the synaptic weights. Substituting $c(t)$ into (5.1), denoting $c_3 = c_2 - c_1$, and taking the average over the input space (at a given afferent location) we obtain

$$\frac{dS(\alpha, t)}{dt} = \lambda A(-\alpha) \left\{ \sum_\beta S(\beta, t) E(a_\sigma(\beta, t) a_\sigma(\alpha, t)) + c_3 E(a_\sigma(\alpha, t)) \right\}$$
$$- \gamma S(\alpha, t) - \varepsilon' A(-\alpha). \tag{5.2}$$

Using $C(\alpha, \beta)$ to represent the correlation of activity $C(\alpha, \beta) \stackrel{\text{def}}{=} E(a_\sigma(\alpha, t) a_\sigma(\beta, t))$ we get

$$\frac{dS(\alpha, t)}{dt} = \lambda A(-\alpha) \left\{ \sum_\beta S(\beta, t) C(\alpha, \beta) - c_1 E[a_\sigma(\alpha, t)] \right\} - \gamma S(\alpha, t) - \varepsilon' A(-\alpha), \tag{5.3}$$

which is a simple case of equation (1) in Miller et al. (1989).

Analogous reformulation is done below for the BCM modification equation (for the purpose of the comparison d is replaced by $a(\alpha, t)$ below)

$$\frac{dm(\alpha, t)}{dt} = \lambda \phi(c(t), \Theta_m) a(\alpha, t), \tag{5.4}$$

for the synaptic weight m (in Miller's notation this is S); for simplicity we omit decay terms and assume a uniform arbor function.

Using a simple form of the modification function, $\phi(c, \Theta_m) \stackrel{\text{def}}{=} c(c - \Theta_m)$ (Bienenstock et al., 1982), substituting $c(t) = \sum_\beta m(\beta) a(\beta, t)$, and taking the average over the input space at a given afferent location we obtain

$$\frac{dm(\alpha, t)}{dt} = \lambda E \left\{ a(\alpha) \sum_{\beta \gamma} m(\beta) a(\beta) m(\gamma) a(\gamma) \right\} - E \left\{ a(\alpha) \sum_\beta m(\beta) a(\beta) \right\} \Theta_m. \tag{5.5}$$

Wherever it is clear, we omit the dependency of $a(\alpha)$ on t (assuming it is a stationary process). Using the same correlation function as defined above, and defining the third-order correlation $\tilde{C}(\alpha, \beta, \gamma)$ of the input activity to be $\tilde{C}(\alpha, \beta, \gamma) = E[a(\alpha) a(\beta) a(\gamma)]$, yields

$$\frac{dm(\alpha, t)}{dt} = \lambda \left\{ \sum_{\beta, \gamma} \tilde{C}(\alpha, \beta, \gamma) m(\beta) m(\gamma) - \Theta_m \sum_\beta m(\beta) C(\alpha, \beta) \right\}. \tag{5.6}$$

Using a definition for the threshold $\Theta_m \stackrel{\text{def}}{=} E[\sum_\alpha a(\alpha) m(\alpha)]^2$ (Intrator, 1990; Intrator and Cooper, 1992) and using the second-order correlation function C, Θ_m becomes $\Theta_m = \sum_{\gamma, \delta} C(\gamma, \delta) m(\gamma) m(\delta)$. Therefore, equation (5.6) becomes

$$\frac{dm(\alpha, t)}{dt} = \lambda \left\{ \sum_{\beta, \gamma} \tilde{C}(\alpha, \beta, \gamma) m(\beta) m(\gamma) - \sum_{\beta \gamma \delta} C(\alpha, \beta) C(\gamma, \delta) m(\beta) m(\gamma) m(\delta) \right\}. \tag{5.7}$$

A few important observations follow. The correlation based models use only the first- and second-order statistical information of the data (in other words, the mean and covariance matrix of input activity), whereas the BCM theory utilizes in addition the third-order statistics of the input activity. Therefore, without going into the details of the limiting behavior, this already suggests that correlation-based models are less sensitive to the input environment. Analysis shows that in many cases the correlation of activity models find the principal components of the input environment (Granger et al., 1989;

Intrator et al.

Kohonen, 1984; Miller et al., 1989; Oja, 1982; Sanger, 1989; Yuille et al., 1989). In the following section we discuss some properties of principal components in information processing.

Comparison Based on the Information Extraction Properties It now becomes relevant to ask what type of structure can be extracted from first and second moments only, and what constitutes an *interesting structure*. The first question is quite old and its answer is well known. First and second moments contain information about the principal components of the input distribution, which are those directions that can minimize L^2 error between the original data and the reconstructed data based only on the first few leading components. Another way to view principal components is to observe that they maximize the variance of the projected distribution, namely the variance of the new random variable that is the projection of the inputs onto the principal components.

Principal Components and Maximum Information Preservation Networks that extract principal components from data are numerous (e.g., Granger et al., 1989; Kammen and Yuille, 1988; Linsker, 1986; Miller et al., 1989; Oja, 1982; Sanger, 1989; Sejnowski, 1977; Yuille et al., 1989). Linsker presented the principles guiding synaptic modification in his layered network and showed that the development rule causes a cell to develop so as to maximize the variance of its output activity, subject to the constraint on the total connection strength and on each synaptic value (Linsker, 1988). Linsker then describes the connection of this rule to principal component analysis and to the principle of *maximum information preservation* taken from information theory. This principle is optimal when the goal is to accurately reconstruct the input, but is not optimal when the goal is classification. This is shown in the following simple example (figure 8.3; see also Duda and Hart, 1973, p. 212). Two clusters, each belonging to a different class, are presented. The goal is to find

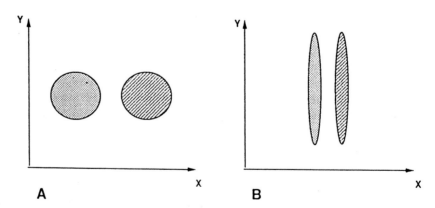

A   **B**

Figure 8.3 Principal components find useful structure in data (A) and fail when the variance of each cluster is different in each direction (B).

a single-dimensional projection that will capture the structure information in the data. In figure 8.3B different clusters have different variance in either direction, whereas in figure 8.3A the variance in both directions is equal. Clearly, the structure in the data is conveyed in the x projection; however, in the first example the variance is maximized in the y projection. This projection also minimizes the mean squared error (MSE) and is therefore superior for maximum information preservation. In the second example, because the variance of each cluster is equal in both directions, the projection that captures the most structure in the data and preserves maximum information is the x projection.

Another way to view what principal components do to the data is to observe that they define a new system of coordinates in which the covariance matrix is diagonal; namely, they eliminate the second-order correlation in the data i.e., correlation between the projections of the input data onto any two principal components. It is important to note here that this procedure does not eliminate higher order correlation in the data.

Finding Other Interesting Low-Dimensional Structure in Data This problem has recently been discussed in the context of a statistical method called projection pursuit (PP) (see Huber, 1985, for review). This method seeks structure that is exhibited by (linear) projections of the data and is therefore relevant to neural network theory, since the activity of a neuron is believed to be a function of the projection of the inputs on the vector of synaptic weights. Diaconis and Freedman (1984), have shown that for most high-dimensional clouds, most low-dimensional projections are approximately normal. This finding suggests that important information in the data is conveyed in those directions whose single-dimensional projected distribution is far from Gaussian. For example, some known measures of deviation from normality are skewness and kurtosis, which are functions of the first four moments of the distribution. These moments contain information about statistical correlations up to fourth order. Intrator (1990) has shown that a BCM neuron (given by equation 4.8) can find structure in the input distribution that exhibits deviation from normality in the form of multimodality in the projected distributions. This type of deviation, which is measured by the first three moments of the distribution, is particularly useful for finding clusters in high-dimensional data (since clusters can not be found directly in the data due to its sparsity) and is thus useful for classification or recognition tasks. Below, we give another interpretation of this projection index in light of the previous discussion.

If we assume that the retina is performing decorrelation of the inputs (Atick and Redlich, 1992) then the covariance matrix $C(\alpha, \beta)$ is diagonal (assuming that the inputs have zero mean) and so for eigen values $e(\alpha)$, equation (5.7) becomes:

$$\frac{dm(\alpha, t)}{dt} = \lambda \left\{ \sum_{\beta, \gamma} \tilde{C}(\alpha, \beta, \gamma) m(\beta) m(\gamma) - e(\alpha) m(\alpha) \sum_{\gamma} e(\gamma) m^2(\gamma) \right\}. \tag{5.8}$$

Table 8.1 Assumptions about Input Environment: Two Models

	Clothiaux et al. (1991)	Miller et al. (1989)
Normal rearing	Patterned input (correlated activity within the eye with addition of noise)	Locally correlated input from both eyes
	Correlation between eyes	No correlation between eyes
Monocular deprivation	Patterned input from the open eye	Same correlation structure as normal rearing
	Uncorrelated noise from the deprived eye	Reduced activity to the deprived eye
Strabismus	Patterned input from each eye	Locally correlated input from both eyes
	No correlation between eyes	Anticorrelations between eyes

This suggests that the BCM synaptic modification equation is performing third-order decorrelation of the inputs subject to some penalty related to the size of the weights. When the second-order statistics of the input data is not decorrelated, then the modification equation can be thought of trying to find some balance between third-order correlation and second-order correlation in the data.

5.2   Comparison Based on Assumptions About the Input Environment

In table 8.1 we summarize the different assumptions about the environment used by our group and by Miller and colleagues to model classical visual deprivation experiments. What is apparent from this comparison is the different emphasis given to between-eye and within-eye correlations in activity. To quote Mastronarde (1984):

Some of the strongest evidence on the importance of correlated firing in development comes from cases where local correlations in activity are induced by sensory stimulation; e.g., formation of binocular cells in visual cortex requires binocularly corresponding visual input to the two eyes (Hubel and Wiesel, 1965). There has been growing interest in a more restricted question: what is the role in development of the correlated activity that occurs in the spontaneous discharge?

In work we have chosen to focus on the influence of activity induced by sensory stimulation on the development of visual cortex.

5.3   Comparison Based on Network Architecture

Although it is possible that network architecture plays an important role in comparison of varuous ideas on visual cortical plasticity, in this chapter we have not done any analysis of different architecture. This will be dealt with in subsequent work.

6 CONCLUDING REMARKS

We have given a short account of the BCM theory of synaptic plasticity, including comparison with experiments in visual cortex and possible cellular and molecular basis for the fundamental modification equations. In addition, we have shown that correlation-based models and BCM theory differ in the type of structure for which they search. Correlation models include first- and second-order statistics of input correlations, while BCM modification also includes third-order statistics.

Evidence exists for a principal component type preprocessing that may be taking place in the retina; we suggest that BCM modification further preprocesses the visual inputs by reducing (extracting) third-order statistical correlations. Extracting third-order statistics from the visual environment is a natural extension and complements the extraction of second-order statistics that may be done in the retina.

Statistical theory tells us that finite-order statistics of the data is not sufficient to uniquely characterize the data distribution. However, the addition of the third-order moment adds important feature such as skewness to the description of the distribution (see Kendall and Stuart, 1977, for review), and in this case, it adds information about multimodality (Intrator, 1992). The method of principal components is sufficient for finding clusters in data when the variance of each cluster is relatively constant in all directions, since then directions that maximize the variance also maximize the information conveyed for the purpose of classification. When this is not the case, the example (figure 8.3) shows that the direction that maximizes the variance of the projection does not necessarily carry information useful for separation of the two clusters although it does find the direction that will minimize the mean squared error between the reconstructed signal and the input; this is dictated by the principle of maximum information preservation.

It is possible that the principle of maximum information preservation is useful in retinal processing, in which an order of magnitude reduction in the number of cells occurs. We suggest that this principle is not general enough to account for processing done in early visual cortex and that such statistical properties provide a convenient framework for comparison of various plasticity theories.

ACKNOWLEDGMENT

Research was supported by the Office of Naval Research, the Army Research Office, and the National Science Foundation.

REFERENCES

Artola, A., Bröcher, S., and Singer, W. (1990) Different voltage dependent thresholds for the induction of long-term depression and long-term potenation in slices of rat visual cortex. *Nature* 347:69–72.

164   Intrator et al.

Atick, J. J., and Redlich, N. (1992) What does the retina know about natural scenes. *Neural Computation* 4:196–210.

Bear, M. F., and Colman, H. (1990) Binocular competition in the control of geniculate cell size depends upon visual corical NMDA receptor activation. *Proc. Natl. Acad. Sci. USA* 87:9246–9249.

Bear, M. F., Cooper, L. N., and Eben, F. F. (1987) A Physiological basis for a theory of synapse modification. *Science* 237:42–48.

Bear, M. F., Gu, Q., Kleinschmidt, A., and Singer, W. (1990) Disruption of experience-dependent synaptic modification in the striate cortex by infusion of an NMDA receptor antagonist. *J. Neurosci.* 10:909–925.

Bear, M. F., Press, W. A., and Connors, B. W. (1992) Long-term potentiation of slices of kitten visual cortex and the effects of NMDA receptor blockade. *J. Neurophysiol.* 67:841–851.

Bienenstock, E. L., Cooper, L. N., and Munro, P. W. (1982) Theory for the development of neuron selectivity: Orientation specificity and binocular interaction in visual cortex. *J. Neurosci.* 2:32–48.

Clothiaux, E. E., Bear, M. F., and Cooper, L. N. (1991) Synaptic plasticity in visual cortex: Comparison of theory with experiment. *J. Neurophysiol.* 66:1785–1804.

Connors, B. W., and Bear, M. F. (1988) Pharmacological modulation of long term potentiation in slice of visual cortex. *Soc. Neurosci. Abstr.* 14:298.8.

Cooper, L. N., and Scofield, C. L. (1988) Mean-field theory of a neural network. *Proc. Natl. Acad. Sci. USA* 85:1973–1977.

Diaconis, P., and Freedman, D. (1984) Asymptotics of graphical projection pursuit. *Ann. Stat.* 12:793–815.

Duda, R. O., and Hart, P. E. (1973) *Pattern Classification and Scene Analysis.* John Wiley, New York.

Dudek, S. M., and Bear, M. F. (1992) Homosynaptic long-term depression in area CA1 of hippocampus and the effects on NMDA receptor blockade. *Proc. Natl. Acad. Sci. USA* 89:4363–4367.

Fox, K., Sato, H., and Daw, N. (1989) The location and function of NMDA receptors in cat and kitten visual cortex. *J. Neurosci.* 9:2443–2454.

Friedman, J. H. (1987) Exploratory projection pursuit. *J. Am. Stat. Assoc.* 82:249–266.

Granger, R., Ambrose-Ingerson, J., and Lynch, G. (1989) Derivation of encoding characteristics of layer II cerebral cortex. *J. Cog. Neurosci.* 1:61–87.

Greuel, J. M., Luhmann, H. J., and Singer, W. (1987) Evidence for a threshold in experience-dependent long-term changes of kitten visual cortex. *Dev. Brain Res.* 34:141–149.

Huang, Y. Y., Colino, A., Selig, D. K., and Malenka, R. C. (1992) The influence of prior synaptic activity on the induction of long-term potentiation. *Science* 255:730–733.

Hubel, D. H., and Wiesel, T. N. (1959) Integrative action in the cat's lateral geniculate body. *J. Physiol.* 148:574–591.

Hubel, D. H., and Wiesel, T. N. (1965) Bimocular interaction in striate contex of kittens reared with artificial squint. *J. Neurophysiol.* 28:1041–1059.

Huber, P. J. (1985) Projection pursuit (with discussion). *Ann. Stat.* 13:435–475.

Intrator, N. (1990). A neural network for feature extraction. In D. S. Touretzky and R. P. Lippmann (eds.), *Advances in Neural Information Processing Systems,* Vol. 2. Morgan Kaufmann, San Mateo, CA, pp. 719–726.

Intrator, N. (1992) Feature extraction using an unsupervised neural network. *Neural Computation* 4:98−107.

Intrator, N., and Cooper, L. N. (1992) Objective function formulation of the BCM theory of visual cortical plasticity: Statistical connections, stability conditions. *Neural Networks* 5:3−17.

Intrator, N., and Gold, J. I. (1993) Three-dimensional object recognition of gray level images: The usefulness of distinguishing features. *Neural Computation* 5:61−74.

Jampolsky, A. (1978) Unequal visual inputs and strabismus management: A comparison of human and animal strabismus. In *Symposium on Strabismus (Transactions of New Orleans Academy of Ophthalmology)*. Mosby, St. Louis, p. 358.

Kammen, D., and Yuille, A. (1988) Spontaneous symmetry-breaking energy functions and the emergence of orientation selective cortical cells. *Biol. Cybern.* 59:23−31.

Kendall, M., and Stuart, A. (1977) *The Advanced Theory of Statistics*, Vol. 1. Macmillan, New York.

Kirkwood, A., Gold, S. M. D. J. T., Aizenman, C., and Bear, M. F. (1993) Common forms of synaptic plasticity in hippocampus and neocortex in vitro. *Science*, 260:1518−1521.

Kleinschmidt, A., Bear, M. F., and Singer, W. (1987) Blockage of NMDA receptors disrupts experience-dependent plasticity of kitten striate cortex. *Science* 238:355−358.

Kohonen, T. (1984) *Self-Organization and Associative Memory*. Springer-Verlag, Berlin.

Linsker, R. (1986) From basic network principles to neural architecture (series). *Proc. Natl. Acad. Sci. USA* 83:7508−7512, 8390−8394, 8779−8783.

Linsker, R. (1988) Self-organization in a perceptual network. *IEEE Comput.* 88:105−117.

Mastronarde, D. N. (1989) Correlated firing of cat retinal ganglion cells. *Trends Neurosci.* 12:75−80.

Miller, K. D., Keller, J., and Stryker, M. P. (1989) Ocular dominance column development: Analysis and simulation. *Science* 240:605−615.

Mioche, L. and Singer, W. (1989) Chronic recordings from single sites of kitten striate cortex during experience-dependent modifications of receptive-field properties. *J. Neurophysiol.* 62:85−197.

Oja, E. (1982) A simplified neuron model as a principal component analyzer. *Math. Biol.* 15:267−273.

Press, W. A., and Bear, M. F. (1990) Effects of disinhibition on LTP induction in slices of visual cortex. *Soc. Neurosci. Abstr.* 16:348.9.

Ramoa, A. S., Paradiso, M. A., and Freeman, R. D. (1988) Blockade of intracortical inhibition in kitten striate cortex: Effects on receptive field properties and associated loss of ocular dominance plasticity. *Exp. Brain Res.* 73:285−296.

Reiter, H. O., and Stryker, M. P. (1988) Neural plasticity without action potentials: Less active inputs become dominant when kitten visual cortical cells are pharmacologically inhibited. *Proc. Natl. Acad. Sci. USA* 85:3623−3627.

Sanger, T. D. (1989) Optimal unsupervised learning in a single-layer linear feedforward neural network. *Neural Networks* 2:459−473.

Scofield, C. L., and Cooper, L. N. (1985) Development and properties of neural networks. *Contemp. Phys.* 26:125−145.

Sejnowski, T. J. (1977) Storing covariance with nonlinearly interacting neurons. *J. Math. Bio.* 4:303−321.

Singer, W. (1977) Effects of monocular deprivation on excitatory and inhibitory pathways in cat striate cortex. *Exp. Brain Res.* 134:508–518.

Yang, X., and Faber, D. S. (1991) Initial synaptic efficacy influences induction and expression of long-term changes in transmission. *Proc. Natl. Acad. Sci. USA* 88:4299–4303.

Yuille, A., Kammen, D., and Cohen, D. (1989) Quadrature and the development of orientation selective cortical cells by hebb rules. *Biol. Cybern.* 61:183–194.

Part II. Neural Networks

A Neural Model for Category Learning

(with D. L. Reilly and C. Elbaum)

Biol. Cybern. **45**, 35 (1982)

The papers of the previous section trace our attempt to relate learning rules to their physiological origins — the path from cellular events to higher level system properties. The papers in this section follow another branch of our work, the attempt to use biologically inspired architectures and learning rules to construct what are now called artificial neural networks — networks that can be realized either in hardware or as machine simulations that can accomplish various real-world tasks.

The human brain, on which such systems are loosely modeled is, of course, an enormously complex system that has evolved over hundreds of millions of years. It is hardly ideally suited for what is likely supposed to do: enable us to adapt quickly in complex changing environments and to communicate with others of our species so that we may better survive and reproduce. From an engineer's point of view this organ is somewhat of a disaster: put together from what was available,rather than from what would have been ideal — too large, too expensive, and much too emotional.

In constructing artificial neural networks we do not attempt to duplicate precisely what is done in the brain. Rather we attempt to use what appear to be brain-like mechanisms and procedures to accomplish useful tasks, the kind of tasks that animals accomplish easily.

Anyone who has thought about this problem is almost immediately struck by the fact that, as mentioned in the very first paper of this series, certain tasks such as recognition of patterns are extremely easy for human children or adults to perform, while extremely difficult for current computers. Other tasks such as rapid execution of instructions, and complex logical and arithmetic operations are done at blinding speed by computers and, if done at all by human beings, done haltingly and after much educational effort.

It seems evident that if one could combine these two capacities one would indeed have machines of the future. In a sense artificial neural networks are an attempt to create this animal-like capacity in machine structures.

The following paper with Charles Elbaum and Doug Reilly (longtime comrades in arms) is our attempt to solve two classic problems in learning theory. One is the problem of constructing perceptron-like systems that can discriminate nonlinear boundaries and the other is the credit assignment problem when one is dealing with multiperceptron systems — two problems mentioned as difficult in the influential book of Minsky and Papert. This is done using what are now known as radial basis functions and the RCE learning rule (officially Restricted Coulomb Energy, but somewhat reminiscent of the initials of the authors). When these ideas were introduced one of the things we had in mind was that actual neural systems can learn very rapidly and that large numbers of neurons are available to accomplish tasks. We made no attempt to be efficient in the use of neurons.

Biol. Cybern. 45, 35–41 (1982)

Biological
Cybernetics
© Springer-Verlag 1982

A Neural Model for Category Learning*

Douglas L. Reilly, Leon N. Cooper, and Charles Elbaum

Center for Neural Science and Department of Physics, Brown University, Providence, RI 02912, USA

Abstract. We present a general neural model for supervised learning of pattern categories which can resolve pattern classes separated by nonlinear, essentially arbitrary boundaries. The concept of a pattern class develops from storing in memory a limited number of class elements (prototypes). Associated with each prototype is a modifiable scalar weighting factor (λ) which effectively defines the threshold for categorization of an input with the class of the given prototype. Learning involves (1) commitment of prototypes to memory and (2) adjustment of the various λ factors to eliminate classification errors. In tests, the model ably defined classification boundaries that largely separated complicated pattern regions. We discuss the role which divisive inhibition might play in a possible implementation of the model by a network of neurons.

I. Introduction

A common concern of neural models has been the problem of relating the function of complex systems of neurons to what is known of individual neurons and their interconnections. In this paper we discuss a neural model that displays a form of learning manifested in human behavior: supervised learning of pattern categories. The terms pattern and event are used here synonymously to refer to a state of the environment that is characterized by a set of measurements. A category of patterns is a set of patterns in the same class. Their members may yield "roughly" the same value for some measurement (or collection of measurements) made on them (e.g. with reference to some feature set). However, one can imagine a category resulting from an association between a collection of

very unlike events and a particular system response (e.g., calling "a" and "A" by the sound of the first letter in the alphabet). In this case, the criterion defining the category is the association itself.

There are several difficulties in the problem of pattern classification that we address here. A given pattern class appears in the primary sensory neurons in a vast variety of manifestations. Consider all of the recognizable distortions of the Arabic numeral "three". All of these must be classified as "three" and at the same time be distinguished from other classes (1, 2, 4, etc.) and all of their distortions. Therefore, the problem of classification involves a separation of "different" classes as well as a grouping together of all distorted members of the same class. Our model is capable of making the separation as well as the grouping with a simple instruction procedure that seems at least roughly comparable to that employed in human learning.

There is a growing body of research dealing with the general problem of learning in an adaptive system composed of neuron like elements. Early work in this field introduced the notion of correlation matrix memories, showing how it was possible for a system to learn associations between pairs of input and output vectors ($\mathbf{x}^i$, $\mathbf{y}^i$) (Kohonen, 1972). Category learning has frequently been viewed as learning an association between $\mathbf{y}^i$ and a set of noisy versions of $\mathbf{x}^i$. Models for such concept formation have been proposed which make use of varying amounts of interaction with an external "teacher" (e.g., Amari, 1977; Grossberg, 1978; Barto et al., 1981; Bobrowski, 1982). Among the various approaches in such systems, learning rules incrementally adjust elements of some weight vector $\mathbf{w}$ whose inner product with the input $\mathbf{x}$ is an important contributing factor to the output of the system.

In our approach pattern classification is accomplished through prototype formation. Evidence from psychological experiments suggests that learning of pattern classes might involve abstraction of a pro-

* This work was supported in part by the Alfred P. Sloan Foundation and the Ittleson Foundation, Inc.

0340-1200/82/0045/0035/$01.40

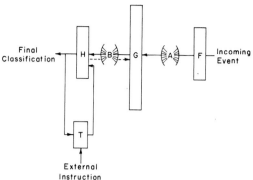

Fig. 1. Architecture of the model. Shown are coding neurons (*F*), prototype cells (*G*), classification cells (*H*), mapping (*A*) from *F* to *G*, mapping (*B*) from *G* to *H*, and the external instructor (*T*). Arrows mark information flow

Table 1. Classification of system responses for various values of α and Q as defined by (2)

Classification	α	Q	$\mathbf{h}$
Correct	1	0	$\mathbf{h}^c$
Unidentified	0	0	0
Incorrect	0	1	$\mathbf{h}^r, r \neq c$
Confusion	0	$\geqq 2$	$\sum_{r \neq c}^{Q} \mathbf{h}^r$
	1	$\geqq 1$	$\mathbf{h}^c + \sum_{r \neq c}^{Q} \mathbf{h}^r$

totype to represent a category of stimuli (e.g., Posner and Keele, 1968, 1970; Franks and Bransford, 1971). Some knowledge of class variance must also be learned. A closely related argument holds that categories are learned by retaining in memory examples of each class (e.g., Brooks, 1978; Medin and Schaffer, 1978). In pattern recognition theory, the technique of nearest neighbor classification is effectively an exemplar learning scheme (Cover and Hart, 1967; Duda and Hart, 1973). The focus of algorithms for such training has been to find and store the example set of minimal size which can guarantee performance within some acceptable error rate.

Here no distinction is made between the single (prototype) and multiple exemplar theory. Any class member stored in memory will be referred to as a prototype for that class. We will discuss learning in a system of neurons and, in particular, a model for prototype formation and development in a class of distributed memory neural networks.

II. Overview

In the architecture we consider, afferents from coding neurons, *F*, project onto prototype cells, *G*, which in turn synapse with classification neurons, *H* (see Fig. 1). Each class of events will be represented by the activity of a unique *H* neuron. An input event is coded by a vector of firing rates (**f**) in the *F* bank. If it causes activity in an *H* cell, it is classified as belonging to the category associated with that cell.

We define four possible network responses to an input pattern. Let $\mathbf{f}(c)$ represent an incoming pattern belonging to the cth class of events, and let **h** be the vector of output firing rates of the *H* neurons. Further, let $\mathbf{h}^x$ be defined as a vector with components

$$(\mathbf{h}^x)_j = \delta_{xj}. \tag{1}$$

The response **h** can be written, in general, as

$$\mathbf{h} = \alpha \mathbf{h}^c + \sum_{r \neq c}^{Q} \mathbf{h}^r. \tag{2}$$

If $\alpha = 1$ and $Q = 0$, then the system has correctly classified the input pattern. A response characterized by $\alpha = 1$ and $Q \geqq 1$, or $\alpha = 0$ and $Q \geqq 2$, we refer to as confusion, since the system is unable to decide upon any of several pattern classifications. The case where $\alpha = 0$ and $Q = 1$ is an outright incorrect response. When $\alpha = 0$ and $Q = 0$, no categorization has been made and the pattern is unidentified. Table 1 summarizes the responses.

The synaptic connections between *G* and *F* are represented in the mapping *A*. In our learning models, a prototype for a class is "imprinted" on the synapses between a *G* cell and the *F* set, thus becoming the most effective stimulus for that cell. For any given class, there may be more than one prototype; each will correspond to a different G_i. The mapping *B*, between cell groups *G* and *H*, develops so that the subset of *G* cells which can cause a given *H* cell to fire consists of prototypes representing the same class. A sufficient stimulus for an *H* cell to fire will be supraliminal activity in any member of its corresponding *G* cell subset.

The *H* set of neurons (and indirectly, *G*) has an additional source of input, that diagrammed by the block *T* in Fig. 1. Through *T*, an external supervisor can correct the network classification responses. The specific form of the mapping *B*, along with some aspects of *A*, will develop as a result of interaction with input patterns and with *T*. Essentially, *T* can cause the commitment of a *G* cell to a prototype and the strengthening of the association between this *G* cell and the proper classification cell. We assume synaptic modification as the vehicle for these network changes. One can imagine a variety of ways in which synaptic changes can result in cell coupling between the *G* and *H* sets. For example, simple Hebbian modification can produce the desired association if the particular *H* cell was receiving concurrent stimulation from *T*. The only requirement of this procedure is that cell commitment never involve a previously committed cell. For simplicity, we further assume that

(1) cell commitment is rapid (i.e., occurring within the duration of event presentation)

(2) only one cell is committed to any one prototype.

In the mapping A, an element A_{ij} represents the logical synapse between G_i and F_j; i.e., it summarizes the total effectiveness of neuron F_j in firing G_i. In accordance with a distributed memory model studied by Anderson and by Cooper, among others (e.g., Anderson, 1970, 1972; Kohonen, 1972, 1977; Cooper, 1973; Nass and Cooper, 1975; Anderson and Cooper, 1978), we take the firing rate of G_i (call it g_i) to be a weighted sum of the firing rates of the F neurons (f_j), gated by some threshold function

$$g_i = \Theta\left(\sum_j A_{ij} f_j\right), \tag{3}$$

where

$$\Theta(x) = 0 \quad \text{if} \quad x \leqq \theta$$
$$= x - b \quad \text{if} \quad x > \theta. \tag{4}$$

Given a prototype $\mathbf{P}(c)$ representing a class c of inputs, the equality

$$A_{ij} = P_j(c), \quad \text{all } j \tag{5}$$

establishes a correspondence between the ith G cell and a particular class of patterns c. The synapse vector of G_i takes on the value of the prototype.

Each prototype cell has a "region of influence" in the input space of events. It is defined as the set of input patterns that satisfies the threshold condition for cell firing. For convenience, assume input events to be normalized ($\mathbf{f} \cdot \mathbf{f} = 1$). The region of influence defined by cell G_i with threshold θ is the intersection of the surface of a unit hypersphere with a cone of angular width γ,

$$\gamma = \cos^{-1} \theta, \tag{6}$$

where γ is the angle between $\mathbf{P}(c)$ and an input $\mathbf{f}$ at threshold.

A class of patterns defines a region or set of regions in the pattern space of input events. Class regions corresponding to different pattern categories are assumed to be strictly disjoint. A priori, we choose not to restrict the complexity that the shape of class boundaries may display. To identify the class of an input event, the neural network must characterize and learn the arrangement of class regions. Our model develops by itself a set of prototypes whose influence regions map out the areas belonging to different categories in the pattern space without prior information of what these areas are. One approach to such prototype organization will be discussed. Several others, differing

in their methods of cell modification and in their assumptions about interaction between G cells, or equivalently, between prototypes stored in memory will be discussed elsewhere.

III. Prototype Formation and Development

For the present, we continue the assumption of normalized input patterns ($\mathbf{f} \cdot \mathbf{f} = 1$). Each committed prototype cell has a synapse vector of the form (for the ith cell),

$$\mathbf{A}^i = \lambda_i \mathbf{p}^i, \tag{7}$$

where $\mathbf{p}^i$ is a normalized ($\mathbf{p}^i \cdot \mathbf{p}^i = 1$) prototype vector and $\lambda_i > 1$. The vector $\mathbf{p}^i$ corresponds to some previously seen input pattern whose presentation failed to excite the H cell of the appropriate class. Modification to prototype cell synapses is governed by the following conditions.

1. New Classification

If $\mathbf{f}(c)$ is presented and

$$\mathbf{h} \cdot \mathbf{h}^c = 0 \tag{8}$$

i.e., the H cell for the cth class does not fire, then a new G cell (call it G_k) is committed to $\mathbf{f}(c)$ and the synapse between G_k and H_c is assigned strength 1. The synapses of G_k with F are modified according to

$$A_{kj} \to P_{kj} = \lambda_0 f_j, \tag{9}$$

where $\lambda_0 > 1$.

2. Confusion

If presentation of $\mathbf{f}(c)$ causes firing rate activity in some H_w where $w \neq c$, then this results in a signal from the T channel to reduce the λ factors of each currently active G cell associated with H_w. The quantity λ is diminished until the response of the cell to $\mathbf{f}(c)$ lies at threshold. If G_r is such a unit, then

$$\lambda_r \to \lambda'_r$$

such that

$$\lambda'_r \mathbf{p}^r \cdot \mathbf{f}(c) = 1. \tag{10}$$

For convenience, we have taken $\theta = 1$.

These two rules for prototype acquisition and modification will enable the network to learn the geography of the pattern classes.

In an untrained network, all G cells are uncommitted. The strengths of the synapses between G and H are all zero or some arbitrarily small number. When a pattern $\mathbf{f}(c)$ is presented to this system, no H cell responds above threshold. Information from the T element enters the system, identifying the correct class of the input. A single G cell is committed to $\mathbf{f}(c)$ as a prototype for that class and, simultaneously, the synapse between this G cell and H_c is set equal to 1. Since this input represents the first example of any pattern class, we can let $c = 1$. If the same pattern were to be presented again to the system, the response of the G cell would be

$$\lambda_0 \mathbf{p}^1(c) \cdot \mathbf{f}(c) = \lambda_0 > 1. \tag{11}$$

The output signal, λ_0, from this G cell would cause H_c to fire.

Suppose a second pattern $\mathbf{f}^2(c')$ is presented to the system. Assume $c' = c$. If

$$\lambda_0 \mathbf{p}^1(c) \cdot \mathbf{f}^2(c) > 1 \tag{12}$$

then H_c will fire and the pattern will be correctly classified. Thus no change occurs. If

$$\lambda_0 \mathbf{p}^1(c) \cdot \mathbf{f}^2(c) < 1 \tag{13}$$

then $\mathbf{f}^2(c)$ will be committed to a new G cell [prototype $\mathbf{P}^2(c)$] and the synapse between this G cell and H_c will be set equal to 1. In this way, a class can be characterized by more than one prototype.

Consider the situation in which $c' \neq c$. Whether or not the existing prototype cell fires past threshold, there will be no active H cells of the class of $\mathbf{f}^2$. The subsequent T signal causes a new prototype cell to be committed to $\mathbf{f}^2$, along with the setting of the synaptic connection between this G cell and a new H cell. If, in addition,

$$\lambda_0 \mathbf{p}^1(c) \cdot \mathbf{f}^2(c') > 1 \tag{14}$$

then λ_0 is reduced to λ_1 such that

$$\lambda_1 \mathbf{p}^1(c) \cdot \mathbf{f}^2(c') = 1. \tag{15}$$

As the system learns, the λ factors associated with any active incorrect class prototypes will be reduced, leaving only the correct H cell to respond to the pattern.

The strategy of this network learning scheme is made clearer by considering the problem geometrically. The size of the influence region of a prototype cell is directly proportional to the magnitude, λ, of the prototype. Class territories in the space of events are defined by covering them with the overlapping influence fields of a set of prototypes drawn from class samples. Should the influence region of a given prototype extend into the territory of some differing class to the point of incorrectly classifying or confusing a member of that class, the λ factor of the prototype is reduced until its region of influence just excludes the disputed pattern. Prototype modification only decreases λ factors. Influence fields of existing prototype cells are never enlarged in an effort to include (classify) an event, since for many of these elements, even slightly larger regions of influence have previously resulted in incorrect identifications. Consequently, a pattern that is excluded from the influence regions of all existing prototypes for its class is an occasion for commitment of a new G cell, with the pattern assuming the role of the new prototype.

Note that the prototype cells in memory are completely decoupled in that there are no mutual inhibitory or excitatory interactions among them. In the network's classification response, there is no vote counting among prototypes. The activity of a single prototype cell counts as heavily as the possibly concerted activity of a set of prototype cells, all specific to some other class.

This model was tested in computer simulations using a design set of input patterns. The patterns were vectors randomly generated in a normalized three dimensional pattern space. Samples were constrained to lie on the top half of a unit sphere ($z > 0$) and represented two classes of patterns labelled A and B. In one arrangement the A region was chosen as a spherical cap centered on the z axis and ringed by the B region, a surrounding band on the sphere's surface. The projection of this design is a pair of concentric circles on the $x - y$ plane. A second geometry pictured the A and B regions as separated by a sinusoidal boundary on the sphere's surface.

Patterns arrived in cycles (trials). A trial consisted in presentation of 200 novel A vectors and 200 novel B vectors, randomly distributed with respect to class. After some number of trials, the distribution of prototypes was graphed together with the effective boundaries between the A and B classes. In this space these boundaries are paths along a spherical surface. They are displayed by graphing projections on the $x - y$ plane.

The graphs in Figs. 2–4 illustrate the performance of the model in resolving class boundaries for the two different geometries. In Fig. 2, the class regions were separated by a gap, i.e., an area of pattern space containing no input patterns. When the angular width of this gap is less than $(\lambda_0)^{-1}$, there can develop prototypes for each class which have influence regions extending right up to the boundary with the other class. Consequently the gap is claimed for both pattern categories. Should a pattern from this region be selected as an input, its contested status (response confusion by the model) would cause the influence region of one or the other class to withdraw from a portion of the gap.

Note that in practice, the model need not develop a single decision surface separating pattern classes. In Figs. 3 and 4, there is no gap between the hypothetical

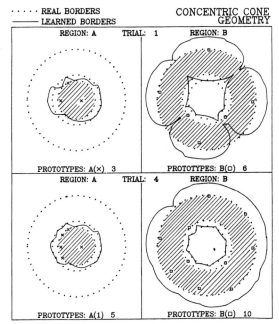

Fig. 2. Prototype regions for the concentric cone geometry with a gap. Region A: shaded area within innermost (first) dotted circle. Region B: shaded annulus defined by second and third dotted circles. Projections of prototype vectors on sphere's surface are plotted as crosses (A) and squares (B). Pictured are graphs of prototype boundaries (solid lines) as they appear after the first and fourth trials. Total numbers of prototypes are given below each graph

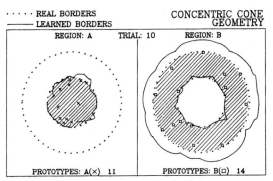

Fig. 3. Prototype regions for concentric cone geometry with no gap. Region A: shaded area within innermost (first) dotted circle. Region B: shaded annulus bounded by first ans second dotted circles. Prototype boundaries (solid lines) pictured after 10 trials

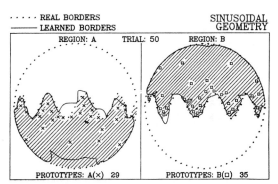

Fig. 4. Prototype regions for sinusoidally separated pattern zones. Region A: bottom scalloped semicircle (shaded area within dotted lines). Region B: upper scalloped semicircle (shaded area within dotted lines). Prototype boundaries (solid lines) pictured after 50 trials

category regions. A single border separates them, yet in the model, this border is approximated by a double line. If either the prototype or the classification cells were coupled by some mutual interaction (e.g., inhibition), this double border could, in places, be replaced by a single boundary. The nature of such a line would be a function of the specific form of the interaction. Excepting such coupling, it is only in the limit of studying a very large number of design samples that the double line category borders could be expected to merge into a single curve lying along the actual class boundary. The response to any input located in an area where the double lines extend beyond each other will be confusion. Patterns falling in regions from which both prototype generated boundaries have retreated will be identified with neither pattern class.

In the case of prototypes committed to inputs near a class border, the initially large influence regions can result in many incorrect or confused responses until the magnitude of the prototype is appropriately scaled. This creates a somewhat unstable learning process which does not converge smoothly to the final pattern

region mapping[1]. Nonetheless, it is clear that this model can resolve pattern classes of arbitrary complexity.

IV. Possible Neural Realization

It is likely that category learning is conducted in different areas of the brain by a variety of cell assemblies. Indeed, one can imagine a number of specific networks of neurons that could implement the important features of our model. We consider a possible

---

1 There are a variety of means of improving this. For example, the magnitude of the initial λ_0 may decrease in time so that prototypes committed late in the process leave smaller initial regions of influence. Alternatively, each new prototype may be automatically tested against each existing prototype (treated as an incoming pattern)

neural substrate whose function could relate to one aspect of prototype development in the model.

It has been calculated that under certain conditions, activity in inhibitory fibers whose synapses are located on or near the cell body can have a divisive effect on the somatic membrane potential (Blomfield, 1974). Inhibitory current across these synapses is postulated to increase membrane conductance, thus shunting off a fraction of the summed post-synaptic potential arriving at the cell from its dendrites. The result is to scale the cell output by some multiplicative factor. Inhibitory synapses occurring amidst the excitatory ones further out along the dendritic spines and shafts would have their normal subtractive effect on cell firing rate. Divisive or shunting inhibition has also been considered elsewhere (Poggio, 1981; Kogh et al., 1982).

Cells have been found in different areas of the brain with significant numbers of synapses on or near the perikaryon that are predominantly characterized by flat vesicles and/or symmetric membrane differentiation (e.g., Davis et al., 1979; White et al., 1980). Such morphology is widely considered to be indicative of inhibitory function. By contrast, synapses located on the dendritic shafts and spines of such cells are both excitatory and inhibitory. This anatomy is consistent with that assumed for divisive inhibition. Indeed, other investigators have observed scaling of cell response as a function of inhibitory transmitter released into the soma (Rose, 1977) and under certain conditions of visual stimulus presentation (Dean et al., 1980).

Divisive inhibition is a candidate mechanism for implementing the λ factor scaling of prototype cell response assumed in the prototype learning model. There its principal effect is to provide for a modifiable cell threshold. The distinction which the model makes between prototype commitment and changes in λ is in the same spirit as the functional distinction which Blomfield's model suggests for synapses. The initial commitment of a prototype might involve changes in the spiny synapses and those in general distal to the soma. Such modification could occur according to any of a number of schemes previously suggested (e.g., correlation learning). Cell tuning, on the other hand, would be controlled largely by adjustments to inhibitory synapses proximal to the soma. Long term changes in somatic membrane conductance might even result from very different inhibitory effects (e.g., chemical deposition within the cell body due to active inhibitory afferents).

The processes of modification to sites distal and proximal to the soma might be mutually interactive in a number of ways. For example, one can imagine the somatic membrane conductance of a cell increased to such a point that the cell rarely fires. (In the model, such was the case for a cell committed to a prototype near a class boundary). Lack of post-synaptic response in conjunction with pre-synaptic activity might cause, as some have suggested (Cooper et al., 1979), the distal synapses of such a cell to lose the information of the stored prototype. This could free the cell to become committed to a new preferred pattern. At the same time, distal modification could be an ongoing process which performs some type of averaging over those inputs able to cause cell firing (the Hebbian requirement). If the environment presented a sequence of smoothly varying events of sufficient duration, the distal modification might cause the cell to "follow" the inputs. In this way, the preferred pattern of the cell could change with only a minimum of change in the degree of cell tuning.

V. Conclusion

Category learning plays an important role in a broad range of mental activity, from learning sequences of task oriented sensori-motor controls to very complex problems in conceptualization. As such it is probably implemented in different ways by different cell assemblies throughout the brain. A successful model for category learning should be consistent with the general features of this host of sub-networks and with their perhaps locally unique architectures. We have presented one such model with properties thought to be characteristic of the neural system as a whole. Among these are: coding of information by neuron firing rates, synaptic transmission of information from cell to cell, excitatory and inhibitory interactions among cells, distributed memory stored over the entire set of synaptic junctions and initially unspecified cell interconnections that are modified by the history of the system's experiences. This model suggests that it is possible to construct plausible neuron networks that incorporate these features and that can display a powerful ability to learn to identify and distinguish categories of events. In a separate publication we will report on the application of this and a related model to a practical problem in categorization (Reilly et al., 1982). The model learning systems were trained to classify examples of unconstrained handwritten numerals. By detecting only very simple information about patterns, the system achieved a high degree of accuracy (approximately 98%) in tests against patterns not viewed during training.

Acknowledgements. We would like to express our appreciation to our colleagues at the Brown University Center for Neural Science for their interest and helpful advice. In particular, we thank Messrs. Paul Munro, Michael Paradiso and Christopher Scofield for several useful discussions.

References

Amari, S.I.: Neural theory of association and concept-formation. Biol. Cybern. **26**, 175–185 (1977)

Anderson, J.A.: Two models for memory organization using interacting traces. Math. Biosci. **8**, 137–160 (1970)

Anderson, J.A.: A simple neural network generating an interactive memory. Math. Biosci. **14**, 197–220 (1972)

Anderson, J.A., Cooper, L.N.: Les modeles mathematiques de l'organization biologique de la memoire. Pluriscience 168–175, Encyclopaedia Universalis, Paris (1978)

Barto, A.G., Sutton, R.S., Brouwer, P.S.: Associative search network: a reinforcement learning associative memory. Biol. Cybern. **40**, 201–211 (1981)

Blomfield, S.: Arithmetical operations performed by nerve cells. Brain Res. **69**, 115–124 (1974)

Bobrowski, L.: Rules for forming receptive fields of formal neurons during unsupervised learning processes. Biol. Cybern. **43**, 23–28 (1982)

Brooks, L.: Non-analytical concept formation and memory for instances. In: Cognition and categorization, pp. 169–211, Rosch, E., Lloyd, B. (eds.). Hillsdale, N.J.: Lawrence Erlbaum Associates 1978

Cooper, L.N.: A possible organization of animal memory and learning. In: Proceedings of the nobel Symposium on collective properties of physical systems, Vol. 24, pp. 252–264, Lundquist, B., Lundquist, S. (eds.). London, New York: Academic Press 1973

Cooper, L.N., Liberman, F., Oja, E.: A theory for the acquisition and loss of neuron specificity in visual cortex. Biol. Cybern. **33**, 9–28 (1979)

Cover, T.M., Hart, P.E.: Nearest neighbor pattern classification. IEEE Trans. Inform. Theor. **13**, 21–27 (1967)

Davis, T.L., Sterling, P.: Microcircuitry of cat visual cortex: classification of neurons in layer IV of area 17, and identification of the patterns of lateral geniculate input. J. Comp. Neur. **188**, 599–628 (1979)

Dean, A.F., Hess, R.F., Tolhurst, D.J.: Divisive inhibition involved in directional selectivity. J. Physiol. **308**, 84p–85p (1980)

Duda, R.O., Hart, P.E.: Pattern classification and scene analysis. New York: Wiley 1973

Franks, J.J., Bransford, J.D.: Abstraction of visual patterns. J. Exp. Psychol. **90**, 65–74 (1971)

Grossberg, S.: Adaptive pattern classification and universal recoding. II. Feedback, expectation, olfaction, illusions. Biol. Cybern. **23**, 187–202 (1976)

Kogh, C., Poggio, T., Torre, V.: Retino-ganglion cells: a functional interpretation of dendritic morphology. Philos. Trans. R. Soc. (to be published)

Kohonen, T.: Correlation matrix memories. IEEE Trans. Comput. **21**, 353–359 (1972)

Kohonen, T.: Associative memory – a system-theoretical approach. Berlin, Heidelberg, New York: Springer 1977

Medin, D.L., Schaffer, M.M.: Context theory of classification learning. Psychol. Rev. **85**, 207–238 (1978)

Nass, M.M., Cooper, L.N.: A theory for the development of feature detecting cells in visual cortex. Biol. Cybern. **19**, 1–18 (1975)

Poggio, T.: A theory of synaptic interactions. In: Theoretical approaches in neurobiology, pp. 28–38, Reichardt, W., Poggio, T. (eds.). London: MIT Press 1981

Posner, M.I., Keele, S.W.: On the genesis of abstract ideas. J. Exp. Psychol. **77**, 353–363 (1968)

Posner, M.I., Keele, S.W.: Retention of abstract ideas. J. Exp. Psychol. **83**, 304–308 (1970)

Reilly, D.L., Cooper, L.N., Elbaum, C.: An application of two learning systems to pattern recognition: handwritten characters (to be published)

Rose, D.: On the arithmetical operation performed by inhibitory synapses onto the neuronal soma. Exp. Brain Res. **28**, 221–223 (1977)

White, E.L., Rock, M.P.: Three-dimensional aspects and synaptic relationships of a Golgi-impregnated spiny stellate cell reconstructed from serial thin sections. J. Neurocytol. **9**, 615–636 (1980)

Received: March 12, 1982

Dr. D.L. Reilly
Center for Neural Science and
Department of Physics
Brown University
Providence, RI 02912
USA

Pattern Class Degeneracy in an Unrestricted Storage Density Memory

(with C. L. Scofield, D. L. Reilly, and C. Elbaum)
Neural Information Processing System, ed. D. Z. Anderson
(American Institute of Physics, USA, 1987) p. 674

Learning System Architectures Composed of Multiple Learning Modules

(with D. L. Reilly, C. Scofield, and C. Elbaum)
IEEE International Conference on Neural Networks **2**, 495 (1987)

In constructing neural network architectures much work is done testing them in various complex situations. It's easy to construct pattern classes that can be separated without any great difficulty. (In fact most real-world data sets can be trivially separated perhaps 60–80% with almost any algorithm – the reason any network seems so promising initially. It is the last few % that distinguish one classification method from another.) The task is to construct systems that identify patterns of speech, writing, and various other complex real-world data that human beings distinguish and classify so easily.

This is somewhat of an engineering problem. It is necessary to test various neural architectures against large real-world data sets. And some of these data sets are extremely useful for commercial applications. Thus, in the mid-seventies, we separated the work that could more easily be carried out in a commercial environment from the academic work and formed what is probably the earliest neural network company: Nestor. Created in 1975 as a limited partnership with four founding partners, Charles Elbaum, Simon Heifetz, Herbert Meeker, and myself, (our original headquarters were Stonington, Connecticut — half-way between Providence and N. Y. by a somewhat flexible yardstick) and incorporated and gone public in 1983.

Various neural network architectures as well as practical applications of neural network systems have been developed at Nestor or in cooperative efforts between Nestor and academic groups.

The first of the following two papers (referring to results of a later paper in this series), describes the evolution of the radial basis function/RCE algorithm in situations where complete class separation is not possible so that some probabilistic measure must be introduced. The second paper describes what we call a multineural network architecture. This has various advantages in increasing resolution and minimizing the problem known as the "curse of dimensionality".

The second paper represents one of our earliest attempts to construct what we now call hybrid neural networks. The basic idea is to use confusion class information, described more fully in the paper "Hybrid Neural Network Architectures: Equilibrium Systems that Pay Attention". The first networks classify roughly, then pass these rough classifications to special networks that separate the confused classes. Thus, for example, the first network might distinguish between A and U or V. The UV confusion is then fed to a second network which is specialized to separate U from V. Obviously, looking at the lower portion of the figure is a very useful separator for UV but not useful for UO. A key feature is that the various networks need not be introduced in advance, but are constructed as part of the learning process. The evolution of the field, it seems to me, will be in thedirection of constructing such interacting feed forward, feed back networks. These eventually will include rules and context.

Neural Information Processing System, ed. D. Z. Anderson. © AIP, 1982

PATTERN CLASS DEGENERACY IN AN UNRESTRICTED STORAGE DENSITY MEMORY

Christopher L. Scofield, Douglas L. Reilly, Charles Elbaum,
and Leon N. Cooper

*Nestor, Inc., 1 Richmond Square, Providence,
Rhode Island 02906, USA*

ABSTRACT

The study of distributed memory systems has produced a number of models which work well in limited domains. However, until recently, the application of such systems to real-world problems has been difficult because of storage limitations, and their inherent architectural (and for serial simulation, computational) complexity. Recent development of memories with unrestricted storage capacity and economical feedforward architectures has opened the way to the application of such systems to complex pattern recognition problems. However, such problems are sometimes underspecified by the features which describe the environment is often non-specific. We will review current work on high density memory systems and their network implementations. We will discuss a general learning algorithm for such high density memories and review its application to separate point sets. Finally, we will introduce an distributions of non-separate point sets.

INTRODUCTION

Information storage in distributed content addressable memories has long been the topic of intense study. Early research focused on the development of correlation matrix memories.[1,2,3,4] Workers in the field found that memories no larger than the number of dimensions of the input space. Further storage beyond this number caused the system to give an incorrect output for a memorized input.

Recent work on distributed memory systems has focused on single layer, recurrent networks. Hopfield[5,6] introduced a method for the analysis of settling of activity in recurrent networks. This method defined the network as a dynamical system for which a global function called the 'energy' (actually a Liapunov function for the autonomous system describing the state of the network) could be defined. Hopfield showed that flow in state space is always toward the fixed points of the dynamical system if the matrix of recurrent connections satisfies certain conditions. With this property, Hopfield was able to define the fixed points as the sites of memories of network activity.

Like its forerunners, the Hopfield network is limited in storage capacity. Empirical study of the system found that for randomly chosen memories, storage capacity was limited to $m \leq 0.15N$, where m is the number of memories that could be accurately recalled, and N is the dimensionality of the network (this has since been improved to $m \leq N$,[7,8]). The degradation of memory recall with increased storage density is directly related to the proliferation in the state space of unwanted local minima which serve as basins of flow.

UNRESTRICTED STORAGE DENSITY MEMORIES

Bachman et al.[9] have studied another relaxation system similar in some respects to the Hopfield network. However, in contrast to Hopfield, they have focused on defining a dynamical system in which the locations of the minima are explicitly known.

In particular, they have chosen a system with a Liapunov function given by

$$E = -1/L \sum_j Q_j |\mu - x_j|^{-L}, \tag{1}$$

where E is the total 'energy' of the network, $\mu(0)$ is a vector describing the initial network activity caused by a test pattern, and x_j, the site of the jth memory, for m memories in R^N. L is a parameter related to the network size. Then $\mu(0)$ relaxes to $\mu(T) = x_j$ for some memory j according to

$$\mu = -\sum_j Q_j |\mu - x_j|^{-(L+2)}(\mu - x_j). \tag{2}$$

This system is isomorphic to the classical electrostatic potential between a positive (unit) test charge, and negative charges Q_j at the site x_j (for a 3-dimensional input space, and $L = 1$). The N-dimensional Coulomb energy function then defines exactly m basins of attraction to the fixed points located at the charge sites x_j. It can be shown that convergence to the closest distinct memory is guaranteed, *independent of the number of stored memories m*, for proper choice of N and L.[9,10]

Equation (1) shows that each cell receives feedback from the network in the form of a scalar

$$\sum_j Q_j |\mu - x_j|^{-L}, \tag{3}$$

Importantly, this quantity is the same for all cells; it is as if a single virtual cell was computing the distance in activity space between the current state and stored states. The result of the computation is then broadcast to all of the cells in the network. A 2-layer feedforward network implementing such a system has been described elswhere.[10]

The connectivity for this architecture is of order $m \cdot N$, where m is the number of stored memories and N is the dimensionality of layer 1. This is significant since the addition of a new memory $m' = m + 1$ will change the connectivity by the addition of $N + 1$ connections, whereas in the Hopfield network, addition of a new memory requires the addition of $2N + 1$ connections.

An equilibrium feedforward network with similar properties has been under investigation for some time.[11] This model does not employ a relaxation procedure, and thus was not originally framed in the language of Liapunov functions. However, it is possible to define a similar system if we identify the locations of the 'prototypes' of this model as the locations in state space of potentials which statisfy the following conditions

$$
\begin{aligned}
E_j &= -Q_j/R_0 \quad \text{for } |\mu - x_j| < \lambda_j \\
&= 0 \qquad\quad \text{for } |\mu - x_j| > \lambda_j.
\end{aligned}
\tag{4}
$$

where R_0 is a constant.

This form of potential is often referred to as the 'square-well' potential. This potential may be viewed as a limit of the N-dimensional Coulomb potential, in which the $1/R$ ($L = 1$) well is replaced with a square well (for which $L \gg 1$). Equation (4) describes an energy landscape which consists of plateaus of zero potential outside of wells with flat, zero slope basins. Since the landscape has only flat regions separated by discontinuous boundaries, the state of the network is always at equilibrium, and relaxation does not occur. For this reason, this system has been called an equilibrium model. This model, also referred to as the Restricted Coulomb Energy (RCE)[14] model, shares the property of unrestricted storage density.

LEARNING IN HIGH DENSITY MEMORIES

A simple learning algorithm for the placement of the wells has been described in detail elsewhere.[11,12]

Reilly *et al.* have employed a 3-layer feedforward network (Figure 1) which allows the generalization of a content addressable memory to a pattern classification memory. Because the locations of the minima are explicitly known in the equilibrium model, it is possible to dynamically program the energy function for an arbitrary energy landscape. This allows the construction of geographies of basins associated with the classes constituting the pattern environment. Rapid learning of complex, nonlinear, disjoint, class regions is possible by this method.[12,13]

LEARNING NON-SEPARABLE CLASS REGIONS

Previous studies have focused on the acquisition of the geography and boundaries of nonlinearly separable point sets. However, a method by which such high density models can acquire the probability distributions of non-separable sets has not been described.

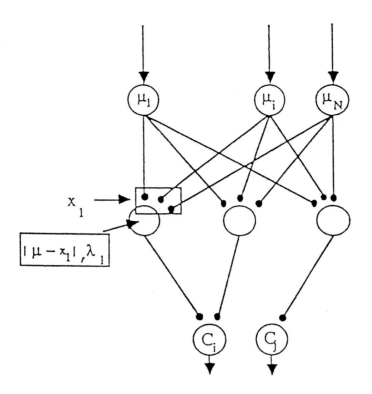

Figure 1. 3-layer feedforward network. Cell i computes the quantity $|\mu - x_i|$ and compares to internal threshold λ_i.

Non-separable sets are defined as point sets in the state space of a system which are labelled with multiple class affiliations. This can occur because the input space has not carried all of the features in the pattern environment, or because the pattern set itself is not separable. Points may be degenarate with respect to the explicit features of the space, however they may have different probability distributions within the environment. This structure in the environment is important information for the identification of patterns by such memories in the presence of feature space degeneracies.

We now describe one possible mechanism for the acquisition of the probability distribution of non-separable points. It is assumed that all points in some region R of the state space of the network are the site of events $\mu(0, C_i)$ which are examples of pattern classes $C = \{C_1, \ldots, C_M\}$. A basin of attraction, $x_k(C_i)$, defined by Equation (4), is placed at each site $\mu(0, C_i)$ unless

$$|\mu(0, C_i) - x_j(C_i)| < R_0, \tag{5}$$

that is, unless a memory at x_j (of the class C_i) already contains $\mu(0, C_i)$. The initial values of Q_0 and R_0 at $x_k(C_i)$ are a constant for all sites x_j. Thus as events of the classes $C_1, \ldots, C_M$ occur at a particular site in R, multiple wells are placed at this location.

If a well $x_j(C_i)$ correctly covers an event $\mu(0, C_i)$, then the charge at that site (which defines the depth of the well) is incremented by a constant amount ΔQ_0. In this manner, the region R is covered with wells of all classes $\{C_1, \ldots, C_M\}$, with the depth of well $x_j(C_i)$ proportional to the frequency of occurence of C_i at x_j.

The architecture of this network is exactly the same as that already described. As before, this network acquires a new cell for each well placed in the energy landscape. Thus we are able to describe the meaning of the wells that overlap as the competition by multiple cells in layer 2 in firing for the pattern of activity in the input layer.

APPLICATIONS

This system has been applied to a problem in the area of risk assessment in mortgage lending. The input space consisted of feature detectors with continuous firing rates proportional to the values of 23 variables in the application for a mortgage. For this set of features, a significant portion of the space was non-separable.

Figures 2a and 2b illustrate the probability distributions of high and low risk applications for two of the features. It is clear that in this 2-dimensional subspace, the region of high and low risk are non-separable but have different distributions.

Figure 3 depicts the probability distributions acquired by the system for this 2-dimensional subspace. In this image, circle radius is proportional to the degree of risk: small circles are regions of low risk, and large circles are regions of high risk.

Of particular interest is the clear clustering of high and low risk regions in the 2-d map. Note that the regions are in fact nonlinearly separable.

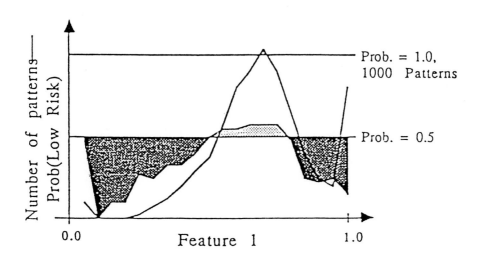

Figure 2a. Probability distribution for high and low risk patterns for feature 1.

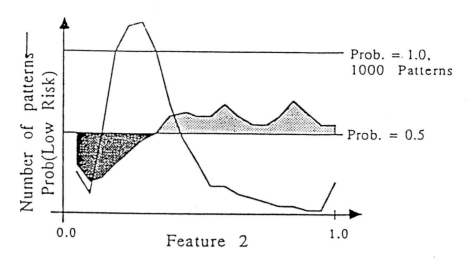

Figure 2b. Probability distribution for high and low risk patterns for feature 2.

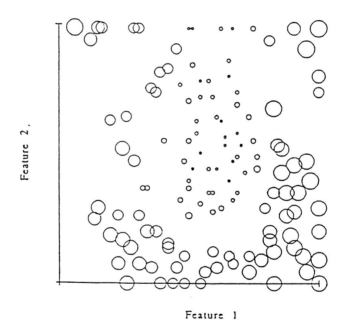

Feature 1

Figure 3. Probability distribution for low and high risk. Small circles indicate low risk regions and large circles indicate high risk regions.

DISCUSSION

We have presented a simple method for the acquisition of probability distributions in non-separable point sets. This method generates an energy landscape of potential wells with depths that are proportional to the local probability density of the classes of patterns in the environment. These well depths set the probability of firing of class cells in a 3-layer feedforward network.

Application of this method to a problem in risk assessment has shown that even completely non-separable subspaces may be modeled with surprising accuracy. This method improves pattern classification in such problems with little additional computational burden.

This algorithm has been run in conjunction with the method described by Reilly et al.[11] for separable regions. This combined system is able to generate nonlinear decision surfaces between the separable zones, and approximate the probability distributions of the non-separable zones in a seemless manner. Further discussion of this system will appear in future reports.

Current work is focused on the development of a more general method for modeling the scale of variations in the distributions. Sensitivity to this scale suggests that the transition from separable to non-separable regions is smooth and should not be handled with a 'hard' threshold.

ACKNOWLEDGEMENTS

We would like to thank Ed Collins and Sushmito Ghosh for their significant contributions to this work through the development of the mortgage risk assessment application.

REFERENCES

[1] Anderson, J. A.: *A simple neural network generating an interactive memory*, Math. Biosci. **14**, 197–220 (1972).

[2] Cooper, L. N.: *A possible organization of animal memory and learning*, in Proceedings of the Nobel Symposium on Collective Properties of Physical Systems, eds. Lundquist, B. and Lundquist, S. **24**, 252–264, London, New York: Academic Press (1973).

[3] Kohonen, T.: *Correlation matrix memories*, IEEE Trans. Comp. **21**, 353-359 (1972).

[4] Kohonen, T.: *Associative memory - a system-theoretical approach*, Berlin, Heidelberg, New York: Springer (1977).

[5] Hopfield, J. J.: *Neural networks and physical systems with emergent collective computational abilities*, Proc. Natl. Acad. Sci. USA **79**, 2554–2558 (April 1982).

[6] Hopfield, J. J.: *Neurons with graded response have collective computational properties like those of two-state neurons*, Proc. Natl. Acad. Sci. USA **81**, 2088-3092 (May 1984).

[7] Hopfield, J. J., Feinstein, D. I., and Palmer, R. G.: *'Unlearning' has a stabilizing effect in collective memories*, Nature **304**, 158-159 (July 1983).

[8] Potter, T. W.: Ph. D. Dissertation in advanced technology, SUNY Binghampton, (unpublished).

[9] Bachmann, C. M., Cooper, L. N., Dembo, A., and Zeitouni, O.: *A relaxation model for memory with high density storage*, to be published in Proc. Natl. Acad. Sci. USA.

[10] Dembo, A. and Zeitouni, O.: ARO Technical Report, Brown University, Center for Neural Science, Providence, RI (1987); also submitted to Phys. Rev. A.

[11] Reilly, D. L., Cooper, L. N., and Elbaum, C.: *A neural model for category learning*, Bio. Cybern. **45**, 35–41 (1982).

[12] Reilly, D. L., Scofield, C., Elbaum, C., and Copper, L. N.: *Learning system architectures composed of multiple learning modules*, to be appear in Proceedings of the First International Conference on Neural Networks (1987).

[13] Rimey, R., Gouin, P., Scofield, C., and Reilly, D. L.: *Real-time 3-D object classification using a learning system*, Intelligent Robots and Computer Vision, Proc. SPIE **726** (1986).

[14] Reilly, D. L., Scofield, C. L., Elbaum, C., and Cooper, L. N.: *Neural networks with low connectivity and unrestricted memory storage density*, to be published.

Learning System Architectures Composed of Multiple Learning Modules

Douglas L. Reilly, Christopher Scofield
Charles Elbaum, Leon N. Cooper
Nestor Inc., One Richmond Square, Providence, RI 02906
Center for Neural Science and Department of Physics,
Brown University, Providence, RI 02912

Published in the Proceedings of the IEEE First Annual International Conference on Neural Networks, June 1987.

IEEE Catalog #87TH0191-7

Learning System Architectures
Composed of
Multiple Learning Modules

Douglas L. Reilly, Christopher Scofield,
Charles Elbaum, Leon N. Cooper

Nestor Inc., One Richmond Square, Providence, RI 02906

Center for Neural Science and Department of Physics,
Brown University, Providence, RI 02912

1. Introduction

The Nestor Learning System™ (NLS-1000) is a supervised learning system for pattern classification which has evolved from study of a general neural model. The system is constructed from several interacting learning modules. We begin by reviewing the construction and operation of a single module. We then discuss how an assembly of modules can learn to classify patterns as a function of their individual internal states. The system can be seen as viewing an incoming event in the context of different representation spaces, each such space corresponding to a module. The system is able to learn which of the several representation (or "code") spaces should be used to discriminate a particular pattern class. The multi-module architecture is well-suited to processing multi-sensor input. Additionally, the structure of the system allows modules to be dynamically incorporated without the need for retraining. Other advantageous features of this multi-module system is its ability to train rapidly and to achieve economies of memory representation.

2. A Single Learning Module

The Nestor Adaptive Module™ is a single-module learning system [1] [3] [4]. Operating in a supervised mode, this module can learn the correct classification(s) of information presented to it. The learning module is constructed from numerous neuron-like assemblies, called cells. Associated with a cell is a set of weights on its input lines, and a threshold on its output line. Both the weights and the threshold are adjustable. The function of a cell (also referred to as a prototype cell) is to produce a response (i.e. to "fire") when the input lines contain a set of signals which are similar to the values of the weights stored on the cell's input lines. The threshold of the prototype cell determines how similar the input signal must be to the weights (the prototype vector) for the cell to fire. All prototype cells see a copy of the module's input signals.

Each prototype cell feeds forward through a fixed strength connection to a particular classification cell. A classification cell implements a logical OR function to indicate the presence of any one firing prototype among all the prototype cells connected to it. Figure 1 presents a diagram of the module's network.

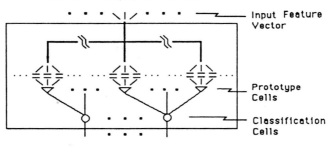

Figure 1

The output of a module is a "response type" (T) and a "class list" (**C**). The response type can be either "identified" ("i"), confused ("c") or "unidentified" ("u"). The class list is a list of the classes represented among firing classification cells.

Memory in a module is stored in the set of weighted connections between each prototype cell and the input signals, the connections between the prototype cells and the classification cells and, finally, in the thresholds associated with the prototype cells. Importantly, the network in a module is not "preconfigured" but dynamically constructs during training. In particular, the module self-organizes to add prototype and classification cells (with the appropriate interconnections) to accomplish the required category learning. The input weight vector and cell threshold of a prototype cell determine that cell's "influence region" in the pattern space defined by the input signals to the module. The influence region of a prototype cell is the neighborhood of events in the pattern space that will cause the prototype cell to fire. Through modifications to the input weight vector and cell threshold, the learning algorithm creates and adjusts prototype cell influence regions to cover class territories in the module's pattern space. The simple learning algorithm outlined in [1] allows a module to learn arbitrarily shaped class territories that may be disjoint and/or nonlinearly separable in the pattern space.

In a simulation of category learning in a two-dimensional pattern space, each prototype cell in the module has two real-valued inputs and the firing equation of a prototype cell implements a simple Cartesian distance metric in the pattern space. During training, class-labelled example patterns, randomly selected from two hypothetical class regions α and β, were presented to the system for identification. Figure 2 shows the influence regions of prototypes that developed in the module as a result of training. The influence fields of prototypes for class α are represented by horizontally striped circles; those for class β, by vertically striped circles.

II-496

277

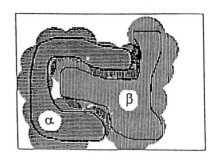

Figure 2

3. Multiple Module Architectures

The ability of learning modules to classify patterns will depend upon the distribution of pattern classes in the module's pattern space (also referred to as the "code" space of the module.) One can imagine a system composed of a set of such modules, each coupled to a different set of input signals describing a given pattern. The properties of the class distributions will then vary from module to module. In particular, the information input to a given module may be well suited for making some class discriminations and not others. Thus, a given module can specialize in learning to discriminate certain types of patterns.

Learning system architectures composed of multiple learning modules can take advantage of the properties of assemblies of such modules to produce more interesting and useful behavior [2]. We will show that a system composed of a set of such learning modules can learn to select which of its several modules is able to uniquely identify a pattern class. We will also demonstrate that through interactions between modules such a system is able to correlate the responses of several modules to identify a pattern class, even when no individual module has learned to be an "expert" at unambiguously identifying the pattern class in question.

In general, a learning system architecture is a specification of the arrangement of modules composing the learning system. Each module is assigned a priority in the system, though different modules may share the same priority. (Modules sharing priority "n" are said to operate at level "n" in the system.) Each module receives as input a different representation or a different subset of the entire input signal.

Let the output of the jth module be (T_j, C^j). We define a higher order mechanism, the Class Selection Device (CSD), that combines module responses (T_j, C^j) as a function of module priority p_j, into a final system output, (T_0, C^o). The CSD can operate in a number of ways to determine the system response. We outline one method below.

Consider a learning system composed of two adaptive modules, U_1 and U_2, with corresponding code spaces C_1 and C_2. Assume that the code spaces C_1

C_2 have class regions as shown below. Clearly U_1 will be able to learn to classify α and β. Since α and β are overlapping or "degenerate" territories in C_2, U_2 will have no possibility of reliably distinguishing them.

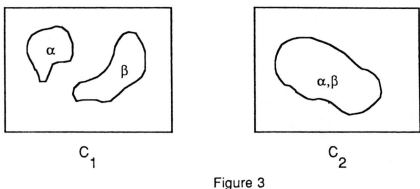

$$C_1 \qquad\qquad\qquad C_2$$

Figure 3

On the basis of computer simulations, we will show that, regardless of module priorities, the system response to incoming examples of α and β will be solely determined by the module response of U_1.

In the first case, assume that U_1 has higher priority than U_2. This means that the CSD will "poll" U_1 before U_2 to determine a system response and that prototype commitment will occur preferentially in U_1 as opposed to U_2. However, module responses that are unambiguous (T_j="i") have a natural dominance over module responses that are ambiguous (T_j="c" or "u")). Thus, if after polling U_1, the system response is ambiguous, the CSD polls U_2.

As examples of α and β are presented to the system, prototypes are committed preferentially in U_1 (because of its higher priority) and a separating mapping develops in U_1 which reliably classifies α and β instances.

Figure 4 shows the final configuration of system memory when trained on this problem. In simulation, samples of classes α and β were presented to the system by randomly selecting points from the α and β regions of both modules. Training feedback consisted of providing the correct classification for comparison with the system classification.

$$U_1 \qquad\qquad U_2$$

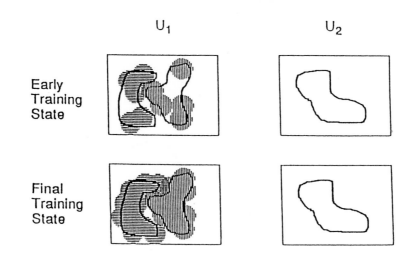

Early
Training
State

Final
Training
State

$$p_1 > p_2$$

Figure 4

Now assume that U_2 has higher priority. This means that prototypes are initially committed in U_2, and U_2 then determines the system response. However, the degeneracy of α,β class distributions causes U_2 prototype influence fields to be rapidly reduced in size to correct module classification errors. (Typically, prototypes with thresholds below some minimum size are "purged" from the memory after some period of time.) Various mechanisms can be introduced to slow the rate of prototype re-commitment in those regions where repeated prototype commitment and shrinkage has occurred. In this locality of the U_2 code space, the effective priority of U_2 for prototype commitment is therefore reduced. Consequently, the CSD begins to deposit prototypes for α and β in U_1. As training in U_1 causes a separating mapping to develop there, its responses become largely unambiguous and correct. As the rate of prototype commitment in U_2 slows, the responses of U_2 become chronically more "unidentified". The dominance of unambiguous responses over such ambiguous ones reverses the initial higher priority of U_2 in driving the system response. Thus, the response of module U_1, despite its initially lower priority, comes to determine the system response for α and β by virtue of its ability to develop a separating mapping. (Note that this is only necessarily true for classes α and β as we have drawn them. For other class distributions which are well-separated in U_2, U_2 begins with an initially higher priority and maintains it.)

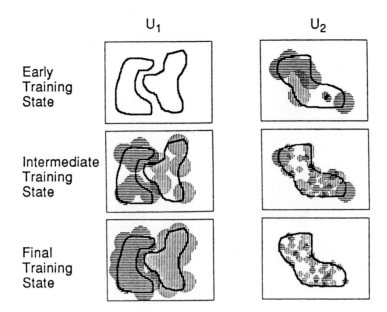

$$p_2 > p_1$$

Figure 5

Finally, consider the case where modules U_1 and U_2 share equivalent priority (exist at the same level in the system). As always, a system response that is confused or incorrect will cause the influence fields of prototypes that are incorrectly firing to be reduced in size. If this results in the system response being updated to "unidentified", then for equi-priority modules, prototype commitment will be attempted for both. If, on the other hand, this updating leaves the system response "identified" and correct, then no prototypes are committed in either module. As training progresses, U_1 will tend towards more unambiguous correct responses while U_2 will increasingly respond with "unidentified". Consequently, U_1 will come to solely determine the system response. As that response becomes unambiguous and correct, prototype commitment in both modules ceases.

II-500

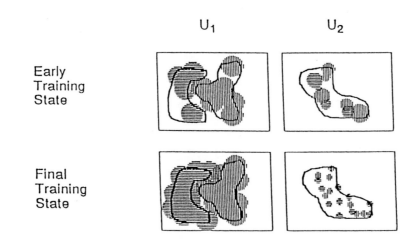

$$p_1 = p_2$$

Figure 6

Thus, regardless of initial module priorities, the system learns to adjust the effective priorities of modules so that the module that is "naturally" suited for a particular class identification determines the system response.

What happens when no module is uniquely capable of identifying a pattern class? In this case, the system combines the ambiguous responses of several modules to correctly identify the pattern.

Consider the case of a pattern class γ that is degenerate with α in unit U_1 and degenerate with β in U_2. Separable mappings for γ cannot develop in either module. Instead, these class degenerate "confusion zones" become layered with prototypes that, when responding to an input pattern from such a region, yield the class degenerate set as module output. (A number of possible mechanisms can implement such coverings; one example is multi-class prototypes.) Consequently, in U_1, the response of the module to an example of γ is the output T_1="C" and $C^1 = \{\alpha,\gamma\}$. In U_2, the response to the same example γ of will be T_2="C" and $C^2 = \{\beta,\gamma\}$. Using a simple vote counting procedure, the CSD can correlate the C^1 and C^2 responses to produce the system response $C^0 = \{\gamma\}$.

In this description of a CSD mechanism, the system is biased to find units that uniquely separate classes before attempting correlations. In other variations, correlations are attempted first, and modules with unambiguous responses are sought later.

4. Features of Multi-Module Learning

An obvious application of this multi-module architecture is in the processing of data from multiple sensors. A system of this sort can learn to base its classification decisions upon data from the appropriate sensor (e.g., infrared camera vs optical range camera) or to correlate partial information from several sensors to reach a conclusion.

An attractive implementation feature of such multi-module systems is their ability to be configured to learn classes with an economy of memory storage. This is achievable by systems in which upper level (high priority) modules are coupled to low resolution representations and in which resolution increases with decreasing unit priorities. Deeper level units in which representations are less memory efficient are only accessed for those classes not identifiable by more efficient upper level representations.

A novel feature of such a system is its ability to incorporate additional modules dynamically, without the need for re-training of the system from "scratch." Modules initially absent can be considered to have essentially very low priority numbers. When these units are added later, their priorities, like that of any module, are candidates for adjustment by the system in an effort to resolve its ambiguous responses.

Finally, we have observed experimentally that training time in a system can be shorter when it implements a multi-module approach as opposed to a single large pattern space. This is especially true for discrimination of those pattern classes which differ in only a few of all the measured features.

5. Applications

We have applied multiple module learning systems in experiments on a number of real world problems. One such system, discussed elsewhere [4], deals with a problem in the machine vision domain, the recognition of "blocks world" objects. Two experimental systems were constructed. One system used three units (modules), one for each of the three largest contours extracted by a region segmentation algorithm. The second system combined layers of units in a hierarchical manner where, at deeper levels of the hierarchy, the pattern is described in more detail. The second system was tested on a simple four object problem and achieved a 99% correct recognition rate. This system was also tested on a ten object problem in which the unique class of some instances of some objects could not be determined from the available information. The system achieved a high confusion rate (29%) and a low error rate (2%). We are now applying such systems to tasks involving detection and classification of three-dimensional man-made structures in aerial images and industrial assembly verification.

II-502

Acknowledgements: The authors wish to acknowledge Raymond Rimey for his helpful review of this manuscript and Linda Nunez for her support of the network computer simulations.

References

[1] L.N. Cooper, C. Elbaum, D.L. Reilly, "Self Organizing General Pattern Class Separator and Identifier," U.S. Patent No. 4,326,259.

[2] L.N. Cooper, C. Elbaum, D.L. Reilly, C. Scofield, "Nonlinear Pattern Class Separation In a Multiple Unit Adaptive System," To be published.

[3] D.L. Reilly, L.N. Cooper, C. Elbaum, "A Neural Model for Category Learning," Biol. Cybem. 45, pp. 35-41 (1982).

[4] R. Rimey, P. Gouin, C. Scofield, D.L. Reilly, "Real-Time 3-D Object Classification Using a Learning System," Intelligent Robots and Computer Vision, Proc. SPIE Vol. 726, 1986.

A Relaxation Model for Memory with High Storage Density

(with C. M. Bachmann, A. Dembo, and O. Zeitouni)

Proc. Natl. Acad. Sci. **84**, 7529 (1987)

Coulomb Potential Learning

(with M. P. Perrone)

The Handbook of Brain Theory and Neural Networks (MIT Press, 1993)

During the seventies working on neural networks was a rather lonely occupation. Learning systems, in general, had fallen out of favor; the artificial intelligence community was convinced that what was required to reproduce the output of human mental activity were more and more elaborate programs such as those that painted walls and stacked blocks. The idea that rule-based systems must be combined with data driven or learning systems was distinctly out of favor.

Two events changed this point of view dramatically. The first was the discovery (or the rediscovery) of the back-propagation network that provided another solution to the credit assignment problem. The second was John Hopfield's paper introducing a spin-glass analogy for neural networks. Hopfield's paper, in particular, spoke to physicists since it was a language they could understand and (as important) provided a delicious toy with which they could play.

Overnight, it seemed, the intellectual climate changed from one in which neural networks, previously believed to be useless, suddenly would do everything. A new ideological party formed around the dogma that, in effect, a learning system without any rules whatsoever should learn everything.

I remember in talks I gave during that period warning against this excess of romanticism, pointing out that there was no biological precedent, that cortex separated from sensory processing did not function very effectively. These warnings were sometimes accompanied by a slide of Georges de la Tour's painting, "The fortune teller". "*Hold on to your wallets,*" I warned.

A problem with Hopfield's model, the subject of a great deal of attention by physicists born again into neural networks, was what became known as the problem of spurious memories. The network would often relax to memories not specifically put in if the memories stored exceeded some number. To those of us familiar with the use of correlation memories for equilibrium recall, this didn't seem too surprising since correlation memories were imperfect in their recall of stored items unless the vectors representing them were orthogonal. The first of the following papers came out of a seminar discussion of this problem at Brown. We show that by improving the storage matrix, one can store as many items as one wishes with no spurious memories. The second paper is a more recent summary of this and related results and their relation to the RCE network.

Proc. Natl. Acad. Sci. USA
Vol. 84, pp. 7529–7531, November 1987
Biophysics

A relaxation model for memory with high storage density

(neural networks/Hopfield model)

CHARLES M. BACHMANN, LEON N COOPER, AMIR DEMBO, AND OFER ZEITOUNI

Department of Physics, Applied Mathematics and Center for Neural Science, Brown University, Providence, RI 02912

Contributed by Leon N Cooper, June 22, 1987

ABSTRACT We present a relaxation model for memory based on a generalized coulomb potential. The model has arbitrarily large storage capacity and, in addition, well-defined basins of attraction about stored memory states. The model is compared with the Hopfield relaxation model.

Equilibrium associative and distributed memories that are content addressable and can recall stored memories more or less imperfectly have been known and studied for years (1–5). Concomitantly, relaxation models have been the subject of much exploration (6). In 1982, Hopfield (7) introduced a relaxation model of memory storage and retrieval that incorporates simultaneously a distributed memory correlation matrix and a relaxation process from a given input to an equilibrium state. Although learning procedures can be included, the model has not emphasized these. Among the problems of this model are poor recall of stored memories when the number of stored items exceeds some percentage of the number of involved neurons.

The correlation matrix originally employed by Hopfield has relatively weak recall properties when employed as an equilibrium distributed memory; it gives perfect recall only when the inputs are orthogonal. When the inputs are not orthogonal, one can still achieve perfect recall by some orthogonal modification procedure such as Widrow–Hoff (8), or what Kohonen calls an *optimal associative mapping* (9). Such procedures work if the number of stored memories is equal to or smaller than the dimension of the system (the number of input synapses on each neuron). A procedure for storing as many memories as desired for a given dimension has also been discussed (10). In this procedure items can be stored at arbitrary points with variable regions of influence on a hypersphere.

In this paper we present a general method for the construction of a relaxation memory in which an arbitrary number of items can be stored. The essence of the problem is to define a function whose minima lie at designated points, corresponding to the items to be stored, and to show that these are the only minima of the function. Then an appropriate relaxation procedure is defined, so that any entering pattern relaxes to one of the stored items.

Hopfield's Model and Some Improvements

In the Hopfield model (7), neurons are binary-valued threshold units and are completely interconnected, with the strength of the connections given by a correlation matrix formed from the memory states to be stored in the system:

$$w_{ij} = \sum_{s=1}^{m} \mu_i^s \mu_j^s, \tag{1}$$

where $\mu_i = \pm 1$ and w_{ij} represents the connection strength

between μ_i and μ_j. Input states are relaxed to local minima of a Liapunov function,

$$\xi = -\frac{1}{2} \sum_{i,j} w_{ij} \mu_i \mu_j \tag{2}$$

by random, asynchronous updating of the neurons in the layer according to:

$$\mu_i \to 2\theta \left(\sum_{j=1}^{N} w_{ij} \mu_j \right) - 1. \tag{3}$$

In its original form, the Hopfield model functions poorly as a categorizer when $(m/N) \lesssim 0.1$, where m = the number of stored states and N = the number of neurons. Given the limitations of the original model, improvements have been sought. "Unlearning," an approach first tried by Hopfield *et al.* (11), employs the relaxation of random states to a stable state (often spurious attractors); a correlation matrix is formed from the relaxed state, and then an amount proportional to this is subtracted from the original matrix:

$$w_{ij} \to w_{ij} - \alpha \mu_i^{\text{relaxed}} \mu_j^{\text{relaxed}}. \tag{4}$$

With "unlearning," the number of stored states that can be correctly recalled approaches N and error correction is improved, but falls to zero as $m \to N$ (12).

Recently, an interesting variation of Hopfield's unlearning has been studied by Potter (12). The algorithm is a hybrid combining elements of Hopfield's unlearning with a modification reminiscent of the Widrow–Hoff algorithm (8):

$$w_{ij} \to w_{ij} - \alpha(\mu_i^{\text{target}} - \mu_i^{\text{relaxed}})\mu_j^{\text{input}}(\mu_j^{\text{input}} + 1). \tag{5}$$

The symmetry of the synaptic matrix is preserved by making the same modification to w_{ji} each time a modification is performed on the element w_{ij}. In simulations for which all of the input states at a radius of one Hamming unit from each stored state were used for the modification procedure, a radius of attraction of one Hamming unit was observed for m just below N. Above N, the radius of attraction and the percentage of stable stored states decreases. In ref. 13, it has been shown that Potter's algorithm may be viewed as an *"effective orthogonalization"* of the input with respect to the nonlinear relaxation process; a more complete discussion of Potter's algorithm is given there.

High-Density Storage Model

In what follows we present a general method for the construction of a high storage-density neural memory. We define a function with an arbitrary number of minima that lie at preassigned points and define an appropriate relaxation procedure. Let $\vec{x}^1, \ldots, \vec{x}^m$ be a set of m arbitrary distinct memories in R^N. The "energy" function we will use is:

$$\xi = -\frac{1}{L} \sum_{i=1}^{m} Q_i |\vec{\mu} - \vec{x}^i|^{-L} \tag{6}$$

where we assume throughout that $N \geq 3$, $L \geq (N - 2)$, and $Q_i > 0$ and use $|\cdots|$ to denote the Euclidean distance. Note

7529

that for $L = 1$, $N = 3$, ξ is the electrostatic potential induced by fixed particles with charges $-Q_i$. If $Q_i > 0_j$ this energy function possesses global minima at $\vec{x}^1, \ldots, \vec{x}^m$ (where $\xi(\vec{x}^i) = -\infty$) and has no local minima except at these points. A rigorous proof is presented in Dembo and Zeitouni (14) together with the complete characterization of functions having this property.

As a relaxation procedure, we can choose any dynamical system for which ξ is strictly decreasing. In this instance, the theory of dynamical systems guarantees that for almost any initial data, the trajectory of the system converges to one of the desired points $\vec{x}^1, \ldots, \vec{x}^m$. However, to give concrete results and to further exploit the resemblance to electrostatics, consider the relaxation:

$$\dot{\vec{\mu}} = \vec{E}_{\vec{\mu}} \triangleq -\sum_{i=1}^m Q_i |\vec{\mu} - \vec{x}^i|^{-(L+2)} (\vec{\mu} - \vec{x}^i) \qquad [7]$$

where for $N = 3$, $L = 1$, Eq. 7 describes the motion of a positive test particle in the electrostatic field $\vec{E}_{\vec{\mu}}$ generated by the negative fixed charges $-Q_1, \ldots, -Q_m$ at $\vec{x}^1, \ldots, \vec{x}^m$.

Because the field $\vec{E}_{\vec{\mu}}$ is just minus the gradient of ξ, it is clear that along trajectories of Eq. 7, $d\xi/dt \le 0$, with equality only at the fixed points of Eq. 7, which are exactly the stationary points of ξ.

Therefore, using Eq. 7 as the relaxation procedure, we can conclude that entering at any $\vec{\mu}(0)$, the system converges to a stationary point of ξ. The space of inputs is partitioned into m domains of attraction, each one corresponding to a different memory, and the boundaries (a set of measure zero), on which $\vec{\mu}(0)$ will converge to a saddle point of ξ.

We can now explain why $\xi_{\vec{\mu}}$ has no spurious local minima, at least for $L = 1$, $N = 3$, using elementary physical arguments. Suppose ξ has a spurious local minimum at $\vec{y} \ne \vec{x}^1, \ldots, \vec{x}^m$, then in a small neighborhood of $\vec{y}$ that does not include any of the $\vec{x}^i$, the field $\vec{E}_{\vec{\mu}}$ points towards $\vec{y}$. Thus, on any closed surface in that neighborhood, the integral of the normal inward component of $\vec{E}_{\vec{\mu}}$ is positive. However, this integral is just the total charge included inside the surface, *which is zero*. Thus we arrive at a contradiction, so $\vec{y}$ cannot be a local minimum.

We now have a relaxation procedure, such that almost any $\vec{\mu}(0)$ is attracted by one of the $\vec{x}^i$, but we have not yet specified the shapes of the basins of attraction. By varying the charges Q_i, we can enlarge one basin of attraction at the expense of the others (and vice versa).

Even when all of the Q_i are equal, the position of the $\vec{x}^i$ might cause $\vec{\mu}(0)$ not to converge to the closest memory, as emphasized in the example in Fig. 1. However, let $r = min_{1 \le i \ne j \le m} |\vec{x}^i - \vec{x}^j|$ be the minimal distance between any two memories; then, if $|\vec{\mu}(0) - \vec{x}^i| \le r/(1 + 3^{1/k})$, it can be shown that $\vec{\mu}(0)$ will converge to $\vec{x}^i$, provided that $k \triangleq (L + 1)/(N + 1) \ge 1$. Thus, if the memories are densely packed in a hypersphere, by choosing k large enough (i.e., enlarging the parameter L), convergence to the closest memory for any "interesting" input, that is an input $\vec{\mu}(0)$ with a distinctive closest memory, is guaranteed.

The detailed proof of the above property is given in ref. 14. It is based on bounding the number of $\vec{x}^j$, $j \ne i$, in a hypersphere of radius R ($R \ge r$) around $\vec{x}^i$, by $[2(R/r) + 1]^N$, then bounding the magnitude of the field induced by any $\vec{x}^j$, $j \ne i$, on the boundary of such a hypersphere by $(R - |\vec{\mu}(0) - \vec{x}^i|)^{-(L+1)}$, and finally integrating to show that for $|\vec{\mu}(0) - \vec{x}^i| \le \theta r/(1 + 3^{1/k})$, with $\theta < 1$, the convergence of $\vec{\mu}(0)$ to $\vec{x}^i$ is within finite time T, which behaves like θ^{L+2} for $L >> 1$ and $\theta < 1$ (fixed). Intuitively the reason for this behavior is the short-range nature of the fields used in Eq. 7. Because of this, we also expect extremely low convergence rate for inputs $\vec{\mu}(0)$ far away from *all* of the $\vec{x}^i$.

The radial nature of these fields suggests a way to overcome this difficulty, that is to increase the convergence rate

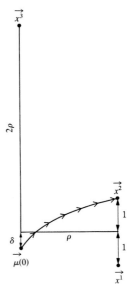

FIG. 1.   Model of memory convergence when $\vec{\mu}(0)$ is closer to $\vec{x}^1$ but converges to $\vec{x}^2$, due to the existence of $\vec{x}^3$ (assuming arbitrary distance $\rho >> 1$ and arbitrary distance $\delta << 1$). $\vec{\mu}(0)$ is the input state; $\vec{x}^1$, $\vec{x}^2$, and $\vec{x}^3$ are stored memories.

from points very far away without disturbing all of the aforementioned desirable properties of the model. Assume that we know in advance that all of the $\vec{x}^i$ lie inside some large hypersphere S around the origin. Then, at any point $\vec{\mu}$ outside S, the field $\vec{E}_{\vec{\mu}}$ has a positive projection radially into S. By adding a long-range force to $\vec{E}_{\vec{\mu}}$, effective only outside of S, we can hasten the movement towards S, from points far away, without creating additional minima inside of S. As an example the force ($-\vec{\mu}$ for $\vec{\mu} \notin S$; 0 for $\vec{\mu} \in S$) will pull any test input $\vec{\mu}(0)$ to the boundary of S within the small finite time $T \approx 1/|S|$, and from then on the system will behave inside S according to the original field $\vec{E}_{\vec{\mu}}$.

Up to this point, our derivations have been for a continuous system, but from it, we can deduce a discrete system. We shall do this mainly for a clearer comparison between our high-density memory model and the discrete version of Hopfield's model. Before continuing in that direction, note that our continuous system has *unlimited storage capacity* unlike Hopfield's continuous system (15), which, like his discrete model, has limited capacity.

For the discrete system, assume that the $\vec{x}^i$ are composed of elements ± 1 and replace the Euclidean distance in Eq. 6 with the normalized Hamming distance $|\vec{\mu}^1 - \vec{\mu}^2| \triangleq (1/N)\Sigma_{j=1}^N |\mu_j^1 - \mu_j^2|$. This places the vectors $\vec{x}^i$ on the unit hypersphere.

The relaxation process for the discrete system will be of the type defined in Hopfield's model in Eq. 3. Choose at random a component to be updated (that is, a neighbor $\vec{\mu}'$ of $\vec{\mu}$ such that $|\vec{\mu}' - \vec{\mu}| = 2/N$), calculate the "energy" difference, $\delta\xi = \xi(\vec{\mu}') - \xi(\vec{\mu})$, and only if $\delta\xi < 0$, change this component, that is:

$$\mu_i \to \mu_{i'}\theta[\xi(\vec{\mu}') - \xi(\vec{\mu})], \qquad [8]$$

where $\xi(\vec{\mu})$ is the potential energy in Eq. 6 and i is the component to be updated. Because there is a finite number of possible $\vec{\mu}$ vectors (2^N), convergence in finite time is guaranteed.

This relaxation procedure is rigid because the movement is limited to points with components ± 1. Therefore, although the local minima of $\xi(\vec{\mu})$ defined in Eq. **6** are only at the desired points $\vec{x}^i$, the relaxation may get stuck at some $\vec{\mu}$ which is not a stationary point of $\xi(\vec{\mu})$. However, the short-range behavior of the potential $\xi(\vec{\mu})$, unlike the long-range behavior of the quadratic potential used by Hopfield (Eq. **2**), gives rise to results similar to those we have quoted for the continuous model (Eq. **7**).

Specifically, let the stored memories $\vec{x}^1, \ldots, \vec{x}^m$ be separated from one another by having at least ρN different components ($0 < \rho \leq 1/2$ and ρ fixed), and let $\vec{\mu}(0)$ agree with at least one $\vec{x}^i$ with at most $\theta \rho N$ errors between them ($0 \leq \theta < 1/2$, with θ fixed), then $\vec{\mu}(0)$ converges monotonically to that $\vec{x}^i$ by the relaxation procedure given in Eq. **8**.

This result holds independently of m, provided that N is large enough (typically, $N \rho \ln[(1 - \theta)/\theta] \geq 1$) and L is chosen so that $N/L \leq \ln[(1 - \theta)/\theta]$. The proof is constructed by bounding the cumulative effect of terms $|\vec{\mu} - \vec{x}^j|^{-L}$, $j \neq i$, to the energy difference $\delta\xi$ and showing that it is dominated by $|\vec{\mu} - \vec{x}^i|^{-L}$. For details, we refer the reader again to ref. 14.

Note the importance of this property: unlike the Hopfield model that is limited to $m \leq N$, the suggested system is optimal in the sense of information theory, because for every set of memories $\vec{x}^1, \ldots, \vec{x}^m$ separated from each other by a Hamming distance ρN, up to $(1/2)\rho N$ errors in the input can be corrected, provided that N is large and L properly chosen.

As for the complexity of the system, we note that the nonlinear operation a^{-L}, for $a > 0$ and L integer (which is at the heart of our system computationally) is equivalent to $e^{-L\ln(a)}$ and can be implemented, therefore, by a simple electrical circuit composed of diodes, which have exponential input–output characteristics, and resistors, which can carry out the necessary multiplications.

Further, since both $|\vec{x}^i|$ and $|\vec{\mu}|$ are held fixed in the discrete system, where all states are on the unit hypersphere, $|\vec{\mu} - \vec{x}^i|^2$ is equivalent to the inner product of $\vec{\mu}$ and $\vec{x}^i$, up to a constant. A detailed implementation can be found in ref. 14.

To conclude, the suggested model involves about $m \cdot N$ multiplications followed by m nonlinear operations and then $m \cdot N$ additions. The original model of Hopfield involves N^2

multiplications and additions and then N nonlinear operations *but is limited to* $m \leq N$. Therefore, whenever the Hopfield model is applicable, the complexity of both models is comparable.

This work was supported by Office of Navy Research Contracts N00014-86-K0041 and N00014-85-K-0607, Army Research Office Contracts DAAG-29-84-K-0262 and DAAG29-84-K-0082, and the Weizmann Postdoctoral Fellowship (A.D. and O.Z.).

1. Anderson, J. A. (1970) *Math. Biosci.* **8,** 137–160.
2. Anderson, J. A. (1972) *Math. Biosci.* **4,** 197–220.
3. Cooper, L. N. (1974) in *Proceedings of the Nobel Symposium on Collective Properties of Physical Systems,* eds. Lindquist, B. & Lindquist, S. (Academic, New York), pp. 252–264.
4. Kohonen, T. (1972) *IEEE Trans. Comput.* C **21,** 353–359.
5. Kohonen, T. (1977) *Associative Memory: A System Theoretic Approach* (Springer, Berlin).
6. Metropolis, N., Rosenbluth, A. W., Rosenbluth, M. N., Teller, A. H. & Teller, E. (1953) *J. Chem. Phys.* **21,** 1087–1092.
7. Hopfield, J. J. (1982) *Proc. Natl. Acad. Sci. USA* **79,** 2554–2558.
8. Widrow, G. & Hoff, M. E. (1960) in *Institute of Radio Engineers, Western Electronic Show and Convention, Wescon Convention Record* (Institute of Radio Engineers, New York), Vol. 4, Part 4, pp. 96–104.
9. Kohonen, T. (1984) *Self-Organization and Associative Memory* (Springer, Berlin), pp. 172–174.
10. Reilly, D. E., Cooper, L. N. & Elbaum, C. (1982) *Biol. Cybern.* **45,** 35–41.
11. Hopfield, J. J., Feinstein, D. I. & Palmer, R. G. (1983) *Nature (London)* **304,** 158–159.
12. Potter, T. W. (1987) Dissertation (State University of New York, Binghamton).
13. Bachmann, C. M. (1986) *ARO Technical Report* (Brown University Center for Neural Science, Providence, RI).
14. Dembo, A. & Zeitouni, O. (1987) *ARO Technical Report* (Brown University Center for Neural Science, Providence, RI).
15. Hopfield, J. J. (1984) *Proc. Natl. Acad. Sci. USA* **81,** 3088–3092.

The Handbook of Brain Theory and Neural Networks. © MIT Press, 1993

Coulomb Potential Learning

Michael P. Perrone and **Leon N Cooper**[*]

Physics Department and

Institute for Brain and Neural Systems

Box 1843, Brown University

Providence, RI 02912

Email: mpp@cns.brown.edu

The Coulomb Potential Learning (CPL) algorithm (Bachmann et al., 1987), which derives its name from its functional form's likeness to a coulomb charge potential, was originally motivated by the short-comings of the Perceptron (Rosenblatt, 1962) and the original Hopfield net (Hopfield, 1979). In the case of the Perceptron, it was clear almost from the outset that the linear-separability provided by the perceptron would not be sufficient to perform complex tasks. In the case of the original Hopfield model, the recall capacity of the network is low due to non-orthogonal memories and the existence of spurious memories. The CPL algorithm addresses both of these problems by providing a simple network which is capable of both storing an arbitrarily large number of memories with perfect recall and no spurious memories; and also constructing arbitrary nonlinear boundaries for classification tasks. In addition the CPL algorithm is easy to implement in hardware and is readily adaptable to parallel computation.

Perfect Memory Recall

The CPL algorithm constructs a network in the following way: Suppose that we are given a set of memories, $\mathcal{M} = \{m_i : m_i \in \mathcal{R}^N \forall i\}$. For each memory construct a neuron, $n_i(x)$, with

---

[*]Research was supported by the Office of Naval Research, the Army Research Office, and the National Science Foundation.

the following activation function $n_i(x) = -||x - m_i||_2^{-L}$, and combine all of these neurons with a single perceptron. The output of the perceptron, $E(x) = \sum_i Q_i n_i(x)$, is given by

$$E(x) = -\sum_i Q_i ||x - m_i||_2^{-L},$$

where $N \geq 3, L \geq N - 2$ and $Q_i > 0$. Clearly, each memory corresponds to a minimum of the network activation Thus we have stored all of the memories. Now, we need to define a method for retrieving memories from this network. We retrieve memories by relaxing from an arbitrary initial state to a minimum of the $E(x)$ function by performing gradient descent on $E(x)$ to find a fixed point of the following differential equation:

$$\dot{\vec{x}} = -\sum_i Q_i ||\vec{x} - \vec{m_i}||_2^{-(L+2)} (\vec{x} - \vec{m_i}).$$

Note that relaxation to a memory may be slow if the initial point is far from all of the memories since in that case, the gradient will be very small. Another practical consideration is that when the gradient descent process is implemented, the gradients near memories will be very large and so care must be taken to terminate computer implementations of the relaxation process before overflow errors occur.

No Spurious Memories

In order to gain more insight into the process of memory recall, we consider the case in which $L = 1$ and $N = 3$. In this case, we can make a direct analogy with physics. The memories can be interpreted as negative electric point charges and the relaxation equation corresponds to the motion of a positive particle in a Coulomb field – thus the name of the algorithm. In this special case, it is easy to see why the network has no spurious memories: Suppose a spurious memory exists then by definition it is distinct from the true memories. In a small neighborhood of the spurious memory, the gradient must point inward towards a spurious minimum; but Gauss' Law (Jackson, 1975) tells us that a net inward gradient implies that there must be charge enclosed in the neighborhood. Thus we arrive at a contradiction; so the spurious memory can not exist. It is possible to prove in general that the CPL network has no spurious memories (Dembo and Zeitouni, 1987).

Note that we can control the size of the basins of attraction for each memory by adjusting the strength of the "charge", Q_i, and the exponential power of the potential, L. This flexibility gives us a natural method for controlling the relative importance of various memories and how likely they are to be recalled. Thus, the network will not necessarily relax to the closest memory in the Euclidean sense but rather it will relax to the most prominent which implies that the network is constructing nonlinear boundaries between memories. Under certain conditions however (Dembo and Zeitouni, 1987), these nonlinear effects can be minimized.

Deterministic Classification

It is also possible to modify the CPL network to function as a classification network. If we have two or more memory sets, $\mathcal{M}_1, \mathcal{M}_2, \ldots$ which correspond to distinct classes and we construct a CPL network for each memory set, then we can compare the outputs of the networks and choose the smallest as the winner. In this way, the CPL networks is acting as a Parzen Window Estimators (Duda and Hart, 1973). Additionally, we can combine the CPL networks and use gradient descent to relax to a memory. The classification is then given by the class of the memory to which the network relaxes.

In practice however, it is not feasible to store all of the memories associated with a given classification task since testing new patterns may be unacceptablely slow for large memory sets. In this case, it is helpful to sub-sampling the memories. A simple yet efficient method for sub-sampling has been outlined by the RCE algorithm (Reilly et al., 1982). The RCE algorithm creates networks of neurons analogous to the CPL algorithm except that $Q_i = 1 \, \forall \, i$ and the neurons have a bounded activity function given by $n_i(x) = 1 - \Theta(||x - m_i||_2 - t_i)$ where $\Theta(\cdot)$ is a step function. Thus the activity of RCE neuron i is 1 if the input is within a distance t_i of m_i and 0 otherwise. The RCE neuron can be viewed as a "clipped" version of the CPL neuron. Classification of a given input is performed by choosing the class corresponding to the RCE network with the largest output. No relaxation process is necessary. In its simplest version, the RCE algorithm builds a network in the following manner. For each memory in the data set:

1) Test the RCE net with the new memory

2) If all neurons are inactive or the classification is wrong:

   2a) Set t equal to the distance to the nearest memory of a different class.

   2b) Add a neuron with the new memory as its center and threshold t.

Once such a network has been constructed, it can be used as is or the RCE neurons can be replaced with CPL neurons where the Q_i are set equal to the number of members from the same class which activate the corresponding RCE neuron. Combining CPL with RCE results in a Radial Basis Function (RBF) network (Powell, 1987) which efficiently uses the data. Note that such a CPL network can also be trained by gradient descent on the Q_i, the t_i and the m_i.

Probabilistic Classification

For probabilistic classification in which class membership is a random variable, we can improve performance by using a variation of the above algorithm. In principle, memories from different classes can be arbitrarily close for probabilistic classification. In practice, this fact results in many neurons having very small thresholds, t_i, and therefore much of the feature space may be left uncovered particularly at class boundaries or anywhere where the class probabilities are nearly equal. This uneven covering can lead to poor classification performance. One way around this problem is the Probabilistic RCE algorithm (Scofield et al., 1987). The probabilistic version is the same as the deterministic version except that a minimum value for the threshold, t, is set. After training, any cell whose threshold is below the minimum value is automatically reset to the minimum value. During testing, the probabilistic regions will give multiple responses from one or more classes. The number of cells active from a given class divided by the total number of cells active can be used as an estimate of the class probability. Also these cells can keep pattern and classification counts that allow them to estimate the probability that a pattern falling with their influence field is an example of the class they represent (Scofield et al., 1987).

Avoiding the Curse of Dimensionality

In high dimensional classification problems, networks such as CPL's, Kernel Estimators and RBF's all suffer from the *"Curse of Dimensionality"* (Bellman, 1961) which implies that the amount of data required to construct an reliable estimate to the true solution increases exponentially with dimensionality. Thus a CPL network which functions well for low dimensional classification problems may perform no better than the level of chance for high dimensional problems unless there is a ridiculously large amount of data. In practice, we rarely if ever have this much data. We run out of physical space to store all of the data long before we can offset the exponential factor!

Hybrid Methods

One way that we can lessen the effects of the curse of dimensionality is to construct hybrid neural networks which can divide a problem into subtasks which are handled in lower dimensional spaces and are therefore less troubled by the curse of dimensionality. In one variation of the RCE algorithm (Reilly et al., 1987), multiple RCE networks are generated to solve subtasks of a large multi-class task in the following way.

1) Train a network on the full task.

2) Identify which classes are being confused.

3) For each pair of confused classes:

   3a) Use the previously trained network to select patterns which are confused.

   3b) Train a new network on the confused patterns.

Note that in both training phases, the networks do not have to be trained with the full dimensionality of the task. In the first phase (Step 1), this fact is justified because we can correct mistakes in the second phase; and in the second phase (Step 3), we are dealing with less complex tasks and therefore the full dimensionality may not be needed. In operation, the hybrid network described above classifies a new pattern using the main network unless there is confusion between classes in which case the pattern is classified by the appropriate subtask network.

In practice, the subtask selection procedure can be iterated as long as there is data available on which to train. However as the data becomes sparse, our estimates become more and more noise and we run into the problem of overfitting. We can lessen this problem by averaging over several hybrid network solutions from several different training runs (Perrone and Cooper, 1993b). In this way, we can reduce the variance of our estimate as much as we like (Perrone and Cooper, 1993a). This method can significantly improve performance.

293

References

Bachmann, C. M., Cooper, L. N., Dembo, A., and Zeitouni, O. (1987). A relaxation model for memory with high storage density. *Proceedings of the National Academy of Sciences*, 84:7529–7531.

Bellman, R. E. (1961). *Adaptive Control Processes*. Princeton University Press, Princeton, NJ.

Dembo, A. and Zeitouni, O. (1987). General potential surfaces and neural networks. ARO technical report # 8, Center for Neural Science, Brown University.

Duda, R. O. and Hart, P. E. (1973). *Pattern Classification and Scene Analysis*. John Wiley, New York.

Hopfield, J. J. (1979). Neural networks and physical systems with emergent collective computational abilities. *Proceedings of the National Academy of Science*.

Jackson, J. D. (1975). *Classical Electrodynamics*. John Wiley and Sons.

Perrone, M. P. and Cooper, L. N. (1993a). Learning from what's been learned: Supervised learning in multi-neural network systems. In *Proceedings of the World Conference on Neural Networks*. INNS. [To appear].

Perrone, M. P. and Cooper, L. N. (1993b). When networks disagree: Ensemble method for neural networks. In Mammone, R. J., editor, *Neural Networks for Speech and Image processing*. Chapman-Hall. [In press].

Powell, M. J. D. (1987). Radial basis functions for multivariable interpolation: A review. In Mason, J. C. and Cox, M. G., editors, *Algorithms for Approximation*, pages 143–167. Clarendon Press, Oxford.

Reilly, D. L., Cooper, L. N., and Elbaum, C. (1982). A neural model for category learning. *Biological Cybernetics*, 45:35–41.

Reilly, R. L., Scofield, C. L., Elbaum, C., and Cooper, L. N. (1987). Learning system architectures composed of multiple learning modules. In *Proc. IEEE First Int. Conf. on Neural Networks*, volume 2. IEEE.

Rosenblatt, F. (1962). *Principles of neurodynamics*. Spartan.

Scofield, C. L., Reilly, D. L., Elbaum, C., and Cooper, L. N. (1987). Pattern class degeneracy in an unrestricted storage density memory. In Anderson, D. Z., editor, *Neural Information Processing Systems*. American Institute of Physics.

Georges de la Tour, painting:
"The fortune teller"

An Overview of Neural Networks: Early Models to Real World Systems

(with D. L. Reilly)

An Introduction to Neural and Electronic Networks, eds. J. Davis and C. Lau
(Academic Press, San Diego, 1990) p. 227

Doug Reilly joined us in the mid-seventies. He was the first of our group to pursue the development of neural networks that could perform various real-world tasks. A simple seeming problem we chose to experiment with was the recognition of hand-written characters as they were being written (online). All of us who have learned to read can more-or-less recognize the individual characters another person writes (not, however, as well as we think we can). Yet to write explicit rules that allow a machine to recognize such writing with all of its variability was (at least at that time) a highly nontrivial task.

So we set out attempting to construct neural networks that could learn from examples (as children do) to distinguish pattern classes from one another. In this effort we worked very closely with Charles Elbaum. One of the results was the RCE algorithm. I recall meeting after meeting with Doug late in the afternoon to go over the day's results – and the frustrations endured in attempting to solve this seemingly easy problem. The following book chapter presents an overview of the neural network situation as Doug and I saw it in 1990.

11

An Overview of Neural Networks: Early Models to Real World Systems

Douglas L. Reilly and Leon N Cooper

INTRODUCTION

Scientific fashion, like the length of women's skirts and the width of men's ties, changes with the seasons. From the old belief that neural networks could do nothing, we have now, among current opinions, the suggestion that they can do everything. Truth, as "the Master of those that know" proposed, might be closer to the golden mean.

We look to biology for inspiration. Neocortex by itself (disconnected from sensory inputs) functions at a distinctly reduced level. (Classifying circuitry from retina through visual cortex as a neural network begs the question since these are largely genetically programmed so that their learning takes place on evolutionary time scales.) Such considerations lead us to conclude that if neural networks can function at all to do useful things it seems very likely that they will do so by being incorporated into systems containing many more or less conventional elements so that they can solve real world problems economically. It is likely that in the future such networks will be components of complex systems involving classification, computation, and reasoning.

Neural networks are inspired by biological systems in which large numbers of nerve cells, that individually function rather slowly and imperfectly, collectively perform tasks that even the largest computers have not been able to match. They are made of many relatively simple processors connected to one another by variable memory elements whose weights are adjusted by experience. They differ from the now standard von Neumann computer in that they characteristically process information in a manner that is highly parallel rather than serial, and they learn (memory weights and thresholds are adjusted by experience). Since neural networks learn, they differ from the usual artificial intelligence systems in that the solution of real-world problems requires much less of the expensive and elaborate programming and knowledge engineering required for such products as rule-based expert systems. Thus, neural network systems seem to some of us to represent the next generation of computer architecture: systems that combine the enormous processing power of von Neumann computers with the ability to make sensible decisions and to learn by ordinary experience, as we do ourselves.

An Introduction to Neural and Electronic Networks. Copyright © 1990 by Academic Press, Inc. All rights of reproduction in any form reserved.

The various neural networks that are currently in fashion differ in their ability to make accurate distinctions. their ability to learn quickly, efficiently, and without extensive retraining, their complexity—the level of interconnectivity (an important consideration for realization in hardware), and in the size of the conventional computer required to simulate the network. An important criterion in differentiating neural networks from one another is whether they learn quickly and accurately enough to be of use in real time situations. A related question is whether they require retraining on an entire data set to learn one new item. (It is necessary for a system to be able to learn new information rapidly and accurately in order to be able to adapt to changes that occur.)

At this point it is perhaps appropriate to make what should be an obvious (but surprisingly not universally appreciated) remark. Neural networks that learn accurately and rapidly are *not* the result of random connections of many elements that learn according to just any rule. Although random networks that adjust themselves in a random manner can be made to learn, this learning, that possibly might be called evolutionary learning, takes place in what might be called evolutionary time. It is therefore important to distinguish between the time required to design an efficient and rapidly learning network and *the time it takes such a network to internalize the rules necessary to deal with a particular environment.* Further, it is necessary to distinguish the generic network architecture and learning rules from the learned internalized rules specific to a particular environment.

Further, one should not be misled into judging the value of neural networks based on their ability to solve certain hard problems such as picking out a mouse in a complicated background fifty yards away. While it is true that some animals can do this, they very likely do this by a complex combination of information processing, pattern recognition, and feedback and -forth between cognitive acts and pure pattern recognition. While this may be an eventual goal for neural network systems, it should not be used as a criterion for their value. Neural networks can be of great value in helping to solve real world problems without duplicating

everything that animals do. There are many situations in which the biological system does better, at present, than anything we can build even though we understand how the biological system functions.

In this chapter we present an overview of neural networks from early models to present systems that can solve real world problems.

Definitions and Notation

A neural network can be defined as a distributed computational system composed of a number of individual processing elements operating largely in parallel, interconnected according to some specific topology (architecture) and having the capability to self-modify connection strengths and processing element parameters (learning). In general a neural network performs some information processing function (pattern recognition, data compression, etc.). Neural networks can learn complex "rules." They differ in their efficiency and rapidity of learning, their ability to make distinctions, their capacity to generalize, and the type of machine and/or hardware on which they can run.

Among the problems that are difficult to solve using conventional rule-based techniques are those that might be characterized as having a high degree of entropy or variability as in Figure 1. A neural network processing element has inputs and synaptic weights that produce an integrated potential as an output as shown in Figure 2. The output is a sigmoidal function of the integrated potential as shown in Figure 3. A neural network is composed of such individual processing elements connected to one another according to some architecture (Figure 4). Input vectors are denoted by d, e, or f. Output vectors by c, g, or t. Transfer or memory storage matrices are A, M, W, or R. $f^1...f^\alpha$ are the N components of the αth vector.

Neural networks are characterized by their *architecture*: what is connected to what. They can be fully connected, sparsely connected, or feedforward. They can also be characterized by their *dynamics* (dynamical or equilibrium systems) and by their *learning rules*, that is, which network parameters (weights, thresholds, number of connections, etc.) change over time and how.

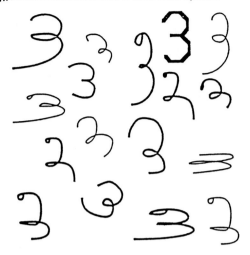

FIGURE 1   Which of the above characters is a three?

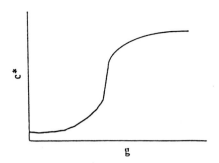

FIGURE 3   Sigmoidal function of integrated potential.
$g_i = \sum_{j=1}^{N} A_{ij}f_j$; $c = c^*(\sum_{j=1}^{N} A_{ij}f_j) = c^*(g)$.

Among the architectural choices, we have to decide between a richly interconnected neural network as shown in Figure 5 or a simple feedforward neural net-

work. Analysis of biological systems (Cooper & Scofield. 1988) suggests that complex completely interconnected neural networks can be approximated by such simple feedforward networks.

EARLY NETWORK LINEAR MODELS

We begin by discussing some of the properties of early linear neural network models. In the linear

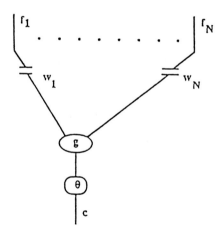

FIGURE 2   A typical neural network element. f_i, inputs: w_n, weights:
$g = \sum_{j=1}^{N} w_jf_j$, integrated potential: θ, threshold function: c, output.

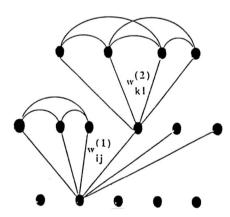

FIGURE 4   Typical neural network.

region let g_i be the cell output

$$g_i^\alpha = \sum_{j=i}^{N} A_{ij} f_j^\alpha \qquad (1)$$

where $f_1^\alpha \ldots f_N^\alpha$ are the components of the αth vector

We may then regard $[A_{ij}]$ (the synaptic strengths of the N^2 ideal junctions) as a matrix or a mapping which takes us from a vector in the F space to one in the G space. This maps the neural activities $f = (f_1, f_2 \ldots f_N)$ in the F space into the neural activities $g = (g_1 \ldots g_N)$ in the G space and can be written in the compact form

$$g = Af \qquad (2)$$

In the earliest neural network models, it was proposed that it was in modifiable mapping of the type A that the experience and memory of the system are stored. In contrast with machine memory, which is, at present, local (an event stored in a specific place) and addressable by locality (requiring some equivalent of indices and files), animal memory is likely to be distributed and addressable by content or by association. In addition, for animals there need be no clear separation between memory and "logic."

It is convenient to write the mapping A in the basis of vectors the system has experienced

$$A = \sum_{\mu v} c_{\mu v} g^\mu \times f^v \qquad (3)$$

Here g^μ and f^v are output and input patterns of neural activity while the $c_{\mu v}$ are coefficients reflecting the degree of connection between various inputs and outputs. The symbol x represents the "outer" product between the input and output vectors. The ij^{th} element of A gives the strength of the ideal junction between the incoming neuron j in the F bank and the outgoing neuron i in the G bank (Figure 6). Thus, if only f_j is nonzero, g_i, the firing rate of the ith output neuron, is $g_i = A_{ij} f_j$. Since

$$A = \sum_{\mu v} c_{\mu v} g^\mu \times f^v \qquad (4)$$

the ij^{th} junction strength is composed of a sum of the

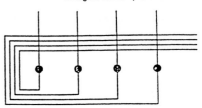

FIGURE 5   A fully interconnected network.

entire experience of the system as reflected in firing rates of the neurons connected to this junction. Each experience or association (μv), however, is stored over the entire array of $N \times N$ junctions. This is the essential meaning of a distributed memory: each event is stored over a large portion of the system, while at any particular local point many events are superimposed.

We have shown elsewhere that the nonlocal mapping, A, can serve in a highly precise fashion as a memory that is content addressable and in which "logic" is a result of association and an outcome of the nature of the memory itself (Cooper, 1973).

The matrix A gives perfect recall if the f^v are orthogonal. If f^v are not orthogonal, they can be orthogonalized by various techniques. This leads to what Kohonen has called an optimal mapping

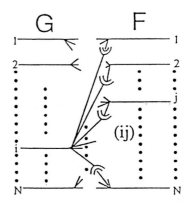

FIGURE 6   A typical neural network containing a distributed memory.

(Kohonen, 1984). The argument in a very simple form is as follows.

For all inputs, f^μ, and desired outputs, g^ν, we want

$$g^\nu = A * f^\nu \quad \nu = 1,2,....K \quad (5)$$

In general, we want

$$G = A * F \quad (6)$$

where G and F are the matrices

$$G = \begin{bmatrix} g_1^1 & \cdots & g_1^K \\ \cdot & \cdots & \cdot \\ \cdot & \cdots & \cdot \\ \cdot & \cdots & \cdot \\ g_N^1 & \cdots & g_N^K \end{bmatrix} \quad (7)$$

$$F = \begin{bmatrix} f_1^1 & \cdots & f_1^K \\ \cdot & \cdots & \cdot \\ \cdot & \cdots & \cdot \\ \cdot & \cdots & \cdot \\ f_N^1 & \cdots & f_N^K \end{bmatrix}$$

If $K = N$ (all matrices are square and inverses exist), then

$$A* = GF^{-1} \quad (8)$$

For nonsquare matrices and for those with no inverses one can obtain $A*$ using the method of pseudoinverses (Kohonen, 1984).

LEARNING

Among the more interesting properties of neural networks is their ability to learn. These are not the only systems that learn, but it is their learning ability coupled with the distributed processing inherent in neural network systems that distinguishes these systems.

Among the various classical learning rules, the oldest and the most famous is that proposed by Hebb. "When an axon of cell A is near enough to excite a cell B and repeatedly or persistently takes part in fir-

ing it, some growth process or metabolic change takes place in one or both cells such that A's efficiency as one of the cells firing B is increased" (Hebb, 1949). In our notation, this could be written:

$$\delta A_{ij} \sim g_i f_j \quad (9)$$

where $g = Af$ is the actual output and f the actual input. We then have

$$A(t+1) = \gamma A(t) + \delta A(t) \quad (10)$$

where

$$\delta A(t) = \eta(g \times f) = \eta A f \times f \quad (11)$$

This yields

$$A(t+1) = \gamma A(t) [1 + \frac{\eta}{\gamma} f(t) \times f(t)] \quad (12)$$

and leads to *exponential growth of recognition of f if f is a repeated input.* Obviously we will need some limit on synaptic strengths. One approach to limiting exponential growth of synapses is "anti-Hebbian" unsupervised learning:

$$\delta A \sim - \eta g \times f \quad (13)$$

$$A(t+1) = \gamma A(t) [1 - \frac{\eta}{\gamma} f(t) \times f(t)] \quad (14)$$

This learning rule projects a repeated incoming vector, f, to zero. The BCM model (Bienenstock, Cooper, & Munro, 1982) combines Hebbian and anti-Hebbian unsupervised learning; it has been applied to many situations in developing visual cortex and appears to explain normal rearing as well as the many deprivation and pharmacological experiments and has been extensively discussed (Cooper, Bear, & Ebner, 1987).

Supervised Learning

In supervised learning, one tries to make adjustments to the set of synaptic weights so that the actual output is guided to a "desired", or "target" output. Such techniques were explored as far back as the late 1950s and early 1960s by Rosenblatt (1958) and Widrow and Hoff (1960). Typically the modification algorithm takes the form

$$A(t+1) = \gamma A(t) + \delta A(t) \qquad (10)$$

where

$$\delta A(t) = \eta(t^a - g^a) \times f^a \qquad (15)$$

and again $g^\alpha(t) = A(t)f^\alpha(t)$ is the actual output for the input $f^\alpha(t)$ while t^α is the target output for pattern α. This yields

$$\delta A(t) = \eta(t^\alpha - A(t)f^\alpha)f^\alpha \qquad (16)$$

Note $\delta A(t) = 0$ if $A(t)f^\alpha = t^\alpha$, so that the correct A is a fixed point (denoted A^*). This A^* corresponds to the *pseudoinverse* discussed above.

We can see the connection with gradient descent if we define an "energy," an "error," or "cost" function:

$$E = \tfrac{1}{2}\sum_\alpha (t^\alpha - Af^\alpha)^2 \qquad (17)$$

The variation of E with respect to A is:

$$\delta E = -(\delta A)\sum_\alpha (t^\alpha - Af^\alpha) \times f^\alpha \qquad (18)$$

If

$$\delta A \sim \sum_\alpha (t^\alpha - g^\alpha) \times f^\alpha \qquad (19)$$

as above, $\delta E \leq 0$ under this variation.

We can picture what is happening in two dimensions as a shift of f^1 and f^2 to f^{1*} and f^{2*} as in Figure 7. With this, mapping A goes to A^* as

$$g^1 \times f^1 + g^2 \times f^2 \longrightarrow g^1 \times f^{1*} + g^2 \times f^{2*} \qquad (20)$$

Note that

$$f^{1*} \cdot f^1 = 1 \qquad f^{2*} \cdot f^1 = 0 \qquad (21)$$

$$f^{1*} \cdot f^2 = 0 \qquad f^{2*} \cdot f^2 = 1$$

Therefore

$$g^1 = A^* f^1 \qquad (22)$$

$$g^2 = A^* f^2$$

Thus the *actual output* becomes the *target output*. With this we can provide separation, recognition *of up to*

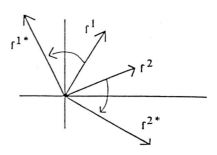

N *vectors* in an N dimensional space. This leads to interesting results but has obvious problems. We get separation only if the number of inputs, K, is equal to or smaller than the dimensionality of the space, N. In general with such linear systems we obtain only linear separability. We must introduce some nonlinearity. Otherwise we get increasing confusion with more and more inputs.

Nonlinear Systems

One way to characterize a neural network is as a nonlinear mapping from input f to output g:

$$g = M[f]$$

M is a nonlinear mapping of the input, f, into the output, g. A theorem of Kolmogorov (1957) indicates that nonlinear threshold devices as arranged in a neural network can yield essentially any nonlinear mapping. Therefore, there is reason to believe that if a problem can be solved by some nonlinear mapping of inputs into outputs, a neural network can provide the solution. Many problems can be solved this way.

Perceptrons

One of the earliest nonlinear neuron-like learning devices was Frank Rosenblatt's (1958) Perceptron

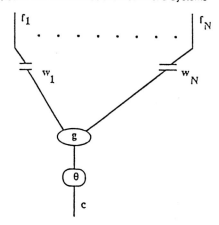

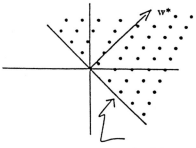

FIGURE 8   A perceptron.

decision   hyperplane

FIGURE 9   Decision hyperplane defined by a weight vector W*.

(Figure 8). Perceptron learning is based on the difference between the actual output and the desired output. Other than such irrelevant features as limiting weights and adjustments to ± 1, it is very similar to gradient descent; it uses a simple threshold so that it incorporates a nonlinearity.

The update rule is

$$c \Rightarrow 1 \text{ if } \sum_j w_j f_j \geq \theta \qquad (23)$$
$$\Rightarrow 0 \text{ if } \sum_j w_j f_j < \theta$$

with c being the output, f_j the inputs, and w_j the synaptic weights. The learning rule is:

$$w_j \rightarrow w_j + \delta w_j \qquad (24)$$

with

$$\delta w_j = \eta(t - c)f_j, \quad (0 < \eta \leq 1)$$

where t is the target response, c the actual response.

An important result is the perceptron learning theorem which states that by this learning mechanism a proper decision hyperplane (Figure 9) will be found if it exists. It was always clear that the Perceptron cannot solve all problems since it is only capable of linear separations (Figure 10). However, multiple layer perceptrons are not limited in this way. If fact, it is now known that with three layers (input, hidden, and output as shown in Figure 11) arbitrary nonlinear separability is possible (see RCE, the network to be discussed later; Lippman, 1987). But multilayer perceptrons suffer from the *credit assignment problem*. How should the weights be adjusted so that the proper decisions are made? In spite of the fact that Rosenblatt proposed a solution like that presently called back propagation of error, the supposed lack of a solution to this problem was a major argument used by Minsky and Pappert (1969) to discredit learning systems.

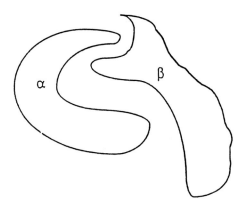

FIGURE 10   The decision regions α and β cannot be separated by a single perceptron.

At present, the credit assignment problem has been solved in various ways. Some of the advantages and disadvantages of the various methods will be discussed next.

A very popular method of dealing with the credit assignment problem in multilayer neural networks is to construct a cost function. This often takes the form:

$$E = \tfrac{1}{2} \sum_{\alpha} (t^{\alpha} - c^{\alpha})^2 \qquad (25)$$

$$= \tfrac{1}{2} \sum_{\alpha} (t^{\alpha} - M[f^{\alpha}])^2$$

with t^{α} the target output and f^{α} the input for pattern α. The object is to find some procedure for adjusting M (a function of all of the weights and thresholds ω^{β}_{ij} and θ^{β}_i $M[(\omega_{ij}).\theta_i])$ so that E is minimized. If $E=0$ exists then a solution exists (Figure 12).

Various Learning Rules

Although learning is one of the primary characteristics of neural networks, the ability to learn is not particularly mysterious. This may be illustrated by what we might call evolutionary learning, a procedure that learns, but rather slowly. Make random variations in the weights ω^{β}_{ij} and the thresholds θ^{β}_i in the nonlinear mapping presented earlier, and retain only those variations that reduce E. Thus we adjust weights and thresholds at random. When such an adjustment results in a lower cost, retain it. When it raises the

cost, discard and try it again. Repeat this procedure on all inputs (the entire environment) or a statistically equivalent sample over and over again. If a solution exists (energy equals zero), eventually one will arrive at it unless one gets stuck in a local minimum (to get out of a local minimum, one can introduce noise). In any case of reasonable complexity, such a procedure takes a very long time.

Backward Propagation of Error, Generalized Widrow–Hoff, or Gradient Descent

To improve the very slow speed of random learning, various methods of directed learning have been proposed. These clearly increase speed of descent, but increase the computation in each adjustment. The most famous is called backward propagation of error (Parker, 1985; Rumelhart, Hinton, & Williams, 1986; Werbos, 1974). In this much discussed procedure, the synaptic weights are modified according to

$$\Delta w_{ij} = \eta d_i c_j \qquad (26)$$

where δ_i is the error signal, η is a small constant, and c_j is the output of cell j, and $\delta_i = (t_i - c_i) \cdot c_i'$ if i is an output unit, and where c_i' is the derivative of transfer function.

The error propagates backward through the network from the last layer (Figure 13). An iterative formula (backward propagation) can be defined relating the

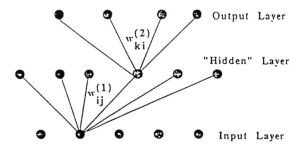

FIGURE 11   Multilayer network.

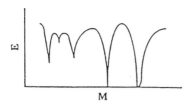

FIGURE 12   The cost as a function of M.

FIGURE 13   Backward propagation of error.

error signal at a given connection to the error signal of the layer after.

$$\delta_k = c'_k \sum_i \delta_i w_{ik} \qquad (27)$$

Although this procedure speeds the learning process, it suffers from a problem common to all methods based on cost reduction. In many applications all data must be represented to learn the solution (Figure 14). In essence, the problem of which minimum is found depends on the starting point (initial conditions). Once in an incorrect minimum it is very difficult. if not impossible, to find a way out.

RELAXATION MODELS AND NEURAL NETWORKS

A network of neurons is a very nonlinear dynamic system. Because of recurrent connections, time delays, etc., a given input will, when regarded in detail, produce a very complicated response that in time may or may not settle down to some stable state. The evolution of the neural network state in such short time intervals (≈ 1 sec) can be and has been described by sets of coupled nonlinear equations (Wilson & Cowan, 1973; Edelman & Reeke. 1982; Grossberg. 1982) or can be approximated by some discrete updating mechanism (Anderson. Silberstein, Ritz. & Jones. 1977) perhaps the best known of which is Hopfield's (1982) model. In what follows. we describe the Hopfield model which illustrates in a fairly transparent manner some of the properties of such systems (Figure 14).

Hopfield's Model

Figure 15 is a schematic representation of the Hopfield model. Note that the architecture is fully interconnected. The updating procedure is random and asychronous. Learning was not emphasized in the original model. although some modification procedures do exist (Hopfield. Feinstein. & Palmer, 1983; Potter, 1987). The updating procedure is

$$c_i \rightarrow \theta(\sum_{i=i} w_{ij} c_j) \qquad (28)$$

The corresponding Hopfield "energy" is:

$$E = -\frac{1}{2} \sum_j w_{ij} c_i c_j \qquad (29)$$

Because the relaxation procedure is the discrete analog of gradient descent. the network will always descend to a local minimum.

$$\Delta E = -(\sum_{i=1}^{v} w_{ij} c_j) \Delta c_i \qquad (30)$$

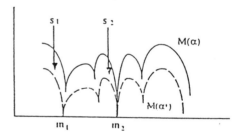

FIGURE 14   If the starting point is S_1. the minimum m_1 may be found for the restricted data set. α'. For the full data set. α the only true minimum is m_2. This would be found with the restricted data set. α'. only if the starting point were S_2.

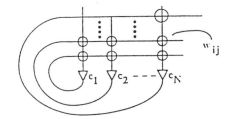

FIGURE 15   Schematic representation of the Hopfield model. w_{ij} weights: $c_i = \pm 1$.

Since $(\sum_{j=1}^{N} w_{ij} c_j)$ and Δc_i always have the same sign, when $\Delta c_i \neq 0$ due to the updating procedure, we have $\Delta E \leq 0$. A frequent problem with relaxation procedures is that local minima may be true memories or *spurious minima* (Figure 16).

Also, the number of spurious minima increases dramatically with the number of stored memories. Some problems of spurious minima in relaxation models may be overcome, however slowly, through the use of Boltzmann machines (Ackley, Hinton, & Sejnowski, 1984). The architecture of the Boltzmann machine is fully interconnected. Hidden units are used. The updating is not only random and asynchronous, but also probabilistic. The system may escape from spurious minima due to noise introduced via the concept of temperature. In the update rule the probability to have c_i go to 1 is:

$$P(c_i \to 1) = \frac{1}{1 + e^{-\Delta E_i / T}} \qquad (31)$$

Relaxation Model with High Storage Density

Whether it is more appropriate to describe the succession of stages in cortex by equilibrium feedforward or dynamic relaxation mechanisms is not yet known. It is quite possible that both have their regions of applicability. However, from the point of view of modeling the succession of states that correspond to recall and association of stored memory states, little seems to be gained by the use of elaborate relaxation mechanisms. All of the results can be obtained in a simpler and more transparent manner using network equilibrium methods that will be described in the following sections.

The problem of spurious memories in the Hopfield model is due to unwanted local or absolute minima in the energy equation (29). One might therefore ask whether it is possible to construct an energy function with minima only at the desired memory sites. This problem was solved in a model proposed by Bachmann, Cooper, Dembo, and Zeitouni (1987).

In this model, memory sites are represented by r^i, $i = 1 \ldots K$. A network state is given by r. An "electrostatic" energy (Figure 17) is defined as a function of the memories and the network state. A simple case in three dimensions is

$$E = -\frac{1}{2} \sum_{i=1}^{K} Q_i |r - r^i|^{-1} \qquad (32)$$

Bachmann et al. (1987) show that this energy has minima only at the designated sites. Such memory

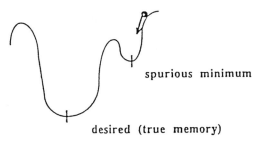

FIGURE 16   Spurious minima in the Hopfield model.

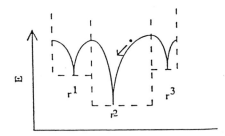

FIGURE 17   Electrostatic energy.

Coulomb Memory Basins

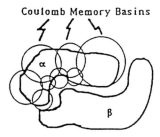

FIGURE 18   Coulomb memory basins.

sites with variable-width attractive basins can be used to outline a class territory as in Figure 18.

Now, we present an *equilibrium system* mapping, R, based on these ideas with *low connectivity*, *high density memory storage*, and *excellent ability to make distinctions* (separations in decision space and rapid learning) in which an arbitrary input *e* is mapped into output *c* by

$$c = R^{-1}e \quad \text{or} \quad Rc = e \qquad (33)$$

RCE Neural Network

We can describe the RCE neural network (Cooper, Elbaum, & Reilly, 1982; Reilly, Cooper, & Elbaum, 1982) in terms of its architecture, the transfer functions of its elements, and the learning laws that govern how the values of the weights and processing element parameters change over time.

The architecture of the RCE network specifies three processing layers: an input layer, an internal layer, and an output layer (Figure 19). Each node in the input layer registers the value of a feature describing an input event. If the application is character recognition, these features might be counts of the number of line segments present at various angles of orientation. If the application is signal classification, the features might be the power in a signal at various frequency bandwidths. If the application is emulating the judgment of a mortgage underwriter, the features would represent information derived from the mortgage application.

Each cell in the output layer corresponds to a pattern category. The network assigns an input to a category if the output cell for that category "fires" in response to the input. If an input causes only one output cell to become active, the decision of the network is said to be "unambiguous" for that category. If multiple output cells are active, the network response is "ambiguous." Though confused about the identity of the pattern, the network nonetheless offers a set of likely categorizations. Cells in the middle or internal layer of the network construct the mapping that ensures that the output cell for the correct category fires in response to a given input pattern.

The internal layer is fully connected to the input layer; each cell on the internal layer is connected to every cell on the input layer. The output layer is sparsely connected to the internal layer; each internal layer cell projects its output to only one output layer

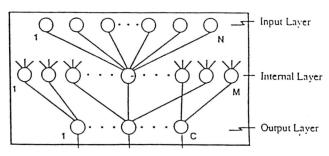

FIGURE 19   RCE network.

cell. Note that there are no recurrent connections in the RCE network, nor are there any lateral connections between cells in any given layer. The RCE is a reduced connectivity, feedforward network. An input pattern on the input layer creates a pattern of activity on the internal layer which in turn produces some set of firing cells on the output layer. All this occurs in a single pass through the network.

The connections between the ith internal layer cell and the input layer cells define a weight vector, w_i. If f represents the activity pattern of the input layer, then the transfer function of the ith internal layer cell in the RCE is given by Equation (34)

$$g_i = \Lambda_\lambda (D (w_i, f)) \qquad (34)$$

$D (w_i, f)$ is some defined distance metric between vectors w_i and f (e.g., a Cartesian distance between real-valued vectors; a Hamming distance between binary-valued vectors; an inner product between normalized pattern vectors, etc.). Λ_λ is some threshold function, chosen appropriately for D. (If D is a Hamming distance, then $\Lambda_\lambda (x) = 1$ $x \le \lambda$: 0, otherwise).

The transfer function of a cell in the output layer of the RCE is such that an output layer cell performs a logical "OR" function on its inputs. Its inputs are unit strength signals from the firing internal cells to which it is connected. The connection strengths between an output cell and its suite of internal cells are all unit valued. Thus, if at least one of the internal layer cells connected to an output layer cell is firing, the output cell will fire.

The transfer function (Equation 34) associates each internal layer cell with a region of the input feature space (the cell's "influence field" or "attractive basin"). The location of the region is specified by the vector of weights coupling the cell to the input layer nodes. The size of the region is determined by the firing threshold of the cell. The geometry of the region is determined by the choice of distance metric. (For example, a Cartesian distance metric in a continuous-valued pattern space, R^N, results in the geometry of an N-dimensional sphere). Any input pattern falling within the influence region of the cell will cause the cell to fire.

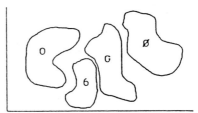

FIGURE 20   Hypothetical category territories for 0. 6. G. 0 in a two-dimensional feature space.

✓ The essence of the function of the network can be seen by regarding a pattern category as a collection of points in the N-dimensional feature space defined by the N cells of the input layer. A pattern of activity among the cells of the input layer corresponds to the location of a point in this feature space. All examples of a pattern category define a set of points in this feature space that can be characterized as a region (or a set of regions) having some arbitrary shape (Figure 20).

Just as a category of patterns defines a region (or regions) in the feature space of the system, a cell in the internal layer of the RCE network is associated with a set of points in the feature space. The geography of this set of points is defined by the transfer function of the internal cell.

For purposes of illustration, consider an RCE network with only two cells on the input layer. (In actual applications, a user defines as many input cells as he needs to represent his input feature vector to the system.) The transfer function can be thought of as defining a disk-shaped region centered at the feature space point w_i (the vector of weights coupling the ith internal cell to the input layer), with a radius λ_i around the w_i (Figure 21). Any point (feature vector) falling within this region will cause this internal, cell to become active.

Thus in an RCE network, the internal layer cells define a collection of disks in the space of input patterns (Figure 21). These disks represent the memories built up in the RCE. Some of the disks may overlap. If a pattern falls within the attractive basins of several internal cells, they will all fire and fire their associated output cells. If all these internal cells are projecting

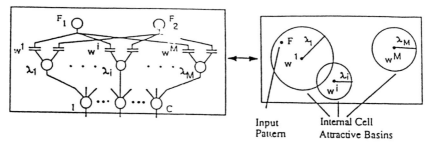

Input   Internal Cell
Pattern   Attractive Basins

Figure 21 Feature space for a two-dimensional input pattern space showing input pattern ($F = F_1, F_2$) and RCE memories as attractive basins.

their outputs to the same output cell, then only that output cell will fire and the response of the network will be an unambiguous classification of the input. On the other hand, if the firing internal layer cells project their outputs to more than one output cell, then all of those output cells will fire, resulting in an ambiguous assignment of several possible classifications to the input pattern.

Learning involves two distinct mechanisms in the network. The first is cell commitment. Cells are "committed" to the internal layer as well as, though less frequently, to the output layer. When cells are committed, they are "wired up" according to the RCE network paradigm. Each cell in the internal layer is connected to the outputs of each of the cells in the input layer. Each cell in the internal layer projects its output to only one cell in the output layer. The second learning mechanism in the RCE network is the modification of the thresholds associated with cells in the internal layer. Thus, each internal processing element has its own weight vector w_i and threshold λ_i and their values are changed under separate modification procedures.

Both the commitment of cells and the adjustment of internal cell thresholds are controlled by training signals that move from the output layer back into the system. If an output cell (representing a given category of patterns) is off (0) and should be on (1), an error signal of +1 is generated for that output cell. If an output cell is on (1) that should be off (0) an error signal of -1 travels from that cell back into the internal layer.

An error signal of +1 traveling from the kth output layer cell into the system causes a new internal processing element to be committed. Its output is connected to the kth cell in the output layer, and its vector

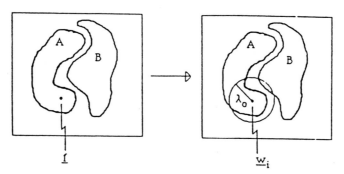

FIGURE 22   Feature space representation of cell commitment in internal layer.

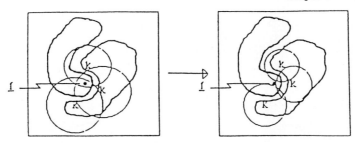

FIGURE 23   Threshold reduction for active internal layer cells associated with K^{th} output cell. After modification, f is no longer covered by any internal cell disk for k^{th} output cell.

of connections w_i to the input layer assumes the value of the current feature vector of the input layer. ($w_{ij} <- f_j$, $j = 1...N$) The threshold of the cell is assigned a value λ, where λ is defined as $\lambda = \max (\lambda_{max}, \lambda_{opp})$ where λ_{opp} is the distance to the nearest center of an internal cell influence field (the cells' weight vector w) for any class different from that of the current pattern, and λ_{max} is the maximum size of an influence field ever assigned an internal cell (a user-defined parameter of learning).

In the feature space of the system, this adds a new disk-shaped region that covers some portion of the class territory for this input's category (Figure 22).

If an error signal of -1 is sent from the kth output unit back into the system, then the system responds by reducing the threshold values (λ_i) of all the active internal units that are connected to the kth output cell. This has the effect of reducing the sizes of their disk-shaped regions so that they no longer cover the input pattern (Figure 23).

Table 1 summarizes the effect on the network of the various error signals traveling back from the output layer.

The RCE network can learn to distinguish pattern categories even when they are defined by complex relationships among the features characterizing the patterns. Figure 24 is taken from a simulation in which two hypothetical geometries were defined to represent two arbitrary pattern classes, α and β.

Training the network consisted of selecting a rar dom set of points from the two class territories an presenting these points as input patterns to the syste: for training. From seeing the training samples, the sys tem builds up a set of disks that covers the territorie for the two hypothetical pattern classes α and β. Eac disk corresponds to the "win region" of an intern: unit and each disk is "owned" by a pattern category b virtue of its unique connection to one and only on output cell. The size of the disk is related to the thres old of the corresponding internal unit. Disk sizes th: are too large cause the wrong internal cell to fire i response to a pattern input. The resultant "-1" trainin signal reduces the disk size to prevent it from firin for that pattern (or patterns like it) again. This proces of committing disks and reducing their sizes allow the system to develop separating mappings even fc

TABLE 1

Error signal from i^{th} output cell	Modification
+1	Commit an internal cell for this event and connect it to i^{th} output cell
-1	Reduce thresholds of all active internal cells connected to this output cell in order to turn the out put cell off
0	No change to any of the internal cells connected to this output cell

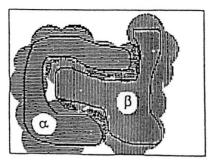

FIGURE 24   RCE network separating classes α and β.

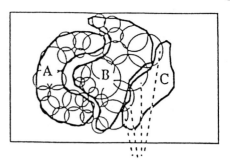

FIGURE 26   Dynamic category learning. New category C can be introduced after categories A and B are learned. Some examples of C will be classified incorrectly as B. Thus, the network needs to be shown examples of C and B, not of A.

nonlinearly separable pattern categories. In principle, an arbitrary degree of nonlinearity can be achieved.

Furthermore, this learning strategy is able to map out category regions even if they are disjoint in the feature space. This allows the system to learn concept formation as required for a problem. For example, the shapes of the images of "A" and "a" have little in common. In a typical feature space, their pattern territories are most likely not connected, or "disjoint" (Figure 25). Nonetheless, the system can map out both of these regions separately and, if so instructed during training, associate them with the same output cell representing the concept "A".

Dynamic category learning is possible with the RCE network. New classes of patterns can be introduced at arbitrary points in training without always involving the need to retrain the network on all of its previously learned data. Assume that the network has been fully trained on classes A and B. Imagine a new

class of patterns C is to be introduced that lies near B in the feature space (Figure 26). To some examples of C the network will respond with B; some of the internal cells for B are generalizing more than they should. To train the network, examples of the new category C must be shown, as well as examples of the categories that are near C in the pattern space. The nearby categories are generally those that the network offers as incorrect answers to examples of the new class. Thus, in the above example, the network must be shown examples of class C and the previously learned class B. However, examples of A will not need to be shown to train the network on the new class C.

The RCE is a partially distributed network in the sense that the information about class A is not distributed over all the weights in the network, but rather over some subset of weights, namely, those associated with the internal cells that have their outputs targeted to the class A output cell. This still provides a reasonable degree of fault tolerance. If one of these internal cells fails to function, only a partial amount of the information about the given class is lost or degraded. This is because a class is generally represented by the overlap of disks associated with a number of internal cells.

The RCE can be trained to sharpen its understanding of a given class by showing it examples of pattern "noise", examples that have no class affiliation. Training on noise input allows the system to better

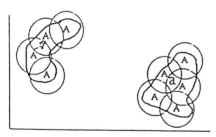

FIGURE 25   RCE network training to map subclasses "A" and "a" onto single category "A".

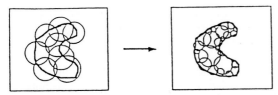

FIGURE 27   "Noise" inputs are any patterns taken from territory outside the class region outline. Training on noisy inputs improves initial approximate class mapping (*left*) by reducing the size of internal cell influence fields (*right*). No internal cells need be committed for noise.

define the extent of the boundary between the class and background noise (Figure 27). Furthermore, the net does not build up an explicit model of the noise, because no internal cells need to be committed for noise inputs.

In some pattern classification problems, there might be overlap between the class territories in the feature space (Figure 28). Within this confusion zone the pattern classes are not, strictly speaking, separable. To patterns falling within that territory, the network should respond with some measure of the likelihood of the most probable class affiliation.

In training, internal cells cannot have their influence field sizes reduced below some critical size, λ_c. Cells with influence field size reduced to this value are termed "probabilistic" cells; all other internal cells are referred to as "deterministic." When a probabilistic cell fires, it fires the output cell to which it is connected, but, weakly, in the sense that the output of the network is now officially "ambiguous." Even if only one output

cell is firing, if it is being stimulated only by probabilistic internal cells, the answer of the network remains ambiguous. The output cell's classification is offered as a possible, but not definite classification of the input.

During learning, probabilistic cells block the commitment of deterministic cells in any space that they occupy. If an input pattern for class B falls only within the influence fields of one or more probabilistic cells for class A, the network will not commit a deterministic cell for this pattern. It will, however, commit a probabilistic cell for class B, centered at the input pattern site, with influence field equal to λ_c. The network allows the influence fields of probabilistic cells for one class to cover the center point of probabilistic cells for another class. This is another way in which they will differ importantly from deterministic cells. The commitment procedure will result in a layering of the confusion zone with probabilistic cells for A and a layering with probabilistic cells for B. In this way, any input pattern that falls within the confusion zone will fire at least one probabilistic cell for A and one for B. This will cause the network to respond with the ambiguous response, A or B. In a later evolution of the RCE, these probabilistic cells keep pattern and classification counts that allow them to estimate the probability that a pattern falling within their influence field is an example of the class they represent (Scofield, Reilly, Elbaum, & Cooper, 1987).

The memories that the system learns during training are stored in the weight vectors between the input layer and the internal layer. The number of cells in the internal layer grows automatically as a function of the complexity of the problem. This is in contrast to other network models in which the size of the network is

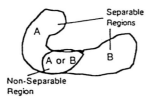

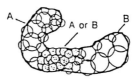

FIGURE 28   On the left, two class territories that overlap in a confusion zone. On the right, separable territories are mapped out with deterministic cells (open circles); the confusion zone, with probabilistic cells (filled circles).

fixed and can only be predetermined through trial and error. The number of memories that the RCE network can develop is limited, in principle, only by the number of available sites in the pattern space. In networks such as the fully interconnected Hopfield network, the number of memories that can be stored is typically limited to 10% of the number of nodes in the network. Various procedures have been studied to relax this memory storage limitation, although none approaches the 1 memory/1 cell ratio of the RCE net. Additionally, as more nodes are added to the Hopfield net, the number of connections grows as the square of the number of nodes. Due to the essentially feedforward connectivity of the RCE net, the number of connections grows linearly with the number of nodes.

We can summarize the features discussed for the RCE network. Its reduced internal connectivity and feedforward architecture reduces the number of computations that need to be performed within the net to compute a response. This has made it possible to develop software applications of the RCE net that run in real time. Furthermore, these architectural properties simplify the problem of designing special purpose silicon for chip level implementation of the net. The network can learn rapidly, generalizing quickly to the notion of a pattern category after only a few examples of the class are presented. This is due to the fact that changes to the weight vectors of internal layer cells are on the same order of magnitude as the input signal, and not some fraction of the signal magnitude. The network is able to resolve pattern classes that are not linearly separable, and even pattern classes represented by disjoint territories in the feature space. The network can store large numbers of memories if required for learning many complex categories. It automatically commits new cells in the internal and output layers as required to accommodate this need. Its partially distributed character allows it to make local changes to memory, thus providing it a capability for true dynamic category learning.

Multiple Network System

In many real-world pattern recognition problems a single representation is not appropriate for all the clas-

sification decisions the system must make. (The representation of the problem is defined by the set of features whose measurements characterize the pattern as an input to the system.) Often, decision making in a problem occurs on the basis of information contained in subsets of all the possible features that characterize the data. The definition of these feature subsets in effect partitions the entire feature space into subspaces. These partitions may arise naturally in a problem as a consequence of different sensor sources for the data. (A feature subset can be associated with each sensor.) Other partitions are suggested by common themes in the characteristics being measured for the pattern (description of a shape in Cartesian versus polar coordinates, measurements in length versus spatial frequency, etc). In other cases, feature subsets are a consequence of general knowledge about the structure of decision making in the given problem domain. Natural partitioning of the feature set can also result from the introduction of new features to the system late in the process of learning.

A multiple neural network system has been designed in which a number of RCE networks are combined together with a Controller module (Cooper, Elbaum, Reilly, & Scofield, 1988; Reilly, Scofield, Elbaum, & Cooper, 1987). The Controller integrates the responses of the RCE networks into a system response and, on the basis of corrective feedback from an instructor, directs the training of the networks in the system (Figure 29). Each RCE network processes a user-defined feature subset. By virtue of the feature subset it is processing, a particular RCE network may be able to be trained to make unambiguous, correct decisions about patterns belonging to certain categories. For other categories, the partitioning may be such that no single RCE network has enough information to reliably classify the event. In such cases, an RCE network can develop category mappings that at least allow it to indicate likely pattern categories for the input. The Controller then correlates the answers for several such networks in an attempt to identify the pattern.

In this multiple neural network system, the RCE nets are arranged in user specified groups or levels. A

Pattern

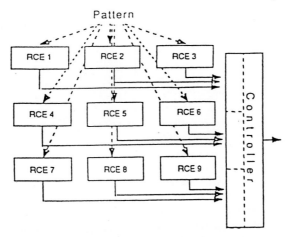

FIGURE 29   Architecture of a 3 x 3 multiple neural network system, showing Controller receiving classification outputs from each of the component 9 RCE neural nets.

level may contain as few as one or as many as all the nets in the system. The Controller polls the RCE networks, level by level, until it has enough information to decide about the pattern's identity. Networks at a certain level in the system are only accessed if the Controller has been unable to decide on an overall system response, given the responses of previous levels.

In a multiple neural network system, the ability of the RCE network to build up knowledge not only of the shapes of separable pattern class territories but also of the approximate shape of overlapping class territories allows the Controller to correlate ambiguous answers of several networks to produce an unambiguous response. For example, network 1 may have enough information to identify class α, but cannot distinguish classes β and γ. On the other hand, network 2 may be able to identify γ, but cannot distinguish between α and β. (Class territories for this problem may look like those shown in Figure 30.) Each RCE network maps out the shapes of the confusion zones in its feature spaces. Consequently, the system can identify examples of β on the basis of responses of both networks. Examples of β will produce a confused response of (β,γ) in network 1 and (α,β) in network 2. The Controller can integrate these ambiguous answers to decide upon β as the pattern's identity.

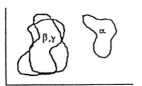

FIGURE 30   Hypothetical class territories for classes α, β, and γ on two different feature spaces. In the feature space on the left, β is largely indistinguishable from γ; in the feature space on the right, β is degenerate with α.

Component networks in the system that have, by virtue of their assigned feature subsets, a certain expertise at deciding for some set of categories must be able to develop that expertise during the course of training. This is ensured by several factors. First, the algorithm that the Controller uses to assemble the system response directs it to weight more heavily a network whose response is unambiguous over one whose response is a set of classes. After training, networks that cannot develop a separable mapping for a class of patterns tend to produce confused responses or no response to examples of that pattern class. Thus, this aspect of the Controller's function ensures that at the least, the confused responses of an inexpert network can be overlooked in favor of the unambiguous responses of an expert network.

Training ensures that the incorrect responses of an inexpert network are "trained out." This results from the fact that the Controller broadcasts its training signals in stages to the various networks. The "-1" signals are sent first to those active RCE networks whose outputs include incorrect classifications. If this modification to system memory does not produce a correct classification of the input pattern, then "+1" signals are sent to networks on a selective basis. If at a given level there is both an expert and an inexpert network (relative to a particular pattern class), this method of training will ensure that the mapping in the inexpert network will not survive.

This is illustrated in Figure 31. There, two hypothetical pattern classes are distinguishable by one network but are totally indistinguishable by another. (In the feature space of the latter network, their class territories are identical. Consequently, only one such territory is pictured.) After initial mappings have developed, training can correct any erroneous or confused system responses simply by cell threshold reduction in the networks. This occurs more often in network 2 than in network 1. Eventually, the internal cells of network 2 have their disk-shaped regions so reduced in size as to be completely ineffective for producing a response from the net. No additional internal cells are committed in network 2 because network 1 has developed a mapping that produces unambiguous and accu-

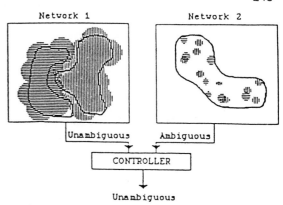

FIGURE 31   Memory states of RCE networks late in training. In network 1, the two pattern classes are represented by two different feature space regions. In network 2, the class regions are degenerate (both classes share a single region). Network 1 trains to dominate system response.

rate responses. One can regard this as a type of "evolutionary" learning in which a given internal cell mapping of a class territory "survives" if it is able to produce accurate classifications on the part of the network for that class. The fracturing of those mappings which produce incorrect responses renders them ineffective for generating a response.

The ability of a multiple neural network learning system to learn pattern categories using feature space partitions has a number of important advantages. First, it allows the system to be configured to what is known about the problem. As an example, consider a problem in which information from two different sensors must be used to identify classes of signals. Some classes of signals may be identified solely on the basis of information from one of the sensors. Other inputs may be classified by appealing to data coming from the other sensor. For still other signals, it may only be possible to identify them by appealing simultaneously to data from both sources. A useful architecture for this problem would use three RCE networks. At the first level would be a network coupled to one sensor; at the next level, a net coupled to inputs from the other sensor. At the third level of the system would be a network coupled to inputs from both sensors.

Hierarchical filtering can be implemented in this multi-network system using a multi-level architecture. Feature subsets may carry the same kind of information, but at different degrees of resolution. For example. in a signal processing application, a representation can be generated by taking the Fourier transform of the input to yield the power spectrum of the signal. The power spectrum samples the energy in the signal at particular frequencies. That sampling can be coarse (gross averages of the power contained in a few frequency bands) or fine (more samples of the energy taken from a larger number of frequency ranges). Networks can be arranged in a hierarchy starting with the coarsest representation in the uppermost level, with successive levels carrying more detailed representations. Those signals which can be identified by appealing only to coarse information will be categorized by the upper levels. Those signal classes which can only be distinguished on the basis of very detailed information will exist as confusion zones in the upper levels. Upper level networks will pass these signals down to the deeper level nets which carry the fine scale information needed to resolve their identities.

With such hierarchical filtering, categorization knowledge can develop with significant reductions in storage space required for implementing the system. Associated with each internal cell is an amount of local storage memory determined, for the most part, by the number of weights per cell. The number of internal cell weights is determined by the number of input features and the precision to which each feature is represented. If upper levels carry coarse feature information, then the mappings that develop in these levels require less memory for storage.

In general, networks are accessed by the Controller only as it needs additional input from them to decide on the identity of a pattern. The order of accessing is determined by their place in the architecture of the system. This makes it possible to incorporate new RCE networks into a system that has already developed memories from some previous training. If the new networks are positioned at the deeper levels of the system, they are accessed only for those pattern

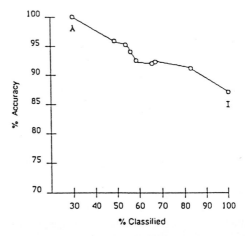

FIGURE 32 Performance of multiple neural network system trained to predict underwriter's Accept/Deny decisions on applications for home mortgages. A. minimum error mode; I. maximum response mode.

classes that the originally configured system could not identify. Importantly, the existing memory of the system is protected and the entire system does not have to be reconstituted by retraining on all previously learned categories. Thus, the system can be upgraded through the introduction of new input features without jeopardizing the current knowledge of the system.

The degree of decision-making or the "tendency for decisions" is adjustable in the system through the definition of Controller parameters. At one extreme, the system can be biased to avoid making an unambiguous decision on a pattern unless there is substantial agreement among its component networks as to the pattern's identity. This "minimum error" mode is desirable in those applications in which there is a high premium associated with each error made. At the other extreme, the system can operate in a "maximum response" mode. This mode is often useful in applications where the system is maximizing throughput, and occasional errors can be tolerated or detected by some filtering system downstream. Thus it is possible to set operating points along a continuum between strict minimum error and strict maximum response (Figure 32).

SUMMARY

In this chapter we have attempted to present an overview of neural networks from the earliest models to a current system.

Multiple RCE neural network systems can efficiently and rapidly learn to separate enormously complex decision spaces and coordinate many neural networks, each a specialist in dividing some portion of the decision space.

To demonstrate that neural networks can solve difficult real-world (as opposed to toy) problems, to show the advantages they offer, and to properly test existing neural network systems as well as to design future systems, these systems must be put to work on real world data. Often the effects we are looking for are small or obscured by noise and variability, so that we cannot be convinced that the system is functioning satisfactorily unless it is in a real world situation.

A number of applications have been developed with multiple RCE neural network systems in problem domains ranging from character recognition to vision (Rimey, Gouin, Scofield, & Reilly, 1986; Reilly et al., 1987) to decision making in financial services (Collins, Ghosh, & Scofield, 1988). Other applications for problems in signature recognition, process monitoring, multisensor fusion, and risk assessment are currently in various stages of exploration or development.

References

Ackley, D. H., Hinton, G. E., & Sejnowski, T. J. (1984). *Boltzmann machine: Constraint satisfaction networks that learn. Carnegie Mellon University.* (Technical Report no. CMU-CS 84-119).

Anderson, J. A., Silberstein, J. W., Ritz, S. A., & Jones, R. S. (1977). Brain state in a box. *Psychological Review*, 84, 413–451.

Bachmann, C. M., Cooper, L. N., Dembo, A., & Zeitouni, O. (1987). A relaxation model for memory with high storage density. *Proceedings of the National Academy of Sciences of the United States of America*, 84, 7529–7531.

Bienenstock, E. L., Cooper, L. N., & Munro, P. W. (1982). Theory for the development of neuron selectivity: Orientation and binocular interaction in visual cortex. *Journal of Neuroscience*, 2, 32–48.

Collins, E., Ghosh, S., & Scofield, C. L. (1988). An application of a multiple neural network learning system to emulation of mortgage underwriting judgements. *IEEE International Conference on Neural Networks*, 2, 459–466.

Cooper, L. N. (1973). A possible organization of animal memory and learning. In B. Lindquist & S. Lindquist (Eds.), *Proceedings of the nobel symposium on collective properties of physical systems*. (pp. 252–264). New York: Academic Press.

Cooper, L. N., Bear, M., & Ebner, F. (1987). Physiological basis for a theory of synapse modification. *Science*, 237, 4248.

Cooper, L. N., Elbaum, C., & Reilly, D. L. (Awarded 1982). *Self organizing general pattern class separator and identifier.* U.S. Patent No. 4,326,259.

Cooper, L. N., Elbaum, C., Reilly, D. L., & Scofield, C. L. (Awarded 1988). *Parallel, multi-unit, adaptive nonlinear pattern class separator and identifier.* U.S. Patent No. 4,760,604.

Cooper, L. N., & Scofield, C. L. (1988). Mean field theory of a neural network. *Proceedings of the National Academy of Sciences of the United States of America*, 85, 1973.

Edelman, G. M., & Reeke, G. N., Jr. (1982). Selective networks capable of representative transformations, limited generalizations and associative memory. *Proceedings of the National Academy of Sciences of the United States of America*, 79, 2091–2095.

Grossberg, S. (1982). *Studies of mind and brain*. Dordrecht, The Netherlands: Reidel.

Hebb, D. O. (1949). *The organization of behavior*. New York: Wiley.

Hopfield, J. J. (1982). Neural networks and physical systems with emergent collective computational abilities. *Proceedings of the National Academy of Sciences of the United States of America*, 79, 2554–2558.

Hopfield, J. J., Feinstein, D. I., & Palmer, K.G. (1983). Unlearning has a stabilizing effect in collective memories. *Nature*, 304, 158–159.

Kohonen, T. (1984). *Self-organization and associative memory*. Berlin: Springer-Verlag.

Kolmogorov, A. K. (1957). On the representation of continuous functions of several variables by superposition of continuous functions of one variable and addition. *Doklady Akademii Nauk. SSSR*, 114, 369–373.

Lippman, R. P. (1987). An introduction to computing with neural nets. *IEEE Transactions on Acoustics, Speech, and Signal Processing*, 4(2), 4–22.

Minsky, M. L., & Pappert, S. (1969). Perceptrons: an introduction to computational geometry. Cambridge, MA: MIT Press.

Parker, D. B. (1985). *Learning-logic* (REPORT TR-47). Cambridge MA: MIT Center for Research in Computation Economics Management Science.

Potter, T. W. (1987). *Storing and retrieving data in a parallel distributed memory system*. Ph.D. thesis. Binghamton: State University of New York.

Reilly, D. L., Cooper, L. N., & Elbaum, C. (1982). A neural model for category learning. *Biological Cybernetics*, 45, 35–41.

Reilly, D. L., Scofield, C., Elbaum, C., & Cooper, L. N. (1987). Learning system architectures composed of multiple learning modules. *IEEE International Conference on Neural Networks*, 2, 495–503.

Rimey. R.. Gouin. P.. Scofield. C., & Reilly. D. L. (1986). Real-time 3-D object classification using a learning system. *Proceedings of SPIE—the International Society of Optical Engineers.* 726, 552–558.

Rosenblatt. F. (1958). *The perceptron. a theory of statistical separab in cognitive systems* (Report no.VG-1196-G-1). Cornell Aeronautical Laboratory, Ithaca.

Rumelhart. D. E.. Hinton, G. E.. & Williams, R. J. (1986). Learning internal representations by error propagation. In D. E. Rumelhart & J. L. McClelland (Eds.), *Parallel distributed processing* (Vol. 1. *Foundations*). Cambridge. MA: MIT Press.

Scofield. C. L.. Reilly. D. L.. Elbaum. C.. & Cooper. L. N. (1987). Pattern class degeneracy in an unrestricted storage density memory. In D. Z. Anderson (Ed.). *Neural information processing systems.* New York: American Institute of Physics.

Widrow. B.. & Hoff. M. E. (1960). Adaptive switching circuits. In *Institute of Radio Engineers, Western Electronic Show and Convention. Wescon Convention Record.* 4. 96–104.

Wilson. H. R.. & Cowan. J. D. (1973). A mathematical theory for the functional dynamics of cortical and thalamic nervous tissue. *Kybernetik.* 13. 55–80.

Hybrid Neural Network Architectures: Equilibrium Systems that Pay Attention

Neural Networks and Applications, eds. R. J. Mammone and Y. Zeevi
(Academic Press, San Diego, 1991) p. 81

In the last few years, neural network research has focused on connections between neural networks and statistical methods as well as the construction of interacting neural networks, each of which perform specialized functions. I believe that eventually we will have groupings of rule-based and learning systems that combine contextual information with pure classification in various complex arrangements in which information is fed forward and backward to the various networks. The following paper is an attempt to outline some of the possibilities; it may be regarded as a descendent of the Gensep idea. In some ways this may be characterized, from the point of view of statistics, as attempting to strike a balance between the bias of a system, that is to say its predilection toward one solution as opposed to another, as compared to the variance that occurs if a system is trained on limited data with no predilection whatsoever, a totally nonparametric solution.

Hybrid Neural Network Architectures: Equilibrium Systems That Pay Attention [1]

Leon N Cooper
Brown University

Attitudes toward Neural Networks have, in the short span of my memory, progressed from skepticism through romanticism to what we have at present: general realistic acceptance of neural networks as the preferred - most efficient, most economic - solution to certain classes of problems.

In this brief paper I would like to present an outline of what seem to me to be the major issues and some of the outstanding problems that confront us. In addition, I would like to present a brief account of how our own thinking has progressed

Neural Networks come in several broad categories:

1. Relaxation neural networks that can be regarded as methods of approximating non-linear dynamics.

2. Equilibrium neural networks that classify or assign probabilities

3. Equilibrium hybrid neural networks that via feed-forward and/or feed-back show some properties of relaxation of dynamic networks and display such phenomena as selective attention.

In what follows, we present hybrid equilibrium neural networks that are designed for high efficiency in classification and/or prob-

[1] The work on which this article is based was supported in part by the National Science Foundation, the Army Research Office, the Office of Naval Research and Nestor Incorporated.

81

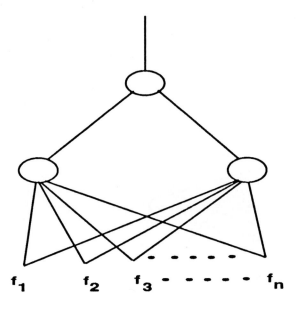

Figure 1

ability ranking and which further have some of the properties of relaxation networks and display selective attention.

Neural networks, in general, do not give optimal solutions. We may regard them in many ways as giving sometimes adequate solutions, sometimes very rapidly. Even training a 3-neuron network as has been shown by Blum and Rivest (Blum 89) is NP-complete. Training a general network is NP-complete, even with only three examples and with two-bit inputs and in some cases they can't even approximate well (Judd 87).

Suppose we attempt to train a general neural network made of the usual individual elements (Fig. 1), let us say one with m levels k_m hidden or internal units per level as in Figure 2, using the generalized gradient descent method (Back Propagation) (Werbos, 74; Rumelhart 86).

The network can be characterized as a non-linear mapping of an input, f, into an output

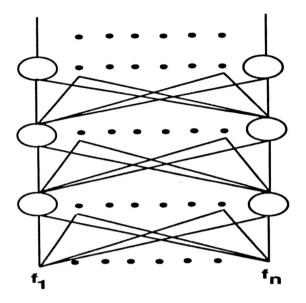

Figure 2

$$M[f] \qquad (1)$$

We can construct an energy function as follows:

$$E = \frac{1}{2} \sum_A (t^A - M[f^A])^2 \qquad (2)$$

Where t^A is the desired (or target) response to the input f^A.

A general picture of the energy as a function of the neural network weights and thresholds is shown in figure 3. It displays many local minima that are far from the desired solution.

The object of the search is to find a solution, if one exists, of zero energy (all of the outputs precisely as desired). The gradient descent method guarantees that at each modification, the energy remains unchanged or decreases. But finding an adequate solution to this problem can be harder than finding the proverbial needle in an actual haystack.

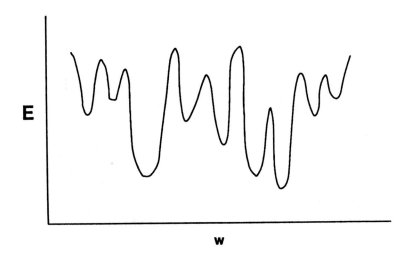

Figure 3

If, for example, the best solution is a network of 'weakly coupled' sub-networks as in figure 4. It is highly unlikely, that by a method of gradient descent, we can find this solution among all the other possibilities.

Referring to Figure 3, as anyone who has used back propagation knows, the starting point is often as important as the method of descent; one starts over and over to try to find an adequate solution. Since the starting point is determined by the data presented and the choice of initial weights, the problem is one of choosing these in an appropriate fashion that will allow one to descend to an appropriate solution.

As most workers in the field have concluded, the general black learning box is a romantic illusion, and what we are now doing is constructing specific architectures for classes of problems. We have recognized for many years is that different types of neural networks work best in different situations. For example, back propagation or charge clustering networks (Scofield 88 a, b) enable us to generate new representations and work extremely well in some situations (such as, for example, the parity problem). They have proved very useful in dimension reduction reducing an original high dimensional

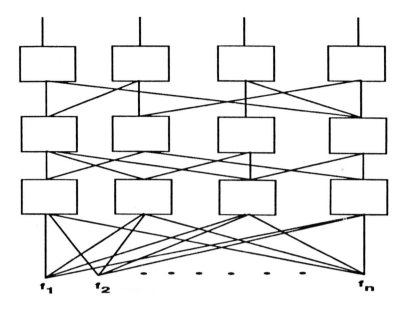

Figure 4

input space, thus helping to overcome what is known as the curse of dimensionality. RCE networks (Reilly 82) are best for classification. They work very well when the feature set is reasonable.

In what follows, we outline procedures we have employed for the past few years in putting together what we have called multi-neural network or hybrid neural network system that consist of different types of neural networks functioning together in a single overall network or system using each to its best advantage. (Reilly 87, 88).

We begin by discussing the problem of classification. To classify a decision space, we must choose an appropriate feature set so that the decision space is reasonably divided. In addition, one has the problem of resolution in each of the features. Classification is possible when the classes are separable, at least in principle, and the appropriate features have been chosen. The RCE solution for classification (newly fashionable as radial basis functions or radius limited perceptron) was introduced to map non-linear boundaries. This method

1. always can do this (produces a Borel covering of the space)

2. always converges

3. learns very rapidly

4. spends most of its learning time at complex boundaries

5. can add classes without retraining since learning is local

6. does not have to specify size of network beforehand

Thus the RCE solution achieves the goal of rapid learning and concentration of effort on difficult regions. Among the problems of RCE, it does not work well if

1. classification is not possible

2. features are badly chosen

Also, the RCE solution is not economical in number of prototypes; it is not efficient for serial computation

If classification is not possible, the RCE neural network has been generalized to assign probabilities. [Scofield 1987]

Dealing with the question of economy, we note that in spaces higher than two dimensions, no computationally efficient method exists to delineate geographies. However, the brain, inspiration for networks, very likely performs this function using many neurons in parallel. Further the problem of computational efficiency, important for simulations on Von-Neuman computers, would likely become much less significant as parallel hardware (neural network chips) become available.

In many cases, features are chosen initially in a manner that allows some classification and results in confusion regions. This problem is attacked using hybrid architectures. In one situation for a given set of features, the decision space appears as in figure 5.

For another set of features, the space might be divided as shown in Figure 6.

To the extent that the feature space is well divided, a method such as that of RCE that has been described in detail elsewhere does

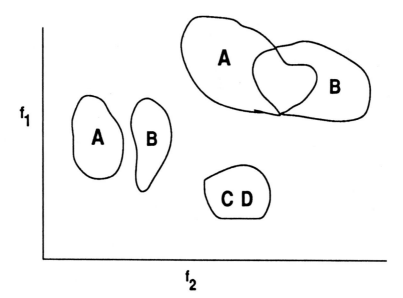

Figure 5

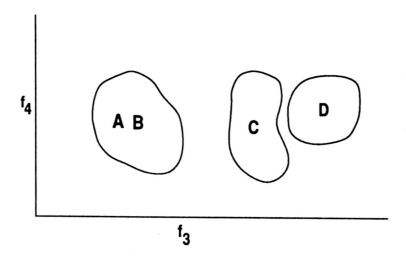

Figure 6

excellent and very rapid classification. (e.g. Reilly 82) Referring to
Figure 5, there will be no problem to separate and A and B regions
for the feature space for f_1, f_2 . Referring to Figure 6, there will
be no problem in classifying the C and D regions for the feature
space f_3, f_4. However, the classification of A and B is not possible
using the features of f_3 and f_4 whereas the classification of C and
D will not be possible using features f_l and f_2. In Figure 5 we see
also another domain in which A and B are partially separated and
partially ov--lap.

Of course, it happens very often that no set of features can completely separate the classes. In such a situation one would need
probablistic estima⁻⁰s (Scofield, 87).

To resolve sucn problems, we constructed what we call multi-
neural network/coupled learning systems. The behavior of such systems has been described in detail elsewhere. [Reilly 1987] Such coupled learning systems automatically assign the classification task to
the appropriate network as indicated in Figures 7and 8. The input
A is assigned to the sub-network one where it is classified, while
the input C is automatically assigned to sub-network 2 where it is
classified. An extremely important point is that one does not have
to know in advance which network will classify which inputs. Coupled learning assigns the appropriate network as part of the learning
procedure.

Why not, we might ask, put the four features f_1, f_2, f_3, f_4 , into a
single network. The simplest way to understand the advantage of the
multi-neural network system is to consider the problem of resolution.

Suppose that one has n features, with N bit total input capacity.
Divided equally, one has N/n bits for each feature. But it might
be necessary in order to make some distinctions to have more than
N/n bits for certain features. Consider, for example, the problem of
distinguishing a U from a V. Some of the early multi-neural network
units might properly classify A's and B's, but deliver a confusion for
U's and V's; that is to say, deliver the judgement that the incoming
entry is either a U or V. Now this is regarded as valuable information,
because this confusion region can be directed to a network that can
separate the confusion. In effect, this network is constructed as a

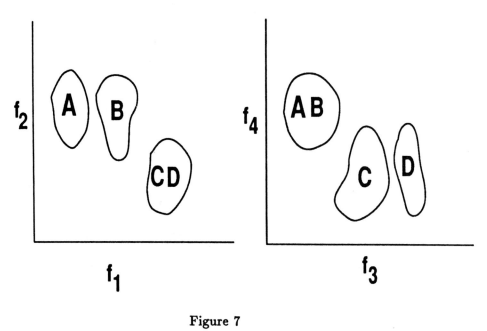

Figure 7

Figure 8

specialist where all of the resolution might be concentrated at the lower part of the figure. This is useful for separating U's and V's, but is not necessarily useful for separating other characters.

What we see here then, is that if one can automatically train a system of networks to perform functions such as this U, V separation, we are, in fact, using equilibrium networks in a manner that is reminicent of relaxation networks. Instead of feeding the output back to the same network, one is, in effect, feeding the output of one network into another. Another way of looking at this is to regard this as building selective attention into an equilibrium network. What the network is doing, is making a preliminary classification, saying that the class is either U or V; then as we do ourselves, it focuses attention on the element that will distinguish U's from V's. What the network does is to pass the provisional classification to a network which is a specialist at separating U's from V's.

It is clear, that in functioning this way, the network can perform much more efficiently than it could if it attempted to make the distinction between U's and V's as part of its overall classification. It can concentrate its entire attention for maximum resolution.on those features that are best at distinguishing U's from V's. In early systems, the network had to choose between the initial feature sets to decide which to emphasize. [Reilly 1988] In later more automated systems, we expect that the networks themselves, as part of the process, will construct the appropriate feature sets: these are what we call hybrid networks. My students as well as my colleagues at Nestor have been working with me on such systems for some time. Some examples are given in the following references. [Zemany 1989, Scofield 1990, Intrator 1990, Scofield 1991].

An early solution to this problem is to

- first construct the multi-neural network architecture, as was done for example in the Nestor system (NDS)

- then to find proper features and their distribution by trial and error

Thus the diagnostics of NDS enabled the user to see which classifications were not being made properly, and to construct a new

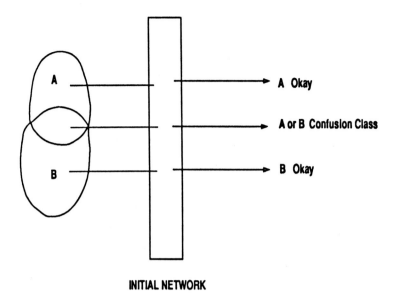

INITIAL NETWORK

Figure 9

network with the appropriate features–this enabled a highly efficient interaction between the user and the system and enabled the users to construct efficient neural networks quickly.

In some cases, such as the ECG analysis system (Carroll 89) the user discovered that the initial features were not sufficient to make the distinctions he desired; he had to add new features. The discovery process is made simpler by a system such as NDS that has diagnostics that allows one to focus quickly on the specific problem.

It is typical of classification problems that a good deal of the input data can be classified easily. What often happens is that it is very difficult to go beyond the initial easy classifications. The systems described above enable one to concentrate on those regions of the classification space that are particularly difficult to separate.

A more sophisticated solution to the problem of constructing a multi-neural network system is to design a system in which the initial multi-neural network system yields some classification and some confusion.

In an architecture we developed called GENSEP [Reilly 1988], the system constructs a separating network specialized to separate the $A - B$ confusion of figure 9 in a very simple way : by taking those features for which $f^A - f^B$ is the largest.

We note that the construction of these separaters is done automatically by the multi-network system and does not have to be put in by hand. A number of networks independently compete to accomplish the required class distinction. The multi-network system learns from experience which networks have the most effective separaters and those determine the final classification.

Hybrid multi-neural network architectures may be regarded as descendents of GENSEP.

- multi-network RCE does initial classification

- confusion regions are fed into sub-networks that can generate non-linear mappings of the original feature set specialized to separate the confusion regions. Back propagation or charge clustering are examples of learning procedures that can perform this function.

- this makes an enormous improvement in training time and accuracy because the time consuming gradient descent learning algorithms need function on very small but critical subsets of the entire data set.

- such a method is related to the advantage of training near boundaries (Ahmad 88).

This last point is illustrated for the perceptron in figure 10.

For Backpropagation using the energy function (2), there is no change in weights and/or thresholds if $t^A = M[f^A]$.

We must select patterns close to the decision region for faster action (figure 11).

Hybrid networks can be thought of as doing this automatically.

- multi - RCE does rapid classification of regions far from boundary

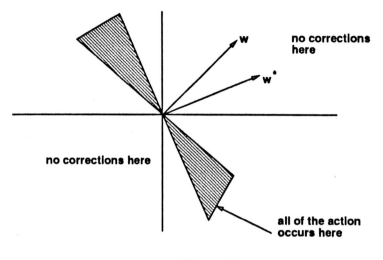

Figure 10

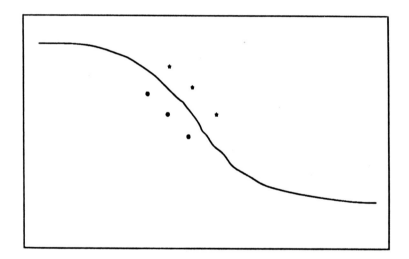

Figure 11

- confusion classes are by their nature near the boundary

- specialized non-linear mappings are fed again into multi-RCE for rapid classification

- note that these specialized mappings can be very different in different parts of the space

We note also that confusion class results can also be fed back to tune the original feature extractors so as to produce better separation.

This very brief summary only suggests the richness and variety of current efforts to enhance the efficacy of neural network architectures. What should be clear is that the question is no longer whether such architectures will prove to be useful. Rather this issue has become, which among these architectures, along with supporting hardware, is fastest, cheapest and/or most accurate. For a specific task, which is best.

Bibliography

[1] [Reilly 88] "GENSEP: A Multiple Neural Network Learning System with Modifiable Network Topology," Douglas Reilly, Christopher L. Scofield, Leon N. Cooper and Charles Elbaum. Abstracts of the First Annual International Neural Network Society Meeting, Boston 1988. Neural Networks, vol. 1, supplement 1, 1988.

[2] [Ahmad88] S. Ahmad and G. Tesauro, "Scaling and generalization in neural networks: A case study," in Proc. of the 1988 Connectionist Models Summer School, D.S. Touretzky, G.E. Hinton and T.J. Sejnowski, eds., Morgan Kaufmann Publishers, San Mateo, CA, 1989, pp. 3–10.

[3] [Blum89] A. Blum and R.L. Rivest, "Training a 3-node neural network is NP-complete," in Advances in Neural Information Processing Systems, D.S. Touretzky, ed., Morgan Kaufmann Publishers, San Mateo, CA, 1989, pp. 494–501.

[4] [Carroll 89] T.O. Carroll, H. Ved, D. Reilly, "Neural Network ECG Analysis" IJCNN, Washington, D. C., June 1989; Technical Report Nestor Incorporated.

[5] [Intrator90a] N. Intrator, "Feature extraction using an exploratory projection pursuit neural network," Ph.D. dissertation, Brown University, May 1990.

[6] [Intrator90b] N. Intrator, "Feature extraction using an unsupervised neural network," to appear in Proc. of the 1990 Connectionist Models Summer School, D.S. Touretzky, J.L. Elman and T.J. Sejnowski, eds., Morgan Kaufmann Publishers, San Mateo, CA, 1990.

[7] [Judd87] S. Judd, "Learning in networks is hard," in Proc. IEEE First Int. Conf. on Neural Networks, San Diego, CA, June 1987, vol. II, pp. 685–692.

[8] [Reilly82] D.L. Reilly, L.N. Cooper and C. Elbaum, "A neural model for category learning," Biol. Cybern., vol. 45, 1982, pp. 35–41.

[9] [Reilly87] D.L. Reilly, C.L. Scofield, C. Elbaum and L.N. Cooper, "Learning system architectures composed of multiple learning modules," in Proc. IEEE First Int. Conf. on Neural Networks, San Diego, CA, June 1987, vol. II, pp. 495–503.

[10] [Rumelhart86] D.E. Rumelhart, G.E. Hinton and R.J. Williams, "Learning representations by back-propagating errors," Nature, vol. 332, 1986, pp. 533–536.

[11] [Scofield87] C.L. Scofield, D.L. Reilly, C. Elbaum and L. N. Cooper, "Pattern class degeneracy in an unrestricted storage density memory," in Neural Information Processing Systems, Denver, CO, 1987, D.Z. Anderson, ed., American Institute of Physics, New York, NY, 1988, pp. 674–682.

[12] [Scofield88a] C.L. Scofield, "Learning internal representations in the Coulomb energy network," in Proc. IEEE First Int. Conf. on Neural Networks, San Diego, CA, June 1987, vol. I, pp. 271–276.

[13] [Scofield88b] C.L. Scofield, "Unsupervised learning in the N-dimensional Coulomb network," in Abstracts of the First Annual Int. Neural Network Society Meeting, vol. 1, suppl. 1, 1988, p. 129.

[14] [Scofield90] C.L. Scofield, "Neural network automatic target recognition by active and passive sonar signals," in Proc. of the Conf. on Neural Networks for Automatic Target Recognition, Tyngsboro, MA, May 1990.

[15] [Werbos74] P. Werbos, Beyond Regression: New Tools for Prediction and Analysis in the Behavioral Sciences. Ph.D. dissertation, Harvard Univeristy, 1974.

When Networks Disagree: Ensemble Methods for Hybrid Neural Networks

(with M. P. Perrone)

Neural Networks for Speech and Image Processing, ed. R. J. Mammone
(Chapman–Hall, 1993)

Learning from What's Been Learned: Supervised Learning in Multi-Neural Network Systems

(with M. P. Perrone)

Proceedings of the World Conference on Neural Networks (INNS, 1993)

Since neural networks are data driven (trained by actual data) and good uncorrupted data are often expensive to obtain (one of the recent advances in this area is the increasing availability of data such as the NIST or post-office sets of hand-printed numerals and letters), it is important to make efficient use of available data in training networks. The following two papers present some recent results concerning efficient uses of data and ensemble averaging when one is using hybrid neural networks.

10

When networks disagree: Ensemble methods for hybrid neural networks

MICHAEL P. PERRONE and LEON N. COOPER
Physics Department, Neuroscience Department, Institute for Brain and Neural Systems, Box 1843, Brown University, Providence, RI 02912, USA

10.1 Introduction

Hybrid or multi-neural network systems have frequently been employed to improve results in classification and regression problems (Cooper, 1991; Reilly *et al.*, 1988, 1987; Scofield *et al.*, 1991; Baxt, 1992; Bridle and Cox, 1991; Buntine and Weigend, 1992; Hansen and Salamon, 1990; Intrator *et al.*, 1992; Jacobs *et al.*, 1991; Lincoln and Skrzypek, 1990; Neal, 1992a,b; Pearlmutter and Rosenfeld, 1991; Wolpert, 1990; Xu *et al.*, 1992, 1990). Among the key issues are how to design the architecture of the networks; how the results of the various networks should be combined to give the best estimate of the optimal result; and how to make best use of a limited data set. In what follows, we address the issues of optimal combination and efficient data usage in the framework of ensemble averaging.

In this chapter we are concerned with using the information contained in a set of regression estimates of a function to construct a better estimate. The statistical resampling techniques of jackknifing, bootstrapping and cross-validation have proven useful for generating improved regression estimates through bias reduction (Efron, 1982; Miller, 1974; Stone, 1974; Gray and Schucany, 1972; Härdle, 1990; Wahba, 1990, for review). We show that these ideas can be fruitfully extended to neural networks by using the ensemble methods presented. The basic idea behind these resampling techniques is to improve one's estimate of a given statistic θ, by combining multiple estimates

Artificial Neural Networks for Speech and Vision. Edited by Richard J. Mammone. Published in 1993 by Chapman & Hall, London. ISBN 0 412 54850 X

of θ generated by subsampling or resampling of a finite data set. The jackknife method involves removing a single data point from a data set, constructing an estimate of θ with the remaining data, testing the estimate on the removed data point and repeating for every data point in the set. One can then, for example, generate an estimate of θ's variance using the results from the estimate on all of the removed data points. This method has been generalized to include removing subsets of points. The bootstrap method involves generating new data sets from one original data set by sampling randomly with replacement. These new data sets can then be used to generate multiple estimates for θ. In cross-validation, the original data is divided into two sets: one which is used to generate the estimate of θ, and the other which is used to test this estimate. Cross-validation is widely used neural network training to avoid over-fitting. The jackknife and bootstrapping methods are not commonly used in neural network training due to the large computational overhead.

These resampling techniques can be used to generate multiple distinct networks from a single training set. For example, resampling in neural net training frequently takes the form of repeated on-line stochastic gradient descent of randomly initialized nets. However, unlike the combination process in parametric estimation, which usually takes the form of a simple average in parameter space, the parameters in a neural network take the form of neuronal weights which generally have many different local minima. Therefore, we cannot simply average the weights of a population of neural networks and expect to improve network performance. Because of this, one typically generates a large population of resampled nets, chooses the one with the best performance and discards the rest. This process is very inefficient. Here we present ensemble methods which avoid this inefficiency, and avoid the local minima problem by averaging in functional space not parameter space. In addition, we show that the ensemble methods actually benefit from the existence of local minima, and that with the ensemble framework, the statistical resampling techniques have very natural extensions. All of these aspects combined provide a general theoretical framework for network averaging, which in practice generates significant improvement on real-world problems.

10.2 Basic Ensemble Method

In this section we present the Basic Ensemble Method (BEM), which combines a population of regression estimates to estimate a function $f(x)$ defined by $f(x) = E[y|x]$.

Suppose that that we have two finite data sets whose elements are all independent and identically distributed random variables: a training data set $\mathscr{A} = \{(x_m, y_m)\}$ and a cross-validatory data set $\mathscr{CV} = \{(x_l, y_l)\}$. Further, suppose that we have used $\mathscr{A}$ to generate a set of functions, $\mathscr{F} = f_i(x)$, each element of which approximates $f(x)$.[1] We would like to find the best approximation to $f(x)$ using $\mathscr{F}$.

One common choice is to use the **naive estimator**, $f_{\text{naive}}(x)$, which minimizes the mean square error relative to $f(x)$,[2]

$$\text{MSE}[f_i] = E_{\mathscr{CV}}[(y_m - f_i(x_m))^2],$$

thus

$$f_{\text{naive}}(x) = \arg\min_i \{\text{MSE}[f_i]\}.$$

This choice is unsatisfactory for two reasons: First, in selecting only one network from the population of networks represented by $\mathscr{F}$, we are discarding useful information that is stored in the discarded networks; second, since the $\mathscr{CV}$ data set is random, there is a certain probability that some other network from the population will perform better than the naive estimate on some other previously unseen data set sampled from the same distribution. A more reliable estimate of the performance on previously unseen data is the average of the performances over the population $\mathscr{F}$. Below, we will see how we can avoid both of these problems by using the BEM estimator, $f_{\text{BEM}}(x)$, and thereby generate an improved regression estimate.

Define the **misfit** of function $f_i(x)$, the deviation from the true solution, as $m_i(x) \equiv f(x) - f_i(x)$. The mean square error can now be written in terms of $m_i(x)$ as

$$\text{MSE}[f_i] = E[m_i^2].$$

The average mean square error is therefore

$$\overline{\text{MSE}} = \frac{1}{N} \sum_{i=1}^{i=N} E[m_i^2].$$

Define the BEM regression function $f_{\text{BEM}}(x)$, as

$$f_{\text{BEM}}(x) \equiv \frac{1}{N} \sum_{i=1}^{i=N} f_i(x) = f(x) - \frac{1}{N} \sum_{i=1}^{i=N} m_i(x).$$

If we now assume that the $m_i(x)$ are mutually independent with zero mean,[3]

---

[1] For our purposes, it does not matter how $\mathscr{F}$ was generated. In practice we will use a set of backpropagation networks trained on the $\mathscr{A}$ data set, but started with different random weight configurations. This replication procedure is standard practice when trying to optimize neural networks.

[2] Here, and in all of what follows, the expected value is taken over the cross-validatory set $\mathscr{CV}$.

[3] We relax these assumption in section 10.4, where we present the Generalized Ensemble Method.

we can calculate the mean square error of $f_{BEM}(x)$ as

$$\text{MSE}[f_{BEM}] = E\left[\left(\frac{1}{N}\sum_{i=1}^{i=N} m_i\right)^2\right]$$

$$= \frac{1}{N^2} E\left[\sum_{i=1}^{i=N} m_i^2\right] + \frac{1}{N^2} E\left[\sum_{i\neq j} m_i m_j\right]$$

$$= \frac{1}{N^2} E\left[\sum_{i=1}^{i=N} m_i^2\right] + \frac{1}{N^2} \sum_{i\neq j} E[m_i]E[m_j]$$

$$= \frac{1}{N^2} E\left[\sum_{i=1}^{i=N} m_i^2\right], \tag{1}$$

which implies that

$$\text{MSE}[f_{BEM}] = \frac{1}{N}\,\overline{\text{MSE}}. \tag{2}$$

This is a powerful result because it tells us that by averaging regression estimates, we can reduce our mean square error by a factor of N when compared to the population performance. By increasing the population size, we can in principle make the estimation error arbitrarily small! In practice, however, as N gets large our assumptions on the misfits, $m_i(x)$, eventually break down (see section 10.5).

Consider the individual elements of the population $\mathscr{F}$. These estimators will more or less follow the true regression function. If we think of the misfits functions as random noise functions added to the true regression function and these noise functions are uncorrelated with zero mean, then the averaging of the individual estimates is like averaging over the noise. In this sense, the ensemble method is smoothing in functional space and can be thought of as a regularizer with a smoothness assumption on the true regression function.

An additional benefit of the ensemble method's ability to combine multiple regression estimates is that the regression estimates can come from many different sources. This fact allows for great flexibility in the application of the ensemble method. For example, the networks can have different architectures or be trained by different training algorithms or be trained on different data sets. This last option – training on different data sets – has important ramifications. One standard method for avoiding over-fitting during training is to use a cross-validatory hold-out set.[4] The problem is that since the network is never trained on the hold-out data the network may be missing valuable information about the distribution of the data particularly if the total data set is small. This will always be the case for a single network using a cross-

---

[4] The cross-validatory hold-out set is a subset of the total data available to us and is used to determine when to stop training. The hold-out data is not used to train.

validatory stopping rule. However, this is not a problem for the ensemble estimator. When constructing our population $\mathscr{F}$, we can train each network on the entire training set and let the smoothing property of the ensemble process remove any over-fitting or we can train each network in the population with a different split of training and hold-out data. In this way, the population as a whole will see the entire data set while each network has avoided over-fitting by using a cross-validatory stopping rule. Thus the ensemble estimator will see the entire data set while the naive estimator will not. In general, with this framework we can now easily extend the statistical jackknife, bootstrap and cross-validation techniques (Efron, 1982; Miller, 1974; Stone, 1974) to find better regression functions.

10.3 Intuitive illustrations

In this section, we present two toy examples which illustrate the averaging principle which is at the heart of the ensemble methods presented in this chapter.

For our first example, consider the classification problem depicted in Figure 10.1. Regions A and B represent the training data for two distinct classes which are Gaussianly distributed. If we train a perceptron on this data, we find that hyperplanes, 1, 2, and 3 all give perfect classification performance for the training data; however, only hyperplane 2 will give optimal generalization performance. Thus, if we had to choose a naive estimator from this population of three perceptrons, we would be more likely than not to choose a hyperplane with poor generalization performance. For this problem, it is clear that the BEM estimator (i.e. averaging over the three hyperplanes) is more reliable.

For our second example, suppose that we want to approximate the Gaussian distribution shown in Figure 10.2a and we are given two estimates shown in Figure 10.2b. If we must choose either one or the other of these estimates we will incur a certain mean square error; however, if we average these two functional estimates we can dramatically reduce the mean square error. In Figure 10.2c, we represent the ensemble average of the two estimates

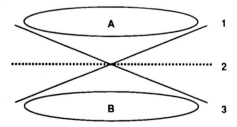

Figure 10.1 *Toy classification problem. Hyperplanes 1 and 3 solve the classification problem for the training data, but hyperplane 2 is the optimal solution. Hyperplane 2 is the average of hyperplanes 1 and 3.*

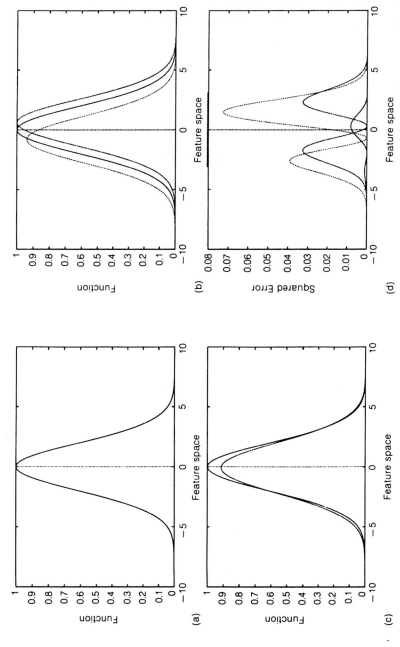

Figure 10.2 (a) Gaussian to be estimated; (b) two randomly chosen Gaussian estimates compared to the true Gaussian; (c) ensemble average estimate compared to the true Gaussian; (d) square comparison of the three estimates. Notice that the ensemble estimate gives the smallest square error. ——— : true function; - - - : estimate 1; ..... : estimate 2.

347

from Figure 10.2b. Comparing Figure 10.2b to Figure 10.2c, it is clear that the ensemble estimate is much better than either of the individual estimates. In Figure 10.2d, we compare the square error of each of the estimates.

We can push this simple example a little further to demonstrate a weakness of the Basic Ensemble Method and the need for a more general approach. Suppose that $x \sim \mathcal{N}(0, \sigma^2)$ and we are given $\mathcal{D} \equiv \{x_i\}_{i=1}^{i=N}$ and σ. We can estimate the true Gaussian by estimating its mean with

$$\mu \equiv \frac{1}{N} \sum_{j=1}^{j=N} x_j,$$

or we can use a modification of the jackknife method (Gray and Schucany, 1972) to construct a population of estimates from which we can construct an ensemble estimator. Define the ensemble estimate as

$$g_{\text{ensemble}}(x) \equiv \frac{1}{N} \sum_{j=1}^{j=N} g(x; \mu_{(-i)}),$$

where

$$\mu_{(-i)} \equiv \frac{1}{N-1} \sum_{j \neq i} x_j$$

and

$$g(x; \alpha) \equiv \frac{1}{\sqrt{2\pi\sigma^2}} e^{-(x-\alpha)^2/\sigma^2}.$$

We can now explicitly compare these two estimates using the mean integrated square error (MISE) of the estimates,

$$\text{MISE}[g(x; \alpha)] \equiv E_{\mathcal{D}} \left[\int_{-\infty}^{+\infty} (g(x; \alpha) - g(x; 0))^2 \, dx \right].$$

Calculating $\text{MISE}[g(x; \mu)]$ and $\text{MISE}[g_{\text{ensemble}}(x)]$ in this special case, it is easy to show that

$$\text{MISE}[g(x; \mu)] < \text{MISE}[g_{\text{ensemble}}(x)].$$

Comparing this result with equation (2), there seems to be a contradiction: the ensemble average is performing worse on average! The reason the ensemble performs worse on average is that two of the main assumptions from section 10.2 are wrong: The misfit functions from the population are not uncorrelated nor are they in general zero mean. Since these assumptions do not hold in general, we present an alternative formulation of the ensemble method in which these assumptions are not made (see section 10.4).

The above example helps to illustrate two important aspects of neural network regression which should be considered when performing ensemble averages: For neural networks, the existence of multiple local minima prohibits the simple parameter averaging we performed when we approximated

$g(x; \mu)$ in the example above. And in general, we do not know whether the function we are trying to estimate is representable by a given neural network as we assumed was true in the example above.

10.4 Generalized Ensemble Method

In this section we extend the results of section 10.2 to a generalized ensemble technique which always generates a regression estimate which is as low or lower than both the best individual regressor, $f_{\text{naive}}(x)$, and the basic ensemble regressor, $f_{\text{BEM}}(x)$, and which avoids overfitting the data. In fact, it is the best possible of any linear combination of the elements of the population $\mathscr{F}$.

Define the Generalized Ensemble Method estimator, $f_{\text{GEM}}(x)$, as

$$f_{\text{GEM}}(x) \equiv \sum_{i=1}^{i=N} \alpha_i f_i(x) = f(x) + \sum_{i=1}^{i=N} \alpha_i m_i(x),$$

where the α_i's are real and satisfy the constraint that $\sum \alpha_i = 1$. We want to choose the α_i's so as to minimize the MSE with respect to the target function $f(x)$. If again we define $m_i(x) \equiv f(x) - f_i(x)$ and in addition define the symmetric correlation matrix $C_{ij} \equiv E[m_i(x)m_j(x)]$ then we find that we must minimize

$$\text{MSE}[f_{\text{GEM}}] = \sum_{i,j} \alpha_i \alpha_j C_{ij}. \tag{3}$$

We now use the method of Lagrange multipliers to solve for α_k. We want α_k such that $\forall k$

$$\partial_{\alpha_k} \left[\sum_{i,j} \alpha_i \alpha_j C_{ij} - 2\lambda \left(\sum_i \alpha_i - 1 \right) \right] = 0.$$

This equation simplifies to the condition that

$$\sum_k \alpha_k C_{kj} = \lambda.$$

If we impose the constraint, $\sum \alpha_i = 1$, we find that

$$\alpha_i = \frac{\sum_j C_{ij}^{-1}}{\sum_k \sum_j C_{kj}^{-1}}. \tag{4}$$

If the $m_i(x)$'s are uncorrelated and zero mean, $C_{ij} = 0 \; \forall i \neq j$ and the optimal α_i's have the simple form

$$\alpha_i = \frac{\sigma_i^{-2}}{\sum_j \sigma_j^{-2}},$$

where $\sigma_i^2 \equiv C_{ii}$, which corresponds to the intuitive choice of weighting the f_i's by the inverse of their respective variances and normalizing. Combining equations (3) and (4), we find that the optimal MSE is given by

$$\text{MSE}[f_{\text{GEM}}] = \left[\sum_{ij} C_{ij}^{-1} \right]^{-1}. \tag{5}$$

The results in this section depend on two assumptions: the rows and columns of C are linearly independent and we have a reliable estimate of C. In certain cases where we have nearly duplicate networks in the population $\mathscr{F}$, we will have nearly linearly dependent rows and columns in C, which will make the inversion process very unstable and our estimate of C^{-1} will be unreliable. In these cases, we can use heuristic techniques to sub-sample the population $\mathscr{F}$ to assure that C has full rank (see section 10.6). In practice, the increased stability produced by removing near degeneracies outweighs any information lost by discarding nets. Since the C we calculate is the sample correlation matrix not the true correlation matrix, C is a random variable as are $\text{MSE}[f_{\text{GEM}}]$ and the optimal α_i's. Thus noise in the estimate of C can lead to bad estimates of the optimal α_i's. If needed, we can get a less biased estimate of C^{-1} by using a jackknife procedure (Gray and Schucany, 1972) on the data used to generate C.

Note also that the BEM estimator and the naive estimator are both special cases of the GEM estimator, and therefore $\text{MSE}[f_{\text{GEM}}]$ will always be less than or equal to $\text{MSE}[f_{\text{BEM}}]$ and $\text{MSE}[f_{\text{naive}}]$. An explicit demonstration of this fact can be seen by comparing the respective MSE's under the assumption that the $m_i(x)$'s are uncorrelated and zero mean. In that case, comparing equations (1) and (5), we have

$$\text{MSE}[f_{\text{BEM}}] = \frac{1}{N^2} \sum_i \sigma_i^2 \geqslant \left[\sum_i \sigma_i^{-2} \right]^{-1} = \text{MSE}[f_{\text{GEM}}],$$

with equality only when all of the σ_i are identical. This relation is easily proven using the fact that $a/b + b/a \geqslant 2 \ \forall a, b > 0$. Similarly, we can write

$$\text{MSE}[f_{\text{naive}}] = \sigma_{\text{min}}^2 \geqslant \left[\sum_i \sigma_i^{-2} \right]^{-1} = \text{MSE}[f_{\text{GEM}}].$$

Thus we see that the GEM estimator provides the best estimate of $f(x)$ in the mean square error sense.

10.5 Experimental results

In this section, we report on an application of the Generalized Ensemble Method to the NIST OCR database. The characters were hand-segmented, hand-labeled and preprocessed into 120 dimensional feature vectors by convolution with simple kernels. The database was divided into three types (numbers, uppercase characters and lowercase characters) and each of these types was divided into independent training, testing and cross-validatory sets with sizes listed in Table 10.1.

Table 10.1 *Database divisions*

Data set	Training set	CV set	Testing set	Classes
Numbers	13241	13241	4767	10
Uppercase	11912	11912	7078	26
Lowercase	12971	12970	6835	26

We trained a population of 10 single hidden unit layer backpropagation networks for a variety of different hidden unit layer sizes for each type of data. Each network was initialized with a different random configuration of weights. Training was stopped using a cross-validatory stopping criterion. For simplicity, we calculated the weights for the GEM estimator under the assumption that the misfits were uncorrelated and zero mean.

Straight classification results are shown in Figures 10.3, 10.5 and 10.7. In these plots, the classification performance of the GEM estimator (labeled 'Ensemble'), the naive estimator (labeled 'Best Individual') and the average estimator from the population (labeled 'Individual') are plotted versus the number of hidden units in each individual network. Error bars are included for the average estimator from the population. In all of these plots there is an increase in performance as the number of hidden units increases. Notice, however, that in all of these results the ensemble estimator was not only better than the population average but it was also as good as or better than the naive estimator in every case.

Typically for character recoginition problems, it is worse for the network to make an error than it is for the network to reject a pattern. This weighted cost

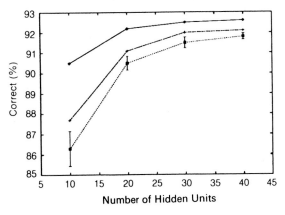

Figure 10.3 *Upper case percentages.* ◆ *: ensemble;* + *: best individual;* ■ *: individual.*

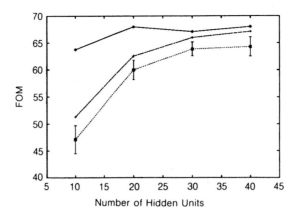

Figure 10.4 *Upper case FOM.* ◆ : *ensemble;* + : *best individual;* ■ : *individual.*

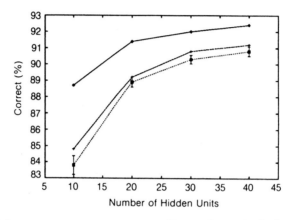

Figure 10.5 *Lower case percentage.* ◆ : *ensemble;* + : *best individual;* ■ : *individual.*

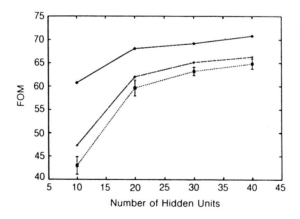

Figure 10.6 *Lower case FOM.* ◆ : *ensemble;* + : *best individual;* ■ : *individual.*

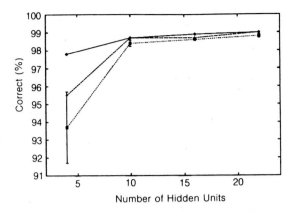

Figure 10.7 *Numbers percentage.* ◆ *: ensemble;* + *: best individual;* ■ *: individual.*

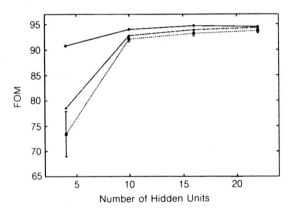

Figure 10.8 *Numbers FOM.* ◆ *: ensemble;* + *: best individual;* ■ *: individual.*

can be taken into account by calculating a Figure of Merit (FOM) instead of a straight performance measure. We define our FOM as follows:

$$\text{FOM} \equiv \%\text{Correct} - \%\text{Rejected} - 10(\%\text{Error}).$$

In our simulations we found an optimal rejection threshold for each network based on the cross-validatory set. FOM results are shown in Figures 10.4, 10.6 and 10.8. Again, notice that in all of these results, just as in the straight classification results, the ensemble estimator was not only better than the population average but it was also better than the naive estimator.

These results for a difficult, real-world problem show that the GEM estimator is significantly and dramatically better than standard techniques.

It is important to consider how many networks are necessary for the ensemble methods presented in this paper to be useful. If we take the BEM result seriously (equation 2), we should expect that increasing the number of networks in the population can only improve the BEM estimator. However as stated in section 10.2, eventually our assumptions on the misfits breakdown and equation (2) is no longer valid. This fact is clearly demonstrated in Figure 10.9, where we show the FOM performance saturate as the number of nets in the population increases. In the figure, we see that saturation in this example occurs after only six or eight nets are in the ensemble population. This is a very interesting result because it gives us a measure of how many distinct[5] nets are in our population. This knowledge is very useful when sub-sampling a given population. This result also suggests a very important observation: Although the number of local minima in parameter space is extremely large, the number of distinct local minima in functional space is actually quite small!

We can make another important observation if we compare Figure 10.9 with Figure 10.4. Consider the value of the FOM on the test data for an ensemble of 4 networks (Figure 10.9). Compare this value to population average FOM for nets with 40 hidden units (Figure 10.4). These values are not

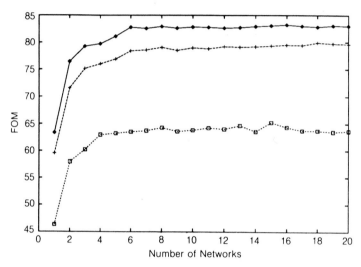

Figure 10.9 *Ensemble FOM versus the number of nets in the ensemble. Ensemble FOM graphs for the upper case training, cross-validatory and testing data are shown. Each net in the populations had 10 hidden units. The graphs are for a single randomly chosen ordering of 20 previously trained nets. No effort was made to optimally choose the order in which the nets were added to the ensemble. Improved ordering gives improved results.* ◆ *: train FOM;* + *: CV FOM;* □ *: test FOM.*

---

[5] By 'distinct', we mean that the misfits of two nets are weakly correlated. It is, of course, arguable what should be considered weakly correlated. For the purposes of this chapter, networks are distinct if the related correlation matrix, C, has a robust inverse.

significantly different; however, training a population of large nets to find the best estimator is computationally much more expensive than training and averaging a population of small nets. In addition, small networks are more desirable since they are less prone to over-fitting than large networks.[6]

It is also interesting to note that there is a striking improvement for an ensemble size of only two.

10.6 Improving BEM and GEM

One simple extension of the ensemble methods presented in this chapter is to consider the BEM and GEM estimators of all of the possible populations which are subsets of the original network population $\mathscr{F}$.[7] All the information we need to perform subset selection is contained in the correlation matrix, C, which only has to be calculated once.

In principal, the GEM estimator for $\mathscr{F}$ will be a better estimator than the GEM estimator for any subset of $\mathscr{F}$, however, in practice, we must be careful to assure that the correlation matrix is not ill-conditioned. If, for example, two networks in the population are very similar, two of the rows of C will be nearly collinear. This collinearity will make inverting the correlation matrix very error prone and will lead to very poor results. Thus in the case of the GEM estimator it is important to remove all duplicate (or nearly duplicate) networks from the population $\mathscr{F}$. Removing duplicates can be easily done by examining the correlation matrix. One can remove all networks for which the dot product of its row in the correlation matrix with any other row in the correlation matrix is above some threshold. This threshold can be chosen to allow a number of nets equal to the number of distinct networks in the population, as described in section 10.5.

An alternative approach (Wolpert, 1990), which avoids the potential singularities in C, is to allow a perceptron to learn the appropriate averaging weights. Of course, this approach will be prone to local minima and noise due to stochastic gradient descent, just as the original population $\mathscr{F}$ was; thus we can train a population of perceptrons to combine the networks from $\mathscr{F}$ and then average over this new population. A further extension is to use a nonlinear network (Jacobs et al., 1991; Reilly et al., 1987; Wolpert, 1990) to learn how to combine the networks with weights that vary over the feature space, and then to average an ensemble of such networks. This extension is reasonable since networks will, in general perform better in certain regions of the feature space than in others.

---

[6] Of course we cannot make the individual nets too small or they will not have sufficient complexity.

[7] This approach is essentially the naive estimator for the population of BEM and GEM estimators. Averaging over the population of BEM or GEM estimators will not improve performance.

In the case of the BEM estimator, we know that as the population size grows our assumptions on the misfits, $m_i(x)$, are not longer valid. When our assumptions break down, adding more nets to the population is a waste of resources, since it will not improve the performance, and if the nets we add have particularly poor performance, we can actually lower the performance of the BEM estimator. Thus, it would be ideal if we could find the optimal subset of the population $\mathscr{F}$ over which to average. We could try all the $2^N - 1$ possible non-empty subsets of $\mathscr{F}$, but for large N this search becomes unmanageable. Instead, we can order the elements of the population according to increasing mean square error[8], and generate a set of N BEM estimates by adding successively the ordered elements of $\mathscr{F}$. We can then choose the best estimate. The BEM estimator is then guaranteed to be at least as good as the naive estimator.

We can further refine this process by considering the difference between the mean square error for the BEM estimator for a population of N elements and the mean square error for the BEM estimator for the same population plus a new net. From this comparison, we find that we should add the new net to the population if the following inequality is satisfied:

$$(2N + 1)\mathrm{MSE}[\hat{f}_N] > 2 \sum_{i \neq \mathrm{new}} E[m_{\mathrm{new}}m_i] + \mathrm{E}[m_{\mathrm{new}}^2],$$

where $\mathrm{MSE}[\hat{f}_N]$ is the mean square error for the BEM estimator for the population of N and m_{new} is the misfit for the new function to be added to the population. The information to make this decision is readily available from the correlation matrix C. Now, if a network does not satisfy this criterion, we can swap it with the next untested network in the ordered sequence.

10.7 Conclusions

We have developed a general mathematical framework for improving regression estimates. In particular, we have shown that by averaging in functional space, we can construct neural networks which are guaranteed to have improved performance.

An important strength of the ensemble method is that it does not depend on the algorithm used to generate the set of regressors, and therefore can be used with any set of networks. This observation implies that we are not constrained in our choice of networks, and can use nets of arbitrary complexity and architecture. Thus the ensemble methods described in this chapter are completely general in that they are applicable to a wide class of problems, including neural networks and any other technique which attempts to minimize the mean square error.

---

[8] The first element in this sequence will be the naive estimator.

One striking aspect of network averaging is the manner in which it deals with local minima. Most neural network algorithms achieve sub-optimal peformance specifically due to the existence of an overwhelming number of sub-optimal local minima. If we take a set of neural networks which have converged to local minima and apply averaging we can construct an improved estimate. One way to understand this fact is to consider that, in general, networks which have fallen into different local minima will perform poorly in different regions of feature space, and thus their error terms will not be strongly correlated. It is this lack of correlation which drives the averaging method. Thus, the averaging method has the remarkable property that it can efficiently utilize the local minima that other techniques try to avoid.

It should also be noted that since the ensemble methods are performing averaging in functional space, they have the desirable property of inherently performing smoothing in functional space. This property will help avoid any potential over-fitting during training.

In addition, since the ensemble method relies on multiple functionally independent networks, it is ideally suited for parallel computation during both training and testing.

We are working to generalize this method to take into account confidence measures and various nonlinear combinations of estimators in a population.

Acknowledgements

We would like to thank the members of the Brown University Institute for Brain and Neural Systems, in particular Nathan Intrator for many useful discussions. We would also like to thank Nestor Inc. for making the NIST OCR database available to us.

Research was supported by the Office of Naval Research, the Army Research Office, and the National Science Foundation.

References

Baxt, W.G. (1992) Improving the accuracy of an artifical neural network using multiple differently trained networks. *Neural Computation,* **4**(5).

Bridle, J.S. and Cox, S.J. (1991) RecNorm: simultaneous normalization and classification applied to speech recognition, in *Advances in Neural Information Processing Systems 3.*

Buntine, W.L. and Weigend, A.S. (1992) Bayesian back-propagation. *Complex Systems,* **5**, 603–43.

Cooper, L.N. (1991) Hybrid neural network architectures: Equilibrium systems that pay attention, in *Neural Networks: Theory and Applications* (eds R.J. Mammone and Y. Zeevi), Academic Press, New York, pp. 81–96.

Efron, B. (1982) *The Jackknife, the Bootstrap and Other Resampling Plans.* SIAM, Philadelphia, PA.

Gray, H.L. and Schucany, W.R. (1972) *The Generalized Jackknife Statistic.* Dekker, New York, NY.

Hansen, L.K. and Salamon, P. (1990) Neural network ensembles. *IEEE Transactions on Pattern Analysis and Machine Intelligence*, **12**(10), 993–1000.

Härdle, W. (1990) *Applied Nonparametric Regression*. University of Cambridge Press, New York, NY.

Intrator, N., Reisfeld, D. and Yeshurun, Y. (1992) Face recognition using a hybrid supervised/unsupervised neural network. Preprint.

Jacobs, R.A., Jordan, M.I., Nowlan, S.J. and Hinton, G.E. (1991) Adaptive mixtures of local experts. *Neural Computation*, **3**(2).

Lincoln, W.P. and Skrzypek, J. (1990) Synergy of clustering multiple back propagation networks, in *Advances in Neural Information Processing Systems 2*.

Miller, R.G. (1974), The jackknife – a review. *Biometrika*, **61**(1), 1–16.

Neal, R.M. (1992a) Bayesian learning via stochastic dynamics, in *Advances in Neural Information Processing Systems*, (eds. J.E. Moody, S.J. Hanson and R.P. Lippmann) Morgan Kaufmann, San Mateo, CA.

Neal, R.M. (1992b) Bayesian mixture modeling by Monte Carlo simulation. *Technical report crg-tr-91-2*, University of Toronto.

Pearlmutter, B.A. and Rosenfeld, R. (1991) Chaitin-kolmogorov complexity and generalization in neural networks, in *Advances in Neural Information Processing Systems 3*.

Reilly, D.L., Scofield, C.L., Cooper, L.N. and Elbaum, C. (1988) Gensep: A multiple neural network learning system with modifiable network topology, in *Abstracts of the First Annual International Neural Network Society Meeting*.

Reilly, R.L., Scofield, C.L., Elbaum, C., and Cooper, L.N. (1987) Learning system architectures composed of multiple learning modules. *Proc. IEEE First Int. Conf. on Neural Networks*, Volume 2.

Scofield, C., Kenton, L. and Chang, J. (1991) Multiple neural net architectures for character recognition. *Proc. Compcon, San Francisco, CA*, 487–91.

Stone, M. (1974) Cross-validatory choice and assessment of statistical predictions (with discussion). *J. Roy. Stat. Soc. Ser. B*, **36**, 111–47.

Wahba, G. (1990) *Spline Models for Observational Data*. SIAM, Philadelphia. PN.

Wolpert, D.H. (1990) Stacked generalization. *Technical report LA-UR-90-3460*, Complex Systems Group, Los Alamos, NM.

Xu, L., Krzyzak, A. and Suen, C.Y. (1990) Associative switch for combining classifiers. *Technical report x9011*, Department of Computer Science, Concordia University, Montreal, Canada.

Xu, L., Krzyzak, A. and Suen, C.Y. (1992) Methods of combining multiple classifiers and their applications to handwriting recognition. *IEEE Transactions on Systems, Man, and Cybernetics*, **22**(3).

Proceedings of the World Conference on Neural Networks. © INNS, 1993

Learning from What's Been Learned:
Supervised Learning in Multi-Neural Network Systems[*]

Michael P. Perrone and **Leon N Cooper** [†]

Physics Department
Neuroscience Department
Institute for Brain and Neural Systems
Box 1843, Brown University
Providence, RI 02912
Email: mpp@cns.brown.edu

Abstract

This paper extends our previous results for averaged regression estimators (Perrone and Cooper, 1993b). We prove that averaged regression estimates always perform as good or better that their unaveraged counterparts. We show that this result derives from the notion of convexity and in this way demonstrate that a wide variety of optimization algorithms can benefit from averaging including: Mean Square Error, a general class of L_p-norm cost functions, Maximum Likelihood Estimation, Maximum Entropy, Maximum Mutual Information, the Kullback-Leibler Information, Penalized Maximum Likelihood Estimation and Smoothing Splines.

1   Introduction

Many neural network algorithms exist for improving network performance by combining two or more networks into one hybrid network (Perrone and Cooper, 1993b; Cooper, 1991; Reilly et al., 1988; Scofield et al., 1991; Baxt, 1992; Bridle and Cox, 1991; Buntine and Weigend, 1992; Hansen and Salamon, 1990; Intrator et al., 1992; Jacobs et al., 1991; Lincoln and Skrzypek, 1990; Neal, 1992; Pearlmutter and Rosenfeld, 1991; Wolpert, 1990; Xu et al., 1992). Perhaps not surprisingly, the simplest approach - averaging network outputs - also appears to be the most general. In this paper we demonstrate the averaging method's generality by showing that it is applicable to a very wide class of optimization costs.

Section 2 is a review of the averaging method. In Section 3, we discuss the notion of convexity and its relation to averaging and use a convexity inequality to generalize the averaging method to a variety of optimization techniques.

---

[*]To appear in "Proceedings of the World Conference on Neural Networks," INNS, 1993.

[†]Research was supported by the Office of Naval Research, the Army Research Office, and the National Science Foundation.

2 Review of the Averaging Method

In this section we review the Basic Ensemble Method (BEM) (Perrone and Cooper, 1993b) which combines a population of regression estimates to estimate a function $f(x)$ defined by $f(x) = E[y|x]$.

Suppose that have generated a set of functions, $\mathcal{F} = \{f_j(x)\}$, each element of which approximates $f(x)$. We would like to find the best approximation to $f(x)$ using $\mathcal{F}$.

Define the average regression function, $\overline{f}(x)$, [1] as

$$\overline{f}(x) \equiv \frac{1}{n} \sum_{j=1}^{j=n} f_j(x).$$

From the *Cauchy inequality* (Beckenbach and Bellman, 1965),

$$\left(\sum_{i=1}^{n} x_i y_i \right)^2 \leq \left(\sum_{i=1}^{n} x_i^2 \right) \left(\sum_{i=1}^{n} y_i^2 \right),$$

we have, by setting $y_i = 1 \ \forall \ i$, replacing the x_i with $f_j(x) - f(x)$ and averaging over the data, that

$$\mathrm{MSE}[\overline{f}] \leq \overline{\mathrm{MSE}[f]}. \tag{1}$$

Equation (1) [2] tells us that the average regressor is always as good or better than the population average. It can also be shown (Perrone and Cooper, 1993a) that the bias of the averaged regressor is equal to the bias of individual elements in $\mathcal{F}$ and therefore the averaging process is performing smoothing by reducing the variance of the estimate. Sufficient averaging will make the variance arbitrarily small.

3 Convexity and Averaging

A function, $h(x)$, is convex on an interval $[a, b]$ if $\forall \ x_1, x_2 \in [a, b] \ h\left(\frac{x_1 + x_2}{2}\right) \leq \frac{h(x_1) + h(x_2)}{2}$. If $\Phi(u)$ is a convex function on the interval $\alpha \leq u \leq \beta$ and $f(x; \omega)$ and $g(\omega)$ defined on $[a, b]$ satisfy $\alpha \leq f(x; \omega) \leq \beta \ \forall \ x$ and $g(\omega) \geq 0$ then *Jensen's inequality* (Gradshteyn and Ryzhik, 1980) states

$$\Phi\left(\frac{\int_a^b f(x; \omega) g(\omega) d\omega}{\int_a^b g(\omega) d\omega} \right) \leq \frac{\int_a^b \Phi(f(x; \omega)) g(\omega) d\omega}{\int_a^b g(\omega) d\omega}. \tag{2}$$

If we use our population of estimators, $\mathcal{F}$, to define $g(\omega) \equiv \frac{1}{n} \sum_{j=1}^{n} \delta(\omega - \omega_j)$, then Eqn. (2) becomes

$$\Phi(\overline{f}) \leq \overline{\Phi(f)} \tag{3}$$

which is just the discrete version of Jensen's Inequality. Thus, we have our fundamental result that *Averaged regressors perform as good or better than their unaveraged counterparts for all convex cost optimization.*

---

[1] Here and in all that follows, an overline indicates an average over the population of estimators.

[2] This result clearly extends to *generalized least mean squares* (Carroll and Ruppert, 1988) where for some non-zero function $g(x)$ we minimize $\sum_{i=1}^{n} (y_i - f(x_i))^2 / g^2(x_i)$.

This result can be extended to various optimization techniques (Perrone and Cooper, 1993a). We list the relevant relations below for L_p-norms, E_1 and E_2 (Gonin and Money, 1989); [3] [4] Maximum Entropy, H (Skilling, 1989); Maximum Mutual Information, I (Galland and Hinton, 1990); Maximum Likelihood Estimation, L (Wilks, 1962); the Kullback-Leibler Information, K (Härdle, 1991); Penalized Maximum Likelihood, P (Hastie and Tibshirani, 1990); and Splines, S (Wahba and Wold, 1975). The goal of each optimization techniques is in brackets next to the appropriate relation.

$$E_1(\{x_j\}) = \sum_{ij} \alpha_i |x_j|^{p_i} \qquad \Rightarrow \qquad E_1(\overline{x}) \leq \overline{E_1(x)} \qquad \text{[minimize]}$$

$$E_2(\{x_j\}) = \sum_{ij}(\alpha_i x_j^{p_{\alpha i}} - \beta_i x_j^{p_{\beta i}}) \qquad \Rightarrow \qquad E_2(\overline{x}) \leq \overline{E_2(x)} \qquad \text{[minimize]}$$

$$H(p) = -\sum_i p(x_i) \ln p(x_i) \qquad \Rightarrow \qquad H(\overline{p}) \geq \overline{H(p)} \qquad \text{[maximize]}$$

$$I(a,b) = H(a) + H(b) - H(ab) \qquad \Rightarrow \qquad I(\overline{p},b) \geq \overline{I(p,b)} \qquad \text{[maximize]}$$

$$L(p) = \prod_i p(x_i) \qquad \Rightarrow \qquad \ln L(\overline{p}) \geq \overline{\ln L(p)} \qquad \text{[maximize]}$$

$$K(f,g) = \int f \ln(\tfrac{f}{g}) \qquad \Rightarrow \qquad K(\overline{f},g) \leq \overline{K(f,g)} \qquad \text{[minimize]}$$

$$P(p) = -\ln \prod_i p(x_i) + \lambda \int (p'')^2 dx \qquad \Rightarrow \qquad P(\overline{p}) \leq \overline{P(p)} \qquad \text{[minimize]}$$

$$S(f) = \tfrac{1}{n} \sum_i (\hat{f}(x_i) - f_i)^2 + \lambda \int (f'')^2 dx \qquad \Rightarrow \qquad S(\overline{f}) \leq \overline{S(f)} \qquad \text{[minimize]}.$$

References

Baxt, W. G. (1992). Improving the accuracy of an artificial neural network using multiple differently trained networks. *Neural Computation*, 4(5).

Beckenbach, E. F. and Bellman, R. (1965). *Inequalities*. Springer-Verlag.

Bridle, J. S. and Cox, S. J. (1991). RecNorm: simultaneous normalization and classification applied to speech recognition. In *Advances in Neural Information Processing Systems 3*.

Buntine, W. L. and Weigend, A. S. (1992). Bayesian back-propagation. *Complex Systems*, 5:603–643.

Carroll, R. J. and Ruppert, D. (1988). *Transformation and weighting in regression*. Chapman and Hall.

Cooper, L. N. (1991). Hybrid neural network architectures: Equilibrium systems that pay attention. In Mammone, R. J. and Zeevi, Y., editors, *Neural Networks: Theory and Applications*, volume 1, pages 81–96. Academic Press.

Galland, C. C. and Hinton, G. E. (1990). Discovering high order features with mean field modules. In *Advances in Neural Information Processing Systems*. Morgan Kaufmann.

---

[3] For E_1 we have $\alpha_i \geq 0, p_i \geq 1$ and $\forall x_1$; and for E_2 we have $x_i > 0$, $\alpha_i, \beta_i \geq 0$, $p_{\alpha i} \in (-\infty, 0) \cap [1, +\infty)$ and $p_{\beta i} \in (0, 1)$.

[4] Of course for $p \in (0, 1)$, the negative of the cost function is not bounded below which may make some people squirm; but we can avoid unboundedness from below be requiring a cost term with sufficiently large p. However there is an even more bizarre characteristic of costs with $p < 0$ and negative costs with $0 < p < 1$. These costs weight large errors more lightly than small errors! The only way around this is to change their respective signs and stop using averaging.

Gonin, R. and Money, A. H. (1989). *Nonlinear Lp-Norm Estimation*. Marcel Dekker, Inc.

Gradshteyn, I. S. and Ryzhik, I. M. (1980). *Table of integrals, series and products*. Academic Press, Inc.

Hansen, L. K. and Salamon, P. (1990). Neural network ensembles. *IEEE Transactions on Pattern Analysis and Machine Intelligence*, 12(10):993–1000.

Härdle, W. (1991). *Smoothing Techniques: with Implementation in S*. Springer-Verlag, New York, NY.

Hastie, H. J. and Tibshirani, R. J. (1990). *Generalized Additive Models*. Chapman and Hall, New York, NY.

Intrator, N., Reisfeld, D., and Yeshurun, Y. (1992). Face recognition using a hybrid supervised/unsupervised neural network. Preprint.

Jacobs, R. A., Jordan, M. I., Nowlan, S. J., and Hinton, G. E. (1991). Adaptive mixtures of local experts. *Neural Computation*, 3(2).

Lincoln, W. P. and Skrzypek, J. (1990). Synergy of clustering multiple back propagation networks. In *Advances in Neural Information Processing Systems 2*.

Neal, R. M. (1992). Bayesian learning via stochastic dynamics. In Moody, J. E., Hanson, S. J., and Lippmann, R. P., editors, *Advances in Neural Information Processing Systems*, volume 5. Morgan Kaufmann, San Mateo, CA.

Pearlmutter, B. A. and Rosenfeld, R. (1991). Chaitin-kolmogorov complexity and generalization in neural networks. In *Advances in Neural Information Processing Systems 3*.

Perrone, M. P. and Cooper, L. N. (1993a). Using averaging to improve convex optimization. [In preparation].

Perrone, M. P. and Cooper, L. N. (1993b). When networks disagree: Ensemble method for neural networks. In Mammone, R. J., editor, *Neural Networks for Speech and Image processing*. Chapman-Hall. [In press].

Reilly, D. L., Scofield, C. L., Cooper, L. N., and Elbaum, C. (1988). Gensep: A multiple neural network learning system with modifiable network topology. In *Abstracts of the First Annual International Neural Network Society Meeting*.

Scofield, C., Kenton, L., and Chang, J. (1991). Multiple neural net architectures for character recognition. In *Proc. Compcon, San Francisco, CA, February 1991*, pages 487–491. IEEE Comp. Soc. Press.

Skilling, J., editor (1989). *Maximum Entropy and bayesian methods*. Kluwer Academic Publishers.

Wahba, G. and Wold, S. (1975). A completely automatic French curve: fitting spline functions by cross-validation. *Communications in Statistics, Series A*, 4:1–17.

Wilks, S. S. (1962). *Mathematical statistics*. John Wiley and Sons.

Wolpert, D. H. (1990). Stacked generalization. Technical report LA-UR-90-3460, Complex Systems Group, Los Alamos, NM.

Xu, L., Krzyzak, A., and Suen, C. Y. (1992). Methods of combining multiple classifiers and their applications to handwriting recognition. *IEEE Transactions on Systems, Man, and Cybernetics*, 22(3).

A Model of Prenatal Acquisition of Speech Parameters

(with B. S. Seebach, N. Intrator, P. Lieberman)

Proc. Natl. Acad. Sci. USA **91**, 7473 (1994)

In the years since we began working in this field there have been many applications of neural networks. Some of these are embedded in larger systems in various industrial processes; some are built into commercial products by the various young companies in this field. Rather than an exotic technology, neural networks have become part of the accepted set of tools, useful for solving certain problems.

In their current state, neural networks are probably best at problems related to pattern recognition. Some existing neural network systems can efficiently and rapidly learn to separate enormously complex decision spaces. The problem of coordinating many neural networks, each a specialist in making classifications in some portion of a decision space, has also been solved. Such neural networks, therefore, are the ones employed in the first commercial applications that have appeared. Products that recognize characters, assembly line parts or signatures, that can perform risk analysis and that make complex decisions mimicking or improving on human experts (such as underwriters) that can diagnose engine or assembly line problems are in the prototype stage and/or are already fielded. One expects further that the pattern recognition ability coupled with, and feeding back and forth to rule-based systems (as has already been done in some simple applications) will finally result in machines that, at least in some measure, share our ability to learn and duplicate our processes of reasoning — machines that might be said to think.

An additional development has been the use of the BCM algorithm (introduced as discussed in the first section of this collection to explain experimental results in visual cortex) for dimensional reduction or feature extraction in preprocessing. In some cases we have combined the preprocessing BCM algorithm with a form of back propagation or RCE type classifiers to improve results.

The following paper is an application of the BCM algorithm to the problem of the acquisition of early speech. The argument is that an algorithm that seems to be functioning in visual cortex, if it is also functioning in the auditory system, could enable categorical perception of speech sounds in the young infant with a reasonable approximation of the prenatal auditory environment.

Proc. Natl. Acad. Sci. USA
Vol. 91, pp. 7473–7476, August 1994
Neurobiology

A model of prenatal acquisition of speech parameters

BRADLEY S. SEEBACH*†, NATHAN INTRATOR‡§, PHIL LIEBERMAN¶, AND LEON N COOPER*§‖

Departments of *Neuroscience, ¶Cognitive and Linguistic Sciences, and ‖Physics, and §Institute for Brain and Neural Systems, Brown University, Providence, RI 02912; and ‡Sackler Faculty of Exact Sciences, Tel Aviv University, Ramat-Aviv 69978, Israel

Contributed by Leon N Cooper, April 18, 1994

ABSTRACT An unsupervised neural network model inductively acquires the ability to distinguish categorically the stop consonants of English, in a manner consistent with prenatal and early postnatal auditory experience, and without reference to any specialized knowledge of linguistic structure or the properties of speech. This argues against the common assumption that linguistic knowledge, and speech perception in particular, cannot be learned and must therefore be innately specified.

Chomsky's view that the "core" features of human linguistic ability are innate (1) is based in part on his assumption that linguistic knowledge, including the processes of speech perception, cannot be learned and thus must be preprogrammed. As this view is not universally accepted (2) and as there is some evidence for early alteration of phonetic perception by linguistic experience (3, 4), it appears useful to examine this assumption. In this paper we show that unsupervised neuron learning, as proposed to account for experimental data in visual cortex (5), can enable learned categorical perception of speech sounds with a reasonable approximation of the prenatal auditory environment. It thus follows that some aspects of early speech perception can be learned and therefore need not be innate.

Some of the strongest evidence for innate "linguistic" brain mechanisms comes from the study of speech. For example, the acoustic signals that differentiate stop consonants such as [b], [p], [d], [t], [g], and [k] are perceived categorically by adult (6) and infant (7) human listeners. The categorical behavioral responses of human adults and infants to these stop consonants has generally been interpreted as evidence for innate neural mechanisms tuned to the acoustic characteristics of speech (2, 7–10). In this view, linguistic development is not a learning process, but a process of selecting the discriminations useful to the maturing infant and forgetting those that are not useful (11). Supporting this belief is the discovery that infants, unlike adults, can discriminate phonetic units of languages they have never heard (12, 13). It is believed that such complex, cognitive behaviors of infants cannot arise from prenatal, experience-dependent modification of neurons.

However, such modifications have been shown to play a critical role in the development of neuronal selectivity. In visual cortex, for example, experience-dependent development proceeds rapidly from the onset of visual function through the so-called "critical period" and is strongly dependent on the visual environment in which the animal is raised. In the following study, we show that the prenatal auditory environment combined with a model of neuronal modification similar to that proposed for visual cortex can account for the acquisition of some basic speech contrasts as well as categorical perception of speech sounds.

The onset of hearing for humans begins as early as the 24th week of gestation (14, 15), raising the possibility that a lengthy "critical period" for auditory development may take place during the last several months of fetal life (16). Clearly, auditory experience in immature animals can alter frequency tuning (17) and spatial mapping (18) in auditory centers of the brain, and cognitive studies of human infants have shown that both prenatal (3) and postnatal (4) experiences may alter aspects of human speech perception prior to language acquisition. The fetus develops in an acoustically rich environment including the mother's voice. Low-frequency sounds dominate (19), whereas pure tones with higher frequencies (from external sources) are more attenuated. A certain amount of masking of low-frequency sounds is to be expected, though, due to the presence of low-frequency intrauterine noise, and tests of fetal hearing commonly use frequencies ranging from 500 Hz to 4 kHz. Low-frequency, broad-band noises are expected to be most efficient in producing responses in such tests (20). No adequate characterization of the transfer functions of the fetal middle ear exists (21).

The auditory periphery is characterized by broad bandpass tuning and poor phase-locking abilities during early mammalian development (22), though the "circuits" passing encoded information to auditory cortex appear to develop as functional units (for example, see ref. 23). Thus, encoded information reaching auditory cortical areas early on seems likely to consist of broad-band frequency information with consistent measures of intensity, but little or no phase information.

Fig. 1 shows an acoustic energy surface of the consonant–vowel (CV) syllable [ta] processed in a manner consistent with these constraints. A high degree of overlap in both time and frequency dimensions produces a "smooth" energy surface. Speech sounds typically display this type of energy surface, with peaks and valleys running in the general direction of the time axis. This stimulus is impoverished in comparison with those often used in speech studies but captures essential qualities of sounds that might be transmitted to the neonate's auditory centers: broad-band frequency information, little or no phase information, and fairly accurate representations of intensity. CV syllables processed in this manner were used as the speech data base for training and testing a neural network model.

We used a neural network based on the work of Bienenstock, Cooper, and Munro (BCM) (5) to determine whether the neural circuitry essential to speech perception might develop inductively in cortical auditory centers. The BCM theory has been used to describe the outcome and kinetics of experience-dependent synaptic plasticity in kitten striate cortex (25, 26). The mechanism for learning used by BCM is one that may be active in immature auditory cortex, as it requires only experience with sensory information.

Consider a neuron with input vector d ($=d_1, \ldots, d_n$), synaptic weight vector m ($=m_1, \ldots, m_n$), both in R^n, and activity (in the linear region) $c = m \cdot d$. The essential properties of the BCM neuron are determined by a modification threshold Θ_m (which is a nonlinear function of the history of activity

---

Abbreviations: CV, consonant–vowel; BCM, Bienenstock, Cooper, and Munro.
†Present address: Department of Neurobiology and Behavior, State University of New York, Stony Brook, NY 11794.

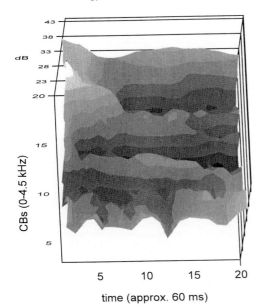

FIG. 1. The energy surface of approximately the first 55 ms of a pronunciation of the syllable [ta] by speaker BR. The ordinate is a perceptually defined scale of frequency known as critical bands (CBs) (24), and due to filter settings, represents frequencies from approximately 70 Hz to 4.5 kHz in a semilogarithmic manner. The abscissa is time, represented by 20 overlapping, 32-ms half-Hamming sampling windows. Each successive window was advanced 2 ms, so that the represented starting times for sampling windows ranged from 0 to 38 ms following the consonantal release. Higher decibel (dB) levels are represented by lighter-gray peaks.

of the neuron) and a ϕ function that determines the sign and amount of modification and depends on the current activity and the threshold Θ_m. The synaptic modification equations are given by

$$\frac{dm_i}{dt} = \mu\phi(c, \Theta_m)d_i, \qquad [1]$$

where in a simple form $\Theta_m = E[(m \cdot d)^2]$ and $\phi(c, \Theta_m) = c(c - \Theta_m)$.

In a lateral inhibition network of nonlinear neurons the activity of neuron k is given by $c_k = m_k \cdot d$, where m_k is the synaptic weight vector of neuron k. The inhibited activity and threshold of the kth neuron is given by $\bar{c}_k = \sigma(c_k - \eta \Sigma_{j\neq k} c_j)$ and $\hat{\Theta}_m^k = E[\bar{c}_k^2]$, for a monotone saturating function σ. The resulting stochastic modification equations for a synaptic vector m_k in such a network are given by

$$\dot{m}_k = \mu\left[\phi(\bar{c}_k, \hat{\Theta}_m^k)\sigma'(\bar{c}_k) - \eta\sum_{j\neq k}\phi(\bar{c}_j, \hat{\Theta}_m^j)\sigma'(\bar{c}_j)\right]d. \qquad [2]$$

This network is actually a first-order approximation to a lateral inhibition network. Its properties were discussed by Intrator and Cooper (27).

A neuronal delay-line mechanism similar to that proposed by Jeffress (28) is assumed to provide the network with a sequence of acoustic events beginning with the syllable onset. Many species, including humans (29), use neuronal delay lines for sound localization.

The phonetic features distinguishing English stop consonants are place of articulation and voicing (6–8, 10). Place of articulation can be viewed as trinary (labial, alveolar, or velar) and voicing as binary (voiced or unvoiced). The problem then

becomes to determine whether neurons can learn to detect these phonetic features in an appropriate environment without explicit preprogramming. The training paradigm we used is aimed at finding subphonemic features distinguishing between the different places of articulation independently of the voicing information. This is conceptually different from learning to recognize consonants as if they are unrelated, indivisible units, and requires a different training paradigm. For example, if one wishes to use a "supervised" model to recognize the stop consonant [k], one might train the network on a data base containing examples of all stop consonants in a variety of contextual situations (for example, ref. 30). In this manner, the net might be "taught" to accept all [k] sounds and reject all non-[k] sounds in many different environments. However, if one's purpose is to develop a neuron's selectivity for a subphonemic feature that distinguishes the consonant [k] from the consonants [p] and [t] (place of articulation), then the best training set is one that contains only that feature distinction.

Such a reduction of available phonetic distinctions in the training data base may typify a real process of development. It may be useful to think of the developing, peripheral sensory system as passing to fetal cortex a very simplified training set in the earliest stages—poorly focused frequency information, no phase locking, etc., in our example—and gradually increasing the complexity of available information, allowing for a progressive, cumulative development and refinement of neuronal selectivities.

A BCM network containing five "cells" was trained on a set of 74 pronunciations of [pa], [ka], and [ta] (unvoiced-stop syllables plus [a]). These syllables were pronounced by a single speaker and were not normalized for loudness or speaking rate. Averages of the three syllable types are shown as gray-scale images in Fig. 2A. The network was trained by random presentations of the 74 tokens until changes in neuronal selectivity became minimal. In the experiments reported here, 5 features were extracted from the 440-dimensional original space. The synaptic weights developed by each of the five cells have been reconstructed graphically (Fig. 2B) with the same axes as the training syllables (Fig. 2A), so that comparisons between the auditory inputs and the selectivities that developed might be made.

After training, cell 4 responded strongly to CV syllables containing labial stops and had excitatory synaptic weights corresponding to a large, low-frequency burst area and the region of the second formant frequency (F2) for [a]. These are the distinctive features of [pa] in the training set (Fig. 2A). Excitatory synaptic weights did not develop corresponding to strong but nondistinctive features of [pa], such as high energy in mid-range frequencies in the earliest time frames. Cells 1 and 5 responded most strongly to alveolar stop CV syllables. Both captured the short, high-frequency burst of [ta] and had negligible or inhibitory synaptic weights in high-frequency regions following the burst. An alveolar stop was identified when both of these cells responded strongly. Cells 2 and 3 together produced a strong response to velar-stop CV syllables. Both cells developed excitatory synaptic weights in extensive high-frequency regions, corresponding to the long duration of high-frequency burst energy for velar stops. The synaptic development of these cells differed in some respects: cell 2 had excitatory weights corresponding to a mid-frequency burst (often associated with [ka]), whereas cell 3 had excitatory weights in high-frequency areas that faded into the region of F3.

The BCM network effectively reduced the dimensionality of the original problem space from 440 dimensions to the 5 dimensions corresponding to the cells' selective responses. These cell selectivities correspond to subphonemic features. In order to interpret the results, a statistical classifier was trained to "phonemically" classify the output of the net's five cells as they responded to a testing data base. Although the

Neurobiology: Seebach *et al.*

Proc. Natl. Acad. Sci. USA 91 (1994)   7475

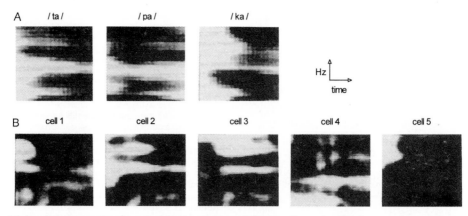

FIG. 2. (*A*) Average energy contour for each of the three training syllable types for speaker BR, shown as gray-scale images. The lighter areas of these images represent the presence of significant energy, with the ordinates of each image representing increasing frequency on the critical-band scale, and the abscissa of each image representing increasing time, which goes from 0 to 38 ms as marked by the start of each sampling window. (*B*) Gray-scale images of synaptic weights for the five cells of a BCM network following training.

BCM network was trained only with the unvoiced tokens of a single speaker, the classifier was trained on (5-dimensional) voiced and unvoiced data from the other speakers as well. The unsupervised feature extraction/classification method was discussed by Intrator (31). Results of the classification of training-set stimuli and testing stimuli from all three speakers are presented in Table 1.

Novel unvoiced-stop syllables from two different speakers, one male and one female, were correctly classified at a 98% rate. In contrast, neural network speech recognizers working on phoneme identification tasks have been highly successful in speaker-dependent tasks on entrained syllable types (30), and utterance recognition systems have been successful with multiple, novel speakers over very limited vocabularies (32, 33). Novel voiced-stop syllables were also successfully classified 96% of the time, despite the absence of voiced-stop syllables in the training data base. The generality of solutions is atypical for speech recognition systems and indicates that the BCM net is discovering features that yield categorical place-of-articulation distinctions.

Such an ability to internalize distinctive features of environmental sounds could explain an infant's ability to discriminate features of languages they have never heard. Phonemic features are defined perceptually. For example, Hindi has types of stop consonants that are distinguished by the presence or absence of aspiration (34). An adult speaker of Hindi can reliably produce and perceive this distinction. Adult speakers of English produce aspirated stop consonants under certain circumstances, though they no longer perceive a difference between an aspirated stop and an unaspirated stop. Because our supposed prenatal auditory neural network learns distinctive features of its sound environment, the child of an English-speaking woman could learn to make phonetic distinctions which his or her mother produces but cannot perceive.

Table 1. Classification results

Speaker	% correct (*n*)		
	Unvoiced stops	Voiced stops	Total by speaker
BR	99 (74)*	95 (73)	97 (147)
LN	98 (45)	91 (44)	94 (89)
JS	99 (75)	100 (75)	99 (150)
Total	98 (194)	96 (192)	97 (386)

*Unvoiced stops from speaker BR were the training set for this sample run.

The network and its training paradigm present a different approach to speaker-independent speech recognition. In this approach the speaker variability problem is addressed by training a network that concentrates on the distinguishing features of a single speaker, as opposed to training a network that concentrates on both the distinguishing and common features, on multi-speaker data.

Although we cannot yet be sure that the features discovered are invariant, the high degree of generalization across states of voicing, loudness, and speakers of both sexes gives reason to believe that neuronal selectivities such as those that develop in this model might provide a basis for perceptual abilities seen in early infancy. It thus appears that the assumption that a complex, cognitive behavior such as categorical perception of speech sounds cannot have its roots in prenatal experience is incorrect.

The questions become empirical. What is the appropriate prenatal auditory experience? Is this sufficient to produce the phonetic perceptual abilities underlying infant phonetic discriminations? Finally, what is really happening in this phase of early auditory development?

We wish to thank the members of the Institute for Brain and Neural Systems for many helpful discussions. Speech preprocessing was done at the Cognitive and Linguistic Science Department of Brown University. Research was supported by the National Science Foundation, the Office of Naval Research, and the Army Research Office.

1. Chomsky, N. (1986) *Knowledge of Language: Its Nature, Origin and Use* (Prager, New York).
2. Lieberman, P. (1991) *Uniquely Human: The Evolution of Speech, Thought, and Selfless Behavior* (Harvard Univ. Press, Cambridge, MA).
3. DeCasper, A. J. & Spence, M. J. (1986) *Infant Behav. Dev.* **9,** 133–150.
4. Kuhl, P. K., Williams, K. A., Lacerda, F., Stevens, K. N. & Lindblom, B. (1992) *Science* **255,** 606–608.
5. Bienenstock, E. L., Cooper, L. N & Munro, P. W. (1982) *J. Neurosci.* **2,** 32–48.
6. Liberman, A. M., Cooper, F. S., Shankweiler, D. P. & Studdert-Kennedy, M. (1967) *Psychol. Rev.* **74,** 431–461.
7. Eimas, P. D., Sisqueland, E. R., Jusczyk, P. & Vigorito, J. (1971) *Science* **171,** 304–306.
8. Blumstein, S. E. & Stevens, K. N. (1979) *J. Acoust. Soc. Am.* **66,** 1001–1017.
9. Lieberman, P. (1984) *The Biology and Evolution of Language* (Harvard Univ. Press, Cambridge, MA).
10. Chomsky, N. & Halle, M. H. (1968) *The Sound Pattern of English* (Harper & Row, New York).

Proc. Natl. Acad. Sci. USA 91 (1994)

11. Piattelli-Palmarini, M. (1989) *Cognition* **31**, 1–44.
12. Eimas, P. D., Miller, J. L. & Jusczyk, P. W. (1987) in *Categorical Perception*, ed. Harnad, S. (Cambridge Univ. Press, New York), pp. 161–195.
13. Kuhl, P. K. (1987) in *Handbook of Infant Perception*, ed. Salapatek, P. & Cohen, L. (Academic, New York), Vol. 2, pp. 275–381.
14. Murphy, K. P. & Smyth, C. N. (1962) *Lancet* **5**, 972–973.
15. Johansson, B., Wedenberg, E. & Westin, B. (1963) *Acta Oto-Laryngol.* **57**, 188–192.
16. Eggermont, J. J. (1986) *Acta Oto-Laryngol. Suppl.* **429**, 5–9.
17. Condon, C. D. & Weinberger, N. M. (1991) *Behav. Neurosci.* **105**, 416–430.
18. Moore, D. R., Hutchings, M. E., King, A. J. & Kowalchuk, N. E. (1989) *J. Neurosci.* **9**, 1213–1222.
19. Armitage, S. E., Baldwin, B. A. & Vince, M. A. (1980) *Science* **208**, 1173–1174.
20. Graniere-Deferre, C., Lecanuet, J. P., Cohen, H. & Busnel, M. C. (1985) *Acta Oto-Laryngol. Suppl.* **421**, 93–101.
21. Rubel, E. (1985) *Acta Oto-Laryngol. Suppl.* **421**, 114–128.
22. Walsh, E. J. & McGee, J. (1987) *Hearing Res.* **28**, 97–116.
23. Brugge, J. F., Reale, R. A. & Wilson, G. F. (1988) *Hearing Res.* **34**, 127–140.
24. Zwicker, E. (1961) *J. Acoust. Soc. Am.* **33**, 248.
25. Bear, M. F., Cooper, L. N & Ebner, F. F. (1987) *Science* **237**, 42–48.
26. Clothiaux, E. E., Bear, M. F. & Cooper, L. N (1991) *J. Neurophysiol.* **66**, 1785–1804.
27. Intrator, N. & Cooper, L. N (1992) *Neural Networks* **5**, 3–17.
28. Jeffress, L. A. (1948) *J. Comp. Phys. Psychol.* **41**, 35–39.
29. Jones, S. J. & Van der Poel, J. C. (1990) *Electroencephalogr. Clin. Neurophysiol.* **77**, 214–224.
30. Watrous, R. L. (1990) *J. Acoust. Soc. Am.* **87**, 1753–1772.
31. Intrator, N. (1992) *Neural Comput.* **4**, 98–107.
32. Lang, K. J., Waibel, A. H. & Hinton, G. E. (1990) *Neural Networks* **3**, 23–43.
33. Tom, M. D. & Tenorio, M. F. (1991) *Neural Networks* **4**, 711–722.
34. Lisker, L. & Abramson, A. S. (1964) *Word* **20**, 384–442.

The Ni1000: High Speed Parallel VLSI for Implementing Multilayer Perceptrons

(with M. P. Perrone)

A background force driving the expansion of neural networks has been the explosive increase in power and reduction in cost of computing hardware. Contrast the present with one example from the recent past: our early attempts to test our on-line recognition algorithms. One of us would write a number (say 3) on a borrowed magnetic bit pad. This would be converted to punch cards duly transported to the mainframe where possibly two days later (depending on the queue) the entered item would be identified (occasionally correctly as 3). The demonstration somehow lacked punch.

For many years computation has been dominated by considerations of speed and efficiency since complex time-consuming calculations had to be accomplished with limited means. We now have to adjust to an era of plenty – an era in which vastly increased memory and processing power make ease of use and convenience of access more important than speed and raw computational power.

In a curiously parallel fashion, neural networks have also been dominated by computational constraints; in some of these networks learning is very slow and in all of them, the intrinsically parallel computations are not efficiently executed on serial machines. This time of constraint may now be ending.

Recently Intel and Nestor under ARPA and ONR contracts have produced the Ni1000, a 3.6-million transistor VSLI chip that has 1024 sparsely connected neurons of 256-input dimensions with 64 outputs. This chip has an onboard RISC processor and onchip learning. It can perform up to 16-billion integer operations or about 33,000 classifications per second and is expected to provide real-time performance in various military and commercial applications.

Since many Ni1000 chips can run in parallel and since future generations of hardware will no doubt increase the number of neurons while decreasing power needs and cost, we appear to be entering an era of hardware plenitude for neural networks. Rather than struggling to make neural networks small and less complex, rather than conserving the number of neurons employed, we may in the future be primarily concerned with ease of use, accuracy, and ability to generalize.

Biology provides us with an example. The brain is a vast ensemble of neurons, each of which can be sluggish, uncertain, and prone to all of the ills that visit us in our journey through this vale of tears. Yet with this instrument we manage with stunning success (at least occasionally) to recognize patterns and make rapid, if sometimes incorrect, decisions in complex real-world situations.

To achieve its speed the designers of the Ni1000 minimized the use of multiplication units. As a result, instead of calculating inner products, what is calculated are what are called city-block distances. In the following paper Mike Perrone and I analyze how good an approximation this is.

The Ni1000: High Speed Parallel VLSI for Implementing Multilayer Perceptrons

Michael P. Perrone
Thomas J. Watson Research Center
P.O. Box 704
Yorktown Heights, NY 10598
mpp@watson.ibm.com

Leon N Cooper
Institute for Brain and Neural Systems
Brown University
Providence, Ri 02912
lnc@cns.brown.edu

Abstract

In this paper we present a new version of the standard multilayer perceptron (MLP) algorithm for the state-of-the-art in neural network VLSI implementations: the Intel Ni1000. This approach enables the standard MLP to utilize the parallel architecture of the Ni1000 to achieve on the order of 40000, 256-dimensional classifications per second. Due to the compact size and affordable price of the Ni1000, this classification speed could be available for the average personal computer.

Preference: Talk

Category: Implementations or Algorithms and Architectures

Keywords: City Block MLP, Ni1000, VLSI, parallel processors, high dimensionality

1 The Intel Ni1000 VLSI Chip

The Nestor/Intel radial basis function neural chip (Ni1000) contains the equivalent of 1024 256-dimensional artificial digital neurons with 5 bit resolution as well as 64 outputs capable of graded responses[Sullivan, 1993]. The chip includes a 20 Mips RISC CPU that supports learning and other operations. In addition it contains a five-stage, pipelined, fixed floating point mathematics unit. The Ni1000 uses CHMOSIV flash technology that permits unpowered, ten-year data retention. It contains 3.62 million transistors in an area of 13.2x15.2 millimeters. The chip can perform 20 billion integer operations per second and is several hundred times faster than conventional microprocessor given the same number of transistors.

The 1024 neurons with 256 dimensional input can perform 1024x256x5 bit comparisons each $2\mu sec$. Thus with full dimensionality, the chip is capable of at least 40,000 classifications per second ($25\mu sec$ per classification). On average, it is expected to be able to perform at a speed of about $15\mu sec$ per classification.

The Ni1000 neurons each have a set of learnable weights corresponding to each of the input dimensions and a learnable threshold value. In training mode, the weight commitment time is about $2\mu sec$ per prototype. Thus to execute a typical RCE [Reilly et al., 1982] or PRCE [Scofield et al., 1987] prototype entry requires about 2μsec. It is expected that in a normal run, one could perform 300-500 training passes per second [Sullivan, 1993]. A normal recognition or classification with RCE or PRCE would be about 1,000 times faster than a Sun Workstation.

To attain this great speed, the Ni1000 was designed to calculate "city block" distances (i.e. the l_1-norm) and thus to avoid the large number of multiplication units that would be required to calculate Euclidean dot products in parallel. Each neuron calculates the city block distance between its stored weights and the current input:

$$\text{neuron activity} = \sum_i |w_i - x_i| \tag{1}$$

where w_i is the neuron's stored weight for the ith input and x_i is the ith input. Thus the Ni1000 is ideally suited to perform both the RCE and PRCE algorithms or any of the other commonly used radial basis function (RBF) algorithms.

However, dot products are central in the calculations performed by most neural network algorithms (e.g. MLP, Cascade Correlation, etc.). Furthermore, for high dimensional data, the dot product becomes the computation bottleneck (i.e. most of the network's time is spent calculating dot products). If the dot product can not be performed in parallel there will be little advantage using the Ni1000 for such algorithms.

In this paper, we address this problem by showing that we can extend the Ni1000 to many of the standard neural network algorithms by representing the Euclidean dot product as a function of Euclidean norms and by then using a city block norm approximation to the Euclidean norm. Section 2, introduces the approximate dot product; Section 3 describes the City Block MLP which uses the approximate dot product; and Section 4 presents experiments which demonstrate that the City Block MLP performs well on the NIST OCR data and on human face recognition data.

2 Approximate Dot Product

Consider the following formula for the dot product

$$\vec{x} \cdot \vec{y} = \frac{1}{4}(\|\vec{x} + \vec{y}\|^2 - \|\vec{x} - \vec{y}\|^2) \tag{2}$$

where $\|\cdot\|$ is the Euclidean length (i.e. l_2-norm).[1] An approximation to this formula which can be implemented in parallel on the Ni1000 is given by:

$$\vec{x} \cdot \vec{y} \approx \frac{1}{4n}(|\vec{x} + \vec{y}|^2 - |\vec{x} - \vec{y}|^2) \tag{3}$$

where n is the dimension of the vectors and $|\cdot|$ is the city block length.

This approximation is motivated by the fact that in high dimensional spaces it is accurate for a majority of the points in the space. In fact, as the dimensionality of the data increases, the approximation is increases in accuracy for an increasingly large region of the space [Perrone, 1993].

Given a set of vectors, V, all with equal city block length, we measure the accuracy of the approximation by the ratio of the variance of the Euclidean lengths in V to the squared mean Euclidean lengths in V. If the ratio is low, then the approximation is good and all we must do is scale the city block length to the mean Euclidean length to get a good fit.[2]

---

[1]Note also that depending on the information available to us, we could use either

$$\vec{x} \cdot \vec{y} = \frac{1}{2}(\|\vec{x} + \vec{y}\|^2 - \|\vec{x}\|^2 - \|\vec{y}\|^2)$$

or

$$\vec{x} \cdot \vec{y} = \frac{1}{2}(\|\vec{x}\|^2 + \|\vec{y}\|^2 - \|\vec{x} - \vec{y}\|^2).$$

[2]Note that in Equation 3 we scale by $1/\sqrt{n}$. For high dimensional spaces this is a good approximation to the ratio of the mean Euclidean length to the city block length.

In Figure 1, we suggest an intuitive interpretation of why this approximation is reasonable. It is clear from Figure 1 that the approximation is reasonable for about 20% of the points on the arc in 2 dimensions.[3] In high dimensions, it is possible to show that the approximation is surprisingly good.

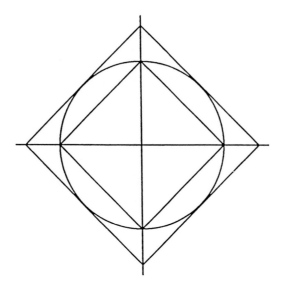

Figure 1: Two dimensional interpretation of the city block approximation. The circle corresponds to all of the vectors with the same Euclidean length. The inner square corresponds to all of the vectors with city block length equal the Euclidean length of the vectors in the circle. The outer square (tangent to the circle) corresponds to the set of vectors over which we will be making our approximation. In order to scale the outer square to the inner square, we multiple by $1/\sqrt{n}$ where n is the dimensionality of the space. The outer square approximates the circle in the regions near the tangent points. In high dimensional spaces, these tangent regions approximate a large portion of the total hypersphere and thus the city block distance is a good approximation along most of the hypersphere.

In particular, it can be shown that assuming all the vectors of the space are equally likely, the following equation holds [Perrone, 1993]:

$$\sigma_n^2 < \left(\frac{2n}{\alpha_n^2 (n+1)} - 1 \right) \mu_{\text{lower}}^2, \tag{4}$$

where n is the dimension of the space; μ_n is the average Euclidean length of the set of vectors with fixed city block length S; σ_n^2 is the variance about the average Euclidean length; μ_{lower} is the lower bound for μ_n and is given by $\mu_{\text{lower}} \equiv \alpha_n S / \sqrt{n}$; and α_n is defined by

$$\alpha_n \equiv \frac{n-1}{n+1} \sqrt{1 + \frac{en}{2\pi(n-1)} \left(\frac{n}{2} \right)^{\frac{1}{2n-2}}} + \frac{2}{n+1}. \tag{5}$$

From this equation we see that the ratio of σ_n^2 to μ_{lower}^2 in the large n limit is bounded above by 0.4. This bound is not very tight due to the complexity of the calculations required; however Figure 3 suggests that a much tighter bound must exist.

A better bound exists if we are willing to add a minor constraint to our high dimensional space [Perrone, 1993]. In the case in which each dimension of the vector is constrained such that the entire

---

[3]In fact, approximately 20% of the points are within 1% of each other and 40% of the points are within 5% of each other.

vector cannot lie along a single axis,[4] we can show that

$$\sigma_n^2 \approx \frac{2(n-1)}{(n+1)^2} \left(\sqrt{\frac{n}{S}} - 1\right)^2 \frac{\mu_{\text{lower}}^2}{\alpha_n^2},$$ (6)

where S is the city block length of the vector in question. Thus in this case, the ratio of σ_n^2 to μ_{lower}^2 decreases at least as fast as $1/n$ since n/S will be some fixed constant independent of n.[5] This dependency on n and S is shown in Figure 2. This result suggests that the approximation will be very accurate for many real-world pattern recognition tasks such as speech and high resolution image recognition which can typically have thousand or even tens of thousands of dimensions.

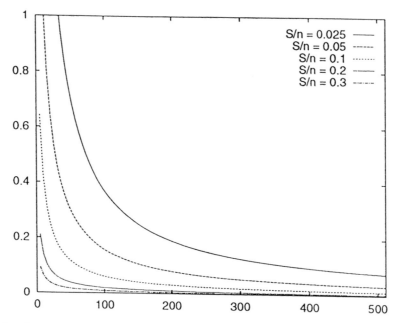

Figure 2: Plot of $\sigma_n/\mu_{\text{lower}}$ vs. n for constrained vectors with varying values of S/n. As S grows the ratio shrinks and consequently, accuracy improves. If we assume that all of the vectors are uniformly distributed in an n-dimensional unit hypercube, it is easy to show that the average city block length is $n/2$ and the variance of the city block length is $n/12$. Since S/n will generally be within one standard deviation of the mean, we find that typically $0.2 < S/n < 0.8$. We can use the same analysis on binary valued vectors to derive similar results.

We explore this phenomenon further by considering the following Monte Carlo simulation. We sampled 200000 points from a uniform distribution over an n-dimensional cube. The Euclidean distance of each of these points to a fixed corner of the cube was calculated and all the lengths were normalized by the largest possible length, $\sqrt{n}$. Histograms of the resulting lengths are shown in Figure 3 for four different values of n. Note that as the dimension increases the variance about the mean drops. From Figure 3 we see that for as few as 100 dimensions, the standard deviation is approximately 5% of the mean length.

3 The City Block MLP

In this section we describe how the approximation explained in Section 2 can be used by the Ni1000 to implement MLPs in parallel.

---

[4]For example, when the axes are constrained to be in the range $[0, 1]$ and the city block length of the vector is greater than 1. Note that this is true for the majority of the points in a n dimensional unit hypercube.

[5]Thus the accuracy improves as S increases towards its maximum value.

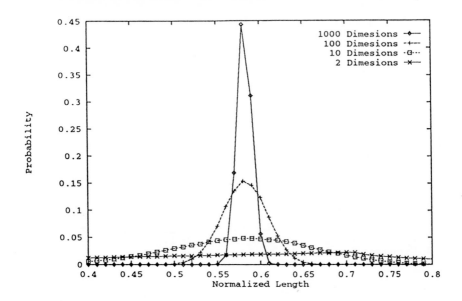

Figure 3: Probability distributions for randomly draw lengths. Note that as the dimension increases the variance about the mean length drops.

The standard functional form for MLP's is given by [Rumelhart et al., 1986]

$$f_k(x; \alpha, \beta) = \sigma\left(\alpha_{0k} + \sum_{j=1}^{N} \alpha_{jk}\sigma(\beta_{0j} + \sum_{i=1}^{d} \beta_{ij}x_i)\right) \tag{7}$$

were σ is a fixed ridge function chosen to be $\sigma(x) = (1 + e^{-x})^{-1}$; N is the number of hidden units; k is the class label; d is the dimensionality of the data space; and α and β are adjustable parameters. The alternative which we propose, the City Block MPL, is given by [Perrone, 1993]

$$g_k(x; \alpha, \beta) = \sigma\left(\alpha_{0k} + \sum_{j=1}^{N} \alpha_{jk}\sigma(\beta_{0j} + \frac{1}{4n}(\sum_{i=1}^{d}|\beta_{ij} + x_i|)^2 - \frac{1}{4n}(\sum_{i=1}^{d}|\beta_{ij} - x_i|)^2)\right) \tag{8}$$

where the two city block calculation would be performed by neurons on the Ni1000 chip.[6] The City Block MLP learns in the standard way by minimizing the mean square error (MSE),

$$\text{MSE} = \sum_{ik}\left(g_k(x_i; \alpha, \beta) - t_{ki}\right)^2 \tag{9}$$

where t_{ki} is the value of the data at x_i for a class k. The MSE is minimized using the backpropagation stochastic gradient descent learning rule [Werbos, 1974]: For a fixed stepsize η and each k, randomly choose a data point x_i and change γ by the amount

$$\Delta\gamma = -\eta\frac{\partial(\text{MSE}_i)}{\partial\gamma}, \tag{10}$$

where γ is either α or β and MSE_i is the contribution to the MSE of the ith data point.

---

[6]The dot product between the hidden and the output layers may also be approximated in the same way but it is not shown here. In fact, the Ni1000 could be used to perform all of the functions required by the network.

DATA SET	HIDDEN UNITS	STANDARD % CORRECT	CITYBLOCK % CORRECT	ENSEMBLE CITYBLOCK
Faces	12	94.6±1.4	92.2±1.9	96.3
Numbers	10	98.4±0.17	97.3±0.26	98.3
Lowercase	20	88.9±0.31	84.0±0.48	88.6
Uppercase	20	90.5±0.39	85.6±0.89	90.7

Table 1: Comparison of MLPs classification performance with and with out the city block approximation to the dot product. The final column shows the effect of function space averaging.

DATA SET	HIDDEN UNITS	STANDARD FOM	CITYBLOCK FOM	ENSEMBLE CITYBLOCK
Numbers	10	92.1±0.57	87.4±0.83	92.5
Lowercase	20	59.7±1.7	44.4±2.0	62.7
Uppercase	20	60.0±1.8	44.6±4.5	66.4

Table 2: Comparison of MLPs FOM. The FOM is defined as the 100 minus the number rejected minus 10 time the number incorrect.

4 Experimental Results

This section describes experiments using the City Block MLP on a 120-dimensional representation of the NIST Handwritten Optical Character Recognition database and on a 2294-dimensional grayscale human face image database. The results indicate that the performance of networks using the approximation is as good as networks using the exact dot product [Perrone, 1993].

In order to test the performance of the City Block MLP, we simulated the behavior of the Ni1000 on a SPARC station in serial. We used the approximation only on the first layer of weights (i.e. those connecting the inputs to the hidden units) where the dimensionality is highest and the approximation is most accurate. The approximation was not used in the second layer of weights (i.e. those connecting the hidden units to the output units were calculated in serial) since the number of hidden units was low and therefore do not correspond to a major computational bottleneck. It should be noted that for a 2 layer MLP in which the number of hidden units and output units are much lower than the input dimensionality, the majority of the computation is in the calculation of the dot products in the first weight layer. So even using the approximation only in the first layer will significantly accelerate the calculation. Also, the Ni1000 on-chip math coprocessor can perform a low-dimensional, second layer dot product while the high-dimensional, first layer dot product is being approximated in parallel by the city block units. In practice, if the number of hidden units is large, the approximation to the dot product may also be used in the second weight layer.

In the simulations, the networks used the approximation when calculating the dot product only in the feedforward phase of the algorithm. For the feedbackward phase (i.e. the error backpropagation phase), the algorithm was identical to the original backward propagation algorithm. In other words the approximation was used to calculate the network activity but the stochastic gradient term was calculated as if the network activity was generated with the real dot product. This simplification does not slow the calculation because all the terms needed for the backpropagation phase are calculated in the forward propagation phase In addition, it allows us to avoid altering the backpropagation algorithm to incorporate the derivative of the city block approximation. We are currently working on simulations which use city block calculations in both the forward and backward passes. Since these simulations will use the correct derivative for the functional form of the City Block MLP, we expect that they will have better performance. Results of these simulations will be presented at the conference.

In practice, the price we pay for making the approximation is reduced performance. We can avoid this problem by increasing the number of hidden units and thereby allow more flexibility in the network. This increase in size will not significantly slow the algorithm since the hidden unit activities are calculated in parallel. Results on these experiments will be presented at the conference.

In Table 1 and Table 2, we compare the performance of a standard MLP without the city block approximation to a MLP using the city block approximation to calculate network activity. In all cases, a population of 10 neural networks were trained from random initial weight configurations and the means and standard deviations were listed. The number of hidden units was chosen to give a reasonable size network while at the same time reasonably quick training. Training was halted by cross-validating on an independent hold-out set [Stone, 1974].

From these results, one can see that the relative performances with and with out the approximation are similar although the City Block is slightly lower.

We also perform ensemble averaging [Perrone, 1993, Perrone and Cooper, 1993] to further improve the performance of the approximate networks. These results are given in the last column of the table. From these data we see that by combining the city block approximation with the averaging method, we can generate networks which have comparable and sometimes better performance than the standard MLPs. In addition, because the Ni1000 is running in parallel, there is minimal additional computational overhead for using the averaging.[7]

5 Discussion

We have shown that we can generate an approximate MLP on the Ni1000 which has comparable performance to a standard MLP but which has the advantages that it is much faster and can have many more hidden units than its serial counterpart with minimal overhead. These results are very promising. They illustrate that it is possible to use the inherent high dimensionality of real-world problems to our advantage. Much more work needs to be done in this area to develop a viable version of the MLP for the Ni1000. For example, it is possible to completely replace the dot product with the city block approximation in both the feedforward and error backpropagation phases. This replacement in the backpropagation phase would allow us to calculate an exact derivative for the network and thereby guarantee that our stochastic gradient descent would converge to a local minimum. This convergence property should improve network performance. We are currently running simulations on this approach which we will present at the conference. In addition, there is currently no proof that such a network would be dense in the set of continuous functions.

In closing we comment that, in light of all of the usual difficulties that high dimensional spaces pose for estimation tasks (e.g. data sparsity, overfitting, local minima, etc.), it is nice to find a situation in which high dimensionality actually helps to improve performance.

Acknowledgements

This research was supported in part by the Office of Naval Research and the Army Research Office.

References

[Perrone, 1993] Perrone, M. P. (1993). *Improving Regression Estimation: Averaging Methods for Variance Reduction with Extensions to General Convex Measure Optimization*. PhD thesis, Brown University, Institute for Brain and Neural Systems; Dr. Leon N Cooper, Thesis Supervisor.

[Perrone and Cooper, 1993] Perrone, M. P. and Cooper, L. N. (1993). When networks disagree: Ensemble method for neural networks. In Mammone, R. J., editor, *Artificial Neural Networks for Speech and Vision*. Chapman-Hall. Chapter 10.

[Reilly et al., 1982] Reilly, D. L., Cooper, L. N., and Elbaum, C. (1982). A neural model for category learning. *Biological Cybernetics*, 45:35–41.

---

[7]The averaging can also be applied to the standard MLPs with a corresponding improvement in performance. However, for serial machines averaging slows calculations by a factor equal to the number of averaging nets.

[Rumelhart et al., 1986] Rumelhart, D. E., McClelland, J. L., and the PDP Research Group (1986). *Parallel Distributed Processing, Volume 1: Foundations*. MIT Press.

[Scofield et al., 1987] Scofield, C. L., Reilly, D. L., Elbaum, C., and Cooper, L. N. (1987). Pattern class degeneracy in an unrestricted storage density memory. In Anderson, D. Z., editor, *Neural Information Processing Systems*. American Institute of Physics.

[Stone, 1974] Stone, M. (1974). Cross-validatory choice and assessment of statistical predictions (with discussion). *Journal of the Royal Statistical Society, Series B*, 36:111–147.

[Sullivan, 1993] Sullivan, M. (1993). Intel and Nestor deliver second-generation neural network chip to DARPA: Companies launch beta site program. *Intel Corporation News Release*. Feb. 12.

[Werbos, 1974] Werbos, P. (1974). *Beyond Regression: New Tools for Prediction and Analysis in the Behavioral Sciences*. PhD thesis, Harvard University.

Thought and Mental Experience: The Turing Test

in press

The ultimate object of all of this work is to discover the material basis of mental activity. The possibility that this can be done inspires a certain paranoia, fear that somehow such knowledge would undermine what we regard as most human — the uniqueness and value of our own experience. I have frequently discussed why I believe this is not the case but, since we have tried to keep this particular volume focused, I will conclude with just one recent essay I wrote that deals with some of these questions. On reading this John Searle commented that he was in general agreement but questioned my sarcastic tone and hostile comments. On a re-reading I do find that I am a bit testy and am not sure why. After all it was written mostly in the month of June, 1993, a month Kay and I spent in a lovely apartment in Paris near the Place St. Sulpice, whose spacious living room, in which I write every morning, overlooks a remarkably calm and radiant inner court where, on a sunny morning, the air is filled with the sound of swooping swallows. Perhaps I went a bit overboard in revenge for the intellectual abuse some of us have suffered during the last twenty years.

In any case, here it is:

Thought and Mental Experience: The Turing Test

Leon N Cooper

Departments of Physics and Neuroscience
and the
Institute for Brain and Neural Systems
Brown University
Providence RI 02912

About the time I had to submit a title for this talk, I came across an article in the New 1York Times, "Can Machines Think? Humans Match Wits". The article describes a not entirely lighthearted match between computers and humans based on Alan Turing's famous test. And it startled me into a reverie on the nature of human consciousness - repressed for many years, in order that I would focus on more concrete questions such as the cellular and molecular basis for learning and memory storage.

You have heard from speakers at this symposium how much progress has been made; we seem to be on the threshold of understanding the physiological basis for learning and for memory storage. But how would knowledge of events on the molecular and cellular level relate to human thought? Is complex mental behavior a large system property of the enormous numbers of units that are the brain?

In an attempt to avoid this question Alan Turing, about forty years ago, proposed his famous test. Put either a machine or a person into a closed room, and communicate from outside this room in a manner that either the person or the machine can understand. (Let us say, by typing on a keyboard.) If one poses questions to whatever or whomever is in the room and if the responses do not allow one to distinguish between human and machine, then according to Turing, we can say that the machine (or the person) thinks.

Suppose, however, the entity in the room answers all questions with "I don't know." , "I don't understand what you are saying.". Human being or machine? Could be a programmed computer. Could be one of our denser colleagues. Suppose the questions concern next best moves in the game of tic-tac-toe. The inhabitant of the room answers with the right move every time. A very easy program to write and not distinguishable from a human being who doesn't have to be genius class. "That's not what Turing meant." you object. "Ask every conceivable question (or some representative subset) about any situation. If the black box answers in a manner indistinguishable from an "intelligent" human then it passes the test".

A new problem arises. Real human beings occasionally come up with somewhat quixotic answers to questions that rub them the wrong way. Suppose then, we confine ourselves to 'reasonable', 'rational' or perhaps 'logical' or 'computable' answers.

Formulated this way we do have a difficult problem but we can, in my opinion, expect that the task will be done. In a certain sense logic and reasoning are easy. Constructing the appropriate 'rational' answer to a question is a process that we already understand conceptually and can presently achieve if the universe of discourse is not too large (e.g. tic-tac-toe). Presumably, for larger problems, patience and more powerful computers are what are required.

The seductive computer metaphor for brain is based on the fact that any logical process can be replicated by a well defined sequence of simple steps, each of which can be executed in a variety of mechanical or electronic devices. Whether the software (the actual sequence of steps) or the hardware (transistors etched into wafers of silicon or living neurons inside a skull) are the same as that of the brain is regarded as irrelevant. As long as inputs and outputs match, the precise internal workings don't count - a point of view, ideally suited to be challenged by Turing's test.

The problem becomes to design an actual sequence of instructions (a program) that can be executed at breathtaking speed on that object of fascination, the computer (misnamed due to its arithmetic origin - however 'instruction executer' is a bit cumbersome) Delicious task for the unshaven persons in the backroom who love to hack. The program of so-called artificial intelligence (a moniker that really stuck) is then to reproduce various human activities by just such sequences of instructions so that the input-output to and from the object of worship becomes indistinguishable from (or even exceeds) that of a human being.

Thus, chess and checker-playing programs, programs that paint walls, stack blocks and so on ad-Ph.d dom.   In the end, we can be assured, all human activity (mental and otherwise) that is equivalent to the execution of a well-defined set of rules can be replicated. The chess program that finally defeats Gary Kasparov or his successor will no doubt be a *tour de force*. But, in fact, it demonstrates only how limited our capacities really are.[1]

How about non-logical or possibly 'non-computable' mental processes.   Recently, Penrose has argued that computers (Turing machines) cannot pass the Turing test because real brains performing real mental processes produce non-computable steps. And, in fact, non-logical processes, leaps (note: 'leaps', never 'steps') of invention or imagination pose somewhat of a problem for the gifted hacker. How does one write a set of rules that produce the unexpected, that compose the magnificent iambic line:

"The multitudinous seas incarnadine"

from two such grotesque, elephantine, English words?

To include such 'illogical' steps into a machine's repertoire of possible responses, requires a deeper understanding of how information is actually acquired, stored and manipulated in the brain.

It has been obvious to many of us for quite a while that the brain is only marginally a computing system. It is no more designed for logic or reason than the hand is designed to play the piano. If it is designed at all, the design concerns survival and, in an ordered

---

[1] The games we play - as perhaps the scientific problems we attack - are chosen from all those possible to fit our actual mental and physical capabilities: difficult enough to be interesting, but not so difficult that we have no chance to succeed. Tic-tac-toe is not serious; two dimensional chess taxes existing minds to the limit, while seven or eight dimensional chess is not usually attempted.

world, survival is enhanced by rapid (even if occasionally incorrect) decision making. If I may be permitted to quote myself. "The animal philosopher sophisticated enough to argue 'the tiger ate my friend but that does not allow me to conclude that he might want to eat me' might then be a recent development whose survival depends on other less sophisticated animals who jump to conclusions".

In recent years progress has been made in understanding how information may be acquired, stored and manipulated in a biological system. Although this work is in its infancy what are designated sometimes as neural network and/or learning systems based on large numbers of processing units that mimic neurons and that can be trained with supervised or unsupervised learning methods, have been extensively studied in the last generation. Systems now exist that recognize patterns, detect fraudulent behavior and can be said to display, at least on a primitive level, features such as recognition, generalization and association and which suggest some of the mental behavior associated with animal memory and learning.

Through years of painful education, somehow, our brain can achieve the ability to reason and execute rules as well as to associate. (It seems clear that association is much easier and more natural than reasoning - the preserve of some of our more cerebral types[2]).

I say somehow because it is not yet all worked out (There are, of course, major areas of current research directed to towards determining precisely how visual and other

---

[2] Among the glories of human intellectual achievement is, from the hazy associations that are the natural capability of the brain, just this creation and successful application to messy real world situations of precise language (to make what is said depend on what was said before) and logical reasoning. In a *fin de siecle* cop out, a California update of dialectical materialism, Hegel come to Los Angeles, we witness a seasonal twist of intellectual fashion that heralds fuzzy logic (a useful engineering tool) in a recent work become fuzzy thinking, as more appropriate than Aristotelian logic to describe the less than sharp boundaries of the real world. Thus it is proposed that we convert the razor sharp distinctions of the trained and athletic mind to the indistinct muttering and imprecise groping of the intellectual couch potato, that to the complexities of the world we add the charm of not knowing what we are talking about. *Sic transit...*

information processing are accomplished.) but, perhaps conceptually, it is there. We can already envisage how interacting associative and rule based systems could function together to reason and associate (at least in simple situations) as we do. We expect that as our experience with rule-based and data-driven learning systems increases, we will more and more be able to reproduce (and finally surpass) human reasoning - or more precisely the output product of human mental activity and thus design a system that passes the Turing test.[3] [We might distinguish human from machine however by noting the time taken to respond to "What is 2579362 x 1279854?"]

Having designed such a system (its practical and commercial value aside), do we have a machine (a generalized Turing machine?) that thinks? In effect, this is the question the Turing test is designed to evade. An evasion due, in part, to an excess of positivism - a fear of and/or aversion to mentalism or assumptions about the internal workings of the mind that cannot be directly verified by experience. An evasion that, to my mind, is totally contrary to the nature and purpose of scientific thinking.

Successful science has given us just such machines (actual or conceptual) that work behind the actual events, not totally unlike the fictions discussed by Professors Fussell and Barth. The greatest include Newton's laws, Maxwell's equations and Schrodinger's equation. Such entities as molecules and/or atoms were assumed to exist (an assumption that was vigorously contested in the 19th century with positivistic-type arguments) long before they were 'seen'. It is not necessarily the case that every element of the 'behind the scenes' machinery can be directly observed. In quantum mechanics, for example, the wave function is not directly observable. The consequences of this sometimes invisible machinery can, however, be put into correspondence with experience. The essence of the positivist argument (as actually employed by Einstein and Heisenberg) is not that we

---

[3] A more reasonable evolution - one that is already taking place - is the design of 'reasoning and/or decision making' systems that handle vast quantities of data with enormous rapidity, but applied to very specialized domains (e.g. airline reservations and seat assignments or fraud detection in credit card transactions)

cannot introduce entities that are not directly observable but, rather, that if an entity is not observable (e.g. absolute time in special relativity or simultaneous position and momentum in the quantum theory) it need not appear in theory.

Thus a satisfactory theory of mind not only is allowed but, in my opinion, *requires* the introduction of mental entities. We will be satisfied only when we see before us constructs that can have mental experience, when we see how they work, how they come about from more primitive entities such as neurons.

Further, in an ingenious construction (known as the "Chinese room"), Searle has argued that purely algorithmic behavior, in this case an English speaker answering Chinese questions about a Chinese story with Chinese answers by applying purely syntactic rules for manipulating Chinese symbols does not necessarily imply understanding of either Chinese or the story. Thus (although he agrees that computers might pass the Turing test) he argues that passing this test is not evidence of 'thinking' or 'understanding'.

All of this can be summarized by saying that the Turing test is not sufficient. At best, it provides us a perfectly responding 'black box" but no knowledge of how it all works - and that, in my opinion, is just the knowledge we want.[4]

What is more surprising, perhaps, is that for what is hardest to understand, the origin of the complex of mental experience: consciousness, awareness of ourselves, feeling, passing the Turing test is not necessary. Nothing in the perfect machine response gives us any indication of how such experience comes about. If I may quote myself again, we have absolutely no idea "how is it that a machine (for unless we accept a Cartesian

---

[4] We could, of course, take the black box apart to investigate the sequence of instructions gave us all of the right answers. It remains an open question whether this would necessarily shed any light on the phenomenon of mental activity.

dualism we are surely machines) can feel?" Or in the words of the late philosopher Hans Jonas, "the capacity for feeling, which arose in all organisms, is the mother value of all."

It is claimed by some (the proponents of "strong artificial intelligence) that mental qualities such as feeling, consciousness and understanding emerge as a consequence of the execution of those algorithms that lead to the appropriate Turing responses. This may or may not be the case. But no one has given any indication of how it would come about. Further, counter examples exist. Searle's Chinese Room, in my opinion, shows that correct algorithmic behavior can occur with no understanding, while every dog wagging its tail tells us that consciousness is there with little or no algorithmic behavior. It is much more plausible to believe that algorithmic behavior and consciousness are independent qualities.[5]

The deepest error and what may be most misleading is the attempt to equate the activity of the brain with a reasoning system. A more appropriate view is that of Albert Szent-Gyorgi:

"The brain is not an organ of thinking but an organ of survival, like claws and fangs. It is made in such a way as to make us accept as truth that which is only advantage. It is an exceptional, almost pathological constitution one has, if one follows thoughts logically through, regardless of consequences. Such people make martyrs, apostles, or scientist, and mostly end on the stake, or in a chair, electric or academic."

For the understanding of such mental behavior as consciousness or feeling, it is thus clear that the Turing test is not only not necessary but is totally irrelevant. One can reason, or

---

[5] Although some aspects of algorithmic behavior can clearly be achieved without conscious awareness, it remains possible that such awareness is required to attain the full power of our information processing abilities.

at least perform logical operations, without feeling: mechanical calculators do it all the time. Dogs and cats (even turtles, probably) feel but can't answer many questions[6]

Let us then put this famous and ingenious criterion to rest and confront again the underlying problem. Can we understand the human mind (all of its components: reasoning, feeling, self-awareness) and its presumed origin in that biological organ , the brain. Could we, in the extreme, construct a machine that was conscious [Whether a machine in this sense could be a Turing machine I will leave for others to answer.]. What are the steps required so that a machine (algorithmic or not) can experience mental activity? The non-sequitur "How would we know?" is an evasion [How do we know anything?]. Whether we can be sure that another creature and/or a machine is conscious is independent of the understanding of how it is that consciousness arises as a property of a very complex physical system. This, not reasoning power, is the profoundest mystery surrounding that biological entity, brain.

What has made this problem even more perplexing is the confusion of the various components of mental activity, the failure to distinguish what we believe we know - at least in principle - from what we do not yet understand at all. Although it is true that we are still far from understanding, for example, how the visual system processes and sorts information, such questions can be precisely formulated and it seems reasonable to believe that answers can be constructed from materials available. The same might be said for reasoning (logical and otherwise). But what is the source of our mental experience - our conscious awareness?

Mental awareness or consciousness themselves have many components, some easy to understand, some still incomprehensible:-an on-off switch; memory in storage, memory

---

[6] In Turing's defense (if he needs a defense) we should note that such mental behavior was not what he had in mind. His concern was the output of human reasoning - not human feeling.

in play; the distinction between ourselves and the external world (still the subject of vast philosophical cerebration) are, to my mind, easily understood.

But what , for example, is desire? No problem to program directives to pursue, functions to maximize or minimize, hot or cold to avoid, warm and cuddly to seek. What is the source of our self awareness of the process, and possibly the mother of all: our ability to feel.

These questions are sufficiently difficult so that we have been subjected to the usual evasions - Cartesian dualisms: variations of homunculus proposals; solutions of one mystery by invoking another: consciousness arises in the quantum measurement process or where gravity meets quantum theory; total refusal to confront the issue: consciousness arises "somehow" when a machine executes the proper algorithmic processes; total retreat under the cover of positivist philosophy: how would we know if a machine were conscious... and so on.

We have heard such arguments before. They seem to be typical responses to the frustration of failure in attacking really difficult scientific problems. First try and fail. Follow this by proving a solution is impossible or irrelevent. Toy with the notion that a new law of nature is involved. Then, when the solution is found, complain that it is really trivial or (even better) that it was suggested in some obscure comment once made in a paper you published a long time ago.

On a personal note, we experienced all of this in the course of developing a theory of superconductivity - also a complex and subtle consequence of an interacting many component system, in this case the quantum mechanics of electrons in a metal. After the fact one, rather well known, physicist expressed his disappointment that "such a striking phenomenon as superconductivity [was]... nothing more exciting than a footling small interaction between electrons and lattice vibrations." - thus missing the point in operatic style.

It could turn out that we must invoke a new "law of nature": pour the conscious substance into the machine - a position not unfriendly to a common view of mankind in its emerging years. But the conservative scientific position is to attempt to construct this seemingly new and surely very subtle property from the materials available - those given to us by physicists, chemists and biologists (as has been done many times before: celestial from earthly material, organic from inorganic substances, the concept of temperature from the motions of molecules - and so on.). If this cannot be done (perhaps one could be patient enough to give us a couple of years to try) then we will genuinely have made one of the profoundest discoveries in the history of thought - consequences of which would shape and alter our conception of ourselves in the deepest way.[7]

The prospect that such a program could be carried out elicits occasional paranoid reactions: cries of reductionism or, as expressed by John Lucas of Oxford, "the rubbishing of human experience".

As for reductionism, I have always been somewhat mystified as to what the fuss is about. Scientists (as mentioned above) have been constructing seemingly new and elevated entities from base material as part of their daily exertions since Thales showed us the way. When such a construction cannot be made, something new must be added. There seem to be no shortage of voices advising us to add that something new before we have had a reasonable chance to construct from the old. My advice is patience, along with the reminder that our brain is programmed to jump to conclusions.

---

[7] A distinction must be made between a new assumption, an entirely new entity (the equivalent of the addition of Euclid's fifth or parallel axiom to the first four that distinguishes between Euclidean and non-Euclidean geometries) and an unexpected and non-inevitable construction from the materials available (e.g. living creatures from chemicals or novels from letters). These latter are often highly dependent on initial conditions. Even temperature, a seemingly straightforward construction from kinetic theory and statistical mechanics, requires equilibrium systems.

The rubbishing of human experience is a greater concern since no small number of the *nouvelle vague* computer and robotic types seem only too happy to do just that. To me this is a reflection of the joy they have experienced and the wisdom they have gained in their passage through this vale of tears. The value we place on our own experience is something we determine ourselves and would never (I hope) forfeit to any machine or in fact (except to a limited extent) to anyone else. This value is completely independent and totally unaffected by any 'reductionist' explanation of how our mental activity comes about. No more than a detailed knowledge of the chemistry of digestion affects our appreciation of the product of a great chef or the bouquet of a fine wine.

Well then how, from materials made available to us by unrepentant reductionists: electrons, protons, atoms, molecules, DNA, RNA, receptors, enzymes, proteins, membranes, neurons, axons, dendrites, synapses ... , can we construct an entity that has mental experience?

We believe that there is a reasonable evolutionary sequence leading from sunshine, lightning and a reducing atmosphere to molecules, the primitive protein soup and to more and more complex structures. Even the simplest cells show reflex-like, chemically directed responses to various stimuli (aversion or attraction.). As has been said, "protoplasm is irritable". (But how does it come to feel irritable?) It seems reasonable to believe that there is advantage for organisms that can communicate from one end to the other, that electrical communication is a very efficient way of doing this and that excitable membranes provide the means.

There is further advantage in the innovation of a nervous system that is plastic. (The transmitter driven synapse provides an excellent option.) For the animal can now learn and store memories of past experience. This animal can adapt to environmental changes in less than evolutionary time.

And there is surely advantage (most of the time) in exercising the option of mental experience - feelings of pleasure and pain, awareness of individuality, instilling a directive to "jump to conclusions" to "accept as truth that which is only advantage" as a means of producing life preserving behavior in complex and somewhat unpredictable real-world situations.

All of this can at least be sketched. We can guess (at least conceptually) how from primitive feelings such as pleasure and pain more complex mental states could be constructed. But how the essential primitive: feeling, arose out of materials such as reflex reactions to hot and cold, how, somewhere in the distinction between those events that produce physical reactions in ourselves and those that do not, in the interplay of present sensory input with memory of past experience, our self-awareness, our mental experience, our consciousness arose, remains a deep mystery.

To paraphrase Shannon:

"Can machines feel?" "Sure they can. We're machines. We feel"

But are we? And if so, how?

A Final Few Words

This series of papers provides a view of the effort we began some twenty years ago: to understand the physical basis, the basis on a cellular and molecular level, for the processes of learning and memory storage as they occur in living animals. To understand various system properties: recall, association, reasoning and finally all of the higher level mental activity that are part of human self-awareness and consciousness. We expect that when we understand these processes in living animals, we will be able to replicate them both in software and hardware. The recent Nestor/Intel chip which contains 1024 processor equivalents of neurons, is, in my opinion, an important early step in the construction of massively parallel hardware systems.

Perhaps most mysterious is how we become conscious. Although there has been a certain discussion of this question (alluded to in the last article) and though its resolution will, very likely, be subtle and possibly complex, the statement of the problem appears, to me, to be wonderfully simple — a truly beautiful scientific problem [reminiscent of the Bourbaki definition of a beautiful theorem: One with a short statement and a long proof (e.g., Fermat's last theorem)]. For the problem of consciousness we can give the statement in a paraphrase of Santayana: All of our sorrow is real but the atoms of which we are made are indifferent, or how do we construct real sorrow from hypothetical indifference?

Although the enterprise is ambitious I believe that it can be accomplished, hopefully in our lifetime. When the many, often technical, details discussed in these papers are contrasted with the noble overall objectives, as with all scientific constructions, I am reminded of Monet's Sunrise. When one stands very close one sees only the brushwork; when one stands far enough away one sees the rising sun over the parliament building. At some appropriate intermediate distance one can see both the overall image and the brilliance of the color and the brushwork. But each person must find that appropriate distance for themself. I hope this collection helps.